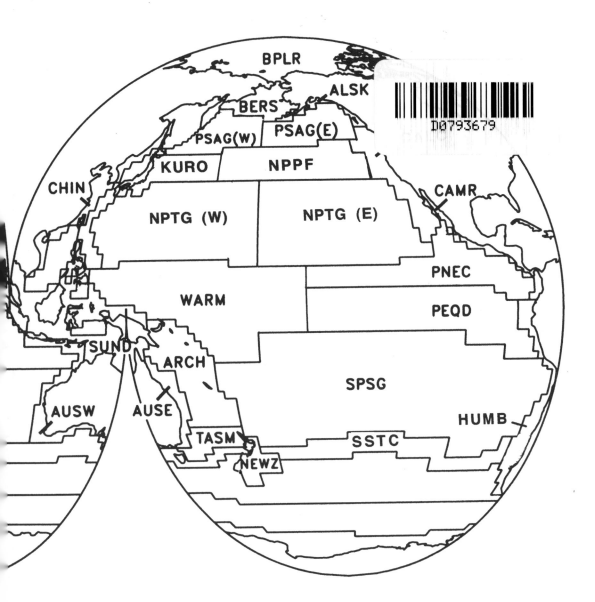

EAFR   Eastern Africa Coastal Province
INDE   Eastern India Coastal Province
INDW   Western India Coastal Province
REDS   Red Sea, Persian Gulf Province

INDIAN OCEAN TRADE WIND BIOME
ISSG   Indian South Subtropical Gyre Province
MONS   Indian Monsoon Gyres Province

PACIFIC COASTAL BIOME
ALSK   Alaska Downwelling Coastal Province
AUSE   East Australian Coastal Province
CALC   California Current Province
CAMR   Central American Coastal Province
CHIN   China Sea Coastal Province
HUMB   Humboldt Current Coastal Province
NEWZ   New Zealand Coastal Province
SUND   Sunda-Arafura Shelves Province

PACIFIC POLAR BIOME
BERS   North Pacific Epicontinental Sea Province

PACIFIC TRADE WIND BIOME
ARCH   Archipelagic Deep Basins Province
NPTG   North Pacific Tropical Gyre Province
PEQD   Pacific Equatorial Divergence Province
PNEC   North Pacific Equatorial Countercurrent
        Province
SPSG   South Pacific Subtropical Gyre Province
WARM   Western Pacific Warm Pool Province

PACIFIC WESTERLY WINDS BIOME
KURO   Kuroshio Current Province
NPPF   North Pacific Transition Zone Province
PSAG   Pacific Subarctic Gyres (East and West)
        Province
TASM   Tasman Sea Province

Ecological
Geography
of the
Sea

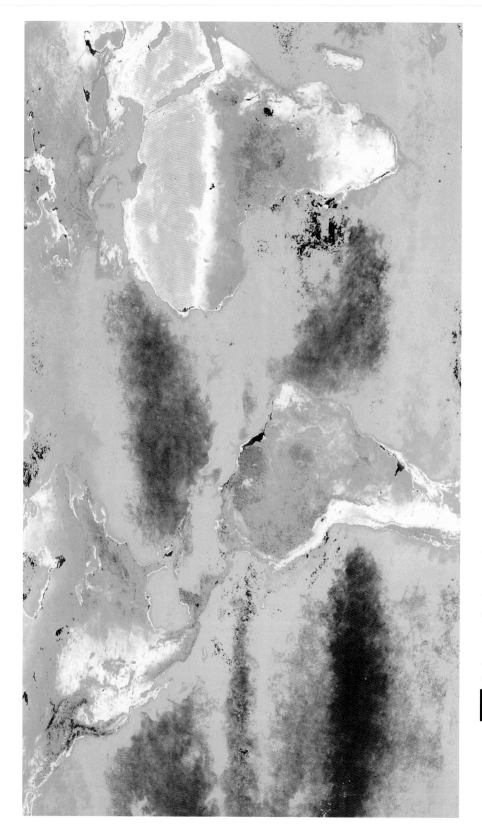

Both terrestrial and oceanic biomes are indicated by relative chlorophyll concentration in this composite SeaWIFS image (see p. 19), representing austral spring and boreal fall, 1997. Chlorophyll concentrations (mg.m^{-3}) at sea surface are about 1.0 (green), 0.1 (blue) and 0.01 (purple). Provided by the SeaWIFS Project, NASA/Goddard Space Flight Center.

Ecological Geography of the Sea

ALAN LONGHURST

ACADEMIC PRESS

San Diego London Boston New York Sydney Tokyo Toronto

Cover photograph: © Jim Brandenburg, Minden Pictures

This book is printed on acid-free paper. ∞

Academic Press
a division of Harcourt Brace & Company
525 B Street, Suite 1900, San Diego, California 92101-4495, USA
http://www.apnet.com

Academic Press Limited
24-28 Oval Road, London NW1 7DX, UK
http://www.hbuk.co.uk/ap/

Library of Congress Card Catalog Number: 98-84363

International Standard Book Number: 0-12-455558-6

PRINTED IN THE UNITED STATES OF AMERICA
98 99 00 01 02 03 MM 9 8 7 6 5 4 3 2 1

"Thus, in these regions at least, observations give better results than theories can provide . . ."

Klaus Wyrtki, *Naga Report*, 1961

CONTENTS

10 The Southern Ocean   339

# PREFACE

How does the ecology of plankton respond to regional oceanography?

This book is an attempt to answer this apparently simple question, not by reference to the distribution of individual species but by analyzing how some basic ecological processes vary in response to the characteristics of regional oceanography. From this analysis, I have tried to draft an ecological geography of the oceans and shallow seas so that, at some level of probability, a prediction may be made about the characteristic ecology of any region.

The main part of the book is arranged geographically by individual ocean, as in classical biogeographies. I have chosen this arrangement because the reader must already have access to several modern texts on marine ecology and the physiology of marine organisms, arranged so as to make it easy to extract information about individual ecological or physiological processes. On the other hand, it may be quite difficult to locate organized information about a particular region of interest other than those few parts of the ocean that have attracted major expeditions in recent decades. Therefore, for the same reasons that Tomczak and Godfrey (1994) published their innovative text on regional oceanography for physicists who lacked a modern geography of the oceans, this book is offered as a geography for marine ecologists.

It will be obvious to the reader that I must have consulted many more original papers than I have been able to cite: This is inevitable in a work such as this, lest the bibliography grow as long as the text. I apologize in advance to those who spot their own uncited contributions to the whole. I have tried to list sufficient key papers to give access to the literature of each province rather than to support each of my contentions with a citation. You will also quickly come to appreciate that I have been unable to locate comprehensive ecological studies for several regions (usually distant from Europe or North America) even if their physical oceanography is quite well-known. This says something about the priorities of oceanographic agencies and also about how physical and biological oceanographers have been trained to look differently at the world.

During the past decade there has been a major cognitive shift among biological oceanographers from an assumption that a simple "nitrate–diatoms–copepods–fish" model adequately summarized the pelagic food chain to more complex constructs that incorporate picoautotrophs, mixotrophic protists, and the role of bacteria in regenerating nitrogen for autotrophs, to say nothing of a more realistic range of possible mechanisms for the flux of major nutrients: Mills (1989) suggests why it took so long to incorporate in our general model of the pelagos the observations of Lohman, made at the end of the nineteenth century, of the relative abundance of nanoplankton cells. Inevitably, most of the region-specific studies I have consulted were done when the simpler model was the accepted norm, and relatively few regional studies incorporate the current models. For some regions I have been able to suggest how the older studies should probably now be interpreted and have extrapolated accordingly, whereas for others I have simply described what I think is known and left you to guess the rest.

I am acutely conscious of the debt I owe to so many people (some now deceased) who taught me enough about the oceans to risk undertaking this book. They are far too numerous to list, but I remember especially the following: Gunnar Thorson, Theodore Monod, and Vernon Bainbridge got me started in the Gulf of Guinea in the 1950s; Bill Thomas and Carl Lorenzen introduced me to oceanic plankton during EASTROPAC in the 1960s; Maurice Blackburn and Ray Beverton showed me the close relationship between fish and temperature off Baja California and Spitzbergen, respectively; Radway Allen and Benny Schaefer helped me understand how fish populations worked; Reuben Lasker and Robert LeBorgne taught me something about physiological rate functions for zooplankton and nekton; Philip Radford introduced me to flux models of ecosystems; and with Bob Williams I explored vertical plankton profiles both in data archives and at sea. Warren Wooster, Karl Banse, and David Cushing always seemed to be there when I needed advice or encouragement.

In recent years, I was strongly supported at the Bedford Institute of Oceanography (Department of Fisheries and Oceans, Canada) by my colleagues of the Biological Oceanography Division with whom I worked at sea in the eastern tropical Pacific, in the Canadian arctic archipelago, and in the North Atlantic from 1977 until 1993. I thank them all for their patience in helping me using their varied expertise in pelagic ecology (Trevor Platt, Shubha Sathyendranath, Glen Harrison, Bill Li, Erica Head, Bob Conover, Paul Kepkay, and Doug Sameoto) and ocean physics (Ed Horne, Fred Dobson, John Lazier, and Allyn Clarke).

We sailed aboard *Hudson*, described to me not so long ago by the late academician Konstantin Fedorov as "the most capable oceanographic ship in the Western world." She is still the only surface ship to have circumnavigated the whole American continent, around Cape Horn, and through the Northwest

Passage in one voyage, but now she teeters on the edge of following our other Canadian research vessels *Baffin* and *Dawson* to scrap. They may be replaced by a Coast Guard hand-me-down.

Finally, this is a good place to apologize to Françoise for all those summers when I was at sea instead of with the family and for all those weekends at the computer.

INTRODUCTION

Ideally, pelagic biogeography should have three components. First, it should describe how, and suggest why, individual species from bacterioplankton to whales are distributed in all oceans and seas. Second, it should tell us how these species aggregate to form characteristic ecosystems, sustaining optimum biomass of each individual component under characteristic regional conditions of temperature, nutrients, and irradiance. Third, and most important for some purposes, it should document the actual areas within which each characteristic pelagic ecosystem may be expected to occur.

The current state of pelagic biogeography is unsatisfactory on all three counts. Biogeographers have concerned themselves almost exclusively with the first question, but a comprehensive account of how species are distributed in the ocean still remains far beyond their grasp. The second question—what are the characteristic ecosystems of the ocean?—has been investigated for almost 100 years and much progress has been made, but there has been little interest in locating the geographical boundaries between ecosystems, which is the third requirement. It is my hope that this book will go a little way toward marrying the work of ecologists, or biological oceanographers, with that of regional oceanographers. It is only by integrating an ecological component into regional oceanography that we can hope to predict the characteristics of the pelagic ecosystem of a given oceanic area.

The simple descriptive phase of biological oceanography occupied almost a full century. Beginning with the straightforward exploration of the marine biota of all oceans by the *Challenger* in 1872–1876, this phase culminated with the multiship, seasonal explorations of large oceanic regions—NORPAC, IIOE, EASTROPAC, EQUALANT, and others—in the middle of the twentieth century. Though some of the very early voyages, notably the Plankton Expedition of 1889, also laid the foundations of our understanding of the biology of oceanic plankton, research on ecological processes was principally pursued in coastal laboratories—especially Kiel, Plymouth, and Woods Hole—where quite rapidly a basic understanding of the cycle of pelagic production and consumption was achieved, a process well described by Mills (1989); as Karl Banse reminds me, a major catalyst in this progress was the introduction into biological oceanography during the 1950s of bulk methods for chlorophyll, seston, proteins, and carbon uptake rates so that samples could be taken that matched the bottle data of the physical and chemical oceanographers.

In recent decades, the detailed study of ecological processes has been taken to the open ocean. Rather than exploring species distributions, the objective of recent large-scale and multiship expeditions has been to quantify the rate constants for physiological processes across a wide range of oceanographic conditions. The catalyst for this change of emphasis was surely the development of solid-state electronics for underwater sensors, for shipboard laboratory equipment, and for data processing. Reliable and compact instruments were critical for biological oceanography especially because of previous difficulties in obtaining data from below the sea surface.

During the past 25 years, and parallel with equivalent progress in the other branches of oceanography, these instruments have delivered nothing less than a revolution in our understanding of ecological dynamics in the oceans and in our ability to archive and process vary large quantities of numerical data, especially concerning the physical environment of the pelagic biota. This new phase of process studies in biological oceanography has now been carried sufficiently far that we can imagine two-dimensional numerical models that simulate characteristic marine ecosystems: coastal upwelling, spring bloom, tropical open ocean, temperate shelf, and so on.

However, there is a catch. Our ability to specify the critical physiological processes under a range of ambient conditions has outstripped our knowledge of the distribution, abundance, and biomass of the organisms themselves. Our accumulated data describing the geographical distribution of plankton still contains insufficient information to partition two-dimensional ecological models among a series of compartments, each having a characteristic ecology, that together might integrate the whole surface of the ocean. Lacking this possibility, global biogeochemical models have usually integrated ecological processes as a parameterized continuum.

This was the situation until very recently. Now, however, as the result of another technological revolution in the natural sciences, the required information to compartmentalize biogeochemical models is at hand. Sensors carried on earth-orbiting satellites can now obtain data representing conditions on the surface of the oceans at very short time intervals and over a grid of many closely spaced points. The NIMBUS satellite carried the Coastal Zone Colour Scanner (CZCS) sensors that for several years obtained data representing the surface color of the oceans and thus, with some reservations, phytoplankton chlorophyll. Other ocean color satellites and other sensor suites are planned to follow or are already orbiting in an experimental mode. Operational and routine chlorophyll sensing should follow later in this decade.

The NIMBUS sensors gave biological oceanographers a tool of unprecedented power. Running through this book is a thread of interpretation of the CZCS data, suggesting how we may use such information to progress from our general knowledge of pelagic ecology to a first approximation of what we have hitherto lacked: comprehensive, region-specific descriptions of seasonal changes in ecological processes over the whole ocean comparable to those we already have for terrestrial ecosystems. In short, it outlines proposals for an ecological geography of the pelagic ecosystem of the surface waters of the ocean.

Why are the proposals restricted to the pelagos, and particularly of the open ocean? Simply because the new satellite sensors tell us nothing about the largest part of the marine biosphere—the interior of the ocean—or about the organisms of the seabed. So, despite their importance in mediating global carbon flux through the respiration of sinking organic material, the biota living below the surface layer remain difficult to study. The same must be said for the benthic communities for which there seems to be no substitute for the slow and steady accumulation of the observations that have occupied a small body of marine ecologists throughout this century. Therefore, you will find little about the benthos or bathypelagic ecology in this book. It is difficult to see how progress in understanding these ecosystems can accelerate in the foreseeable future.

Nor could the CZCS sensors resolve biological properties shorewards of the turbidity front that occurs along almost all coasts at some depth less than 50 m and where, in any event, the fractal nature of coastlines will continue to frustrate logical and comprehensive mapping of their ecological characteristics. Thus, I have had to be very cautious in what I have said about the neritic zone.

Philosophically, if we finally seem to have the right kind of information to define an ecological geography of the pelagic realm, then it should be attempted if only to test the quality of the new data. However, more practically, maps of the ecological characteristics of the ocean must bring aid and comfort to those who use biogeochemical models to quantify the effects of climate change on oceanic biota and, inversely, of the mediation of climate change by oceanic biota. Perhaps a geographic synthesis may also help those who monitor fisheries production to quantify the effects of environmental variability on renewable resources.

In this chapter I examine the principles that might be used to partition the pelagic ecosystem into objectively defined geographic compartments. It is far from evident how this can be done since ecological boundaries in the ocean (even if it is agreed that such boundaries exist) must change their location seasonally and between years. Therefore, how can they possibly be mapped? How can the coordinates of the boundaries be usefully defined? How can they accommodate the consequences of the seasonal meridional march of the radiation and wind fields or the effects of the El Niño–Southern Oscillation events that modify global wind fields and circulation patterns every few years? Even more, can they possibly capture the effects of longer, decadal-scale changes in weather patterns?

I cannot emphasize too strongly that the maps of biogeographic provinces offered in this study are not intended to represent boundaries that are fixed in space and time but are only intended to indicate the approximate time-averaged spatial relations between provinces. How to actually locate the boundaries between provinces when required for global integration of a set of data representing a real period of time is a separate and complex question. For this to be possible, the boundaries must be defined by features that can be observed by remote sensing, for example, discontinuities in the slope or elevation of the sea surface or in the sea-surface thermal or color fields. Only from earth-orbiting remote sensors can we hope to have a flow of regional and global data sufficiently comprehensive and timely to predict the coordinates of variable boundaries between ecological provinces when required.

First, however, we have to define the discontinuities which may subsequently be located, perhaps by reference to proxy information, in real time. To do this, the first question we should address is whether the state of knowledge about the distribution of individual species of plankton (the present corpus of classical biogeography) is sufficient to suggest where we should place discontinuities in an ocean biogeography, and even whether such boundaries exist.

THE INADEQUACY OF CLASSICAL BIOGEOGRAPHY IN ECOLOGICAL ANALYSIS

I have already suggested that the results of 150 years of study of the distribution of the marine flora and fauna are so meager that they permit us to predict comprehensively the characteristic assemblage of species likely to occur in no region of the ocean. In fact, currently, the taxonomic biogeography of the sea belongs to the family of intractable scientific problems. I am not alone in this opinion: Dunbar (1979) and Rosen (1975) have commented that biogeography is an uncritical branch of marine science with neither a useful factual basis nor an agreed methodology: Dunbar stated, "The biogeographic method does not exist, or there are as many methods as biogeographers" and Rosen noted, "Biogeographers of today follow their own preferred and sometimes bizarre premises without casting more than a troubled glance at . . . conflicting theories and ideas." What follows tends to support such criticism, especially for oceanic biogeography, which has even more serious inherent difficulties than terrestrial biogeography (e.g., see de Beaufort, 1951): the cost of collecting samples on the high seas, the relative lack of isolation between natural regions, the high levels of expatriation among plankton species, and three-dimensional distributions that vary in both space and time. These are among the most serious problems, but not the only ones.

Despite the general difficulty of the discipline, terrestrial biogeographers quickly located the important discontinuities between the great faunistic divisions on land, but marine biogeographers still have only relatively sketchy accounts of faunistic or floristic discontinuities in the ocean. What came to be known as "Wallace's line" in the Indo-Pacific archipelago, separating the Australian from the Oriental terrestrial faunas, was identified by Alfred Russel Wallace in almost the same year that the *Challenger* set her sails for the first global exploration of the still-unknown marine biota. The general outlines of the taxonomy and biogeography of the terrestrial flora and fauna were sufficiently well described by the first half of the twentieth century that integration was already achieved between biogeographic regions (inhabited by organisms related in the taxonomic sense) and their constituent biotopes (inhabited by organisms forming characteristic ecosystems) as described by Pitelka (1941). In this way, it was possible to recognize ecologically similar but taxonomically distinct biotopes (e.g., see de Beaufort, 1951): for instance, the rain forests of Brazil, Papua-New Guinea, and Zaire are ecologically quite similar but taxonomically very different. However, we are nowhere near being able to make the same kind of analysis from the accumulated knowledge of the taxonomy and biogeography of oceanic biota, both aspects of which are still explored only very superficially.

This situation is paradoxical because there are probably more than 1.0×10^6 species of terrestrial animals, but no more than 1.0×10^4 species of marine plankton and nekton in all the animal phyla. Since the earliest voyages, a small part of the total effort of oceanography has been devoted to biogeography and even as late as the 1960s strong teams, expressly devoted to this task, were being recruited at some major oceanographic institutions. Even now, however, 150 years after the *Challenger* voyage, the total number of species in each major group of pelagic organisms is not even approximately agreed on and we have descriptions of the seasonally variable distribution of no more than a small proportion of them. This is really not surprising: Although a quantitative comparison cannot be made, the number of stations at which marine plankton have been collected, identified, and enumerated must be orders of magnitude smaller than that for the terrestrial invertebrates, though the area of the oceans is more than twice that of the continents.

It has been remarked that the basic unit of biogeography is the distribution of individual species, though even this simple adage is questionable because few taxonomists of the plankton have faced up to the relationship between binomial Linnaean "species" distinguished simply by some morphological criterion and the individual self-sustaining populations that occupy individual home ranges, are presumed to have some genetic isolation, and may (or may not) be distinguishable on the sorting tray of a microscope. The surface of such problems has scarcely been scratched.

There is some agreement among taxonomists that the number of species (however defined) of mesozooplankton is around 2000 so that McGowan was able to write in 1971 that "holoplanktonic zooplankton are taxonomically well-known at the species level." In retrospect, this was surely an overoptimistic statement. Consider the taxonomic status of copepods of the genus *Calanus* during the ecological exploration of the oceans of recent decades: Two North Atlantic copepods of the genus are surely the best known of all zooplankton species, but *C. finmarchicus* (Gunnerus 1770) and *C. helgolandicus* (Claus 1863) were not clearly distinguished until the work of Fleminger and Hulseman (1977) and were, for a long time, thought to be geographical races (or subspecies) of a single cosmopolitan species. Recent work (quoted in Bucklin *et al.*, 1995) suggests that there are 14 species of *Calanus*, of which 3 (*C. hyperboreus*, *C. simillimus*, and *C. propinquus*) are morphologically distinct from the remainder, which are themselves distinguishable only by expert analysis of fine differences in secondary sexual characters of the exoskeleton. These comprise two species groups: a northern hemisphere arctoboreal trio (*C. finmarchicus*,

C. glacialis, and *C. marshallae*) and a larger group occupying midlatitudes in both hemispheres (*C. helgolandicus, C. pacificus, C. australis, C. orientalis, C. euxinus, C. agulhensis, C. chilensis,* and *C. sinicus*). Recently, molecular systematics using the DNA base sequences of the mitochondrial 16S rRNA gene (Bucklin *et al.,* 1995) has been used to confirm the reality of the group *C. finmarchicus* + *glacialis* + *marshallae* and also of the more genetically diverse *C. helgolandicus* group as listed previously, though genetic information on *C. orientalis* is still lacking.

However, this analysis also identifies significant differences between the individuals of *C. pacificus* from the California coast, from OWS "PAPA," and from Puget Sound so that "subspecies" level is accorded to individuals from these three locations, thus recalling Fleminger and Hulseman (1977), who referred to morphologically distinguishable forms of *C. helgolandicus* from the eastern and western Atlantic. In neither case do we know if the characters— morphological or genetic—lie along gradients or are discontinuous, and we understand very little about how species of *Calanus* may be sympatric. This unhappy situation will be found to extend to other important zooplankton genera: A similar molecular analysis of the difficult but abundant genus *Pseudocalanus* is currently in preparation by the same team.

I have reviewed this case in perhaps greater detail than it warrants, except as an example of how the present state of plankton taxonomy is a hindrance to ecological geography. However, this case is not the only one, and another indication of how very far we are from understanding the geography of pelagic species is given by recent ribosomal RNA analysis of 41 specimens of the small mesopelagic fish *Cyclothone alba,* which is ubiquitous and abundant in all subtropical and tropical oceans (Miya and Nishida, 1997). This analysis identified five monophyletic populations with low levels of mutual gene flow under conditions in which there appears to be no discernible barriers to prevent complete dispersion and intermingling of stocks: The central North Pacific population is genetically closer to those of the North Atlantic than to the other three populations inhabiting the Indo-Pacific ocean. Miya and Nishada reach the same conclusion as Dunbar (1979) and Rosen (1975): They write "we are a long way from knowing what species really exist in the oceanic pelagic realm."

Unfortunately, because RNA procedures are expensive and highly specialized, genome analysis will be very slow in helping us interpret the existing base of knowledge and leaves the biogeographic literature of many important groups of species still in rags and tatters and open to different interpretations. The critical reader of this book will quickly realize that in most cases in which the differential distribution of "species" of plankton is discussed, I have been quite unable to follow the taxonomic trail between the quotation of a specific name by a specific author and how the same specimens would be classified today.

Consider also the following statement (Dodge and Marshall, 1994) concerning the dinoflagellate genus *Ceratium:*

> Since the earlier work of Jorgenson . . . taxonomy within the genus has been fairly consistent and species can be readily determined. To date just over 100 species have been described, but there are problems with the large number of varieties which have been recognized although, over the years, many of these have been raised to species rank.

One might be forgiven for questioning the validity of the authors next comment: "For the preceding reasons *Ceratium* would seem an excellent, if not the best, dinoflagellate to use for biogeographic study"!

However, the identity of species in such well-known genera, though fertile ground for wrangling, is only the tip of the iceberg of this problem. Shih (1979) reminds us of the case of the thecosome mollusc *Cavolinia tridentata,* which exists as nine morphological types. Shih comments that such phenotypes may be allotted to species, subspecies, or forma according to apparently arbitrary decisions. Since such types also commonly appear to be arranged along a latitudinal (or thermal) gradient, it is not surprising that there often appear to be latitudinal discontinuities in the distribution of taxa. Between oceans, populations of apparently the same species frequently show morphological differences that to a future taxonomist may well suggest specific distinction. Shih cites the example of two species of *Rhincalanus*—one described from the Atlantic and one from the Indo-Pacific—of which the males are indistinguishable though there are minor differences between females from the two oceans: These have variously been considered to be two separate species or two forms of a single species.

Phenotypic variations in many phyla of plankton are the rule rather than the exception: Successive generations during a season may be morphologically distinguishable, nonreproducing expatriates may be different from individuals in the home range, individuals from warmer and colder, shoaler and deeper environments may differ, and so on. No wonder that five geographical forms of a common euphausiid *(Stylocheiron affine)* have been identified from the Indo-Pacific.

If one adds to such taxonomic problems those arising from the three-dimensional transport of biota, then certainty becomes stretched very thin indeed. Not only are boundaries between currents or water masses leaky, but the surface layers in which we are interested lie above deeper water masses that may have been subducted from the surface at convergent fronts even thousands of kilometers distant, together with at least some of their planktonic organisms. In this way, the characteristic surface organisms of a subtropical water mass may lie only a few hundred meters above organisms expatriated from a subarctic environment. In a simple plankton tow (and the vast majority of biogeographical data are obtained with such tows) organisms from the two environments may appear with great regularity in the same sample. About ½ km below the tropical copepods in the EASTROPAC zooplankton profiles, I often found quite large numbers of *Eucalanus bungii,* a copepod of the subarctic Northeast Pacific obviously subducted at the subarctic front; they were translucent, inactive, nonreproductive, and existed—one might think—only to confuse the biogeographer. Finally, we must remember the diel migrations which are performed by many species of mesozooplankton between the surface mixed layer and depths (300–500 m) at which we most commonly find expatriates from other water masses: These "interzonal" species are just that—they spend day and night in different depth zones and even different water masses.

Despite all these difficulties, Sinclair (1988) has reviewed the evidence for the existence of distinct, self-sustaining populations of oceanic zooplankton and is able to list about 20 studies which he believes successfully identified the general location of such populations. In several cases, the individual populations inhabiting different areas of the Pacific Ocean are morphologically distinct and, though direct evidence is lacking, may be assumed to be genetically isolated. It is usually assumed that in the highly dispersive oceanic environment, such plankton species are able to maintain a self-sustaining population, and consistently close their life history cycle within a specifiable region of the open ocean, often by means of appropriate seasonal vertical migrations between two depth zones with opposing mean flows. Some possible examples of this strategy will be discussed in the special regional descriptions in later chapters.

If there was a sufficient commonality between their home ranges, the boundaries of the areas occupied by self-sustaining populations might serve our purpose. Unfortunately, the total accrued information on such boundaries is insufficient to do more than offer some support—sometimes for conclusions we may have reached by other methods. This situation is unlikely to change rapidly because progress in marine biogeography, and the accumulation of new data on the distribution of pelagic organisms, will surely remain slow. Systematic biogeography is now perceived as a "filling-in" activity rather than an innovative branch of marine science, and this perception is unlikely to change despite current concerns for loss of biodiversity and habitat degradation by pollution and other forms of exploitation.

Nor can we realistically expect that a technological innovation will replace the slow and steady approach, which has changed little since the early days of oceanographic exploration. Although some progress has been made in mechanizing the slow chore of sorting and counting plankton samples, this is limited mostly to speeding the transfer of data to electronic media. Instrumentation to replace net sampling by counting and identifying organisms *in situ* with electronic and optical sensors is very far from replacing the need for time-consuming deployment of nets and the employment of specialists to identify the captured organisms. The most that has been done is to deploy imaging equipment in towed instruments that can obtain images of the larger organisms, from which a limited level of taxonomic identification may be inferred.

For all these reasons, there are still very few cases in which it has been possible to survey large regions of the oceans with a station spacing and frequency sufficient to specify individual distributions of the hundreds of species of plankton, nekton, and fishes occurring there, together with their seasonal and interannual distributions.

The California Cooperative Fisheries Investigations (CalCOFI) surveys are perhaps closest to the required model, having covered the California Current from about 1950 to the present day. There are two lessons to be learned from CalCOFI, however: (i) even though supported by both federal and state agencies (note, of the richest state in the richest union), the CalCOFI teams have been able fully to enumerate the plankton organisms from the surveys of only a single year and for the remaining years have had to be content with bulk measurements from the samples and accurate counts of a few selected taxa, mostly fish eggs and larvae; and (ii) even here the surveys have frequently been threatened with closure and have had to be performed on a progressively reduced frequency—currently only every third year. The other major large-scale survey of pelagic biogeography, the Continuous Plankton Recorder (CPR) surveys of the North Atlantic (e.g., Colebrook, 1982; Warner and Hay, 1994) along shipping routes from New England to Iceland and the North Sea, has also obtained samples since about 1950 and is almost the sole source of data on long-term changes in species distributions in the open ocean. These surveys, however, have been hard to maintain and currently have no assured long-term funding, being relegated to the status of a private foundation.

Some progress, of course, has been made in the analysis of how species distributions are associated with natural features of the ocean environment. The results are probably more reliable and useful for our purposes when a relatively small region has been analyzed rather carefully: A good example of this kind of approach is use of planktonic foraminifera as water mass indicators along a meridional transect from 30–55°N in the central North Atlantic (Ottens, 1991). Statistical species clustering of surface-collected foraminifera showed four distinct faunal assemblages associated with subpolar water (No. 1), the North Atlantic Current (No. 2), and the Azores Current (No. 4), with an additional assemblage (No. 3) lying zonally in the transition between the latter pair. These fit very well with the boundaries proposed for three ecological provinces in Chapter 7: Atlantic Subarctic Province is inhabited by No. 1, North Atlantic Drift Province by Nos. 2 and 3, and North Atlantic Subtropical Gyral Province by No. 4. Rather, I should have written that those small parts of these three provinces that were crossed by the transect were inhabited by these assemblages at one season of 1 year: I am not confident that we should extrapolate from this single transect to three wide zonal provinces of the North Atlantic and make assumptions about their characteristic assemblages of foraminifera.

There are very few analyses as promising as this one, but despite the tiny proportion of the surface of the ocean (or of the diversity of its biota) that has been included in these studies, attempts have been made since the earliest days to map biogeographical provinces in all oceans. These maps tend to have two characteristics in common: (i) The provinces do not have contiguous boundaries and (ii) the provinces resemble a series of latitudinal zones from poles to tropics, stretching across the ocean basins. Even though they usually conform to this general arrangement, there will be significant differences between almost any two examples that you might choose to compare.

Reviews of historical progress in biogeography generally start with the map of Steuer (1933) that is based on copepod distributions; this divided the oceans into circumpolar arctic, antarctic, and subantarctic regions and then divided each ocean basin (where appropriate) into subarctic and northern and southern subtropical and tropical regions. This is not very different from Sven Ekman's proposal at approximately the same time for a "pelagic warm-water fauna, northern and southern cold-water plankton, and a neritic plankton." From here to Beklemishev's (1969) map is not a far step, though he uses 20 instead of 11 regions. However, when compared, the two maps show limited congruence between the coordinates of comparable regions even in the same oceans. The most extensive compilation of such distribution maps is the comparative atlas of zooplankton of van der Spoel and Heymen (1983), but this does little more than remind us how far we are from achieving a comprehensive, species-based geography of the pelagic ecosystem and how little we have progressed since the

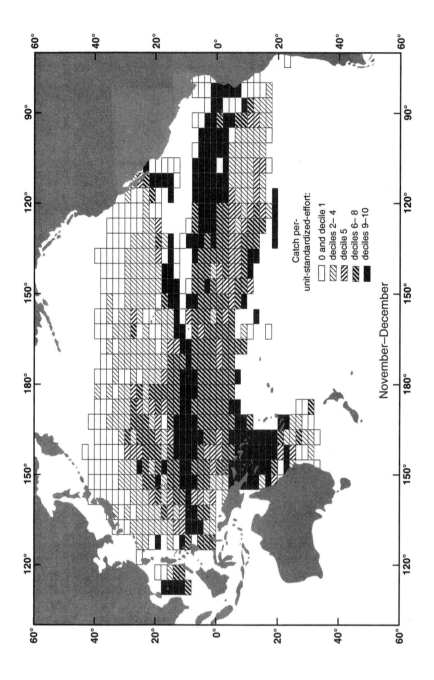

November–December

Catch per-
unit-standardized-effort:

☐ 0 and decile 1
▨ deciles 2– 4
▧ decile 5
▨ deciles 6– 8
■ deciles 9–10

10

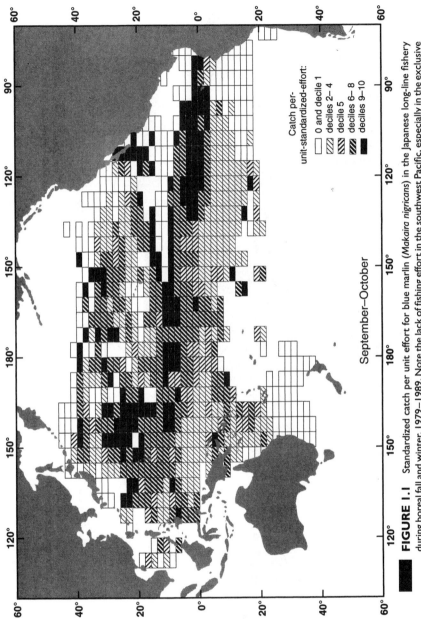

Catch per-
unit-standardized-effort:

☐ 0 and decile 1
▨ deciles 2–4
▧ decile 5
▤ deciles 6–8
■ deciles 9–10

September–October

FIGURE 1.1 Standardized catch per unit effort for blue marlin (*Makaira nigricans*) in the Japanese long-line fishery during boreal fall and winter, 1979–1989. Note the lack of fishing effort in the southwest Pacific, especially in the exclusive economic zones of the Federated States of Micronesia and neighboring administrations, though adjacent catch rates suggest that marlin are nevertheless abundant there (from Hinton and Nakano, 1996, Fig. 5, with kind permission from Inter-American Tropical Tuna Commission).

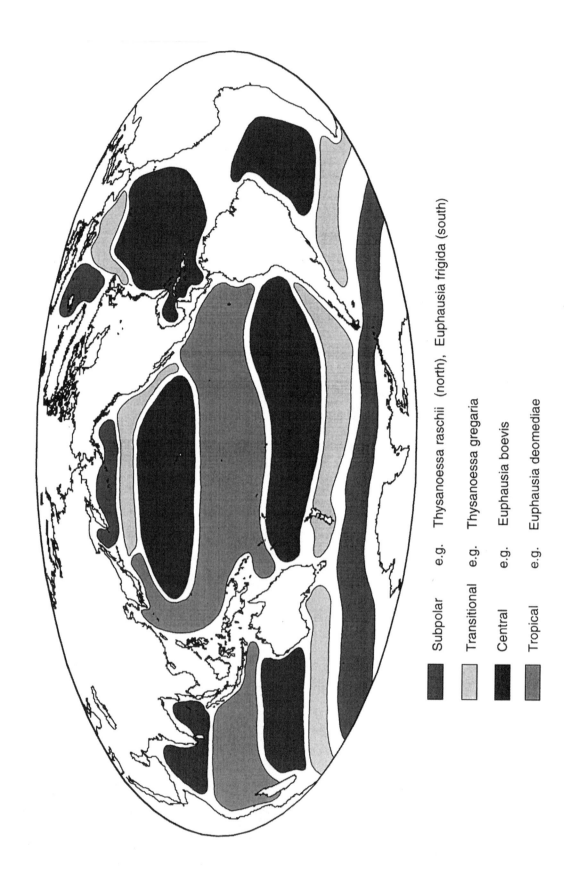

Subpolar   e.g.   Thysanoessa raschii (north), Euphausia frigida (south)

Transitional   e.g.   Thysanoessa gregaria

Central   e.g.   Euphausia boevis

Tropical   e.g.   Euphausia deomediae

early maps. After reviewing the approximately 60 maps that illustrate this atlas, I find it difficult to see any commonality between the distributions of different species.

Things appear even more confusing when proposals are made for partitioning the ocean into geographic units based on the distribution of a single group of organisms or even of single species; in this case, so many are useless or unconvincing because the author fails to show clearly, or even at all, the grid of samples on which the map is based from which we could see where a taxon occurred and where it is thought to be absent. It is worth noting how common this problem is, sometimes for quite unexpected reasons such as the exclusion of the sampling pattern from a particular zone of national jurisdiction (Fig. 1.1). Many (perhaps even most) published maps of species distributions show only those stations where the species was encountered in the samples; it is very common that the stations where samples were taken but the species was not found are unrecorded, which renders the map almost valueless. This kind of intellectual lapse reminds one that biogeography has been a very uncritical subject.

Of all the published maps of oceanwide distributions, few match those of Pacific euphausiids of Brinton (1962), the Pacific species distribution envelopes of McGowan (1971), or the maps from the North Atlantic CPR surveys (Anonymous, 1973). One is convinced by these presentations of the reality of some recurrent patterns of species distributions in the Pacific, the subarctic, central, equatorial, and eastern tropical species, together with those specializing in the transition zones at about 40° of latitude (Fig. 1.2); and in the North Atlantic, neritic, northeast intermediate, northeast oceanic, northwest oceanic, western intermediate, southern oceanic, and southeastern intermediate (Fig. 1.3). These maps provide information which will serve to confirm the reality of some of the ecological boundaries proposed in this work. Otherwise, the more synthetic global biogeographic maps (such as those of Beklemishev) will serve to indicate changes in the pelagic species of algae and zooplankton across some of the principal oceanographic discontinuities: I shall use some of these findings in Chapters 7–10 when I discuss the individual provinces. Such material is most useful in the southern ocean, in relation to the subtropical transition zones and perhaps at the boreal polar fronts. Apart from the neritic–oceanic transition, these appear to be the most important taxonomic discontinuities within the pelagos.

Several conclusions can be made from this brief review of the current state of oceanic biogeography. Most important, it is clear that the available description of the distribution of the species of the pelagic ecosystem is woefully inadequate to predict the distribution of the characteristic pelagic biomes which are the subject of this book. We shall therefore have to approach the matter differently.

FIGURE 1.2   The global distribution of groups of oceanic epipelagic euphausiids showing a general relationship with the biomes and provinces described in later chapters. The lack of a specialized group of tropical species in the Atlantic is attributed to the direct flux from the South Atlantic into the roots of Gulf Stream combined with the small zonal extent of the ocean. Groups of coastal species also occur, and match the meridional distribution of the oceanic species, but are not shown (derived from Brinton, 1962).

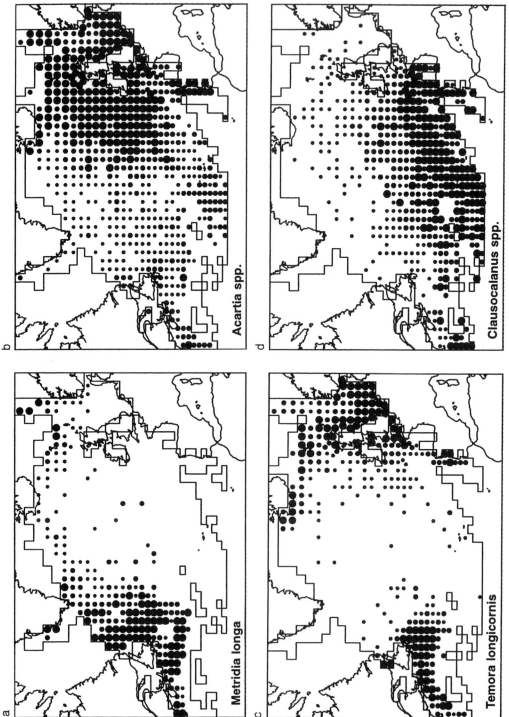

a   *Metridia longa*

b   *Acartia spp.*

c   *Temora longicornis*

d   *Clausocalanus spp.*

We may also concur that "pelagic biogeography has somehow not shared in the renaissance of interest that has occurred in the last decade for shallow-water and terrestrial systems" (summary report of the 1985 International Conference on Pelagic Biogeography), and at least one reason for this lack of energy seems obvious. Biogeographers could not participate in the solid-state electronic revolution of the 1970s that offered such intriguing new possibilities for biological oceanographers, who turned away from descriptive work to studies of ecological and physiological processes with tools that rapidly came to resemble those of physical oceanographers, at least in their ability to capture data.

It is only the exploration of the hitherto relatively unknown microplankton that has been greatly advanced by the new technology. Instruments transferred from the medical field, including laser optics in continuous flow cytometers and the use of fluorescence microscopy, have demonstrated the existence everywhere in the oceanic photic zone of novel micrometer-scale cells, the photosynthetic cyanobacteria and prochlorophytes. These cells are now known to be a significant component of the phytoplankton and critical in hitherto unsuspected ways for the flow of energy and material through the pelagic ecosystem.

During the same period, a conscious effort (Fasham, 1984) has been made to generalize critical physiological rates of algae and zooplankton. Agreement seems to be converging on values within a factor of less than 2, whereas prediction of regional numbers of organisms or their biomass could hardly be done with error terms of less than an order of magnitude. That this new ability to predict physiological rate functions characteristic of different environmental conditions should coincide with our new ability to measure at least one biomass index comprehensively at the surface of the oceans makes it all the more appropriate to revisit the question of an ecological geography of the ocean and to attempt to devise a scheme which will be valid for the entire surface of the ocean in all seasons and under all conditions.

THE NEW AVAILABILITY OF TIMELY, GLOBAL OCEANOGRAPHIC DATA

Novel data have become progressively available to oceanographers in the past 10 years from sensors carried on earth-orbiting satellites. Though only limited information is obtained at each data point on the ocean surface, the flow of simple data obtained every few days or even hours, and at a resolution of only a few kilometers over the whole surface of the ocean, has been revolutionary.

If these early developments mature, oceanographers of the twenty-first century will have everyday access to unprecedented information about the variability of ocean circulation and atmospheric forcing, and to a limited extent the future is already here. Though the sensors have been flown by government

FIGURE 1.3   Climatological distribution of four species of near-surface copepods in the Atlantic from a 60-year time series of Hardy Continuous Plankton Recorder samples from within the indicated area. (a) *Metridia longa* (western cold-water species); (b) *Acartia clausii* (northeastern intermediate species); (c) *Temora longicornis* (coastal species); (d) *Clausocalanus* spp. (southern oceanic species group) (courtesy of Plymouth Marine Laboratory).

agencies, frequently with military objectives, sufficient data have been made available to the oceanographic community as to open a new window on ocean physics. Primary users are rapidly starting to gain direct access to the incoming and stored data through the Internet, and this progress is unlikely to be reversed or halted. The data revolution which is in progress will certainly transform our ability to analyze regional oceanographic patterns, not only in physics but also in biology.

Real-time data on sea surface temperature are currently obtained from radiometers (AVHRR, VISRR, and VIRR) carried on earth-orbiting satellites and, in the coming decades, we can expect to have routine access to fields of surface winds (>13 existing or projected satellites through the end of this century), surface waves (7 satellites), and surface currents (14 existing or planned satellites). The success of TOPEX-POSEIDON 1992, which is currently measuring sea surface elevation with an accuracy of 2 or 3 cm and so is capable of mapping ocean circulation globally at short intervals, provides a glimpse of the future. One of the immediate and (relatively) unsophisticated benefits of the data from the new sensors has been an unprecedented ability to locate and map thermal fronts at sea down to the kilometer scale; in this way, there has already been a rapid increase in our knowledge of the locations of individual fronts, their evolution, and the physical processes that maintain them. Such information has been particularly valuable in understanding the nature of fronts at the shelf edge and also those associated with mesoscale eddies in the open ocean. By inference, and by comparison with chlorophyll images, the ecological significance of these features is now much better understood.

Near-surface chlorophyll fields for all oceans are also potentially available from satellite radiometers, based on the wavelength and intensity of backscattered visible light from the sea surface. Partial coverage of this field was obtained from 1978 to 1986 by the CZCS sensors carried aboard a NIMBUS satellite. Even at the first inspection, it was evident that the CZCS images contained entirely novel information concerning the global, seasonal distribution of phytoplankton chlorophyll. These distributions had previously been known only in broad outline from ship data. I shall frequently refer to the images of the global seasonal chlorophyll field which are shown in Color Plates 1–4.

The indicated chlorophyll values in the CZCS images represent not only surface chlorophyll but also integrate over a large fraction of the first optical attenuation depth for the relevant wavelengths, biased surfaceward where the chlorophyll profile is nonuniform. The sensed depth is approximated by $Z_e/4.6$, where Z_e is the euphotic depth (Morel and Berthon, 1989). Using the attenuation analysis of R. C. Smith (1981), we can infer that sensed depths range from about 25 m in clear oligotrophic water (0.1 mg chlorophyll m^{-3}) to about 5 m in eutrophic ocean water (10 mg chlorophyll m^{-3}).

Parenthetically, it will be convenient (in using these terms for the first time) to define how I shall use them. Originally referring to water containing low or high nutrient concentrations, "oligotrophic" and "eutrophic" are used more loosely these days respectively for (i) clear oceanic water perpetually with few biota and (usually but not always) low nutrients and (ii) green water with many biota and an originally abundant nutrient supply. If we can agree that "nutrient" means the Liebigian-limiting element (which may be a trace metal, as shall

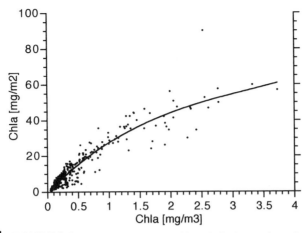

FIGURE 1.4   Relationship between chlorophyll values at the surface and values integrated down the profile: The surface value represents the chlorophyll observed by the CZCS sensors and the profile integration is performed with the profile parameters obtained for each province from the profile archive described in the text. The relationship is best described by a third-order polynomial.

be shown) and not always a macronutrient (as nitrate, silicate, and phosphate are coming to be termed), then the original meaning is almost preserved in my usage, which will refer to water clarity rather than anything else.

Returning to the CZCS images: As a rule, we may assume that the CZCS-indicated chlorophyll values represent mixed layer pigments and that subsurface chlorophyll maxima deeper than 25 m are not detected directly (Fig. 1.4). In this case, therefore, where uplift and illumination of the nutricline lead to a subsurface bloom, this may not be observed in water-leaving radiation. However, in most such cases we can expect that such a bloom will enhance mixed layer "background" chlorophyll values by the same vertical mixing process that maintains the mixed layer itself. This enhanced background may then be regionally detectable in the CZCS chlorophyll field, though this may not represent the true intensity of the subsurface bloom. Moreover, as will be discussed later, we do have a sufficient archive of observations of chlorophyll profiles to make a reasonable prediction of the seasonally variable subsurface chlorophyll field in all parts of the oceans.

It is difficult now to remember how ignorant we were of the extent, variability, and seasonality of algal blooms prior to their global visualization by the CZCS (Banse and McClain, 1986; Brock *et al.*, 1993; Müller-Karger *et al.*, 1989; Longhurst, 1993; McClain, 1993). CZCS data were at first available simply as a brochure of a few striking global and regional images, but the files are now available in a variety of formats; the usual starting point for study is the climatology of monthly means for 1978–1986 on a 1° grid covering all oceans and seas, comprising 42,732 points. This archive of "level 3" monthly composite images was generated as the arithmetic mean of the pigment data stored in the matching level 3 files for each 1° bin. This was the archive used, for instance, by Banse and English (1993) in their study of seasonality of near-surface chlorophyll in the ocean. Ease of access to this archive is continually improving. It should also be noted that many seasonal CZCS images, especially those that are now called "classics," may be downloaded from the SeaWIFS home page (URL at *http://seawifs.gsfc.nasa.gov*). These may help you interpret the text of this book. The individual, 4-km resolution level 2 scenes are available, and were used for this study, on a set of CD-ROMs.

For present purposes, I assume that a region of enhanced CZCS chlorophyll sustained between two monthly images represents an open-ocean algal bloom that should be explicable by reference to the oceanographic and ecological situation in that region and season. To locate the occurrence of such blooms I scanned all level-3 images (both monthly and seasonal) for the tropical Atlantic, eastern Pacific, and Indian Ocean. Many individual level-2 images were also scanned to ascertain the patch form of the seasonal blooms, detail that is obscured in the monthly composites. The chlorophyll values were computed from the water-leaving radiation (measured by the orbiting radiometer) with a NASA algorithm which uses the blue/green ratios at 433/550 and 520/550 nm respectively for low and high chlorophyll concentrations, and for which an accuracy of 35% is claimed in open-ocean areas (Feldman *et al.*, 1989). This algorithm accommodates a range of atmospheric humidity but performs less well in the presence of atmospheric dust veils or of humic substances and suspended sediment in near-surface water. In coastal regions, very high latitudes, and areas lying under the West African or Asiatic dust veils, other algorithms are more appropriate though still of limited reliability.

However, although the radiometer was unable to discriminate the chlorophyll signal from radiation emitted at the sea surface due to suspended sediments or dissolved organic substances in coastal waters, the CZCS images have been used in demonstrations of the feasibility of calculating primary production of whole ocean basins (e.g., Eppley *et al.*, 1985; Sathyendranath *et al.*, 1995; Longhurst *et al.*, 1995; Antoine *et al.*, 1996) and an image-based method has been devised for separating production fueled by nitrate from production fueled by biologically regenerated ammonium and urea, a vital distinction in biological oceanography (Sathyendranath *et al.*, 1991). It should be noted that a small number of seasonal graphs representing ocean-basin scale regions were derived from the CZCS data by Yoder *et al.* (1993), and some were verified by reference to *in situ* chlorophyll measurements: The correspondence was always in the correct sense and accurate enough for our purposes.

The new multiple-spectral radiometers which are coming on line will also discriminate between some taxon-specific combinations of individual chlorophyll pigments. It is already possible to distinguish coccolithophore blooms (by white reflectance from their coccoliths) in AVHRR images (Aiken *et al.*, 1992), and taxonomic mapping of this kind should become available for other major groups of phytoplankton as experience grows with the new sensors and better algorithms are formulated to correct for atmospheric turbidity, water vapor, and nonbiological turbidity of sea water.

After the CZCS sensors failed in the mid-1980s there was a long interval before further sea surface color sensors were flown: Paradoxically, this interval was useful in the sense that it enabled the CZCS data to be properly digested and widely circulated. Biological oceanographers were then better prepared to assimilate data from the next generation, the first of which (the Japanese Advanced Environmental Orbiting Satellite with its 12-band Ocean Colour and Temperature Scanner, having a 6-km resolution) was launched in 1996 and produced some interesting imagery, especially of the northwest Pacific Ocean, but unfortunately for only a limited period.

Just before this book was completed, the next major step was taken when the SeaWiFS/SeaStar satellite of NASA was placed in a 705-km altitude orbit after launch from an L1011 aircraft in August 1997, carrying spectraradiometers sensitive to eight SeaWiFS wavebands. Tests of all systems were successful, and useful sea surface color images, including chlorophyll values down to the first optical depth, were received in mid-September when the satellite became fully operational (see Frontispiece). Local area coverage (LAC) data are at 1.1-km resolution; these are transmitted directly to ground stations, with only small amounts being stored aboard the satellite. Global area coverage (GAC) data have a resolution of 4.5 km and are acquired daily as 14 GAC swaths, each 1500 km wide, starting in the Arctic Ocean and following the orbital path to Antarctica. The GAC data are acquired by the SeaWiFS project daily and are then available to users in a variety of formats, binned and unbinned, and in browse and archival mode. The reader is referred to the Internet site:

http://seawifs.gfsc.nasa.gov/seawifs_scripts/seawifs_subreg.pl.

Beginning in early October 1997, regular products have become available and, subjectively at least, appear to rival the CZCS images in quality and to better resolve the coastal zone. The files are imaged representations of five geophysical parameters: chlorophyll-a concentration, CZCS-type pigment levels, band 5 water-leaving radiance, band 8 aerosol optical thickness, and the band 3 diffuse attenuation coefficient. If the flow of data continues as planned, this is indeed a new beginning in biological oceanography.

Chlorophyll at the ocean surface, and some of its contained information, is the only biological field likely to be routinely available in the foreseeable future on the scale of whole oceans but with relatively fine resolution. Fields of other critical biological variables will surely remain inaccessible to direct measurement by sensors borne by earth-observing satellites. However, perhaps inferences can be drawn from the chlorophyll field in the same way that terrestrial ecologists might infer general distributions of the fauna associated with different biomes whose extent and distribution is obtained from maps of terrestrial vegetation. This is one possibility which is explored in this book.

THE USE OF CZCS IMAGES IN THIS STUDY

In addition to the archive of individual 4-km resolution scenes mentioned in the previous section, an archive of monthly means for 1978–1986 on a 1° grid was obtained from the CZCS surface chlorophyll fields; this file was generated as the arithmetic mean of the pigment data stored in the matching level-3 files for each 1° bin (Kuring, personal communication). From these, the actual chlorophyll values used in this study were obtained by the method of Sathyendranath and Platt (1989).

A sufficient archive of chlorophyll profiles, thought to be a particularly potent descriptor of regional biological oceanography since the early and groundbreaking review of Yentsch (1965), should enable us to extrapolate from the surface chlorophyll field to a three-dimensional representation of global chlorophyll distribution in the ocean. For this purpose, a new archive is available of 26,232 multidepth chlorophyll profiles, obtained from more than 60 individual sources, that had been gathered so as to compute global primary production from CZCS surface chlorophyll data (Longhurst *et al.*, 1995), a task which was the catalyst for the ecological geography proposed in this book.

Each profile in this archive was parameterized by fitting to a shifted Gaussian distribution as described by Platt and Sathyendranath (1988). This procedure delivered a set of parameters uniquely defining the shape of each profile: the depth of the chlorophyll maximum (Z_m, m), the standard deviation around the peak value (s, m), the total pigment within the peak (h, mg chlorophyll m^{-2}), the ratio (r') of peak height to total pigment, and the background chlorophyll value (B_o, mg chlorophyll m^{-3}). Specifying that a minimum of six depths were required for a successful fit of the model, we are left with 21,872 sets of profile parameters to use in the search for boundaries between ecological provinces.

Finally, and most important, primary production rate had also been integrated for the photic zone (mg C m^{-2} month) by a local algorithm (Platt and Sathyendranath, 1988; Sathyendranath *et al.*, 1995) which utilizes

- Surface chlorophyll from the CZCS data
- Parameters descriptive of the chlorophyll profile
- An assumption for the photosynthesis/light relationship
- Surface irradiance from sun angle and a cloud-cover climatology

Thus, for all the areas for which monthly images were available, we have mean values for sea-surface chlorophyll (mg chlorophyll m^{-3}), integrated profile chlorophyll (mg chlorophyll m^{-3}), and the integrated primary production rate (mg C m^{-2} day^{-1}).

For each of the provinces this seasonal cycle of production and accumulation of phytoplankton biomass can be associated with some of the environmental variables likely to force the biological cycle. On the same spatial and temporal scales we have mixed layer depth defined as appropriate for each region either by temperature or by density and also a computation of photic depth based on cloudiness, sun angle, and water clarity as described previously. By comparing the production obtained with an assumption of a uniform chlorophyll profile and that indicated by the profile parameters described previously, an estimate was obtained of primary production in the deep chlorophyll maximum as a fraction of total production.

These variables are assembled into an internally consistent series of seasonal graphs for each individual province, presented over a 1.5-year cycle so that boreal winter and austral summer seasons can be readily examined. These are, of course, less reliable for those coastal provinces where suspended sediments seasonally render the chlorophyll field unreliable and a few have had to be rejected on that count. With some graphs I have been able to associate a regional mean seasonal cycle of zooplankton biomass obtained from various sources.

The graphs are presented individually in Chapters 7–10 and are to be considered an extension to the descriptions of the ecological seasons of each prov-

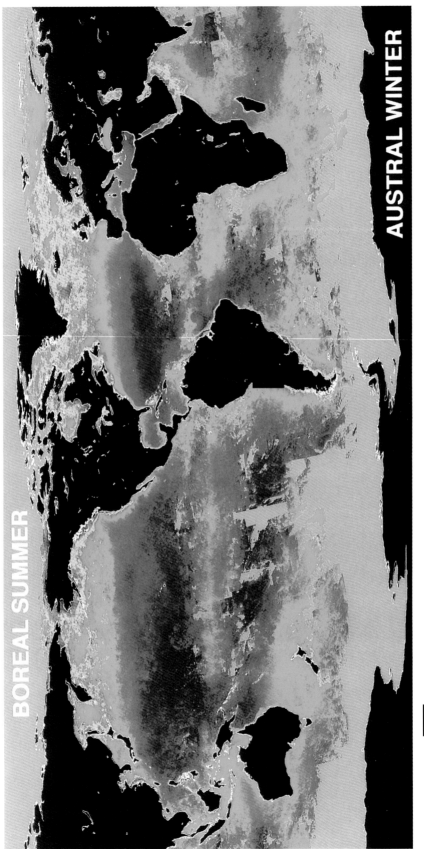

COLOR PLATE 1 Climatological (1978–1986) seasonal sea surface chlorophyll field obtained with Coastal Zone Color Scanner sensor for boreal summer (June–August). Color is a log scale for chlorophyll: purple = <0.06 mg Chl m^{-3}, orange-red = 1–10 mg Chl m^{-3}. Provided by NASA/Goddard Space Flight Center.

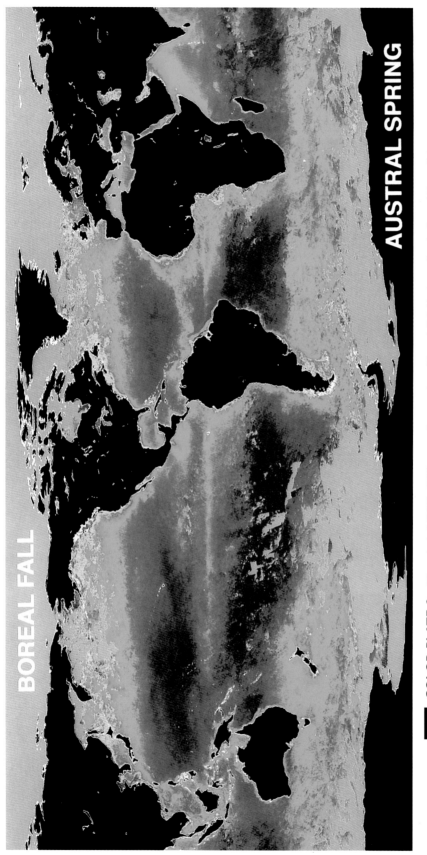

COLOR PLATE 2 Climatological (1978–1986) seasonal sea surface chlorophyll field obtained with Coastal Zone Color Scanner sensor for boreal fall (September–November). Color is a log scale for chlorophyll: purple = <0.06 mg Chl m^{-3}, orange-red = 1–10 mg Chl m^{-3}. Provided by NASA/Goddard Space Flight Center.

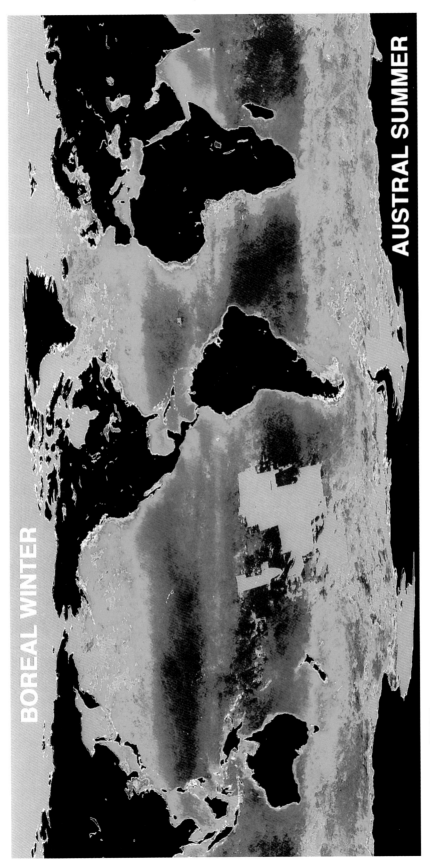

COLOR PLATE 3 Climatological (1978–1986) seasonal sea surface chlorophyll field obtained with Coastal Zone Color Scanner sensor for boreal winter (December–February). Color is a log scale for chlorophyll: purple = <0.06 mg Chl m^{-3}, orange-red = 1–10 mg Chl m^{-3}. Provided by NASA/Goddard Space Flight Center.

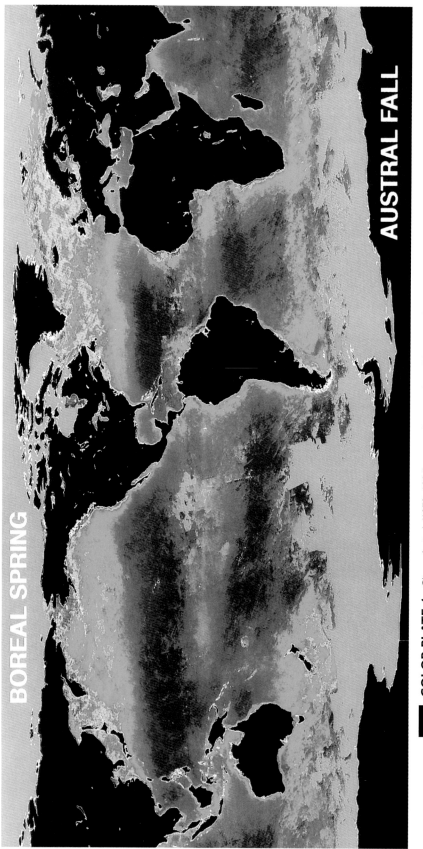

COLOR PLATE 4 Climatological (1978–1986) seasonal sea surface chlorophyll field obtained with Coastal Zone Color Scanner sensor for boreal spring (March–May). Color is a log scale for chlorophyll: purple = <0.06 mg Chl m^{-3}, orange-red = 1–10 mg Chl m^{-3}. Provided by NASA/Goddard Space Flight Center.

ince. You will find that each section describing an individual province ends with a brief, formal "synopsis" which I intend in each case to serve as an extended figure legend for the relevant seasonal graph. Where possible, I have compared these inferred seasonal cycles against seasonal data obtained at sea by conventional means.

THE CHOICE OF SCHEMES TO PARTITION THE PELAGIC ECOLOGY OF THE OCEANS

Terrestrial ecologists have little difficulty in partitioning their studies between natural compartments because the discontinuities and ecological differences between forest, savanna, and grasslands are so profound and easily observed as to be represented in our common vocabulary. Terrestrial ecosystem models intended to compute (for example) global production (e.g., Melillo *et al.*, 1993) are usually partitioned among 20–25 vegetation types mapped on the land surface from satellite images. Different assumptions are appropriate in modeling carbon assimilation within each vegetation type, though all are linked by common assumptions concerning photosynthetic biochemistry and the physiological response of plants to rainfall, sunlight, temperature, and nutrients.

At sea the situation is very different. Even to an oceanographer, one part of the ocean's surface looks remarkably like all the rest, and relatively few oceanographers have sought to partition its apparent continuity. For the most part, it has been thought necessary to identify a real discontinuity only between coastal and oceanic regions. However, we should be able to do better than that, and this book proposes specifications for a set of pelagic biomes compatible with the well-established sets of terrestrial biomes. It is further suggested that much can be inferred about the ecological structure of each oceanic biome from a simple set of environmental variables, comparable to the minimal set (e.g., latitude, altitude, exposure, rainfall, and geological substrate) from which terrestrial biome distribution may be predicted (Atjay *et al.*, 1979). The discussion is simplified as far as possible by avoiding the use of complex terminology. Only two terms are used from ecosystem studies and biogeography: biome and province. These are intended to denote the primary and secondary hierarchical areas of the upper ocean for which definable and observable boundaries are suggested and within which it is suggested that unique ecological characteristics may be predicted. Biome has the same use as in terrestrial plant geography, where it is used to denote a characteristic type of vegetation (tundra, wet tropical forest, dry grassland, etc.), and is probably a better term for the primary spatial biological divisions of the ocean than "domain," which has previously been used (e.g., Longhurst, 1995; Banse and English, 1993), because it emphasizes that these divisions are similar to the major terrestrial vegetation types.

It was a suggestion of Yentsch and Garside, made to a congress of pelagic biogeographers in 1985, which most nearly foreshadowed the proposals made in this book for an ecological classification of the seas. They asked if the major patterns of pelagic biogeography were a response to seasonal and spatial patterns of primary production of algae. If so, could they be reproduced by analysis of the pattern of physical forcing of algal growth rather than (they might have added) by tedious and expensive analysis of net samples? In fact, this was such a novel suggestion that it will be useful to quote briefly from their actual remarks:

> Are the large-scale biogeographic distributions of oceanic biology a response to seasonal and spatial patterns of primary production? . . . Geostrophic currents dictate the shape of the density structure and what we term the 'degree of baroclinicity' is a global mirror of primary production. The fact is that for at least fifty years we have recognized that . . . light, wind, temperature . . . regulate phytoplankton growth. We recognize that seasonality in the primary processes is controlled by stirring of the upper layers of the ocean, and the nutrient-density field. . . . Why, then, are we so hesitant to tie the climatological primary process together with concepts concerning biogeography? One answer might be that we do not have enough knowledge of primary productivity over large areas of the worlds oceans.

This was written before the general availability of images from the CZCS sea surface color sensors and the maps of global primary production computed from them.

It was partly the availability of the global CZCS chlorophyll field that reactivated the search for a satisfactory way of defining ecological provinces. Platt and Sathyendranath (1988) suggested that calculations of primary production should be partitioned between biogeochemical provinces (BGCPs) within which photosynthetic parameters, and the form of the chlorophyll profile, might be seasonably predictable. Their proposal rests on the fact that regional differences exist in the photosynthetic response of phytoplankton to changes in environmental conditions, probably associated with changes in the composition of the phytoplankton. Following this proposal, but using different biooptical criteria, Mueller and Lang (1989) suggested how the northeast Pacific may be partitioned objectively into provinces compatible with the concept of the BGCP.

At the same time, not all biological oceanographers think that ecological models must be partitioned to achieve satisfactory global integration. Morel and Berthon (1989), for instance, suggested that surface chlorophyll specifies subsurface chlorophyll distribution sufficiently well that a one-dimensional model of primary production may be scaled to the surface chlorophyll field and applied globally. This suggestion is not so far from the concept of ecological continuity that is often assumed when one-dimensional models of biological processes are incorporated into general circulation models or other simulations. Ecological continuity requires either that the parameters of a biological model remain constant everywhere or that they always respond in the same way to changes in the environmental covariables such as temperature and nutrients. However, biological responses to changes in environmental conditions are often species dependent and, in nature, further complicated by species succession. Therefore, they may be highly nonlinear. In fact, the debate between partitions and a continuum is somewhat sterile because both concepts will be required for different purposes, and they are not mutually exclusive.

Even in the absence of a definition of a rational set of boundaries for biomes at the ocean basin scale, it is often assumed that such a definition must exist. A recent example is that of Dickey *et al.* (1993), who discussed the different physical conditions controlling algal growth in two "different oceanographic provinces" of undefined extent: in fact, they were comparing conditions at two isolated stations in the western Atlantic [Station "S" of Menzel and Ryther (1960) and their own BIOWATT station] where long-term observations had been made. The assumptions behind this approach, which is common in the literature of biological oceanography, are unhelpful to global integrations of any biological process.

Parallel with the continuing search for biogeographic patterns based on aggregated distributions of individual pelagic species, other proposals have been made for a biological geography of the oceans based on other criteria: surface water mass characters (e.g., Emery and Meincke, 1986), nutrient distributions (Fanning, 1991), features of the surface circulation (Ware and McFarlane, 1989), and the depth of the mixed layer (Longhurst and Harrison, 1989). More economically, it has been suggested that the primary ecological classification of the ocean should simply bifurcate into strongly and weakly stratified areas (Cushing, 1989) or that three domains based on relative seasonal variation of chlorophyll and relative nutrient depletion should be recognized: (i) low-latitude oligotrophic gyres with little seasonality in phytoplankton productivity, (ii) regions where phytoplankton growth does not exhaust available nitrate at any season, and (iii) regions where a spring bloom is followed by summer oligotrophy (Banse and English, 1993).

In the same vein, Barber (1988) identified six characteristic ecosystems (coastal upwelling, low-latitude gyre, equatorial upwelling, subarctic gyre, southern ocean, and eastern boundary current) though he did not suggest that the whole surface of the ocean could be allocated to one of them. Barber also addressed the difficulty of isolating physical cause and biological effect within the unique boundaries of any region of the ocean, emphasizing the effects of distant physical forcing by, for instance, wind stress on the far side of an ocean basin. We must bear both these difficulties in mind when devising specifications for oceanic biomes. A system on a somewhat similar scale but having a much simpler, indeed simplified, ecological approach has been used by fisheries scientists in the analysis of long-term change in fish stocks and in their biological support systems in each of a comprehensive global series of large marine ecosystems (LMEs), which are rather loosely defined but recognizable coastal regions having characteristic fisheries resources (Sherman and Alexander, 1989; Sherman *et al.*, 1996). The analysis of each LME includes a simple consideration of its biological oceanography which attempts only to state the characteristic biomasses of phytoplankton and zooplankton with their decadal-scale variability. Because, as will be discussed later, the partitioning of the coastal regions used in this book comprises, in many cases, provinces that could with profit be subdivided, I recommend that you seriously examine the LME entities if you are principally interested in partitioning the coastal zone.

An elegant scheme for understanding the ecological geography of clupeid fish was proposed by Rosa and Laevastu (1959), foreshadowing the suggestions of Yentsch and Garside. These authors partitioned the oceans into 3 primary, 6 secondary and 37 tertiary compartments, each with specified boundaries, and they systematically listed some characteristics of the oceanography, meteorology, and biology of each. Unfortunately, this far-sighted proposal was ahead of its time. At that time, there was totally inadequate information about conditions in many of the proposed compartments. There are many points of agreement between what is proposed here and the scheme of Rosa and Laevastu. In particular, their basic division of the ocean into three compartments—polar, boreal, and tropical—resembles the primary biomes argued here on oceanographic grounds. Another early system, that of Beklemishev (1969), can also be interpreted as supporting a division of the pelagos into a small number of primary biomes: here, approximately 25 subregions of the whole ocean are grouped into polar, coastal, gyral, and transitional classes, though the coastal class is more generously defined than the coastal biome introduced in Chapter 4. A number of maps have been based on the characteristics of the surface water masses of physical oceanographers: This category suffers from a common failing such that the maps thereby ignore the ecotones lying at the boundary between surface water masses. Ecotones are ecologically important and some form regional entities which must be recognized.

What is attempted here is a framework for a regional ecology of the ocean that brings together our knowledge of relevant physical features of regional oceanography with what we know of the response of planktonic algae to seasonal forcing by physical processes. The proposed scheme is not doctrinaire and does not depend on the establishment and application of a set of rigid criteria; rather, each ocean has been examined independently to determine how best it may be partitioned as part of a global scheme.

2

ECOLOGICAL GRADIENTS

Fronts and the Pycnocline

We naturally expect that boundaries between biogeographic or ecological provinces, if indeed these are definable, will be sharpest where there are the strongest discontinuities in the physical environment. Even a brief acquaintance with the biogeographic literature will reveal a preoccupation with the performance of fronts as barriers to distribution or as ecotones with special ecological characteristics, although recent advances in understanding the physical dynamics of fronts have demonstrated the extent to which parcels of water, containing biota, pass from one side to the other by a variety of mechanisms which will be reviewed in this chapter. However, fronts can indeed represent a leaky boundary between different ecological regimes, particularly where tidally mixed shelf water meets stratified water at a sharp discontinuity in shelf sea fronts. On a much larger scale, the discontinuities at the convergent frontal zones of the Southern Ocean, where subantarctic water passes below the subtropical surface water mass, are well-known to be observable at sea because of the relatively high biomass of plankton that are supported at these ecotones.

However, as writers on oceanic biogeography from the earliest time have been aware, it is in the vertical plane that the most significant environmental gradients and discontinuities occur. The epipelagic zone, whose ecological geography is the subject of this study, is a thin layer of light, lighted, and wind-mixed water lying atop the mass of the ocean; the change in physical and chemical properties between the two depth zones is more frequently abrupt than gradual. When abrupt, the ecological changes across a few tens of meters depth are greater than those across any known vertical front which intersects the sea surface.

This phenomenon is most striking in the eastern parts of the tropical oceans. In the Gulf of Guinea, at a depth not much greater than the height of a tree in the tropical forest, and across a depth not much greater than the height of the shrubs on the forest floor, the temperature drops from 28°C to about 16°C and a different planktonic fauna is encountered. The benthic fauna of the shelf edge and upper slope in the same region strikingly resembles (species for species) the fauna at similar depths off the west coast of Europe, and the temperature of their environment differs by less than 2°C. On the other hand, the environment and biota of the epipelagic zone differ fundamentally between the two regions. Consequently, as has been noted by many authors, a diel migrant copepod or euphausiid moving across this boundary in the tropics at dawn and dusk passes makes an environmental adjustment as great as traveling several thousand kilometers equatorward or poleward.

It is therefore useful at this juncture to discuss the ecological significance not only of vertical fronts, such as the north wall of the Gulf Stream which intersects the sea surface, but also the global horizontal front that occurs at the shallow or seasonal pycnocline, usually at a depth between 30 and 200 m.

OCEANIC AND SHELF SEA FRONTS ARE ECOTONES AND LEAKY BIOGEOGRAPHIC BOUNDARIES

The ecotone, or transition zone between two ecological communities, is an established concept in terrestrial ecology (Odum, 1971) but has been little discussed in biological oceanography. Ecotones are, by definition, linear and less extensive than the communities they separate. They are associated with a gradient either in the physical environment or in an external stress in which, for example, grazing pressure is more or less constrained to one side of the ecotone. It is often observed that ecotones exhibit special ecological characteristics which differ from either of the separated communities, and they may be the habitat of specialized "edge-effect" species. For all these reasons, Odum notes, "we would not be surprised to find the variety and density of life greater in the ecotone." A short examination of ecotones at sea will demonstrate a similar effect there–an effect that must be acknowledged in any overview of the ecological geography of the surface waters of the ocean.

Linear features with accentuated gradients in the properties of their surface waters can be found at all scales in the ocean and should be regarded as ecotones; as already noted, a significantly enhanced ability to map and understand such features has been one of the early benefits of satellite imagery. In the sections devoted to individual regions in the chapters which follow, I shall use this information in a descriptive sense, but it will be useful now to examine three kinds of ecotones more generally: (i) those associated with continental shelves and the effect of the semidiurnal tide, (ii) those associated with mesoscale eddies and filaments in the open ocean, and (iii) those that may occur between two major surface water masses or current systems.

Perhaps most important for shelf and shelf-edge fronts, we must remember that these dynamic fronts are leaky boundaries at best. Physical processes associated with the formation of fronts mediate the transfer of water from one side to the other and are probably responsible for the higher biomass and growth rates of phytoplankton which mark the front. We should, therefore, briefly review the physics and ecology of shelf and shelf-edge fronts so as to understand their global distribution. The following account owes much to recent investigations and reviews of the North Atlantic coastal fronts and is stimulated by the VHRR imagery that became available to map such fronts during the 1970s (see, for example, the review by Le Fèvre, 1986, and research papers by Simpson, 1981; Pingree and Mardell, 1981; Loder and Platt, 1985; Horne *et al.*, 1989).

Shelf Sea or Tidal Fronts

When the semidiurnal tide encounters shoaling water, as over the continental slope, the amplitude of the tidal wave and its horizontal velocity progressively increase. At some depth, usually shoaler than the continental edge, vertical turbulence produced by friction between the tidal stream and the seabed is sufficiently enhanced (when added to turbulence produced by wind stress at the sea surface) as to overturn seasonal thermal stratification of the water column, giving rise to the tidally mixed regions of the shelf. As an aside, John Simpson likened the frictional effect of tidal streams to "hurricane force winds blowing regularly twice each day": no wonder, then, that stratification is so regularly broken down on continental shelves. The tidally mixed and stratified regions of the shelf are separated by a frontal region that migrates semidiurnally and also seasonally because, of course, the area of vertically mixed water over the shelf is significantly larger in winter because of increased wind mixing. There exist other and more complex formulations, but it is convenient to use maps of the parameter $\log_{10} h/u_s^3$ (where h is water depth and u_s is the tidal current at the surface) to locate the transition between mixing and stratification: for most shelf areas, the critical value of this parameter to locate a tidal front is between 1.8 and 2.0. More simply, a tidal stream of 1 m sec^{-1} in 100 m of water falls within this range and therefore mixes (Figs. 2.1a and 2.1b).

Though it has been suggested that high biomass of phytoplankton observed at fronts results from accumulation rather than growth *in situ*, this is probably not usually the case at shelf sea fronts. Nitrogen uptake within the Georges Bank front was 0.36 nM m^{-2} sec^{-1} compared with 0.09 on the mixed side and 0.18 on the stratified side of the front; in nearby oceanic water, the demand was only 0.02 μM m^{-2} sec^{-1} (Horne *et al.*, 1989). Phytoplankton demand at the front was partitioned between nitrate and ammonium (using ^{15}N tracer techniques) showing that 61% of the daily production within the front was new production, based on nitrate introduced into the front by the processes discussed previously, compared with only 41 and 27% in stratified and mixed water, respectively. We can therefore have confidence that the remarkably precise coincidence that has been observed between shelf sea thermal fronts and chlorophyll, as observed on the Plymouth–Roscoff section in the western channel using towed undulating instrumentation (Aiken, 1981), represents *in situ* growth rather than accumulation. This conclusion represents a partial confirmation of the concept of fronts as ecotones, though at the scale of shelf sea fronts the exchange between adjacent waters will be sufficiently dynamic that we will not expect that edge-effect species could maintain self-sustaining

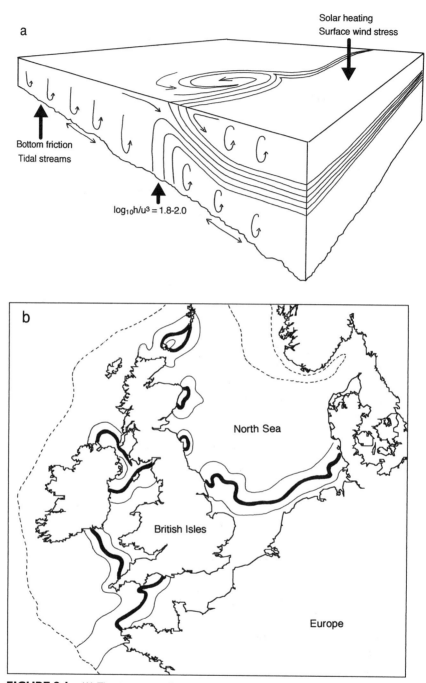

FIGURE 2.1   (A) The structure and circulation within a convergent, prograde tidal front illustrating how bottom friction in tidal streams overturns the pycnocline of offshore water to form a front between tidally mixed and stratified water where the critical value of the h/u^3 parameter occurs. Enhanced value of chlorophyll may be expected at such a front, most likely in the stratified water just seawards of the front, for the same reason that the canonical spring bloom must be constrained below by stratification. (B) The European continental shelf, showing the average locations of tidal fronts between stratified and mixed water. Mean position of the critical value of the h/u^3 parameter (heavy lines) and extreme positions (thin lines) are shown.

populations within the front, except perhaps for species at high trophic levels. Thus, there is some evidence that even shelf sea fronts are attractive feeding zones for fish and seabirds.

Several mechanisms have been advanced to account for the enhanced availability of nutrients in shelf sea fronts, always on the assumption that here phytoplankton growth is nutrient limited. The simplest explanation is that as the lunar month advances, tidal friction increases and the front advances toward its stratified side, progressively incorporating nutrients from below the stratification. Other, more general models were proposed independently in 1981: Holligan suggested that on the stratified side of fronts, phytoplankton growth is released from constraints of light limitation that would occur below the stratified layer and from nutrient limitation that would occur on the mixed side of the front; Tett suggested that vertical eddy diffusion from bottom friction should be stronger in the front than on either side, thus constantly supplying nutrients from the near-bottom layer.

Other potential mechanisms exist to explain linear zones of enhanced plankton biomass in shelf seas; for example, in the North Sea, where the velocity of tidal streams falls below a critical value, organic material sinks from suspension and forms linear zones of soft mud at midshelf depths, of which the Friesian Front off Holland (Baars *et al.*, 1991) is a typical example. In summer, above such benthic fronts the flux of remineralized nitrogen from the organic sediments is thought to be responsible for a zone of enhanced phytoplankton production in the water column. However, in practice it may be difficult to separate this effect from that of tidal fronts which may be aligned with the linear benthic remineralization zone.

Shelf-Edge Fronts

It has long been known that a linear zone of cool surface water, supporting a plankton bloom, overlies the upper slope and shelf edge off western Europe. Satellite imagery shows this to be a quasi-permanent feature, at least 800 km in length and wider at the end of summer than in spring; also, equivalent features occur over other shelf edges elsewhere in the world. For this often retrograde shelf edge front, explanations have evolved more slowly than for tidal fronts since the early suggestions by Leslie Cooper that westerly gales might overturn (he actually used the more colorful term "capsize") the water column. Not all agree on the details of the mechanism by which deeper, cooler, and nutrient-rich water is entrained surfaceward, but earlier explanations involved wind-induced upwelling or overturning of the water column by gales. Now, the most widely accepted explanation involves the generation of internal standing waves on the thermocline at the shelf edge where the tidal stream encounters rough topography on the seabed. These are thought to originate principally during the seawards, off-shelf tidal stream when maximum velocities occur, and the existence of long standing waves has now been confirmed by the satellite observation of linear, sun-glint features at the sea surface having a wavelength of approximately 40 km. Together with wind mixing, such features are sufficient to explain the observed surface cooling and nutrient enhancement.

At shelf edge fronts, a variety of mechanisms have been invoked to account for cross-frontal mixing and, hence, for the enriched phytoplankton biomass

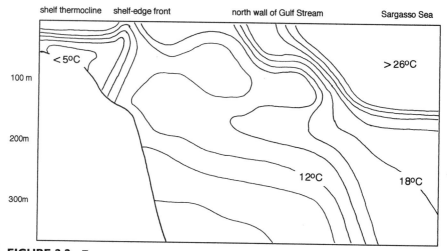

shelf thermocline   shelf-edge front   north wall of Gulf Stream   Sargasso Sea

FIGURE 2.2 Temperature section across the shelf and slope region of the northwest Atlantic (at about 40°N, in summer) to illustrate the relationship between the convergent shelf edge front associated with the break of slope and the oceanic front at the north wall of the Gulf Stream. Note the bubble of shelf water (<10°C), calving at the shelf edge; these features, called "bourrelets froids" on the European shelf, occur where the stratification parameter takes maximal values and bottom water is thus shielded from summer solar heating (from Bowman and Esaias, 1978, Figs. 2, 3, and 5).

observed to occur there (Fig. 2.2). Most important, perhaps, is vertical shear within the tidal stream, together with the effect of baroclinic eddies having a semidiurnal frequency. Horne and Petrie (1986) described "shear-flow" dispersion as another potential mechanism for transfer across the front. In such cases, nutrient-rich subsurface water passes into more shallow areas with the incoming tide, to retreat a half-tidal cycle later with a diminished nutrient content. Whatever physical mechanisms may dominate at each location, rates of cross-frontal transfer may be quite high and apparently sufficient to maintain the observed phytoplankton biomass. It should also be noted that shelf edge fronts may also occur seawards of coastal upwelling regions, in which case they appear in the chlorophyll field as a secondary, outer zone of biological enhancement.

Oceanic Fronts

For the open oceans, I shall distinguish between two kinds of frontal systems: (i) linear convergent zones, usually at the confluences of two major current systems, and (ii) linear features or filaments around, or shed from, mesoscale eddies whether or not these are associated with topographic features. Observations of enhanced levels of biomass are commonplace for both categories, but it is at oceanic confluent fronts which are also convergent that the evidence for the accumulation of "wrecked" biota is greatest. These are the "siomes" of Japanese oceanographers and fishermen, where turbulence at the confluence may disturb the wind–wave system and aggregations of biota may concentrate pelagic fish and birds and attract fishermen (Sournia, 1994). At such fronts, accumulation of biota may depend on their flotation/sinking rates or swimming ability to avoid submergence in the descending, convergent water mass (e.g., Owen, 1981).

At fronts where vertical turbulence associated with shear deformation is sufficiently strong, the consequent entrainment of nitrate from below the nutricline may induce localized zones of enhanced phytoplankton growth and the accumulation of biomass. Weaks and Shillington (1994) have studied the consistently higher-than-background (by an order of magnitude) chlorophyll biomass that is seen in the Coastal Zone Colour Scanner (CZCS) imagery in the southern Subtropical Convergence extending zonally for great distances to the southeast of South Africa; they attribute this effect to the stability induced by the passage of cold, nutrient-rich subantarctic water sliding below the subtropical water. This suggestion is supported by the observation that the feature weakens during winter, when vertical wind mixing is greatest.

We shall discuss several of the major frontal regions (e.g., at the Kuroshio/Oyashio and South Equatorial/Peru current confluence regions) in later chapters. We shall find that some are so sharply defined at the ocean surface as to continue to surprise oceanographers who come across them at sea, whereas others are sufficiently broad and permanent that we shall be tempted to identify them as specific ecological provinces. Thus, some will be entities and others will be boundaries between entities.

The linear fronts and filaments associated with mesoscale eddies could only be understood once satellite imagery of their real form and evolution became available, and this new kind of information has been rapidly exploited so that we now have a good understanding of their distribution in the ocean, how they modify the chlorophyll field, and, hence, their significance in ecological geography. The enhanced chlorophyll concentrations confirmed by CZCS data to occur consistently around the margins of warm-core (anticyclonic) mesoscale eddies are the result of geostrophic forces in the high-velocity zone around the edge of such features (Yentsch and Phinney, 1985; Lohrenz *et al.*, 1993). The required vertical motion will occur if vortex contraction on the flank of such eddies is sufficiently strong and sustained (Woods, 1988; Woods *et al.*, 1977) and usually this effect is limited to patches on the 10-km scale, which serves to explain the beaded string of high chlorophyll often observed around anticyclonic eddies in CZCS images. The same effect is observed on the filaments shed by mesoscale eddies, especially on their curved rather than their straight sections (Tranter *et al.*, 1983). It is also suggested, in relation to the Agulhas eddy field, that production within warm-core rings is limited by convective instability, but around the edge of such rings stability (and hence enhanced productivity) is conferred by the warm water of the ring overlying the cooler water in which it is embedded (Dower and Lucas, 1993).

However they are formed, frontal zones are prominent features not to be neglected in any ecological geography of the ocean; we shall frequently return to this theme in later chapters. It should also be noted that they form only leaky boundaries in the biogeography of individual species; although it is commonplace—as it has been since the beginning of descriptive biological oceanography—to observe that different groups of species and different ecological conditions dominate on either side of oceanic fronts, it is also commonplace to observe that individuals of many, perhaps most, of the relevant planktonic and nektonic species can also be found in small numbers on the other side of the front. Given the dynamic exchange of water across fronts at all scales, it could hardly be otherwise. The problem of expatriation (in three dimensions) is a problem which bedevils all attempts at a tidy biogeography of the pelagic biota of the oceans.

THE UBIQUITOUS HORIZONTAL "FRONT" AT THE SHALLOW PYCNOCLINE

The fronts discussed in the previous section are all visible where they intersect the surface, but the most ubiquitous front in the ocean lies hidden below the surface. This is the seasonal or tropical pycnocline and is globally associated with the change from epipelagic to deeper ecosystems, so it is the most significant feature in the three-dimensional ecological geography of the oceans. It is useful to remind ourselves of some of its main characteristics: most important, what determines its depth and the strength of its density gradient and whether or not it occurs within the lighted zone.

Though the depth at which the pycnocline occurs is determined principally by baroclinicity associated with the ocean circulation, local processes also modify mixed layer depth at least seasonally: turbulence induced by wind stress at the sea surface, shear-stress turbulence induced at its base by inertial oscillations of the mixed layer water mass induced by impulsive wind stress, local heating at the surface by short-wave solar radiation, and the local supply of fresh or brackish water. It will, of course, be useful to remember that density of seawater is greatly more influenced by temperature than by salinity, which (with some important exceptions) is only dominant in evaporative basins or estuaries.

All these factors must be considered in the next chapter, so here it is sufficient to say that the surface wind-mixed layer is differentiated from the interior of the ocean—to which mixing induced by surface processes cannot reach—across a permanent pycnocline whose depth varies from 25 m in the eastern tropical oceans to 250 m in the center of subtropical gyres. It may also be noted, parenthetically, that the baroclinic upsloping of nitrate isopleths toward the edges of the anticyclonic subtropical gyres is a necessary condition of the bowl-shaped mixed layer of these gyres and is reflected in the surface chlorophyll field. For this reason, this has consistently higher values around the edges and the lowest values in the center of the subtropical gyres of each ocean.

At higher latitudes, where seasonality of wind stress and solar heating is stronger, there is a clear discontinuity between the mixed layer of one year and that of the next. After winter mixing and cooling has produced a surface water mass of uniform density down to the depth of the permanent pycnocline, a new shallow seasonal mixed layer develops during spring and progressively deepens during the summer, only to be eroded again by wind mixing during the subsequent winter. Peak development of the summer mixed layer is typically several months after midsummer, and greatest penetration of deep convection occurs several months after midwinter. In some regions, a halocline may somewhat obscure the simple model.

You should be aware, when reading seasonal graphs of mixed layer depth which are based on archived time series of monthly mean values (such as those used in this book), that these may not capture the establishment of a near-surface stratification in spring, far above the winter pycnocline, by solar warming, and hence the induction of buoyancy in the near-surface layer. Instead, they may show a progressive shoaling from a very deep to a very shallow mixed layer depth—a process which simply cannot happen, or at least not by the advent of spring sunshine and weaker winds. This is due to the averaging of shoal and deep values during a single month (or even during two months) indicating a false mean value somewhere between the two. In fact, you will find in many (perhaps most) published models of the seasonal phytoplankton growth that the mixed layer depth is allowed to shoal progressively in early spring as winter wind stress is relaxed. This assumption, of course, violates Dodimead's first rule for thermoclines: that they can deepen by mixing warmer surface with cold deeper waters but they cannot shoal by unmixing the same! It appears as if a confusion has arisen between the diel depth of wind mixing, which does of course follow the pattern often used by modelers, and the depth to the seasonal pycnocline. It is the latter, of course, which has the greater biological significance because it carries the seasonally variable nutricline with it.

At all latitudes, vertical velocity (Ekman's W_E) is imparted to the water column by the curl of the local wind stress at the sea surface, and this motion shoals or deepens the pycnocline. Cyclonic stress imparts positive values to W_E; anticyclonic stress imparts negative values. In this book I follow the notation of Isemer and Hasse (1987): Therefore, positive values of W_E (upward motion) shall be labeled "Ekman suction" and negative values (downward motion) shall be labeled "Ekman pumping." You should be aware that in many papers the latter term is loosely used without definition and often in the opposite sense to that used by Isemer and Hasse or by Tomczak and Godfrey (1994).

The pycnocline, whether seasonal or permanent, lies at a depth generally predictable for each region and has a gradient predictable from its depth: in short, the shoaler the pycnocline, the sharper the gradient. From its depth (and some knowledge of regional oceanography) the vertical gradients in other properties of ecological significance that are associated with it may also be predicted, especially nutrients, light, and the biomass of both phytoplankton and zooplankton. The interactions between nutrients, light, turbulence, and the biota in the mixed layer and pycnocline have been a central theme of biological oceanography in recent decades: for a good discussion of all this, see Mann and Lazier (1991) or Banse (1994).

The simplest expression of the ubiquitous submerged, horizontal front occurs in the typical tropical profile (TTP) of Herbland and Voituriez (1977) which is most clearly expressed in the eastern parts of the tropical oceans. More generally, the TTP is a description of an oligotrophic profile, either the end member of plankton succession in midlatitudes from spring to summer or the permanent tropical condition (Fig. 2.3). During oligotrophic conditions, the depths of nutricline, pycnocline, deep chlorophyll maximum (DCM), and productivity maximum do not differ by more than a few meters, and the vertical distribution of zooplankton and micronekton also conform in a predictable manner (Longhurst and Harrison, 1989). Below the pycnocline, bacterial oxidation

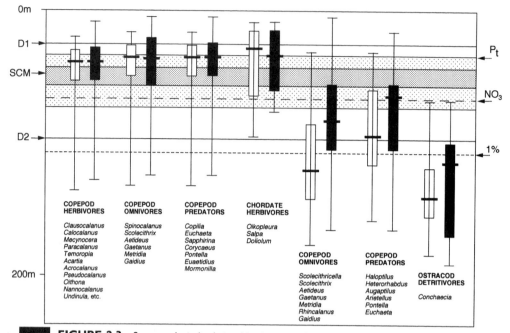

FIGURE 2.3 Some ecological relationships in a typical tropical situation, representing the pelagic ecosystem at the BIOSTAT station in the eastern tropical Pacific (Longhurst and Harrison, 1989). Day and night distributions of groups of plankton organisms having similar feeding strategies, showing depth of maximum numbers, depth range of 50% of the population, and depth range of all except extreme outliers. D1 and D2 are the top and bottom of thermocline as defined by Wyrtki (1967); the DCM lies within the depth zone of relatively high chlorophyll, P_t is the depth of maximum production rate, NO_3 is the depth above which nitrate is undetectable, and 1% is the depth of 1% surface irradiance.

of sinking organic material commonly leads to the development of an oxygen minimum zone and a characteristic anomaly in the usual vertical distribution of plankton profiles (Longhurst, 1967; Saltzman and Wishner, 1997a,b).

It is now generally accepted that in many situations the appropriate model for an oligotrophic profile has a two-layered euphotic zone. Where the mixed layer water is sufficiently clear (few algal cells and little suspended particulates) that light attenuation is dominated by seawater absorption, the 1% isolume often lies within the pycnocline so that the upper (mixed layer) zone is well lit and nitrate poor, while the deeper (pycnocline) zone is poorly lit and nutrient rich. Here we have a sufficient model for the steady-state DCM in which algae are larger, shade adapted, and receive a constant flux of new nitrate across the nitracline so that new production is high relative to production based on regenerated ammonium. On the contrary, the well-lit and nitrate-poor upper mixed layer has low rates of new production relative to ammonium-based, regenerated production. Production in the DCM relative to the production in the mixed layer is complex to compute but probably usually lies in the range 5–50%. Though the shade-adapted cells of the DCM may have access to adequate nitrate, their P_{max} (photosynthetic rate per unit of chlorophyll at light saturation) may be 10 times lower than that for near-surface cells. The plots of seasonal cycles for individual provinces (see Chapters 7–10) show that while an illuminated pycnocline is the normal condition in low latitudes, it is ephemeral in polar seas. The same graphs illustrate the increase in the absolute rate of primary production in the deep chlorophyll maximum during seasons when the pycnocline is illuminated.

In short, all this describes an ecosystem vertically ordered about a discontinuity in the density, nutrient, and biotic gradients. Nitrate diminishes to very low values upwards across the pycnocline as both chlorophyll biomass and the rate of primary production increase. Zooplankton herbivores are often concentrated as dense layers at the depths of maximum rate of algal growth, though the total vertical distribution of zooplankton is complicated by the vertical migrations of some species, usually with diel or seasonal frequency, across the pycnocline. In this way, the migrant species utilize to their advantage the contrasting ecological conditions of both euphotic and bathypelagic zones.

Ocean basin-scale baroclinicity defines a pycnocline topography of troughs, ridges, bowls, and domes associated with the geostrophic flow, and everywhere the pycnocline itself and the other gradients associated with it remain essentially in the same depth sequence. Generally, the deeper the pycnocline, the greater the depth interval over which the features of the oligotrophic profile are spread. In later chapters, I discuss many regional variations on this general theme, forced by regional topography and climate, but this introduction is sufficient to demonstrate the ubiquity of this most significant ecotone and the most important ecological boundary in the ocean excepting only the sea surface, the sea floor, and the shoreline.

BIOGEOGRAPHY OF THE SHALLOW PYCNOCLINE: BOTH HABITAT AND BOUNDARY

Over the 4 or 5 km from the deep sea floor to the surface, the ordered change in the characteristic biota is both predictable and equivalent to 4000 or 5000 km meridional distance at the surface. It is, as suggested in the previous section, at about 25–200 m below the surface that the rates of change in everything we might care to measure are greatest. This is an ecotone and should have the characteristics of one; that is, not only should ecology be different above and below it but also there should be special ecological conditions within it.

If there is a characteristic flora at the pycnocline as distinct from shade adaptation of phytoplankton also occurring in the upper euphotic zone, we could regard it as the shade flora of the ocean and comparable with the shade flora of the forest floor. In fact, for each major group of photosynthetic cells, there is evidence that an oceanic shade flora does indeed exist (Longhurst and Harrison, 1989). In discussing this evidence, let us begin with the smallest cells.

The ultraplankton are now known to occur ubiquitously and contribute significantly to total autotrophic production. Of the ultraphytoplankton production of the subtropical North Atlantic, the 0.7-μm prochlorophyte (*Prochlorococcus*) and the 1.0-μm cyanobacteria (*Synechococcus*) contribute, respectively, 75 and 10%, whereas the 0.7- to 5.0-μm eucaryotes contribute another 10%, though the carbon biomass of the latter is about three times greater than that of the other two fractions combined. Although *Synechococcus* is capable of growth over a wide range of light intensities, the time required for photoadaption at typical rates of vertical mixing is longer than its generation time. This may explain why cyanobacteria and prochlorophytes are most abundant in the mixed layer of the North Atlantic from the Gulf Stream to Morocco, whereas peak abundance of small eucaryotes, mostly chlorophytes, prymnesiophytes, and chrysophytes, occurs at the DCM (Li and Wood, 1988; Li, 1995). In the North and northwest Pacific, very small eucaryotes, especially *Micromonas* (1–3 μm), also dominate the DCM. Thus, within the smallest photosynthetic cells, there does appear to be evidence for the existence of a specialized shade flora.

It is the larger cells, especially dinoflagellates and diatoms, for which we have the best profiles and it is in the open ocean, rather than on the shelf, that we would expect to find the clearest evidence that some species of algae may be able to overcome vertical turbulence and so to occur only at certain restricted depths in the layered ecosystem of the euphotic zone.

Most of the profiles which have been examined for the existence of a shade flora have come from the North Pacific. Early in the season in temperate or subtropical regimes, during periods of relatively high turbulence and diapycnal mixing, the taxa at the DCM resemble those in the mixed layer above (Venrick *et al.*, 1973; Taniguchi and Kawamura, 1972). However, at 26°N in summer, after a pycnocline and DCM has developed, there are two distinct diatom assemblages, meeting at the top of the nutricline: The shallower assemblage is nutrient limited and the deeper is light limited (Venrick, 1982, 1988). Each assemblage has the characteristics of a mature, predation-controlled assemblage and community diversity increases to a maximum near the depth of the DCM.

From the more extensive general literature on phytogeography, some conclusions may be drawn which seem to support the few available floristic profiles. For example, the widespread existence has been noted of a diverse "shade flora" of four Bacillariophycae, 10 Dinophycae, one Prasinophycae, and three Prymnesiophycae. Some of these shade species are very large organisms, such as the diatom *Plantoniella sol* (Furuya and Marumo, 1983) and the widespread prasinophyte *Halosphaera viridis*. In the Kuroshio region another 11 species, in addition to *P. sol*, are shade species (including the diatoms *Asteromphalus sarcophagus*, *Oolithotus fragilis*, *Thorosphaera flabellata*, and *Thalassionema* spp.).

Thus, the hypothesis that the ordered ecosystem of the euphotic zone is partitioned among different assemblages of algal cells, one of which constitutes a shade flora, appears to be supported.

For zooplankton, the pycnocline represents a special depth zone that not only has unique characteristics but also lies close to the separation between the two principal life zones of the ocean: lighted and dark (Longhurst and Harrison, 1989). The increase of zooplankton biomass (five or six orders of magnitude) over the vertical distance (5 or 6 km) from the ocean floor to the sea surface would represent an unprecedented degree of variability if translated into horizontal patchiness in the mixed layer and would have an unprecedented predictability (Longhurst, 1985). In the interior of the ocean, the vertical rate of change of biomass is very small, and the gradient is greatest over a few tens of meters of the pycnocline, where the sparse bathypelagic plankton is separated from the much more abundant epiplankton across a planktocline. The epiplankton and the deeper acoustic scattering layers of diel migrants are the most prominent features in full-depth profiles of pelagic biota and, like the DCM and the pycnocline, are features that may be traced across ocean basins.

At some depth within the epiplankton, and most often also within the pycnocline, a maximum of zooplankton abundance (ZM) usually occurs. Where the water column has stabilized, the ZM lies somewhat shallower than the DCM, especially at night, and closer to the depth of the productivity maximum (PM). Where there is a very shallow mixed layer, as in upwelling regions or at the start of a spring bloom, the ZM occurs very close to both PM and DCM, which are coincident in these circumstances. The depth difference between ZM and DCM is positively correlated with the absolute depth of DCM. Such observations lead one to enquire if all taxa aggregate at the ZM, or is it only certain species that are specialized for life in the ecotone we are considering?

It is convenient to discuss this question by reference to special investigations of vertical distribution of zooplankton species made at the BIOSTAT station in the eastern tropical Pacific (Longhurst, 1985), where there was a shoal pycnocline and all the features of the TTP. Groups of species could be identified which feed similarly and which occur within common depth horizons, though a rearrangement of the vertical pattern occurs at dawn and dusk because some of these groups are diel migrants (Fig. 2.3).

The following characteristic groups of copepods were identified by species-specific depths of maximum abundance ("preferred depths") and depth ranges ("layer depths") of the central 50% of the populations of the 72 most abundant species:

- Small herbivores (<2.0 mm): All 12 species (*Calanus, Clausocalanus, Calocalanus, Undinula, Nanocalanus, Acrocalanus, Paracalanus, Ischnocalanus, Acartia, Eucalanus, Oncaea,* and *Corycaeus*) had preferred day depths on the upper shoulder of the DCM and thus close to the PM. Some of these shifted a few meters upwards into the lower mixed layer at night. *Oithona* lay deeper than the other genera—closer to the DCM. However, this is the most narrowly specialized group in its selection of preferred depths.
- Larger herbivores (2.5–6.5 mm) and most omnivores (1.5–4.6 mm): These occupy a wider layer depth than the small herbivores. Most exploit the lower shoulder of the DCM down to and even below the bottom of the thermocline, with some extending as deep as 250 m. These are species of *Eucalanus, Rhincalanus, Scaphocalanus, Lucicutia, Gaetanus,* and *Neocalanus.*
- Interzonal migrant omnivores: Mostly *Pleuromamma,* these occurred at night both in the mixed layer and at the DCM; diel migrant *Euphausia,* small fish, and siphonophores were similarly distributed. By day, all diel migrants occurred well below the pycnocline.
- Predatory species: Preferred depths include both the mixed layer and depths down to 250 m, although there is much individual specialization. For example, the two smallest (2.5–3.3 mm) species of *Euchaeta* preferentially occurred with the small herbivorous copepods on the upper shoulder of the DCM.

Obviously, this analysis supports the hypothesis of a specialized fauna of the ecotone associated with the pycnocline and emphasizes how herbivorous copepod taxa cluster in the upper part of the open ocean pycnocline.

Other taxa were similarly ordered down the eastern Pacific profiles. Ostracods, which are all detrital feeders, occurred preferentially just below the pycnocline. Noncrustacean herbivores (e.g., *Oikopleura*) tended to aggregate like herbivorous copepods on the upper shoulder of the DCM, though doliolids exploited a wider depth range; predatory chaetognaths, like predatory copepods, had overlapping depth ranges covering the whole mixed layer. This vertical distribution of feeding groups is recognizable elsewhere in the Pacific Ocean, where the mixed layer is both extremely shallow (Costa Rica Dome) or very deep (at the CLIMAX station in the North Pacific central gyre).

We should not expect to find such a consistent pattern of vertical distribution in latitudes with deep winter mixing, where plankton profiles are driven alternately by diapausing and feeding strategies. However, available summer profiles from all oceans suggest that, as the summer progresses, plankton profiles approach the vertical pattern of the tropical climax community described previously.

The North Atlantic from 44 to 62°N in August exemplifies this situation; at this time, many genera have depth distributions remarkably similar to their occurrence in tropical profiles. Small herbivorous copepods (*Acartia, Paracalanus, Pseudocalanus, Oithona,* and *Clausocalanus*) and cladocerans occur most abundantly in the upper thermocline, near the DCM, with only scattered individuals deeper. Already in this season, ontogenetic migrations had carried diapausing individuals of the large herbivores to 1000 m. At night, the interzonal copepods and diel migrant euphausiids occurred at the same depths as the epiplanktonic filter-feeding copepods (Longhurst and Williams, 1979).

Therefore, there seems to be abundant evidence from observations of plankton organisms from the smallest autotrophic cells to herbivorous mesozooplankton to support the original statement: that we should consider the horizontal front of the pycnocline not only as a way station where sinking particles may aggregate but also as an ecotone that is the preferred living space of a characteristic group of planktonic organisms.

3

PHYSICAL FORCING
OF BIOLOGICAL PROCESSES

In terrestrial ecology, knowledge of a simple set of factors (latitude, altitude, rainfall, slope, and rock type) enables one to predict which of 20–30 vegetation types or biomes (xerophyllous forest, grassland, tundra, boreal forest, and so on) is likely to occur at any site. Such predictions are easily verified on land from existing vegetation maps and Landsat images. If the same could be done at sea, biogeography would be provided at a blow with seven-league boots and we could expect rapid progress in a new and more useful direction. This could be achieved if we could identify a comparable set of factors in marine ecology by identifying a minimum suite of oceanographic factors which control the growth of plant cells in the open ocean. It might be possible to utilize regional oceanography to give a more fundamental understanding of regional differences in ocean ecology than can be drawn either from water mass analysis or from classical species-based biogeography alone.

To the extent that this approach is successful, and yields a set of biomes defined by their characteristic modes of phytoplankton ecology, forced by ambient oceanographic factors, it will also yield by inference some information on organisms of higher trophic levels likely to occur in each biome. As for terrestrial vegetation types, the general relations between phytoplankton–herbivore-predator relationships in characteristic pelagic habitats are well-known at least in outline, differ significantly between habitats, and should be generally predictable for each biome defined by its environmental forcing of algal growth. Here, "biome" is used in the sense of Odum (1971): "The largest (land) community unit which it is convenient to recognize. In a given biome the life form of the climatic climax vegetation is uniform. Thus the climax vegetation of the grassland biome is grass." Here, it is assumed that a knowledge of the characteristic seasonal phytoplankton cycle for any region is equivalent to a description of its "climax vegetation" and that much more can be inferred about the regional ecology from this knowledge.

The seasonal shift of values between Northern and Southern hemispheres that we can see so strikingly in the Coastal Zone Colour Scanner (CZCS) global images of sea-surface phytoplankton chlorophyll seems to confirm the generalization that phytoplankton seasonal maxima may simply be modeled as a zonal response to changing sun angle. Theoretically, at 50–70° latitude, surface chlorophyll in the two hemispheres should be out of phase so that summer and winter minima occur simultaneously in each; the same should occur at 10–40°, though with winter rather than summer maxima. However, at <10°, where two maxima of solar radiation occur as the sun twice crosses the equator, there should thus be two seasonal maxima of phytoplankton growth (Rudjakov, 1997). In an ideal ocean, this might be sufficient to define the primary ecological zones. However, the real oceans are not ideal: The shape of ocean basins, the distribution of wind stress, the topography of the mixed layer, and other blunt facts of geography combine to damp these ideal seasonal oscillations.

While geographical asymmetry is a general characteristic of ocean basins, the current distribution of land masses causes the observed seasonal cycles in the eastern parts of tropical oceans to depart most strongly from the simple model. The bulge of West Africa, and the NE–SW trend of the American continent from Canada to Panama, damp out the two maxima expected to result from the dual-irradiance maxima of the solar cycle: In the Gulf of Guinea and the eastern tropical Pacific, surface wind regimes, sea-surface temperature, and currents exhibit only single-seasonal maxima (Tianming and Philander, 1996). The same is also true of biological cycles in the pelagos, as shall be shown in later chapters. This effect of asymmetric geography can be generalized even further to explain why the atmospheric intertropical convergence zone, with its maxima of convective cloud cover, sea-surface temperature, and rainfall, should generally lie north of, and not above, the equator in the Pacific and Atlantic Oceans. This asymmetry is less pronounced in the Indian Ocean, where mixed layer depths are deeper in the east and ocean–atmosphere interactions correspondingly less effective (Philander *et al.,* 1996). Such considerations suggest that in analyzing the regional characteristics of the pelagic ecology of ocean basins, we must bear in mind the quotation from Klaus Wyrtki printed at the front of this book.

MINIMAL SET OF PREDICTIVE FACTORS FOR ECOLOGY OF THE PELAGOS

A bloom of phytoplankton, inferred from an increase in the greenness of oceanic water, is usually assumed to represent a seasonal or episodic increase in algal growth rate, but this assumption may not be justified. An easily forgotten consequence of the rapid, almost daily, replacement rate of phytoplankton biomass compared with the slow seasonal turnover of biomass in terrestrial plants is the close coupling between growth and loss rates that must exist for phytoplankton. Therefore, an observed increase in biomass on even a scale of days may be caused by a relative increase in growth or a decrease in loss rates. A small and temporary change in either may cause a "bloom," observed in the CZCS data as an increase in the green coloration of ocean surface water. It is essential that we clearly distinguish between the two processes leading to a bloom: increased growth rate or a decrease in the loss term.

It is not difficult to suggest a parsimonious set of factors which might determine the basic pattern of phytoplankton growth characteristic of any region of the ocean. These must be sensitive to observations of the vertical profiles discussed in Chapter 2 and must include the factors invoked by Sverdrup (1953) to explain the onset of the North Atlantic spring bloom: the interaction between light, nutrients, mixing, and stability in the upper part of the water column. Sverdrup's model rapidly became central to the theory of plant production in the ocean and remains so today, but we usually forget that his important contribution was to encapsulate, in a simply stated theorem, several concepts that had been discussed earlier by Nathansohn (1909), Gran (1931), Riley (1942), and others as the nitrification theory of plankton production came to be replaced by modern understanding of the nitrogen cycle: already known was the concept of a compensation depth at which photosynthesis balances respiration and the significance of stratification in defining a critical depth for vertical turbulence that allows the algal cells of the mixed layer to be in a positive photosynthesis/respiration balance. Sverdrup's model has recently been critically reexamined by Smetacek and Passow (1990), Sathyendranath and Platt (1994), and Platt et al. (1994), and the following section owes much to their contributions.

A simple dynamic model based on Sverdrup will compute growth and loss terms for phytoplankton biomass by reference to illumination and stratification of the surface water mass. Growth is assumed to be a function of depth, responding to available light, whereas loss is depth-independent and related to biomass; the loss term includes respiration, herbivore consumption, and sinking. Illumination includes self-shading by phytoplankton cells, and stratification is the result of surface heating and wind mixing. At some level of irradiance, photosynthesis and loss rates are balanced and the depth at which this occurs is the compensation depth. In a perfectly stable water column, growth rates would be positive above this depth but negative at greater depths.

Sverdrup (1953) considered the effect of vertical mixing on this model and better defined the concept of a critical depth (D_{cr}), above which mean irradiance experienced by a cell is equal to that at the compensation depth. If the mixed layer is deeper than D_{cr} then cells will experience light limitation (because, being mixed randomly deeper than D_{cr}, they will experience a time-integrated mean irradiance which is lower than that at the compensation depth) and their time-integrated growth will have a negative sign. If, on the other hand, mixed layer depth (MLD) is shallower than D_{cr} then cells will be light replete and their time-integrated growth will have a positive sign.

Thus, when the actual mixed layer depth is shallower than the critical depth, the net growth rate for phytoplankton must be positive, and when the mixed layer is deeper, the loss term dominates and net growth must be negative. These premises, remark Sathyendranath and Platt (1994), are so fundamental that they have to be true. Since MLD is usually rather variable, then the dynamic balance between growth and loss terms for phytoplankton biomass must also be inconstant.

Unstated in the Sverdrup model is the necessary assumption that deepening of the wind mixed layer in winter recharges surface layers with inorganic nutrients, and that the biomass loss term includes the export of nutrients in organic form to the interior of the ocean, progressively stripping the mixed layer nutrient pool. Since a chlorophyll-rich layer in the water column must induce local warming by absorption of incident radiation at that depth, this will modify local stratification (e.g., Sathyendranath and Platt, 1994); the consequent changes in the photosynthetic parameters for algal growth will then add new complexity to the overall biological response to irradiance. The Sverdrup model, therefore, is a component of a coupled biophysical system.

The simple model is capable of simulating the spring bloom of the North Atlantic remarkably well though the North Atlantic is—in Karl Banse's words—an odd-ball ocean in having such a prominent spring-bloom feature. Near-surface stratification is induced in the very deep winter mixed layer by increasing sun angle and incident radiation during spring, and a bloom develops after a characteristic delay as the physiology of algal cells matures; when this delay period is shorter than the intervals between spring gales (wind-mixing events) blooms will likely occur. This evolution continues with increasing solar radiation and decreasing gale frequency until stratification overcomes mixing and the vernal bloom is fully established. Obviously, the principles of the Sverdrup model must also apply to the declining phase of a bloom.

Implicit in Sverdrup's assumptions, as is appropriate for North Atlantic midlatitudes, is that changes in mixed layer depth should be forced by local wind mixing and irradiance. This assumption is not appropriate when the model is used to interpret phytoplankton seasonality in lower latitudes, as has been done by Obata *et al.* (1996), who investigated the interaction over a $1 \times 1°$ grid covering all oceans of relative changes in MLD and D_{cr} on the concomitant increase or decrease of surface chlorophyll using a climatological data set similar to that described below: Levitus MLD, CZCS surface chlorophyll, and the surface light field. As expected, when and where MLD shoals up through D_{cr}, a bloom follows in the next month. The reverse effect is also observed: When and where MLD becomes deeper than D_{cr}, a decrease in surface chlorophyll is observed in the following month.

A striking observation from Obata *et al.*'s (1996) maps is that MLD is consistently shoaler than D_{cr} in the permanently stratified parts of the oceans, and it is only where significant winter overturn occurs (polewards of approximately 45°N and 45°S) that MLD is deeper than D_{cr} at the season when irradiance begins to increase. Thus, although we must therefore seek an explanation different from the simple Sverdrup model for seasonal blooms in lower latitudes, such as those in the oceanic eastern tropical Atlantic and Pacific, the requirements of this model are not violated. We shall need to examine in greater detail the effects of distant forcing on the nutrient field, and we shall generally have to consider that plant cells even deeper than MLD may be light-replete.

Therefore, more general effects must also be accommodated if our small set of critical factors for algal ecology are to have global significance and to be relevant to regions where other processes may override the effects of local winds and sunshine in determining the depth of the surface mixed layer. There are two principal effects to be accommodated as anomalies to the original Sverdrup concept: (I) In low latitudes, and thus over much of the ocean, seasonal changes in mixed layer depth may primarily be a response not to local but to distant wind forcing, and (ii) especially in tropical and polar latitudes, a surface layer of low-salinity water, not always induced entirely by local rainfall, may dominate the near-surface stratification calendar.

Unfortunately, therefore, any model used to subdivide the ocean into primary biomes should be sensitive to a somewhat larger set of factors than those required by Sverdrup though still not much longer than the list required by terrestrial ecologists. It is a principal argument of this book that this will be a sufficient set of factors to predict the first-order sequence of the seasonal phytoplankton growth cycle and, hence, the nature of pelagic biome. It is suggested that the following information is required to predict the course of the biological calendar for any region of the ocean:

- Latitude (determines if wind stress induces mixing or momentum)
- Depth of water (stratification may be overturned in shallow seas by tidal stress)
- Proximity of coastline (effects of river effluents and coastal upwelling)
- Seasonal irradiance cycle (forces photosynthesis, stratification, and ice)
- Local wind regime (forces mixing or vertical Ekman transport of nutrients)
- Local precipitation regime (may induce near-surface stratification)
- Distant forcing of pycnocline depth (modifies effects of local wind mixing)
- Nutrients in intermediate water mass (modifies upward nutrient flux)
- External source of Fe (lack of which may constrain nutrient uptake)

It is therefore processes such as these that we direct our attention in reviewing regional characteristics in ecological oceanography.

WHAT DO WE REALLY MEAN BY AN ALGAL BLOOM?

This is one of the most frequently used terms in biological oceanography and one which is often given very little thought. I have already used it several times, always referring to an increase in the standing stock of chlorophyll. Before going further, it is important to note the falseness of this parallel. As biological oceanographers, our thinking about algal blooms is probably still colored by the title of Sverdrup's (1953) classical paper "On the Vernal Blooming of Phytoplankton" and by terms such as the "spring outburst" or "vernal flowering" of diatoms commonly used not so long ago. We may also remember that Sverdrup discussed his critical depth concept by analysis of what oceanographic conditions would permit "an increase in the phytoplankton population" to occur—adding only, and somewhat as an afterthought, that "if grazers are present . . . the phytoplankton population may remain small in spite of heavy production." Despite this caution, it is all too easy to interpret an increase in chlorophyll (as we easily observe in serial CZCS images) as having some simple proportionality to production rate and a change in the sign of biomass change as necessarily reflecting a similar change in sign of the production rate. None of which, as we should all know, is necessarily so.

One of the earliest findings in the study of plankton dynamics was that phytoplankton cells could be consumed almost as rapidly as they were produced. In 1935, Harvey and others thought that 98% of the production in the English Channel of a single spring bloom was grazed down by copepods in a few weeks, and similar calculations were made in coastal waters by later workers, including those by Fleming, Riley, Gauld, and Cushing. By 1942, Hart had extended such calculations to the Southern Ocean, where he found that standing stock was only about 2% of the production rate. The dynamic balance between production and disappearance of phytoplankton cells and the relationship between size-fractionated consumption and the passive sedimentation of cells should subsequently have been the central theme of biological oceanography; however, as Banse (1992) points out, this has, most disconcertingly in retrospect, not been the case. Perhaps, he suggests, because of the difficulty of research on consumers (great functional diversity and dimensional range) and the easier access to simple demonstration of biological principles in phytoplankton research, teachers have given priority to the latter so that research on production and consumption has been only very loosely coupled.

Without reviewing all that has nevertheless been revealed since the work of Harvey, which has of course confirmed the ephemeral nature of phytoplankton and demonstrated the biological coupling between the growth dynamics of algal cells, their consumers, and the bacterioplankton recyclers, it is instructive to ask what the CZCS imagery can tell us about the nature of algal blooms. The answer, I think, is quite a lot.

You will recall that we have, for each province and for each month, mean values specifying surface chlorophyll, integrated chlorophyll, and integrated primary production rate. By comparing rates of actual and potential chlorophyll increase from these, it was possible to propose a plankton calendar for each province. This was done by computing the change in average algal biomass for each province (expressed as g C m^{-2}) in month N minus biomass in month $N - 1$; the sign of this value indicates loss or gain. From the rate of change of algal biomass and the rate of primary production, both on a monthly basis, it was possible calculate how much of the production of algal biomass does not accumulate but must be lost to consumption, sinking, advection, or entrainment.

The result of this investigation is shown in Fig. 3.1. To make the point as obvious as possible, I have computed the relationship between standing stock, its accumulation, and its productivity in as simple terms as I can devise: monthly accumulation of biomass in terms of the daily production rate and standing stock in terms of the monthly production rate. In each case, you may be surprised by the result if you have never thought seriously about the relationship. Look first at the open ocean areas lying under the trades and westerlies—monthly accumulation usually represents less than a quarter of 1 day's primary production, and standing stock represents less than one-tenth of the production of a whole month. These values are higher in polar regions, but not by very much.

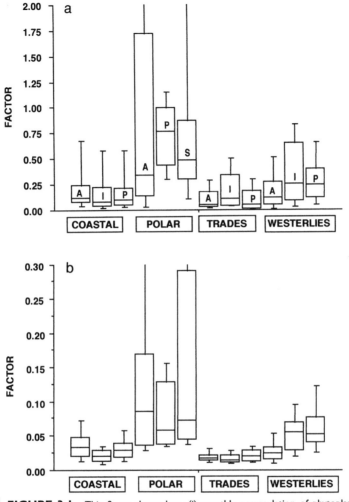

FIGURE 3.1 This figure shows how (i) monthly accumulation of phytoplankton carbon (top) is generally about one-half of 1 day's primary production and (ii) that standing stock (bottom) is equivalent to 0.02–0.10 of the monthly primary production or about one-half to 3 days productivity. The figure is based on CZCS data as described in the text; the analysis is split between oceans (A, I, P, and S) and biomes.

If we examine individual months, rather than the mean values, we get the same result. The June–July bloom induced by the southwest monsoon in the Arabian Sea accumulates no more than 3.0 and 2.3%, respectively, during these months of the production that apparently occurs during each. The spring bloom in the North Atlantic in April and May accumulates even less—0.26 and 0.58%, respectively. These percentages are almost matched in May and June in the subarctic province to the north.

The series of seasonal graphs for production rate and chlorophyll standing stock presented in Chapters 7–10 illustrate the lack of confidence we should have in interpreting underlying ecological dynamics from indications of plant biomass (in most cases, chlorophyll concentration) alone. Phytoplankton standing stock may track seasonal trends in primary production rate quite closely or may significantly diverge from them. Therefore, our graphs show that chlorophyll may accumulate when the production rate is decreasing or may decline when the primary production rate is apparently increasing. One must remember how slight a fraction of total productivity seems to accumulate. Such deviations should be welcomed as opportunities to infer something about the dynamic biology which forces the observed pattern. Where possible, I have suggested in Chapters 7–10 what the reasons may be for each important example. A different presentation of the CZCS data as global maps showing the relation of the critical depth to the mixed layer by Obata *et al.* (1996) assumes that a doubling of the surface chlorophyll concentration indicates a bloom; however, as I shall discuss, this information may not be as useful as it seems.

What are we to make of all this? Obviously, that the production and loss rates for plant material in the ocean are everywhere fiercely coupled and do not resemble those pertaining to the spring outburst of terrestrial plants where caterpillars, for example, in no way keep up with oak trees. Loss cannot be accurately partitioned between sinking and biological consumption, but the former term is usually accepted to be the smaller. Diatoms, after all, unless aggregated have sinking rates of no more than 1 m day^{-1}, whereas the small cells which inhabit the realm of low Reynolds numbers in which inertial forces are negligible in comparison to viscous forces, and which usually comprise by far the largest biomass component, do not sink. The measured precipitation of organic material from the photic zone comprises, for the most part, carbon which has already been consumed. It was early recognized that diatom frustules obtained from the sediments are mostly broken as if by copepod grazing, and sediment traps capture mostly fecal pellets. The sinking flux (the *j*-flux of Berger *et al.*, 1987) is generally believed to comprise around 5% of total primary production in oligotrophic areas, 10–15% in productive oceanic situations, and 25–30% in coastal waters. Given the already recycled nature of much of this material, and the degradation of chlorophyll in it, these figures are not out of line with the conclusions just reached concerning the accumulation of living plant cells.

Demonstration of the control of algal growth by herbivores must be based on observed or modeled budgets for the flow of material through the pelagic food chain. This is a difficult thing to do conclusively, given the complexities of the pelagic food web, not only taxonomically but also spatially. A simple grazing control of algal accumulation might seem to be somewhat illogical; it might seem strange that herbivore control of the uptake of nitrate by algal cells should so perfectly balance the externally forced flux of nitrate across the pycnocline as to result in a predictable and characteristic concentration of mixed layer nitrate and chlorophyll observed in data fields. The only stable and predictable balance would appear to apply to the case (as it is for most of the ocean) in which nitrate is driven to undetectable levels.

The answer to this riddle must lie in the dynamics of grazing organisms, including protists. For copepods, there is a wide literature on the response of grazing rate to the concentration and size of available phytoplankton cells and much attention has been given to grazing thresholds:. the response of grazers to increasingly dense and increasingly attenuated concentrations of food particles. The question, somewhat simplified, has been the following: Do they eat until they burst and (at the other end of the scale) can food be so scarce that they quit and go on a fast? It is the lower threshold that is of interest here. Plankton ecosystem modelers have found it necessary to introduce such a threshold to stabilize their models, and it has in fact been demonstrated experimentally for copepods on a number of occasions (see Conover, 1981, for examples) and recently also for protists. I have been informed (Lessard, personal communication) that protists in an oligotrophic sea have a threshold for successful feeding which is close to the average chlorophyll value outside the brief winter bloom. If this is so, writes Karl Banse (1992), perhaps small cells are kept near this concentration by protistan grazers, whereas the macronutrients that are measurable in the oligotrophic phase do not control plant growth but are—in his words—left over.

One of the conclusions that I draw from these musings is that it is no longer very useful to examine only one of the two terms in the production/consumption equation as, for example, was done in the study by Obata *et al.* (1996); their analysis, based on the global CZCS data and the Levitus mixed layer depth climatology, appears to confirm some aspects of Sverdrup's model—that the North Atlantic and northwest Pacific spring blooms occur after "euphotic conditioning" (i.e., after the mixed layer becomes shoaler than the critical depth), and that some blooms are terminated by the opposite condition. However, I suggest that this tells us very little about what causes the observed chlorophyll accumulation: Simple inferences from maps of change of biomass can be entirely erroneous.

CAN WE SPECIFY MAJOR BIOMES IN THE PELAGOS?

Wherever the wind blows and the sun shines, turbulent diffusion and thermal stratification will be imposed locally and will interact locally, as in all one-dimensional models of the spring bloom. Wherever distantly forced geostrophic adjustment of mixed layer depth occurs, the elements of algal growth models will be shoaled or deepened (Yentsch, 1990).

That these two statements apply singularly to different characteristic latitudes, and serve to distinguish two fundamentally different and characteristic latitudinal zones in all oceans, has only recently become clear to biological oceanographers. That both local and distant forcing must be accommodated in analyses of local algal dynamics is perhaps still not generally understood, though the relative importance of local and distant forcing is what principally distinguishes algal dynamics at low and high latitudes. This is the principal distinction between the two most extensive biomes of the pelagos: the region generally lying below the westerly winds and that lying below the trades or tropical wind regimes.

Wherever the winter sun lies so low on the horizon that the ocean freezes, brine is extracted from the freezing seawater. Wherever sea ice melts with the climbing sun in spring, fresh water is released to form a brackish surface layer above the deeper layers of normal salinity, with the two water bodies separated by a sharp density gradient. The low-salinity surface layer retains its identity while spreading in the wind-driven surface circulation far beyond the margins of the ice fields, and the oceanic polar fronts of both hemispheres define the extent of its dispersal. During this transport the pycnocline constrains wind mixing and conserves the polar brackish-water layer, so this process is associated with a characteristic expression of the Sverdrup model and serves to locate and specify the third primary biome of the pelagos.

Finally, the general oceanic circulation and stratification of surface water masses is modified along the margins of the oceans where the continental topography is encountered and where the influence of continental masses modifies the regional wind regime. These factors, for Mittelstaedt (1991), define a coastal zone often limited seawards by a shelf-break front above the upper part of the continental slope. It is convenient to consider this coastal zone as the region occupied by a fourth major biome, characteristically more diverse than the others. For present purposes, we exclude from it the landward neritic zone where the optical characteristics of the shallow water column are dominated by the effect of suspended sediments because this narrow longshore strip is not amenable to a generalized definition. It will only be in a few exceptional regions, particularly the coasts along which the turbid and powerful plumes of the Amazon and Yangtse rivers pass, that the coastwise turbid zone is sufficiently wide in relation to the width of the shelf itself that we must consider it as coastal zone for present purposes. The logic of this can be challenged, but it will be a useful exception: Nature does not always agree to conform to a pleasantly tidy classification system!

It is a principal argument of this book that the processes which force stratification of the surface layers thereby also determine characteristically different phytoplankton regimes which are the primary biomes of pelagic ecology comparable to the biomes we recognize collectively as forest, savannah, grassland, and desert in terrestrial ecology. It will be useful to consider in greater detail the mechanisms and processes which specify the four primary biomes before proceeding to consider the secondary division of these biomes into the provinces of an ecological biogeography.

Critical to this investigation will be to what extent the characteristic physical regimes supporting the four primary biomes are discontinuous: Can we really distinguish four entities, or compartments, or are we dealing with a continuum between four characteristic situations? Only to the extent that we can specify discontinuities to serve as boundaries will our biomes be a useful advance on previous suggestions, such as the typical six pelagic ecosystems of Barber (1988). To examine this critical criterion of the biomes, it will be necessary to review some elementary physics of the ocean circulation.

STRATIFICATION AND MIXING IN THE OPEN OCEAN: THE CONSEQUENCE OF LATITUDE

Where a brackish surface layer does not impose stratification, we may assume that stability and mixing of the upper ocean are the opposing consequences of solar irradiance and wind stress at the sea surface.

The characteristics of the incident radiation field over the oceans can be encapsulated in two relatively simple propositions: (i) In low latitudes the seasonal cycle has a smaller amplitude than the diel cycle, whereas the annual cycle exceeds the diel at high latitudes (Hastenrath, 1985), and (ii) the irradiance field is dominated by a simple and predictable meridional gradient, which migrates with the seasons and is modified by variable cloudiness and atmospheric clarity.

The wind stress field is more complex, and any ecological geography of the seas must accommodate the complex regional, seasonal effects of local and distant winds. In midlatitudes, seasonal changes in wind stress modify mixed layer depth locally (Niiler, 1977; Lighthill, 1969) and, interacting with the baroclinicity of the nutrient field, determine the regional nutrient level available for phytoplankton growth as stability returns at the end of winter. In the tropics, however, seasonal changes in wind stress induce not only seasonal changes in flow pattern within the Ekman layer but also, through geostrophic adjustment, depth changes in the zonal thermocline ridge/trough system (Philander, 1985; Katz, 1987; Hastenrath and Merle, 1987) often in distant regions.

Although wind stress in low latitudes is generally thought to be weak compared with the westerlies, this is not a correct generalization except in the central tropical–subtropical gyral regions (Isemer and Hasse, 1987) and especially in the South Pacific Ocean. In the Atlantic, for example, resultant wind stress in the trade wind zone during boreal winter approaches that of the westerlies at midlatitudes (the "Roaring Forties"); in the summer when the westerlies are quieter, the trades still remain relatively strong. The monsoon region of the Arabian Sea experiences some of the strongest and steadiest winds anywhere: Off Ras Asir, Somalia, 40% of observations in July are stronger than Beaufort Force 7 (Knox and Anderson, 1985).

There is an essential, but often overlooked, difference between the effect of wind stress at the sea surface at high and low latitudes which is critical to any generalization of the Sverdrup model. Simply stated, wind stress is translated mostly into potential energy at mid- to high latitudes but into kinetic energy at low latitudes. Thus, moderation of wind stress at the sea surface over the North Atlantic in summer has a negligible effect on the flow of the Gulf Stream, whereas seasonal reversal of monsoon winds in the Arabian Sea reverses the direction of the Somali Current in only a few weeks to almost 1000 m depth. Reversal of the North Atlantic westerlies would take about 10 years to have the same effect on the Gulf Stream.

Such are the consequences of the changing balance between the Coriolis force and the pressure-gradient force in achieving geostrophic balance at different latitudes. This balance changes, as is well-known, because the Coriolis force diminishes equatorward as a direct function of latitude, and it does so without discontinuities. This is not the place for a formal discussion of the consequences of this effect, for which a good modern source is Tomczak and Godfrey (1994). Simply stated, however, since flow is forced by the pressure gradient induced by the slope of the sea surface and because geostrophic balance is maintained between this and the Coriolis force, the slope of the sea surface required to induce motion must diminish equatorwards (Lighthill, 1969; Philander, 1985). At the equator (or, in practical terms within about 2° of latitude from it) geostrophic balance ceases to function, and motion is determined by nonrotational fluid dynamics; the most striking manifestation of this effect is the undercurrent flowing eastwards in the pycnocline below the equator in the Pacific and Atlantic Oceans.

That seasonal changes in wind stress should quickly modify the flow in major barotropic currents, such as that along the coast of Somalia, is a fundamental characteristic of tropical oceans. Such seasonal changes in flow of the tropical zonal currents result in geostrophic adjustment of mixed layer depths on a temporal scale of months. Moreover, since the spatial scale of the adjustment of barotropic currents is similar at all latitudes, these seasonal geostrophic adjustments in the mixed layer depths can be identified across whole ocean basins and have important consequences for considering the Sverdrup model: Mixed layers can deepen because of far-field effects and can be forced by local irradiance and windiness. I shall return to this problem later.

Another consequence of the equatorward vanishing of f (the Coriolis parameter, with dimensions of 1/sec) is that the values taken by the Rossby radii of deformation are latitude dependent (Emery *et al.*, 1984; Houry *et al.*, 1987). The external radius (R_e) determines the length scales for barotropic (geostrophically balanced) phenomena; it is the length scale over which gravitational phenomena balance the tendency of f to deform a surface and may be computed as (gH/f), where H is bottom depth and g is acceleration due to gravity. The internal radius (R_i) also depends on stratification and determines the horizontal dimensions of quasi-geostrophic features—mesoscale eddies and Rossby waves. It may be approximated as (1/first eigenvalue N) $\times$ f and, because f vanishes at the equator, R_i takes a value of infinity. At 5° latitude, R_i takes values of around 300 km, whereas at 50° it is only 25 km. We may expect larger and fewer mesoscale eddies as we approach the equator.

There are also latitudinal, regional, and seasonal variations in the density gradients (and thus in their resistance to wind mixing) of the summer pycnoclines of midlatitudes and the permanent shallow pycnocline of low latitudes. Equatorwards, pycnoclines are sharper and have greater resistance to mixing, enhancing the effects discussed previously. Furthermore, the heat balance at the surface of low-latitude oceans is such that there is a mean downward heat flux across the sea surface (balanced by horizontal transport to higher latitudes), and seasonal changes are sufficiently small that at no season is there sufficient loss of heat to produce convective deepening of the permanent tropical thermocline. Finally, there is often an excess of precipitation over evaporation at the surface of the tropical ocean, so mixed layer water at low latitudes is characteristically not only warmer but also a little fresher than deeper water. The tropical Ekman layer, therefore, lies above a pycnocline having greater stability and resistance to mixing than elsewhere; this is expressed as a relatively high subsurface maximum of the Brunt–Väisälä buoyancy frequency (N, in cycles hr^{-1}; see below).

Consider also the effect of a seasonal increase in the westward stress of the trade winds across the tropical oceans within about 15° latitude of the equator; it is this stress which maintains an upward slope of the sea surface and downward slope of the pycnocline to the west in each ocean. However, the tropical Atlantic (5000 km wide at the equator) responds to the seasonal cycle of trade wind stress with a seasonal basinwide geostrophic adjustment of mixed layer depth to maintain equilibrium. Because the adjustment time of an ocean basin to wind stress is related to the time taken for planetary waves to propagate across the ocean, the much greater width of the tropical Pacific (15000 km) is too large for an equilibrium seasonal response (like that of the Atlantic) to occur within a single season. Additionally, over this great distance, wind stress in the west is not in phase with that in the east so the response of the ocean cannot be simple, as in the Atlantic. Consequently, it is only when wind stress changes for longer periods, on the interannual El Niño Southern Oscillation scale, can an equilibrium response in Pacific circulation occur.

All this may be very fundamental and serves to distinguish very clearly the characteristics of low- and high-latitude oceans as biological habitats but does not deliver the essential discontinuities we seek. However, if we look more closely at one feature—relative pycnocline stability, controlling resistance to vertical mixing—from the previous discussion, I believe sufficient discontinuity can be demonstrated.

REGIONAL DISCONTINUITIES IN RESISTANCE TO VERTICAL MIXING IN THE WARM OCEANS

This section attempt to locate such discontinuities, starting with those that may occur between the strong pycnocline of low latitudes and the weaker pycnoclines of higher latitudes. To do this the distribution of N was examined in all oceans, through computations of "BVF 1" with OceanAtlas 2.0.1 for the Macintosh (Osborne *et al.,* 1992), along ocean-basin scale sections from the extensive data archives, especially for the Atlantic Ocean, that are available with this software. In this way, a maximum value for N (and hence of maximum resistance of the pycnocline to turbulent mixing) and the depth at which it occurred in the water column were obtained at intervals of 1° of latitude or longitude along each of 19 long sections (11 meridional and 8 zonal) from oceanographic voyages and from 10 meridional sections synthesized from the accumulated profile data at the U.S. National Oceanographic Data Center in Washington, DC (Levitus, 1982; Levitus *et al.,* 1994; Levitus and Boyer, 1994a,b). Most of the 19 sections specially obtained by research vessels show features of the Atlantic Ocean: The sections synthesized from archived data (in this case, usually termed Levitus data) enable us to examine all oceans at 30° intervals of longitude.

The *Oceanus-Melville* Atlantic section of 1988–1989, which was worked along 25°W from Iceland to the Southern Ocean (Fig. 3.2), will serve to illustrate this approach to the problem of locating discontinuities in the resistance of the pycnocline to mixing. Obtained during summer in each hemisphere, it shows clearly the distinction between the permanent tropical pycnocline, where N takes values of about 12–14 cycles hr^{-1}, and the shoaler summer thermocline (7 or 8 cycles hr^{-1}) of the subtropical gyres. The transition occurs across the equatorward limbs of the subtropical gyres in each hemisphere; the North Equatorial Countercurrent and the main flow of the South Equatorial Current both lie above the tropical pycnocline, which thus lies somewhat asymmetrically to the north in relation to the equator. The actual depth and slope of the

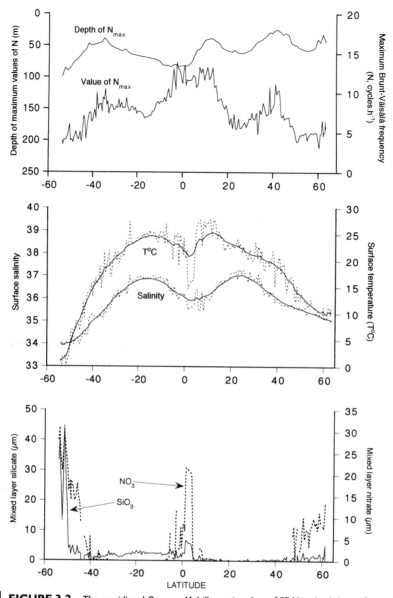

FIGURE 3.2 The meridional *Oceanus-Melville* section down 25°W in the Atlantic Ocean for summer in each hemisphere, showing the maximum values for stability (as buoyancy frequency, N) in the pycnocline and the depth at which N_{max} occurs. The relative slope of the pycnocline reflects the locations of the zonal current systems. The lower panels show the relatively continuous meridional fields of mixed layer temperature and salinity and the highly discontinuous fields of available macronutrients. Computed by OceanAtlas 2.1 software and data sets (Osborne et al., 1992).

maximum value of N reflects (as does the depth of the pycnocline itself) the flow of the zonal currents within the Ekman layer.

The same features are readily identifiable in climatological sections derived from archived data along 30°W, though the seasonal summer thermocline is weaker (as it should be in the archived data representing all seasons), taking a value of only about five cycles hr^{-1}: The strong tropical pycnocline from 20°N to 20°S is clearer (as it should be) in the archived data and takes the same values of N as in the *Oceanus-Melville* section; furthermore, it lies in the same relationship to the zonal tropical current systems, inferred from the depth of the pycnocline. Because the archived data extend further polewards they better show the effects of the low-salinity polar surface layer. At 60°N, as the section passes over the Irminger Basin, and at 40°S the South Subtropical Convergence Zone, the value of N increases progressively polewards as the pycnocline approaches the surface toward the ice edge.

This comparison gives us confidence that the Levitus sections may be used to explore systematically the strength and depth of N at 30° latitude intervals in all oceans and such an investigation confirms that, by reason of its stability and depth, the tropical pycnocline is a unique feature in all oceans and, moreover, that the transition to this feature tends to usually occur at about 20° of latitude (Longhurst, 1995).

What else can can be discerned about the tropical pycnocline? Certainly, the data suggest that the relatively abrupt transition between the weak subtropical and the strong tropical pycnocline (which is a prominent feature in most meridional sections) does not occur always at the edge of the equatorial zonal currents, as one might expect it to do. Consider again the *Oceanus-Melville*: In the North Atlantic, the transition occurs clearly at about 22 or 23°N, coincident with the northern margin of the North Equatorial Current (NEC) flowing west around the equatorward limb of the subtropical gyre. However, in the South Atlantic no such relationship exists, perhaps because the South Equatorial Current (SEC) is wider and more diffuse than the NEC.

Clearly, the area of the tropical pycnocline with high values of N follows the distribution of low-salinity tropical surface water more closely than the margins of the zonal currents. This effect is more clear in the southern than the Northern Hemisphere and is seen not only in the *Oceanus-Melville* section at 25°W but also in *Knorr, Melville* and *Meteor* sections, which run down the Greenwich meridian, and in the "Atlantic Western Basins" sections. In all these, part of the SEC (4°N8°S) and the entire South Equatorial Countercurrent (8–12°S) lie equatorwards of the abrupt transition from weak subtropical to strong tropical pycnoclines; the poleward part of the SEC (from 12–23°S) exhibits a relatively weak pycnocline.

The Brunt–Väisälä frequency takes higher values in the east than in the west of each ocean (Fig. 3.3). Thus, the evolution of the values for N_{max} along zonal sections shows that the pycnocline is most stable ($N = c.15$ cycles hr^{-1}) in the eastern tropical Atlantic and Pacific, illustrating the well-known westward deepening of the tropical thermocline in response to westward trade wind stress at the sea surface. This slope will be about two orders of magnitude greater than the associated wind-driven upsloping of the sea surface. The discontinuity between the relative strengths of the tropical and subtropical pycno-

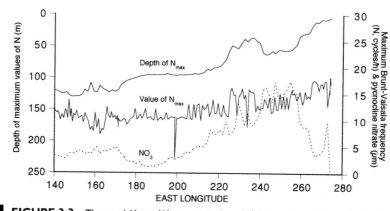

FIGURE 3.3   The zonal *Moana Wave* section along 10°N from Costa Rica to the Philippines showing the deepening and weakening (smaller values of N_{max}) of the pycnocline toward the west and the relative lack of discontinuities compared with the meridional section of Fig. 3.2. Computed by OceanAtlas 2.1 software and data sets (Osborne *et al.*, 1992).

clines is also strongest in the east. It is in the western Indo-Pacific (180°W to 90°E) that the value of N between 20°N and 20°S is lowest (<10 cycles hr^{-1}), and transition to the southern subtropical condition is most continuous. Along the Greenwich meridian, south of Ghana, typical values for N_{max} are the following: SEC, 12–20 cycles hr^{-1}; equatorial zone, 10–17 cycles hr^{-1}; and the Guinea Current, 12–14 cycles hr^{-1}. In the western tropical Atlantic, in comparable regimes, the Western Basins Section shows typical values in the range of 8–12 cycles hr^{-1}.

Because it forms an exception to arguments just made, it will be useful at this juncture to introduce the special case of the western part of the tropical Pacific, the "warm pool" of high surface temperatures lying below the heavy cloud cover of the low-pressure cell at the conjunction of the Intertropical and South Pacific atmospheric convergence zones (Tomczak and Godfrey, 1994). This region has a near-surface halocline within the deeper isothermal mixed layer, the two features being separated by a "barrier layer" (Yan *et al.*, 1992; Lukas and Lindstrom, 1991) between thermo- and haloclines. Meridional sections along 130 and 137°E show that the strongest inflection in the density profile occurs at about 50 m, near the thermocline. The highest values of N (in the range 10–14 cycles hr^{-1}), and therefore the greatest resistance to mixing in the pycnocline, occur rather deep, and well into the thermocline, at 100–150 m. These conditions extend almost to the data line, across a zone of rapidly changing values for N in the tropical and subtropical pycnoclines. This is an extreme case of the situation in the tropical Atlantic and Bay of Bengal, where the halocline does lie shallower than the thermocline but only by about 10 m (Sprintall and Tomczak, 1992) so that both contribute to the high stability associated with the density gradient. I shall return to the ecological consequences of the barrier layer later.

The discontinuities discussed in this section are perhaps sufficiently solid to suggest that it is not unreasonable that the warmer parts of the ocean need not be treated as an ecological continuum, but what of the cooler regions where wind stress in winter may deeply mix the water column? In the next section, we examine one example in which I believe sufficient discontinuity can be demonstrated—at the boundaries of the polar regions.

THE LIMITS OF THE CHARACTERISTIC CONDITIONS OF POLAR SEAS

The polar regions, defined by the polar fronts, are relatively small and comprise only about 6% of the whole ocean. Because it is the critical defining feature of the polar seas, I first discuss if there are definable limits to the imposition of resistance to mixing by the polar brackish surface layer. If no discontinuity can be located, the effect may not be useful for our present purpose and it may be difficult to define a boundary for polar seas. We will first look for evidence of this in some relevant oceanographic sections which can be visualised with OceanAtlas.

A hydrographic section worked by *Hudson* into the Norwegian–Greenland Sea from Atlantic into polar water serves to illustrate the winter situation in the open waters of Atlantic polar regions. This section crossed the Iceland Gap Front and the Oceanic Polar Front (Johanessen, 1986) encountering (i) arctic water, north of the polar front, which is a cold (<2°C), brackish (34.6%) surface water mass with a weak N_{max} (1 or 2 cycles hr^{-1}) at about 75 m; (ii) subarctic water, between the two fronts, which is isothermal (>3°C) and isohaline (>35), with a weak pycnocline at 200 m with N_{max} of 0.5–2.0 cycles ht^{-1}; and (iii) Atlantic water south of the Iceland Front which is warmer (7°C), saltier (35.2%), and lies above a deep pycnocline having its upper inflection at about 250–275 m and an N_{max} of <2.0 cycles hr^{-1}. Here, in winter, the deep mixing typical of the northern part of the North Atlantic subtropical gyre extends into the open, stormy waters of the polar seas. The Western Basins Section (25°W) shows the transition to the surface low-salinity layer in the Denmark Strait.

In the Pacific, the *Washington* (150°W) and *Hakuho Maru* (170°W) sections clearly show the Polar Frontal Zone (32–45°N, depending on longitude) to be the southern boundary of the surface low-salinity water of the subarctic gyre. In the Southern Ocean, the near-surface salinity gradient across the South Subtropical Convergence Zone, is especially well seen in the *Knorr* section along the Greenwich meridian and the *Oceanus-Melville* 25°W section; South of this feature, shallow low-salinity layers are frequently encountered in the sections. A discontinuity occurs in the Southern Hemisphere within the South Subtropical Convergence Zone and in the Northern Hemisphere at the polar fronts of each ocean. Here, an abrupt increase in the values of N_{max} and a shoaling of the pycnocline occur at the boundaries of the polar biome.

However, specification of the geographical limits of the polar biome is complicated by seasonal changes. Even beyond the polar fronts, the surface mixed layer may be overturned by thermal convection and wind mixing during winter, when sea ice cover is not established. It is only in spring that a shoal pycnocline becomes reestablished below the polar layer of low-salinity water and it is only during that season that we may expect to see a distinction between the strong polar and the weak gyral pycnoclines.

It might be protested that a surface brackish layer, constraining vertical mixing, is not a unique characteristic of high-latitude seas. After all, as been notedpreviously, the western Pacific warm pool also has a near-surface halocline in a deeper isothermal mixed layer over an area equivalent to 81% of the polar oceans as defined previously. However, this is really a special case because the nutricline occurs as much as 100 m deeper than the halocline. The surface brackish layer is thus ineffective in controlling the exchange of water across the nutricline.

THE SEAWARD BOUNDARY OF PROCESSES ALONG OCEAN MARGINS

The seaward edge of the coastal boundary biome is often marked by a simple retrograde (isotherms sloping up offshore) front, such as occurs at the edge of many, perhaps most, continental shelves (Mooers *et al.*, 1978) as noted in Chapter 2. Seasonal factors moderate the occurrence of these fronts because shelf water is most strongly mixed in winter or during monsoon conditions and during this season most clearly separated from stratified slope water. During other seasons, stratification of shelf water may obscure the position of the surface front. The shelf-break front off Europe is induced by the amplification (by a factor of 30) of the incoming barotropic oceanic tide at the shelf edge (Le Fèvre, 1986) and injection of nutrients into the photic zone occurs with the period of the M2 tide (Le Fèvre and Frontier, 1988). Similar examples are known from other oceans and at all latitudes.

Other processes commonly serve to set a convenient limit to the coastal zone: Where there is much freshwater runoff, the separation between brackish and oceanic water may form a sufficiently well-defined front to be observable in satellite thermal or color imagery; where coastal upwelling occurs, as in the eastern boundary currents, the cool, green upwelled water may be sufficiently well separated by an upwelling front, discussed in Chapter 4, from blue oceanic water so that this front is a satisfactory limit to the coastal zone. These processes are locally and regionally differentiated to such an extent that it will be more convenient to include them later as part of our discussions of the individual coastal zone provinces.

CONCLUSION: THERE ARE FOUR PRIMARY BIOMES IN THE OCEAN

The primary classification of the ecological geography of the ocean suggested by this analysis (see also Longhurst, 1995; Longhurst *et al,*. 1995) is into four biomes:

- Westerlies biome: where the mixed layer depth is forced largely by local winds and irradiance
- Trades biome: where the mixed layer depth is forced by geostrophic adjustment on an ocean-basin scale to often-distant wind forcing
- Polar biome:where the mixed layer depth is constrained by a surface brackish layer which forms each spring in the marginal ice zone
- Coastal biome: where diverse coastal processes force the mixed layer depth

We shall examine the general properties and boundaries of these biomes in Chapter 4; in doing so, we must bear in mind that only the trades biome, as we shall define it, represents a continuous body of water. The polar and westerlies biomes are each expressed in two separate boreal and austral zones, and because land masses are not uniformly distributed in each hemisphere, we should not expect to find boreal and austral polar biomes expressed identically. The boreal polar biome is characterized by a central ocean largely surrounded by continental coasts and those of a major archipelago, whereas the austral polar biome comprises open ocean surrounding a central continent. Despite such asymmetries, it will be argued, the degree of ecological commonality is sufficient to support the biome concept. It should be clearly stated at this point, however, that I shall not be dogmatic in my application of these definitions in what follows; in particular, the definition for the coastal biome will be violated quite often for convenience. This will especially be the case in enclosed seas such as the Mediterranean in which it is much more logical to include the whole in one biome. This is acceptable because we are not considering the inshore, neritic zone in any detail.

4

BIOMES

Primary Compartments in Ecological Oceanography

A DEFINITION OF THE PRIMARY BIOMES

The definition of ecological biomes and provinces offered here rests principally on observed or inferred regional discontinuities in physical processes, particularly those which affect stability of the upper kilometer of the ocean or the mixing of deeper water up into the photic zone. Regional differences in other ecologically significant variables, such as the characteristics of the light field and the illumination of the nutricline, are also considered. In short, as discussed in Chapter 3, the distinction between the oceanic biomes is based largely on the factors required for the Sverdrup model.

There is obviously a wider set of ecological factors that, if they were available and if we could identify discontinuities in their global fields, could be used to define the ecological characteristics of each biome. For instance, it would be useful to know if there are discontinuities in the abundance of herbivorous zooplankton or in the values attributable to the photosynthetic parameters of phytoplankton, in addition to the geographical distribution of individual species that were discussed in Chapter 1. To the extent that such discontinuities corresponded with the proposed boundaries, there would be additional confirmation that the spatial limits of the biomes have been reasonably defined. Where there was disagreement, further examination of the boundary could be undertaken. Unfortunately, these are unreasonable expectations. For none of the desirable observations or measurements is there a global data field in which we might observe discontinuities.

Consider a very simple case: How does plankton diversity (and hence the complexity of trophic networks) change across the ocean? Diversity is clearly an attribute that is likely to take characteristic values for individual ecosystems (or biomes), but the practical answer to the question must be that for very few water bodies is there a complete enumeration of the species present, let alone a comprehensive measure of ecosystem diversity. Any analysis of relative regional diversity must perforce be based on partial data from which we must then generalize. A good illustration of this problem is provided by one of the most complete and recent enumerations of pelagic biota—that of Margalef (1994), who compares species lists of phytoplankton in the size range 12 μm^3 to >1 mm^3 (i.e., from small flagellates to very large dinoflagellates and diatoms) taken at two stations, one each in the Caribbean and the western Mediterranean. These stations yielded 353 and 257 named species, respectively, and there were about 50 more unidentified taxa in each. To these counts, we would have to add the flagellate flora of the nano- and ultraplankton and the cells of the photosynthetic picoplankton (cyanobacteria and prochlorophytes) to approach a complete inventory of the autotrophic organisms. It is very unusual to be able to locate even single samples of the heterotrophs (micronekton, meso- and micrometazoans, and the protists) that have been analyzed taxonomically as well as Margalef's phytoplankton.

To obtain regional fields of true diversity estimates, within which we might hope to discern discontinuities, we would need observations such as those described by Margalef (1994) (but completed by inclusion of all taxonomic groups) for all points on a suitable grid across the ocean and repeated seasonally. It is not necessary to remark that such information exists for no group of organisms. Even for mapping the simple presence/absence of marine organisms, our information is much less complete than that for terrestrial organisms as was noted in Chapter 1. In any case, we must remember that in the current state of knowledge even a complete list of species present is an incomplete statement of ecological diversity because of the dispersed nature of autotrophy and heterotrophy among protists. All groups of classical autotrophic "phytoplankters" *excluding only diatoms* are now known to include species that are also capable of heterotrophy, and many protist groups include species that have autotrophic protoplastids within their cells (Longhurst, 1991). I think that for no location in all the oceans is there a complete description of the kind required and of the kind which exists for many sites ashore.

Much of what can be inferred about the characteristic ecology of the proposed set of biomes must therefore come from indirect sources. A good case in point is the general relationship between phytoplankton biomass, given as chlorophyll concentration, and the characteristic dimension of phytoplankton cells. As Chisholm (1992) notes: "The one thing that can be unequivocally stated about phytoplankton size is that the fractional contribution of small cells to the standing stock increases as total chlorophyll decreases." She might have put it more simply: cells are smaller in blue than in green water. Furthermore, each size fraction of cells seems restricted in its ability to contribute to total chlorophyll biomass. The <1, <3, and <10 μm cell fractions seem limited to a biomass of about 0.5, 1.0, and 2.0 mg chlorophyll m^{-3}, respectively, and above these thresholds an increase in total chlorophyll can only occur as a contribution from a larger cell fraction. This is what occurs during the course of an episodic bloom.

However, on this apparently simple observation hangs a rather complex tale. Because of the size-dependent capability of cells to acquire nutrients by diffusion, a 5-μm cell becomes diffusion limited below 100 nM (of either nitrogen form), whereas, at the same relative doubling rate, a 0.35-μm cell is limited only below 5 nM. This fact, in turn, explains the characteristic forms taken by large cells in oligotrophic situations which often increase their surface area/volume ratio by evolving either complex cell shapes to enhance surface area or a symbiotic relationship to enhance nitrogen fixation by the action of the symbionts.

Such knowledge provides some confidence that a simple global chlorophyll field has the potential to inform us about a rather wide range of ecological characteristics of the organisms whose pigment is observed. We can, at least in very general terms, infer from a chlorophyll map something about the contribution of the small cell fraction to the phytoplankton community. From this, we can infer something about the nature of the heterotrophic organisms present in the ecosystem. Moreover, we can infer that discontinuities in one field (that of surface chlorophyll) probably indicate discontinuities in secondary ecological characteristics theoretically derivable from it.

Another way of proceeding would be to seek information on the global distribution of biota at higher taxonomic levels than species, hoping to aggregate these into ecologically meaningful groups. It is reasonable to hope that such data might be accessible in a more uniform format than the almost hopelessly diverse descriptions of species lists at individual stations that can be found in the biogeographic literature. Fortunately, the Smithsonian Institution in Washington, D.C. has for many years utilized a standardized first-order sorting technique and record sheet for the plankton samples it has archived, and these protocols have been adopted by other sorting centers. Therefore, it is possible to acquire an archive of these sheets from several different plankton sorting centers and from the data recorded on the sheets to perform a first-order analysis of the composition of the zooplankton; in one case this was done with data from 4166 stations in all oceans, where nets were worked between the surface and 250 m (Longhurst, 1985).

At each station, all organisms (either in a subsample or in the whole) had been identified at least to the level of phylum and class and this made it relatively simple to allocate the counts to an approximation of ecological groupings. For example, the counts could be stratified into the following groups: gelatinous predators, raptorial predators, micro- and macroparticle herbivores, omnivores, and detritivores. An estimate of the relative biomass in each of these groups could then be made. Alternatively, the information could be stratified between groups such as medusae, siphonophores, chaetognaths, polychaetes, ostracods, copepods, mysids, euphausiids, and pteropods. Such grouped information could then be stratified among the 34 sampled regions, and four seasons could be represented and they enabled first-order differences between oceans and continental shelf faunas to be quantified for the polar, temperate, and tropical zones. The general results of that inquiry are shown in Table 4.1, which illustrates the relative first-order differences in biomass of phyla

TABLE 4.1 Differences in the Composition of Mesozooplankton Between Biomes and Between Oceans[a]

Biomes	Oceans	Medusae	Siphophora	Chaetognatha	Ostracoda	Copepoda	Euphausidae	Pteropoda	Appendicularia	Thaliacia	Coelentrate predators	Raptorial predators	Micro-herbivores	Macro-herbivores	Omnivores	Detritivores
					Taxonomic groups								Trophic groups			
POLAR	Arctic	0.07	0.00	9.15	3.30	69.31	0.04	10.08	0.30	0.00	0.08	19.23	0.32	64.57	12.36	3.45
POLAR	Antarctic	0.39	1.49	8.08	2.79	65.86	15.29	3.38	0.06	0.06	1.69	24.22	0.11	47.69	23.79	2.51
WESTERLIES	North Atlantic	1.24	0.94	2.89	1.21	53.07	31.57	3.66	0.07	1.22	1.68	31.75	0.99	20.86	43.79	0.93
WESTERLIES	North Pacific	5.35	4.40	8.21	1.61	40.19	29.01	3.43	0.15	0.39	8.02	33.05	0.45	22.08	35.08	1.32
WESTERLIES	Southern Indian	3.25	0.00	7.41	4.72	34.36	39.24	1.75	0.03	0.23	2.55	36.15	0.60	16.47	40.53	3.70
TRADES	Atlantic	0.86	8.90	6.56	0.60	32.96	23.91	4.00	0.04	6.03	13.61	29.35	8.46	22.48	25.27	0.84
TRADES	Pacific	1.27	10.90	10.46	2.18	29.13	40.78	1.20	0.20	0.93	17.20	38.26	1.63	17.18	22.65	3.08
TRADES	Indian	3.81	5.73	9.97	5.06	32.28	25.84	3.37	0.07	1.72	12.08	31.52	2.27	21.78	25.95	6.40
COASTAL	North Atlantic	0.14	0.45	7.51	0.17	74.84	7.96	1.24	0.00	0.26	0.62	16.96	0.27	44.49	37.49	0.18
COASTAL	North Pacific	1.71	5.36	11.29	0.67	35.37	27.39	1.66	0.22	2.00	7.99	18.99	2.51	24.18	45.57	0.76

[a] Taxonomic and aggregated trophic groups of plankton are expressed as percentage of carbon biomass. Data from 1500 samples, representing all oceans and latitudes, processed by the Smithsonian Museum Plankton Sorting Center, Washington, D.C.

and trophic groups. This enables the quantification of some latitudinal trends and seasonal changes in plankton composition that were well-known for only a few study sites but also suspected to occur more generally from qualitative observations.

To illustrate the possibilities offered by these data, note how they quantify the change in dominance of copepods—for instance, from 67% of polar zooplankton biomass to 33% in the tropical oceans—and how predators (both gelatinous and raptorial) increase from 22% in high latitudes to 47% in the tropics. One of the most striking aspects of these zonal differences is the increasing contribution of gelatinous predators equatorwards: from 0.9% of total zooplankton biomass in polar latitudes to 14.3% in the tropics. The more equal distribution of relative biomass among both the taxonomic and trophic groups in the tropical, compared with the polar, seas is clearly shown in Table 4.1.

At this point, another look at the maps of euphausiid distribution discussed in Chapter 1 may be instructive: These suggest that, if the information base of pelagic biogeography was sufficiently strong, it might be possible to assemble lists of species characteristic of each biome. You may see, as I do, a congruence between the biomes as expressed by the definitions used here and the distributions of some species of pelagic crustacea—species having limited horizontal mobility compared with their predators, the large ocean-ranging and warm-blooded tuna which we may assume are more capable of transgressing boundaries between biomes seasonally. Such a concept should not be novel: That many birds are both tropical and temperate species according to season in no way nullifies the partitioning of terrestrial vegetation into tropical and temperate biomes.

Within the constraints discussed previously, some of the characteristic features of the ecology of each biome will be reviewed in as general terms as possible to introduce the regional descriptions which follow in subsequent chapters.

POLAR BIOME

A comprehensive Sverdrup model must accommodate all factors which will affect the stability of the water column; the low salinity surface layer common to polar seas is such a factor. Winter ice cover and moderate wind speeds at extremely high latitudes constrain winter deepening of the mixed layer to shallower depths than in the ice-free, windy zone of the westerlies between 40 and 60° latitude. Areas of shallow brackish surface water, originating in meltwater from sea ice, may be encountered anywhere poleward of the oceanic polar fronts and may have important local consequences for the timing of the spring bloom and for its initial nitrate conditions.

After the spring thaw, a superficial low-salinity layer stabilizes the upper water column so that seasonal change in mixed layer depth is relatively small during the open-water season. Despite these constraints, winter mixing is sufficient to recharge the surface layers with inorganic nutrients. This is a critical characteristic of polar seas because the stability imposed by the near-surface halocline (particularly strong in spring) may be sufficient to allow algal growth to occur as soon as there is sufficient sunlight. In this connection, note that because of its low salinity, equivalent changes in the density of polar water are forced by changes of 0.1% in salinity or of 5°C in temperature. Values of N_{max} (commonly two to eight cycles hr^{-1}) within the shallow polar pycnocline are not as high as those in the tropical pycnocline, but the timing of the spring chlorophyll increase confirms that the presence of a shallow surface layer, somewhat resistant to wind deepening, has a significant effect on the processes described by the Sverdrup model.

Seasonal ice cover in the open ocean is necessarily accompanied by a front where pack ice meets open water. These seasonally migrating, "marginal ice zones" (MIZs) may extend across tens or hundreds of kilometers, within which mesoscale variability is imposed by the interaction of wind, current, and the ice edge (Smith, 1987). Retreat of the ice edge may occur rapidly (10 km day^{-1} in the Bering Sea), and in 3 months it may shift several hundred kilometers. Ice melt produces a stable vertical density gradient and rapid algal growth is induced in a shallow (often <25 m) mixed layer. Retreating MIZs are then the focus of biological activity at all trophic levels; local "biological spring" may, of course, occur throughout the summer months. Available nitrate is progressively depleted, though renewal may occur by ice-edge upwelling (either density or wind driven); without such input, the algal bloom at a rapidly retreating ice edge may be sustained for as little as 10 days (Smith and Sakshaug, 1990). There is an important ecological connection between epontic algal communities and the phytoplankton. As frazil ice forms in the fall, crystals floating upward scavenge algal cells and incorporate them in the forming ice pack, whereas algae released from the melting ice in spring may either sink as large floc to the sediments or seed the burgeoning phytoplankton community. The polar biome has strong near-surface stratification both because it lies polewards of the global westerlies which induce deep winter convection in lower latitudes and because of the effects of brine rejection when surface water freezes. For this reason a pycnocline below a brackish surface layer commonly occurs in areas which are both ice covered and ice free in winter. The occurrence, even patchily, of a surface layer of low-salinity water within or poleward of the zonal bands of winter westerlies, defines the area in which this biome occurs. In the North Pacific sector of the polar biome an excess of precipitation over evaporation is an additional cause of an unusually extensive polar brackish layer.

There is a singular difference between algal dynamics in the arctic and antarctic polar biomes that is related to the asymmetry between the boreal and austral polar regimes noted in Chapter 3. In boreal regions, the initial nitrate conditions prior to spring are much lower than those in austral regions. Furthermore, the boreal "spring" bloom may rapidly exhaust ambient nitrate, whereas in the austral region, depletion occurs relatively rarely and only close to the continent. The explanation for this distinction is uncertain and probably complex. Suggestions are many, including that nitrate may not be the limiting molecule, that strength of convective mixing prevents sufficient stability for a full bloom to develop, that light is limiting, and that grazing limits algal growth and hence uptake of nitrate (Nemoto and Harrison, 1981; Smith and Sakshaug, 1990).

The polar biome is characterized by low taxonomic diversity at all trophic levels. The total number of species reported for all groups is, relative to low latitudes, extremely small. Associated with the short pulse of primary production that occurs during the brief period of open water and high sun, planktonic algae resemble those of the spring bloom of lower latitudes. Large diatoms dominate the "net" phytoplankton of the nitrate-based bloom and their frustules form a diatom ooze on the deep ocean floor below this biome; this is especially marked in the Southern Ocean in which the northern limit of the diatom ooze corresponds very closely with the location of the Antarctic Divergence, which itself defines the equatorward boundary of the austral polar biome. When a brief oligotrophic phase follows the spring bloom, as it may do in open water in Baffin Bay, smaller cells dominate as they do everywhere when nitrate is no longer available and biologically regenerated ammonium is utilized.

The herbivore response to the availability of phytoplankton may not be as direct as in lower latitudes because of relatively long generation times; some of the larger species of copepods may require more than a single season to complete their growth and reproductive cycles. Characteristically, the larger herbivorous copepods perform deep seasonal migrations, passing the winter in water as deep as 1000 m. The timing of their descent is apparently not controlled directly by availability of food, as might appear most efficient. Rather, the stratagem of overwintering at depth is associated with the fact that the larger copepods cannot complete a full life cycle in a single season: Having completed a certain number of instars, they cease to feed and migrate to overwintering depths, most usually as fifth-stage copepodites (the preadult instar). As will be discussed in a later chapter, this migration/consumption pattern of herbivores has significant consequences for the standing stock of phytoplankton cells.

The annual cycle of the large copepods has another, often neglected, geographical consequence. Where water depth is insufficient for the seasonal ontogenetic migration to occur, large copepods do not complete their life cycles successfully and exist principally as expatriate populations. A significant difference between Northern and Southern hemispheres is that in the south there are important populations of herbivorous, planktonic euphausiids which, unlike the large calanoid copepods of both north and south, do not perform seasonal migrations: they remain relatively near the surface during winter, usually associated with sea ice. The differences between Northern and Southern hemispheres can be to some extent generalized (Table 4.1); the clearest difference is the much greater contribution of euphausiids in the antarctic polar seas, in which they comprise 16% of total zooplankton biomass compared with <1% in arctic seas. In the arctic, amphipods and pteropods are much more abundant.

Also at higher trophic levels, diversity is relatively low, though the marine mammal fauna is more diverse than in lower latitudes and includes several whale species that feed directly on plankton and micronekton; <1% of the approximately 20,000 species of marine fish occur south of the Antarctic Convergence, and there are no Antarctic species of densely schooling pelagic fish. Small shoaling fish occur only in the southerly parts of the boreal polar biome: for instance, capelin (*Mallotus villosus*) are significant components of the ecosystem of the Labrador Sea and north of Norway. In higher boreal latitudes, lacking abundant schooling euphausiids, the mesopelagic polar cod *Boreogadus saida* is a key species in the transfer of material from planktonic production to birds and mammals. In the austral polar biome, another gadoid (silverfish, *Pleurogramma antarcticum*) undertakes the same role, although here the dominant transfer organism between plankton and mammals is the "krill" of Norwegian whalers, the pelagic euphausiids *Euphausia superba* and *E. crystalophorius*.

WESTERLIES BIOME

The mixed layer depth is usually strongly seasonal, deepening in winter with increased wind stress and convective cooling forced by the seasonal changes in sun angle, cloudiness, and strength of the zonal westerly winds. However, in the North Pacific, and especially in the Alaska gyre, winter mixing is constrained by a shallow halocline, though convective cooling extends much deeper. The westerlies extend equatorwards to the Subtropical Convergence Zone; because the globe-encircling westerlies are stronger (and more consistent in location) in the Southern Hemisphere, the Subtropical Convergence there is better developed. Wind stress is most seasonal across the North Atlantic (winter <30 and summer $<5 \times 10^{-1}$ dyn cm^{-1}) and least seasonal in the Southern Ocean (<20–40×10^{-1} dyn cm^{-1} at all seasons). Westerlies have storm tracks along the atmospheric Polar Front. In the North Atlantic in winter this lies southwest–northeast (Florida to the United Kingdom) and, in the North Pacific, zonally from south of Japan to the west coast of Canada. In the Southern Ocean in winter, it is circumpolar, crossing Patagonia at 45°S.

Shoaling of the pycnocline occurs in spring, and in response to development of water column stability and increasing solar radiation, an algal spring bloom may be induced, especially in the North Atlantic. One of the revelations of the global Coastal Zone Colour Scanner (CZCS) images was the striking demonstration to what degree the North Atlantic spring bloom is a singular feature of the oceans. Be that as it may, the discipline of biological oceanography is based historically on studies of the North Atlantic spring bloom and, in fact, what we know about the induction and evolution of this bloom serves us well as a starting point for discussing the other, mostly less well researched, regions under the midlatitude westerlies. Who knows how our subject would have evolved if Sverdrup had not had a spring bloom to try to understand off the coast of his native Norway?

The classical vernal sequence may be summarized as follows: Initial pre-bloom conditions for nutrients at the end of winter are related to the depth of winter mixing and the baroclinicity of the subsurface nitrate field. A bloom is initiated when the mixed layer, as it approaches the surface because of increased warming and reduced wind stress, passes up through the critical depth for net algal growth.

The bloom becomes nutrient limited when the initial charge of inorganic nitrate (or, in some cases, silicate) in the mixed layer is exhausted, and the summer growth that follows is an oligotrophic, nutrient-limited ecosystem fueled by biological regeneration of nitrogen as NH$_3$, with some minor contribution of NO_3 from entrainment (wind and eddy events) and weak Ekman suction (curl of wind-stress field). An autumn bloom, usually weaker than the spring bloom, may follow when the mixed layer deepens so that vertical nutrient flux increases, driven by increasing wind stress and reduced surface warming.

In the remainder of the oceans lying below the westerlies, the CZCS images suggest that a moderate increase in surface chlorophyll values frequently occurs during winter in each hemisphere, especially equatorward of the strongest zonal bands of westerly wind stress, at the time of the local seasonal wind maxima. Because of the baroclinicity of the main oceanic gyres, nitrate values are low at depths to which wind mixing is carried, but Dandonneau and Gohin (1984) suggest that because incident radiation levels are high at the surface, growth of phytoplankton will result even in the absence of near-surface stabilization. The cells which respond most rapidly to transient advection of nutrients in the subtropical oceans are likely to be picoplanktonic (G. Harrison, personal communication) because upgrowth of larger cells is likely only after sufficient stratification is established to retain them in the euphotic zone. As shall be shown in subsequent chapters, although the North Atlantic spring bloom is a singularity of the chlorophyll field, the same phenomenon can be observed elsewhere (e.g., in the Tasman Sea).

The westerlies biome is biologically transitional between polar and trade wind biomes, showing some characteristics of each, both in its alternation between eutrophic and oligotrophic phytoplankton seasons and in its taxonomic composition at all trophic levels. Here we expect to find spring blooms dominated by diatoms and large dinoflagellates and a summer oligotrophic season when the phytoplankton comes progressively to resemble that of tropical seas. Significant phytoplankton biomass is attributable to ultraplankton: prochlorophytes and cyanobacteria (0.7–1.0 μm) and small eukaryotes (0.7–5.0 μm). During summer in the North Atlantic, the former contribute about 30% and the latter about 60% of the primary production from the ultraphytoplankton (Li, 1995); it is supposed that the cells larger than about 5 μm contribute about the same amount again.

Herbivore ecology is also intermediate. Both the seasonal ontogenetic migration of the polar seas and the diel vertical migration of trade wind seas occur in this biome. At higher latitudes, at least some of the larger calanoid copepods descend in fall to overwintering depths, usually in excess of 500 m, and rise again in early spring. The timing of the fall migration once again appears to be ontogenetic rather than a response to lack of food. At all latitudes within this biome, during at least some part of the oligotrophic phase, some large calanoids perform diel vertical migrations, rising to feed at night from daytime residence depths of around 200–500 m, in this way presumably escaping predation in the clear water during daylight hours. During the oligotrophic phase, both phyto- and zooplankton have much in common with biota of warm seas, though overall diversity is lower: This is to be expected from the canonical relationship between ecosystem stability and taxonomic diversity.

Many of the pelagic carnivores of this biome are fish, though these are not as diverse as they are in warmer seas. Migratory stocks of herring and other obligate shoaling clupeids, which spend much of the year in the open ocean, return to the coastal zone to spawn. Others, such as the saury of the North Pacific, spend most of the year over deep water. Higher level carnivores are also well represented by fish in this biome: Both North Atlantic and North Pacific salmon, though spawning in freshwater, spend almost their whole lives in the open ocean. Some species of tuna perform meridional migrations that enable them to exploit the westerlies biome even to quite high latitudes during summer in all oceans. Here, too, we encounter the poleward limits of the small black or silvery bathypelagic fish, exemplified by *Myctophus,* that perform diel migrations comparable to those noted previously for some copepods and presumably for the same reason.

TRADES BIOME

This biome represents what is more informally known as the tropical ocean, defined by the trade wind regime. This extends polewards to about the same latitudes in each hemisphere (30°N–30°S) and in each ocean. The Subtropical Convergence Zone (STC) represents convergence between trades and westerlies, so the poleward boundary of this biome is constrained by the STC in each ocean and each hemisphere.

Because of the different shape of ocean basins and their relation with the STC, central gyres of each ocean do not all have the same relation with trades and westerly winds and hence with the biomes associated with each. This is an important distinction between the current proposals for an ecological geography based on phytoplankton dynamics and previous systems. Here, for instance, it is recognized that almost the whole South Atlantic gyre lies entirely within the trades but that the northern half of the North Atlantic gyre lies below the westerlies.

The trades regime covers about 45% of the total area of the ocean, and for that reason alone it will be useful here to review very briefly its principal physical characteristics that determine its singular ecology:

• The radiation balance is such that there is a mean annual positive downward heat flux across the surface. The area so defined has its widest latitudinal extent in the Atlantic.

• The seasonal radiation flux is such that surface mixed layer is maintained continuously; at no season is there sufficient loss of heat to produce convective deepening of the permanent tropical thermocline.

• The level of solar radiation is such that autotrophs are less commonly light limited than they are at higher latitudes.

• The Rossby radius of internal deformation increases into the tropical zone as a consequence of diminishing Coriolis force; therefore, eddies are increasingly larger but fewer toward the equator.

• Baroclinic time scales are weeks in the tropics rather than years at higher latitudes. The equatorwards-diminishing Coriolis term (f) means that the slope of the sea surface required to force horizontal motion diminishes toward the equator.

• The pycnocline is coincident with permanent tropical thermocline and has very high stability (N; the Brunt–Väisälä frequency takes high values) equatorward of latitude at which winter convective mixing becomes trivial.

• The tropical Ekman layer $[Z_{ekman} = $ friction of imposed stress$/(Nf)]$ lies above a very sharp density discontinuity where turbulence is weak. Most energy from wind friction becomes kinetic and little is used to further deepen Ekman layer.

Although for all these reasons, the locally forced mixed layer depth Sverdrup model for seasonal algal growth is insufficient to account for observed algal blooms in the tropical ocean, it is still often invoked for this purpose (e.g., Wrobleski *et al.*, 1988; Yentsch, 1990; Obata *et al.*, 1996). In fact, changes in the mixed layer depth in the tropical ocean are dominated by distantly forced geostrophic responses to seasonal variations in the strength and location of the trade wind zones; the resultant seasonal changes in the depth of the wind-mixed layer are relatively smaller than those caused by deep winter convection in some regions of the westerlies biome. Some of these distantly forced changes result directly in a bloom, as in the case of the spin-up of geostrophic domes and where basin-scale thermocline tilt shoals the nutricline into the photic zone. It is, of course, the seasonal meridional migration of the north and south trade wind belts and of the intertropical convergence zone (ITCZ), marking the doldrum of calm winds between the trades, that is the source of the variability in wind stress that forces seasonality in the tropical ocean. The migration of the ITCZ modifies the sign and intensity of Ekman vertical motion over large regions (Isemer and Hasse, 1987).

A variety of secondary forces are imposed on the distantly forced seasonal changes in mixed layer depth which induce local and regional blooms: principally, Ekman suction, eddy upwelling, and particularly strong local wind mixing at the event scale. Vertical Ekman velocities in the tropical ocean are similar to rates at high latitudes (Isemer and Hasse, 1987) but may act over larger areas and may be effective in sustaining relatively high chlorophyll at the surface for extended periods, as occurs in the Arabian Sea (Brock *et al.*, 1991).

Winter deepening of the mixed layer may be traced (at least in the Atlantic) to within about 20–25° of the equator or well within the tropical gyres and in a zone which lies below the trade winds for much of the year. We may predict that a very brief, early spring bloom occurs equatorwards of the Subtropical Convergence at least in some years, but that these blooms are soon nutrient limited.

In this biome, the pelagic ecosystem is at its most taxonomically diverse. This is an expression of the general ecological diversity of tropical seas, in which the number of trophic relationships that must be accommodated in ecosystem models is higher than that elsewhere in the open ocean. This is, in fact, the climax community of the pelagos. All other pelagic ecosystems can be viewed as derivatives from it, constrained or stressed in different ways to suppress some of the complexities of the tropical pelagos. As shown in Table 4.1, it is in the tropical oceans that the overall dominance of copepods, both numerically and as relative plankton biomass, is weakest; copepods are reduced to only 33% of total biomass, with the difference between this and the 68% of polar oceans being accounted for by relatively greater biomass of several groups in tropical seas.

Here, associated with relatively very low ambient levels of the limiting nutrient, usually nitrate (although trace elements may be the Liebigian-limiting nutrient in some regions), small phytoplankton cells take their greatest relative importance (in accordance with the general principles mentioned previously) and individual cells tend to depart from simple spherical or discoid form. Dinoflagellates frequently dominate the net plankton and replace the large diatoms of higher latitudes.

Although nitrogen fixation occurs in the pelagos of the westerlies biome, especially by epiphytic cells on *Sargassum* of the North Atlantic and as endosymbionts within diatoms, the blooms of N_2-fixing cyanophytes in the oligotrophic warm oceans of the trades biome have become the paradigm for N_2 fixation among phytoplankton cells: The coincidence of these blooms with trajectories of aeolian dust may be related to the high demand for cellular iron during N_2 fixation (de Baar *et al.*, 1997) which, because of its low fallout rate in aeolian dust, must be retained by the cells in order for N_2 fixation to be as important in the global nitrogen flux as suggested by Carpenter (1989). Four pelagic species of the cyanobacterium *Trichodesmium* (= *Oscillatoria*) contribute macroscopic cellular mats to these blooms, which occur seasonally during the most windless period of the year and may color the sea surface red as far as the eye can see from a research vessel. The global distribution of actively growing *Trichodesmium* populations is limited by the seasonal 20°C sea-surface isotherm (Carpenter, 1989) and it is an almost ubiquitous organism in tropical and subtropical seas. Its remarkable ability to fix gaseous nitrogen and release oxygen under aerobic conditions as well as to offer a substrate for a wide range of consorting organisms within its loose colonial mats certainly are significant factors in the pelagic ecology of warm seas (Capone *et al.*, 1997). It should also be noted that some strains of the ubiquitous picoplanktonic *Synechococcus,* taxonomically related to *Trichodesmium* though ecologically very different, are also known to be N_2 fixers.

Although the complex cell shapes of many phytoplankters may be (as noted previously) a mechanism to increase the surface/volume ratio and hence nutrient diffusion rates, this argument cannot explain the complexity of form exhibited by many tropical zooplankton species. In fact, for both plants and animals, one can probably invoke a mechanism to increase the "perceived size" of an organism for a potential predator, whose capture mechanism may be dimension limited. Though diel vertical migration of planktonic organisms is a global phenomenon, occurring in all kinds of waters from small freshwater ponds to the deep ocean, it is in the trades biome that it extends over greater depth intervals and is most ubiquitous. At all seasons and in all areas, we may expect that a substantial fraction of the zooplankton and nekton, especially copepods (*Pleuromamma, Metridia,* and *Euchaeta*), euphausids, myctophid fish, and squids, to arrive at or close under the surface soon after dusk and to descend again to 200–500 m at dawn. Watching the swarming of deep red bathypelagic squid and black or silvery myctophid fish under the lights of a research ship station at night, often feeding on migrant euphausids, is one of the great pleasures of tropical oceanography.

It is also in this biome that pelagic fish reach their greatest development, and a further pleasure of tropical oceanography is to stand on the foredeck under way and watch the scatter of flying fish from under the bows. A great variety of shoaling clupeids, loosely schooling tuna and other scombroids, and solitary sharks also inhabit the tropical pelagos so that there are multiple food chains, each more complex and longer than those in high latitudes. This is reflected in the temporal stability of the tropical biome and the climax nature of the ecosystem. For tropical tuna, the variance in recruitment between years is a factor of less than 3, whereas for northern cod it may be a factor of about 30 or even more; this renders it much easier to maintain populations in the long term in tropical seas, and for each group progressively to evolve ways to exploit every possible ecological niche.

COASTAL BOUNDARY ZONE BIOME

The coastal boundary biome is topographically far more diverse than the biomes of the open ocean and is therefore relatively unsatisfactory to generalize. For some purposes, inevitably, it will be necessary further to subdivide this compartment and some suggestions as to how this could logically be done will be made in the subsequent chapters. An extreme example of how the fractal nature of the coastal zone (remembering that a coastline is a common paradigm of fractal geometry) can lead to an apparently logical yet very fine subdivision is that of the Gulf of California, which is here treated as a single potential subdivision of the California Current Province. This narrow gulf was divided into 14 biogeographic regions by Santamaria-del-Angel *et al.* (1994) on the basis of seasonal cycles of chlorophyll at 33 locations from the CZCS files—a methodology very close to that used in this book. However, because of the fractal coastline geography, it would be just as logical (if more complete chlorophyll data were available) for a subdivision of each of these 14 regions to have been made.

In this section, some of the generalizations that can be made concerning the diversity of ecological situations within the coastal boundary zone, where the oceanic circulation and tides are modified by interaction with continental topography and the associated coastal wind regime, are reviewed. This useful concept was introduced by Mittelstaedt (1991) and defines what many biological oceanographers regard as the critical regional distinction between processes over continental shelves and those in the open oceans. The coastal boundary biome as defined here is not coincident with the whole of the area of flow of the major eastern and western boundary currents, but only with their inshore parts where flow is modified by the coastline. The width of the coastal boundary will tend to be a function of latitude, if topography permits, because the Rossby radius of deformation is a function of the relative strength of the Coriolis force, itself a function of latitude.

This is not the only consequence of the fact that the coastal biome extends over a far greater range of latitudes, from polar to equatorial, than the other biomes; the associated temperature gradient and the differences in seasonality between high and low latitudes have fundamental consequences for ecological processes so that it will be very difficult to formulate generalizations about coastal ecology: The most extreme consequence of this problem is the distinction between tropical coasts (where water temperature and clarity permit), which are dominated by the work of reef-building coral organisms, and all other coasts. In considering the coastal biome, we must evaluate not only the width, depth, orientation, and topography of the continental shelf but also the geomorphology of the coastline itself. Rias and fjordlands, shallow sedimentary regions (with or without fringing mangroves), and the existence of major rivers are some of the characteristics which will determine the nature of the regional coastal ecosystem. Finally, the discharge of silt by rivers onto the shelf is much greater in low than high latitudes: Three-fourths of all silt delivered by all rivers comes from the Amazon/Orinoco (11%) and by the rivers which enter the Bay of Bengal, the South China Sea, and the basins of the Indo-Pacific archipelago (65%).

The concept of shelf and slope water regimes, originally applied only to wide midlatitude shelves, is probably relevant at all latitudes. Shelf-break fronts are not only the most useful defining feature for the boundaries of this biome but (as discussed in Chapter 2) also have consequences for algal growth that are likely to be of importance in the interpretation of surface chlorophyll fields and require special comment. Such fronts may be associated with elevated levels of algal biomass, either by growth *in situ* because of nutrient enrichment at the front or by aggregation when convergence occurs. Shelf sea fronts inshore of the break of slope at the edge of the continental shelf have major implications for coastal ecology and may be the focus of physical aggregation of biota, the site of enhanced biological growth, or the line of separation between two different biological regimes. Although most of the research on fronts has been done in the North Atlantic, similar processes occur on continental shelves in all oceans and at all latitudes; these will be addressed in the following chapters relevant to each province, and some have been reviewed by Le Fèvre (1986).

In the eastern boundary currents that pass equatorward along the west coasts of the continents there are upwelling regions that are the sites of important (though ephemeral) pelagic fisheries and have attracted much attention from oceanographers. Their characteristics are briefly reviewed here, though I deal with each in detail in subsequent chapters.

Over eastern boundary currents, the equatorward trade wind field has two maxima: One offshore, at 100–300 km from the coast, is also the line of zero wind curl so that cyclonic curl (upward Ekman motion) occurs landward and anticyclonic curl (downward Ekman motion) seaward. The offshore current velocity maximum is also coincident with this offshore maximum in the wind field. A second equatorward wind velocity maximum often occurs at the coast as a boundary-layer response of the eastward component of the oceanic wind field when it encounters coastal topography. Coastal winds also have a greater diel component than the oceanic wind field, forced by the opposition of daytime solar heating and evening wind mixing, and this phenomenon may give rise to diel changes in near-shore water column stability.

Theoretically, offshore and onshore drift of upwelling water should occur in the surface and bottom Ekman layers, respectively, until balance is established between bottom drag and wind stress, but this is not always the case and, in any event, individual upwelling wind events may be of too short duration for balance to be established. Dimensions of upwelling are a function of latitude, water depth, and width of the shelf. Computations of the Rossby internal radius of deformation suggest that latitude is less important than other factors within the eastern boundary currents in determining the location and strength of upwelling events.

A surface density discontinuity is frequently observed between upwelled water, transported offshore by Ekman drift, and offshore surface water, especially during the "spin-down" phase of an upwelling event; this is the frequently discussed "upwelling front" (e.g., Brink, 1983). The strength of upwelling fronts may be related to how strongly stratification is developed within the water mass over the shelf since (at its simplest) the front represents the surfacing of the shallow subsurface pycnocline. Downwelling may occur on the landward side of an upwelling front whenever cross-shelf and vertical transport is organized as two upwelling cells: in this case, upwelling occurs not only at the coast but also on the seaward side of the front.

Though upwelling in each eastern boundary current is forced by analogous wind regimes (Bakun and Nelson, 1991), and though this occurs against analogous geostrophic backgrounds, significant regional differences are nevertheless forced by coastal orientation and by the small-scale topographic features characteristic of each coastline (e.g., Huyer, 1976; Smith, 1981a,b; Brink, 1983). Along a straight "theoretical" coastline, upwelling would occur in a linear zone and something resembling this is seen in multiyear fields of surface density or temperature. Instantaneous reality is quite different, as is obvious from individual satellite images of the temperature and chlorophyll fields, which show that upwelling occurs in isolated coastal cells as well as in a variety of offshore cyclonic meanders and eddies and cool filaments; these often exhibit vorticity, often bearing a pair of terminal eddies—one cyclonic and the other anticyclonic. The observed complexity owes much to the topography of the coastline and shelf; although upwelling centers often appear to be topographically locked to prominent capes, this persistence is probably forced more directly by bottom relief than by the geometry of the coastline.

Although coastal upwelling is classically associated with the coasts of eastern boundary currents, it may also occur along other coasts and in all latitudes, seasonally forced by interaction between local topography, coastal currents and wind, and Kelvin waves induced by distant wind stress. However, with the exception of the Somalia upwelling, all these are minor compared with upwelling in the California, Humboldt, Canary, and Benguela current systems. The wide range of conditions in upwelling sites invites study by the comparative method, which has been a feature of research on this phenomenon.

Coastal divergence upwelling, especially if the continental shelf is relatively narrow, may bring nitrate-rich water to the euphotic zone and deliver conditions for an episodic bloom. Such blooms are associated with an increase in size of the phytoplankton cells, and large diatoms are characteristic components of upwelling blooms (discussed in greater detail in the following chapters).

It is characteristic of both major and minor upwelling sites that the specialized zooplankton herbivore should be a seasonally migrant calanoid copepod: In low latitudes this is frequently *Calanoides carinatus* or a closely related taxon. Their strategy is a modification of that of oceanic calanoids of high latitudes: a descent to depths of 500–1000 m during the nonupwelling season and a rapid ascent when upwelling commences. Unlike their boreal analogs, however, generation time is rapid and several generations may be achieved during a single upwelling season. Nevertheless, in all sites investigated, the resting population comprises only stage 5 copepodites, the final stage before sexual maturity. In this way, reproduction occurs rapidly after the rise to the surface. In the coastal biome at higher latitudes, adjacent to oceanic regions in which some of the large copepods perform seasonal vertical migrations, parts of these populations are advected over the shelf during summer and are unable to descend to their full overwintering depths at the end of the season. These individuals may gather in deep basins on the shelf in some regions, where they may form apparently permanent expatriate populations.

More generally, plankton, and especially the zooplankton, of the coastal biome at all latitudes has characteristics distinguishing it from the plankton of the open ocean. The most characteristic bulk property of coastal plankton is a significantly higher proportion of meroplankton—the planktonic larvae of benthic and littoral invertebrates—than in the open ocean. Once again, I must insert a note of caution: In high boreal and austral latitudes, benthic invertebrates tend toward direct development, so the contribution of meroplankton to the coastal biome is another variable which is modified by latitude. For the rest, species comprising the coastal holoplankton are most frequently congeneric with species of the oceanic plankton; only a few higher taxonomic groups of holoplankton, such as the cladocerans, are coastal specialists. These crustacea, most species of which occur in freshwater, occur as seasonal swarms, especially after phytoplankton blooms, in coastal regions at all latitudes (Longhurst, 1985).

EXCEPTIONAL REGIONS: BOUNDARY LAYERS AND THE HIGH-NUTRIENT, LOW-CHLOROPHYLL AREAS

The conclusion that there are four fundamentally different biomes in the ocean, and the logic for their subdivision into ecological provinces, was reached in this and Chapter 3 by considering how the canonical two-dimensional Sverdrup model of phytoplankton growth might be perturbed by regional changes in the stability of the water column. It must be emphasized that this model, like most other generalizations, is not always appropriate. Some exceptions must be noted now because of their great importance, but they will be discussed in detail in the chapters devoted to each ocean.

These exceptional areas may be grouped into two (or perhaps three) families: (i) where the water column is capped by a surface layer of extremely light water of low salinity; (ii) where nutrient limitation may be imposed not by nitrogen or phosphorus supplied across the nutricline at the density discontinuity but by a trace element having a distant source and sometimes supplied across the sea surface; or perhaps (iii) where herbivores are always sufficiently numerous to contain the accumulation of algal biomass and prevent the occurrence of a "bloom," where this is defined as an observable increase in the standing stock of algal cells. These families are simply stated, but the last two hide competing hypotheses of considerable complexity, as I shall discuss in the following chapters.

The first of these exceptional areas is best exemplified by the western tropical Pacific Ocean, west of the date line, where rainfall over the sea is exceptionally heavy; these conditions specify the Western Pacific Warm Pool Province. Here, surface salinity is significantly lower than that at similar latitudes east of 180°W. An isohaline layer, 40–50 m thick, lies above a strong halocline within the much deeper thermostad which extends down to the thermocline at 100–120 m, coincident with the main pycnocline and (most important) the nutricline. The deeper part of the thermostad is thus a "barrier layer" between the halocline and the pycno- and nutriclines: "barrier" should be understood in the sense that since there is no vertical temperature gradient across the halocline, vertical heat flux cannot occur across it (Lindstrom *et al.*, 1987). For similar reasons, the shallow halocline is unlikely to support an algal bloom when it is wind deepened because nutrient concentration does not change unless, exceptionally, wind deepening extends down to the pycno- and nutriclines. We shall encounter analogous but less extreme situations elsewhere, including parts of the Bay of Bengal, of the northeast Pacific, and of the polar seas.

The second exceptional area is best exemplified by parts of the eastern Pacific, mostly included in the Pacific Equatorial Divergence Province. Here, nitrate appears to be supplied across the nutricline faster than it is taken up by algal cells so that there is always a measurable amount of unutilized nitrate in the mixed layer. Given that there is no limitation by light and that the herbivorous plankton appears to be normally constituted, a different Liebigian-limiting molecule has long been sought for these high-nutrient, low chlorophyll (HNLC) areas. As will be recounted in Chapter 9, large-scale experiments at sea have recently shown that Fe, probably associated with wind-blown aeolian dust, is the missing nutrient. I shall also discuss the more controversial HNLC regions of the Southern Ocean where, far from land and below a lower atmosphere of unsurpassed optical clarity, excess nitrate is also permanently present in the mixed layer: Here, it has been shown that the Fe-rich water (originating from shelf water entrained at the Drake Passage) of the jet current along the Polar Frontal Zone supports an algal bloom which is strictly restricted in dimension (see Chapter 10, Subantarctic Water Ring Province) and distinct from adjacent Fe-poor surface water masses unable to sustain a bloom.

The third case, that of herbivore constraint on the accumulation of algal biomass, is more controversial. It is, however, part of the best candidate model to explain why, in the eastern subarctic Pacific (see Chapter 9, Pacific Subarctic Gyres Province), chlorophyll does not accumulate in spring and there is remnant mixed layer nitrate in summer. In this case, the currently accepted model is that the supply of aeolian Fe is inadequate to support the free, nitrate-based growth of large diatoms, which are replaced by small cells better able to absorb sufficient Fe in dilute suspension and which are held in balance by microplanktonic herbivores. I have already discussed some of the consequences of the fierce coupling between production and consumption which are very pertinent to this case.

5
OCEANS, SEAS, AND PROVINCES

The Secondary Compartments

Subdivision into four biomes neither satisfies the requirement for a regional geography of the ocean nor captures the fine detail shown in satellite images of the surface chlorophyll field. Where relevant comparisons can be made, as in the Gulf Stream eddy field, the observed details of the Coastal Zone Colour Scanner (CZCS) chlorophyll fields closely match the meanders, eddies, and squirts observed in simultaneous sea-surface thermal fields. We can be confident, therefore, that the observed details in the global chlorophyll field are not random noise. Where we can also be confident that the details of the chlorophyll field are not artifactual (because of muddy coastal water or atmospheric dust and humidity), we can assert that what is observed must be the consequence of physical forcing in the surface water mass and secondary biological interactions. This is equally true for patches of high chlorophyll as it is for regions of uniformly low surface chlorophyll. It is for the interpretation of this detail that we require comprehensive information concerning the probable spatial, temporal distribution of values for the wide range of physical forcing factors which process studies have shown to induce or constrain algal blooms.

The primary ecological biomes of the pelagic realm were specified rather simply in Chapter 4, and though the defining factors do not form a spatial continuum over all oceans, the discontinuities—or boundaries—are obviously more fuzzy than precise. Therefore, if we are interested in finer geographical partitioning of the ocean we must have recourse to a wider set of factors having greater precision, especially those apt to define boundaries or interfaces between ecologically distinct regions. These will certainly include the factors which determine the characteristics of regional oceanography at all scales, of which the resistance of the mixed layer to deepening (discussed in Chapter 3) is only a single example. Regional bathymetry, surface circulation, winds, islands, and the distribution of land masses must all be reviewed.

In some parts of the ocean, important regions of chlorophyll enhancement cannot be accommodated within provinces that are defined either by circulation features or by water mass types because the bloom is itself induced by boundary conditions between components of the near-surface circulation. The extreme case of this situation is the linear zone of high chlorophyll which is a prominent feature in CZCS images of the Southern Hemisphere, coincident with the austral Subtropical Convergence. Logically, if circulation features are allowed to dominate the setting of boundaries between provinces, this zonal band of high chlorophyll values must either be split between two adjacent provinces meeting at the convergence or assigned arbitrarily to one of them: Either solution would be unsatisfactory, so in this case it is better to allow the chlorophyll field to determine the boundary. The convergence should be a province, in this view, rather than a boundary.

HOW TO GROUP THE MAJOR OCEANS, THE MARGINAL SEAS, AND COASTAL REGIONS

It is important to accept, at the outset and as reiterated at several points in this book, that partitioning the ocean into compartments is a problem in fractal analysis because homologous physical phenomena can be observed across a wide range of dimension. Though some phenomena have a characteristic scale, many others which will be critical to our analysis do not. An example frequently used to illustrate this concept is vorticity in the atmosphere, which occurs across a range of horizontal dimensions from dust devils (10 m) to storms (100 km) and weather systems (>1000 km). Similar observations of vorticity may be made in ocean circulation: from small and mesoscale eddies to ocean gyres. Fronts between opposing flow in water bodies can occur alongside a jetty or across an ocean basin. Islands, and disturbance to the flow past them, also occur as entities across a wide fractal scale, from anchored buoys to islets to small continents.

We cannot expect, therefore, that in an objective (and, it is hoped, quantitative) review of the processes likely to be relevant to pelagic ecology we shall be constrained to converge on an objective number of compartments. Rather, it may be better to pursue our investigations with a subjective ideal in mind: it has already been suggested that about 50 compartments globally would be convenient for computation of global primary production (Sathyendranath *et al.*, 1995), and this suggestion has been allowed to guide these proposals for a global system for the ecological partitioning of the pelagic realm. This is not only a matter of practical convenience but also of necessity because we are constrained by the available number of observations of any ecological variable or parameter. Where few observations are binned among many compartments, we may have insufficient numbers in each to judge whether the characteristics of adjacent compartments are significantly different and therefore if the boundaries between them are real.

Obviously, as discussed in Chapter 1, the idea of partitioning the ocean into provinces is not new, and many schemes to do this already exist. Why, you might ask, do we need yet another scheme? Apart from the problem that the compartments defined by most existing schemes have inadequately defined boundaries, the compartments of those actually used in practical applications have little basis in oceanography. Consider the partitioning of global fishery statistics among subareas of the ocean by the United Nations Food and Agricultural Organisation (FAO) for the past many decades. These compartments have been used (Gulland, 1971) as if they represented natural areas of the ocean, though a glance at the FAO map will show that they conform to no possible oceanographic reality. One of the most extreme examples is FAO area "H2," which runs from the Bay of Bengal to Tasmania and includes the whole eastern part of the Indian Ocean. To put a part of the Southern Ocean together with the Bay of Bengal for fishery purposes is a breathtaking denial of the natural order of the ocean and also of political reality.

This suggests that biological oceanographers face the same difficulty as did Tomczak and Godfrey (1994) in their recent review of regional oceanography at the global scale. These authors comment that previous regional systems fail to match our current understanding of ocean circulation. They point out that the subdivision of the ocean basins adopted and widely used by the International Hydrographic Bureau (IHB) is not optimal for scientific description of natural processes: the IHB map assumes that the Atlantic, Pacific, and Indian oceans extend polewards down to Antarctica and it was drafted before the true nature of the circulation of the Southern Ocean was understood. In fact, the polar oceans are a good starting point to introduce the current proposals for a system of oceanic biogeographic provinces.

Since the days of the seminal investigations of Alistair Hardy and colleagues in the *Discovery* expeditions to the Southern Ocean of the 1920s, oceanographers, including biologists, have found overwhelming evidence for the ecological, chemical, and physical unity of the Southern Ocean. This ocean is unique because of the annular currents which flow continuously around the globe and because of the importance of convergence and divergence within these flows. Therefore, contrary to the view of the International Hydrographic Bureau, an oceanographer must place the southern boundaries of the three principal oceans—Atlantic, Pacific, and Indian—at the Subtropical Front. This lies between Patagonia and Tasmania and just to the south of Australia and South Africa. Further poleward is the Southern Ocean.

At the other pole, the Arctic Ocean must be treated differently. This is a relatively small basin, almost land-locked and with narrow connections to the Atlantic and Pacific oceans, and is almost perpetually ice-covered, except marginally along the Asiatic coast. It will be appropriate to follow the physical oceanographers and treat this as a Mediterranean, land-locked sea such as the Baltic, Mediterranean, Caribbean, Okhotsk, and Red Seas. Such enclosed seas could either be discussed collectively or allocated to the adjoining ocean as a more or less open "marginal sea." The latter option is greatly more informative and will be followed. The Arctic shall be considered a marginal sea of the Atlantic, to which it is connected through the Fram Straits and the Norwegian–Greenland Sea. Tomczak and Goodall consider the Arctic Ocean, the Barents

Sea, and the Norwegian–Greenland Sea to be a useful unit (the "Arctic Mediterranean Sea") when describing regional circulation, but we shall find it more useful to treat them separately.

Obviously, the defining characteristic that we have used to locate the coastal zone must include all marginal seas; their circulation and ecology must be strongly influenced by topography and coastal wind regimes. The coastal biome, as we have defined it in Chapter 4, shall therefore comprise open and reasonably linear continental coastlines, marginal seas and archipelagoes where the islands are sufficiently large and numerous to modify the ocean circulation, and important areas of shallow water. For instance, we shall include in this category the Arctic, Indonesian, and Philippine archipelagoes but not the Marquesas or Cook Islands.

Our analysis will profit from careful comparisons between the three major oceans and their marginal seas. We shall use the powerful comparative method of geographical analysis and utilize apparent effects of differing dimension and boundary conditions between different oceans and seas as natural experiments to understand regional oceanography and ecology. Such apparently simple matters as the difference between the latitudes to which Patagonia and South Africa extend polewards have profound effects on the ecology of the tropical Atlantic and Pacific oceans.

A METHOD FOR SPECIFYING ECOLOGICAL PROVINCES IN THE OCEANS

It is important to understand that even at the scale of the current scheme, we cannot possibly attain the ideal that boundaries of provinces should be set with perfect knowledge that the biogeochemical characteristics (that we agree are useful in understanding pelagic biology) within each boundary are actually unique and predictable. As I have already discussed, the quantitative data required for this level of certainty are neither available now nor will they be in the foreseeable future. What can be done, however, is to use existing knowledge to propose a comprehensive global series of provinces that are compatible with what has been recently learned of the surface chlorophyll field. Then, for some parts of the ocean, we can test to what extent critical conditions (subsurface chlorophyll profile, values for photosynthetic parameters, depth of mixed layer and nutricline, etc.) within each compartment are seasonally predictable and whether conditions do indeed change across the boundaries between provinces. I shall suggest to what extent such a test is practicable.

Ideally, global fields of relevant data would carry information that would set objective boundaries between provinces or would demonstrate that a set of provinces postulated subjectively could not be supported objectively. Available data archives, for both biological rate processes and the distribution of biota, are inadequate for this task and even the most elementary physical observations, such as mixed layer depths, are adequate only for some parts of the ocean. Elsewhere, as in the South Atlantic and South Pacific, there are too few data even to contour seasonal mean mixed layer depths (e.g., Levitus, 1982).

Despite such problems, it is possible to make some progress. What follows is a description of the steps that were taken toward defining a global set of biogeochemical provinces and their boundaries. This description implies a logical progression rather different from that actually taken: Most of the activities described in this section were undertaken more or less simultaneously, and it was only the testing of boundaries that logically had to follow the other work and did so. Things are often tidier in the telling than in the doing.

The first step was an examination of all available regional and seasonal images of the surface CZCS chlorophyll field in a variety of formats for characteristic, observable, and repetitive regional patterns, both spatial and temporal, of surface chlorophyll enhancement at the surface of the ocean. Where necessary, the individual scenes for critical regions were scanned to clarify the nature of blooms observed in the monthly and seasonal composites. This subjective technique of interrogating the images to locate boundaries was a proxy for the objective and numerical technique that would be the method of choice.

The second step was to examine the regional oceanography of all parts of the ocean not only by bibliographic search but also by consulting data archives. The descriptive physical oceanographic literature for each ocean basin was reviewed extensively to compare the seasonal and regional distribution of chlorophyll values found in the CZCS images with current concepts of surface circulation, together with atlas and other data on mixed layer topography, seasonal wind stress and wind stress curl, heat flux across the surface, and the distribution of observable oceanic frontal zones. Also considered in this analysis were critical studies of plankton ecology, particularly those describing the growth, consumption, or sinking of phytoplankton blooms.

Apart from this extensive literature review of regional oceanography, the data that were used to define biomes and provinces (and locate realistic boundaries between them) are listed below; some have already been referred to in Chapter 4. Even this large data resource failed to describe all parts of the ocean uniformly and adequately: In particular, the South Atlantic and the South Pacific were relatively data poor. A review of the literature of regional oceanography revealed the obvious: data-poor areas are also relatively poorly understood or described. This is undoubtedly reflected in how well the Northern and Southern Hemisphere subtropical regions have been partitioned in this review and will be borne in mind by the cautious reader.

In addition to the global data on the surface (and subsurface) chlorophyll field obtained from the CZCS sensors (see Chapter 1) and the estimates of primary production rates derived from them—both forming a global time series at monthly intervals—the following primary data sets were found to be useful:

• *Global climatologies of mixed layer depth:* Obtained from NOAA-NODC (updated 1994 archive) on-line data and based on temperature (0.5°C) and density (0.125 σ_t) criteria as appropriate for each province. These data were arranged as monthly mean depths on a 1° latitude and longitude grid.

• *Brunt–Väisälä frequency:* Profiles were computed from OceanAtlas data (Osborne *et al.*, 1992) along many archived zonal and meridional oceanographic sections and along synthetic sections derived from archived density profiles at standard intervals of 30° latitude and longitude.

• *Rossby internal radius of deformation:* Obtained along meridional sections for the main oceans from tables and maps of Emery *et al.* (1984) and Houry *et al.* (1987).

• *Photic depth (m):* Calculated from the light field and the chlorophyll profile thought to be typical of the season and region derived from the profile archive. The estimate is therefore valid only for Jerlov oceanic type I–III waters (Jerlov, 1964), in which phytoplankton comprise the only significant source of turbidity, and it is expressed as the monthly mean depth of the 1% isolume (i.e., 1% of integrated surface light, itself based on sun angle and regional cloudiness archives) for each 1° grid point (Sathyendranath *et al.*, 1995). An independent measure of water clarity was afforded by an archive of Secchi disc depths obtained from the NODC, and the comparability of seasonal cycles of this value for Atlantic open ocean regions may be made between Fig. 5.1 and the seasonal cycles of CZCS chlorophyll discussed in Chapters 7–10.

• *Surface nutrient fields:* Obtained from seasonal multidepth climatologies (NOAA-NODC 1994 archives) for all regions of the ocean, issued originally in atlas format but available also as a data file for manipulation (Levitus *et al.*, 1993; Conkright *et al.*, 1994).

The steps outlined previously, together with a detailed consideration of all previous proposals for partitioning the oceans into a global set of ecological, oceanographic, or fisheries provinces, led, by a process of trial and error, to a proposal for a series of 51 provinces whose global distribution is shown on the endpapers of this book and whose mean boundaries are defined in Chapters 7–10, in which a section is devoted to describing each province. Some of the steps taken to test the realism of these boundaries against the available archives of oceanographic data and CZCS images are discussed.

Because the boundaries between provinces must be allowed to vary seasonally and between years, they must be generalized on a map which does not represent any specific season or year. This can either be done with noncontiguous shapes representing the average area covered by each province, but without specifying precise boundaries (as Banse did for his three oceanic domains of productive systems and as is often done in biogeographic analysis), or the areas of each province can be coded onto a geometric grid to represent mean conditions. The final selection of coordinates for the boundaries can be made sensitive to the data field, as will be described.

A STATISTICAL TEST OF THE PROPOSED BOUNDARIES

The final step was to test the reality of the proposals for a global set of boundaries between ecological provinces. The only practicable technique was to distribute the individual data from a suitable global archive into the compartments representing provinces and to investigate statistically the similarities and differences between the data sets that then represented the attributes of the individual provinces.

This was done with the 21,872 sets of profile parameters obtained from the global chlorophyll profile archive described previously. These values were

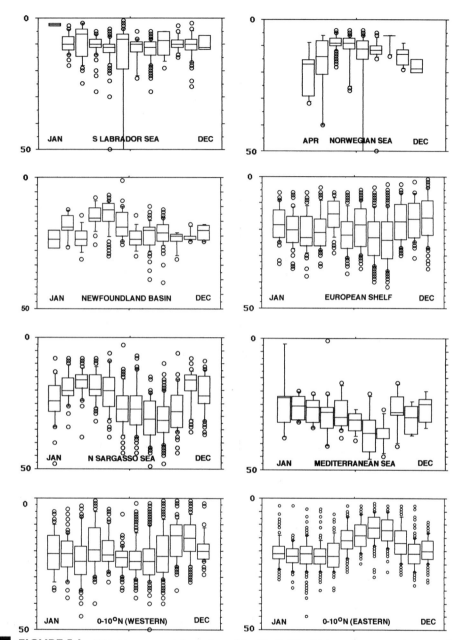

FIGURE 5.1 Water transparency for regions of the North Atlantic shown to support the seasonal CZCS interpretations. The European shelf data confirm that transparency is not always dominated by seasonal changes in chlorophyll biomass. Elsewhere, compare (i) the extremely clear Mediterranean with the always turbid southern Labrador Sea, (ii) the seasonal differences between eastern and western equatorial regions, and (iii) the winter bloom of the Sargasso Sea and the summer bloom of the Norwegian Sea. Analysis was based on 6720 Secchi disc observations from the archives of the NODC.

aggregated seasonally, centered on the 15th day of January, April, July, and October to represent the boreal (austral) winter (summer), spring (fall), summer (winter), and fall (spring) in each province. Of the resulting 204 possible cases (four seasons in each of 51 provinces), for 145 there were >25 profiles, for 38 there were >25 profiles, and for 19 cases there were no profiles at all, though some of these represented polar provinces during winter darkness. This seemed a much better result than one might have expected.

Because the depth and relative strength of the deep chlorophyll maximum layer is an integral of the nutrient, light, and density environments of phytoplankton and because it is the most consistent feature of chlorophyll profiles, it seemed to be a good criterion for testing the proposed boundaries. The depth of the chlorophyll maximum is described by the parameter Z_m. Where this is a positive integer, it represents the subsurface depth of the layer; where Z_m is a negative integer, the chlorophyll maximum lies at the surface. It is not surprising, therefore, that of the parameters describing the chlorophyll profiles, Z_m consistently showed (as I shall discuss) the greatest differences between regions and seasons when partitioned by the proposed set of boundaries.

Some of the resulting seasonal frequency distributions of Z_m were bimodal, suggesting nonuniform conditions for that season with a province. The reason for this was resolvable by identifying the data creating the minor peak. In most cases, this showed that the anomalous data were a discrete set of observations, concentrated in one part of the province, often over a restricted period. In such cases, it was usually possible to obtain a unimodal set of data to represent a province by a minor adjustment of its boundaries. Obviously, if the historical record of oceanographic expeditions and experiments had been different, the final solution would also have been somewhat different.

Using this process, a set of unimodal distributions of parameter values describing the chlorophyll profiles was obtained that represented each case (province and season) and could be used to check that the boundaries between provinces were realistic. This was done by investigation of the differences between seasonal mean values (for each parameter) across the boundaries between each pair of adjacent provinces using analysis of variance and the Bonferroni–Dunn post hoc test. This investigation showed that measures of chlorophyll biomass and depth of its maximum value (Z_p, $B(0)$, B_{int}, and Z_m) differ between adjacent provinces more significantly than the parameters describing the form of the chlorophyll peak itself (s, h, and R), which show weaker regional differentiation; this is evident for both annual and seasonal means for each province.

For all parameters, the most significant boundaries are between provinces of the coastal and oceanic biomes (Sathyendranath *et al.*, 1995), whereas within the oceanic provinces the most distinct boundaries lie to the north and south of the subtropical gyres and thus between the provinces of the westerlies and trades biomes. Most boundaries between oceanic provinces proved to have some significance for the more variable measures (Z_p and Z_m), though even these do not discriminate convincingly between boundaries within the most northerly and southerly groups of provinces. These same parameters confirm the reality of the east–west partition of both the subtropical gyre and the tropical regions of the Atlantic.

Currently, this appears to be as far as one can go in testing that the boundaries proposed in this book (or some other set that might take your fancy) have some objective reality since, as has been discussed previously, the data fields needed for a purely objective ecological partitioning of the ocean neither exist nor are likely to in the foreseeable future. In practice, the proof of the pudding must be in the eating. If the provinces proposed here, or something like them, prove to be useful for practical purposes, such as the partitioning of fishery statistics to investigate the biological basis of fisheries production, then they probably resemble those that might have been objectively drafted had the needed data been available. If not, they will rapidly join the heap of already discarded biogeographic systems. I take some comfort from the fact that two recent proposals to partition the northeast Pacific into "bio-optical" and "fisheries production" provinces have come to conclusions not unlike those presented here (see Chapter 9).

6

TEMPORAL VARIABILITY AND THE ADJUSTMENT OF BOUNDARIES

So far, we have discussed ecological boundaries in the pelagic ecosystem as if their coordinates only varied seasonally and otherwise could be confidently predicted in space and time. It has been assumed that the average location of features in the composite 1978–1986 Coastal Zone Colour Scanner (CZCS) sea-surface color field and in the various oceanographic atlases and data sets that have been referred to are a sufficient description for our purposes. As is surely obvious to most readers, this is no more than a useful first approximation.

In terrestrial ecological geography one can effectively ignore temporal changes in the location of ecotones, or other ecological discontinuities, because unless man intervenes, the tree line on a mountain or the passage between grassland and savannah remains approximately static over periods of centuries. It is only on the millennial scale that such boundaries migrate significantly or disappear. Suburban sprawl, deforestation, and overgrazing are admittedly accomplishing in a few decades what nature cannot do in centuries, but that sad fact does not alter the argument. The human population explosion can produce pressures which rapidly shift ecological boundaries ashore, but paradoxically it is more difficult to directly modify the average locations of the ephemeral and shifting ecological boundaries of the seas which are the subject of this study. We can only accomplish this indirectly by atmospheric modification, resulting in a changed global climate and shifted ocean circulation.

For many, perhaps most, purposes it will be the climatic average conditions that will interest us, and for this the set of coordinates suggested in Fig. 5.2 for the boundaries of the primary biomes and secondary provinces will suffice. For other purposes, it may be necessary to predict or hindcast the coordinates of the provinces for a particular set of years for which we know (or can predict with simulation models) the relevant pattern for ocean circulation or global weather systems or both. Useful knowledge will be of two kinds: (i) the observed location of oceanographic features, both physical and biological, obtained by satellite sensors that serve to locate boundaries between provinces as we have defined them (see Chapter 1), and (ii) sufficient data on meteorological conditions to force general circulation models (GCMs) with which we can hope to predict the locations of the relevant physical features to locate the boundaries of interest. It was with these considerations in mind, therefore, that as far as possible the boundaries of provinces have been located at oceanographic features whose location can be observed or predicted.

We shall have to consider two effects of changes in atmospheric forcing at the sea surface: (i) The boundaries between our provinces may migrate in response to changed circulation patterns and (ii) the ecological characteristics (used to specify a province) may be so modified that adjacent provinces come to be indistinguishable for longer or shorter periods. The latter effect may be expected if, for instance, a change in distant forcing modifies the depth of the thermocline of a tropical region. In such a case, the second-order effect is likely to be a regional modification of the characteristic vertical relationships between light, nutrients, and phytoplankton.

SCALES OF EXTERNAL FORCING: FROM SEASONS TO CENTURIES

This is not the place for a full account of the complex interactions between the variability of weather systems and ocean circulation, which to an important degree are mediated by heat exchange across the sea surface. Nor will it be necessary to explore in depth the possible effects on atmosphere–ocean interactions of subtle changes in solar radiation due to variations in orbital geometry or of the solar constant. It will be enough to recognize that such changes are well documented and constrain the scales of variability that we observe in the stability of the water column within predictable limits.

We shall deal with three scales of variability in regional oceanography induced by (or associated with) changes in the planetary light and wind fields: (i) the seasonal scale, (ii) the El Niño–Southern Oscillation (ENSO) scale, and (iii) the more obscure decade-to-secular scale changes in global weather patterns. This, of course, will be yet another simplification since it is becoming increasingly clear that changes at all scales and all latitudes are causally linked and that the processes by which solar heating is redistributed between ocean, atmosphere, and soil are interacting parts of a single complex mechanism. Although climatology is moving very rapidly to integrate distant phenomena within this complex whole, we cannot hope to achieve a satisfactory integration here of all environmental changes we want to discuss. Many well-observed and regionally important changes are surely part of a greater whole, even though the distant teleconnections have often not been identified; our recognition of these three scales should not, therefore, be regarded as more than a convenience and some phenomena that must be mentioned will be difficult to assign to a single scale.

LINKING SEASONAL TO ENSO-SCALE EVENTS

As will become evident in later chapters, it will not be necessary to modify the locations of borders to accommodate seasonality in ocean conditions; each province accommodates a characteristic seasonal production cycle within its boundaries. This should be equally true at high latitudes, where stability is forced by increasing solar radiation in summer and dispersed by increasing wind stress in the autumn, as at low latitudes in (for instance) the eastern boundary currents, where it is the polewards march of anticyclonic wind-stress curl that leads to maxima in coastal upwelling and algal growth in summer.

However, seasonality is not identical from year to year. Consider the variability in the seasonal march of the atmospheric intertropical convergence zone (ITCZ), which follows the apparent movement of the sun from about 2°S to about 9°N in the Atlantic Ocean. There are differences between years not only in the date that its march toward the north begins but also in the northing attained during boreal summer; in some years the ITCZ reaches only to 8°N, whereas in others it reaches to 10°N. These variations are matched by the northing achieved by the ITCZ of the Indian Ocean. The amplitude of the between-year differences may appear to be minor, but these differences are associated with distant changes in weather patterns that are very significant— for example, the continental wind and rainfall fields over sub-Saharan Africa. The effect of anomalous seasonality in the location of the ITCZ on the surface temperature anomalies in the eastern tropical Atlantic may be complex and related to ENSO events. The 1984 anomalously warm season in the Gulf of Guinea was apparently caused by unusually intense trade wind stress during the previous summer and fall (as the strong 1982–1983 ENSO event was winding down), leading to an unusually deep mixed layer of warm water in the western basin. On the seasonal relaxation of the trades, this mass of warm water surged eastwards along the equatorial waveguide to form an unusually deep and warm mixed layer in the Gulf of Guiana that persisted for many months (see Merle, 1980; Citeau *et al.*, 1988; Carton and Huang, 1994; Delacluse *et al.*, 1994). This event occurred early in a decadel-scale period during which Atlantic climatology responded to unusual ENSO-like conditions in the tropical Pacific, as I shall discuss later.

In any case, we may generally expect that between-year variability in the ITCZ, which is the principal convergence zone in the planetary atmospheric circulation, will be associated with major anomalies in sea-surface temperature (SST), characteristically of scale 3–4000 km and observable for periods of up to 12 months, and hence with variability of planetary weather systems (see, for example, the discussion in Cushing and Dickson, 1966; Cushing, 1982). That such anomalies are coherent over even longer periods, spanning more than a single decade, will be discussed later.

There are also significant effects on pelagic ecology over large areas. The curl of the wind stress in the ITCZ is associated with Ekman suction in the

North Equatorial Counter Currents (NECCs) of both the Pacific and the Atlantic Oceans and, hence, with a zonal band of enhanced chlorophyll in each ocean. Two of the provinces of the trade wind biome (North Pacific Equatorial Countercurrent and North Atlantic Tropical Gyral Provinces of the Pacific and Atlantic, respectively) are partially defined in relation to this process so that their actual locations are probably determined by annual variation in the meridional position of the ITCZ. Operationally, this might be observed if sea-surface color images are routinely available; unfortunately, the coverage of the CZCS images is insufficient to match changes in the countercurrent regimes with changes in the meridional location of the ITCZ.

The tropical monsoon rains are reliable from year to year but notoriously fickle in the longer term, and this unreliability is associated with the significant interannual variability that occurs in ocean circulation in tropical seas. Soon after the Great Famine of 1899, the Indian meteorological observatories undertook to predict the strength and timing of the monsoon; during this work, Gilbert Walker's analysis of surface pressure records over the Indo-Pacific uncovered the Southern Oscillation and gave us the critical tool to understand interannual change in global weather patterns and ocean variability. Walker's "Southern Oscillation" (SO) refers to the observation that when atmospheric pressure is high over the Pacific Ocean (the southeast Pacific subtropical high-pressure cell) it is low over the Indian Ocean (the Australian–Indonesian low-pressure trough) and vice versa. The state of the SO is quantified by the SO Index (SOI), which, in its simplest form, is the difference between sea-level pressure at Darwin and Tahiti. The seesaw effect of the Southern Oscillation has attracted a major research effort in recent decades because it involves a great deal more of the global weather pattern than was at first apparent. Low values of the SOI foreshadow ENSO events which will merit more than passing attention, though for a deeper understanding of the physics involved the reader is referred to the now abundant literature, for which my preferred starting point would be Chapter 19 of the regional oceanography of Tomczak and Matthias (1994).

Because ENSO events bring remarkable and even disastrous rains to the South American coastal deserts and other major climate anomalies (Fig. 6.1), direct historical records exist of the return interval of the stronger events since 1525, near the end of the Little Ice Age, and of all events since 1803 (Enfield and Cid, 1990). The period of the SO is determined by the rate of propagation of planetary waves across the Pacific Ocean, modified by orbital and solar constants: from 1803 to 1987, ENSO events recurred every 3.7 years during periods of high solar constant and every 3.2 years when the solar constant was low. Extreme return intervals ranged from 1 to 8 years, and there has been no long-term trend in the frequency of ENSO events in the five centuries of warming climate since the Little Ice Age . As I shall discuss later, in the past two decades the return interval has shortened significantly for reasons that are not at all clear. Our understanding of the phenomenon has strengthened significantly since the attention of oceanographers and fisheries scientists was drawn to it by the ENSO events of the 1960s and especially that of 1972–1973.

It was recognized early that the Southern Oscillation was associated with the strength of the trade winds and tropical rainfall patterns and it was also

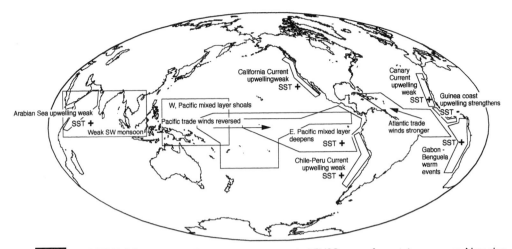

FIGURE 6.1 Low-latitude responses to a canonical ENSO event, from various sources. Note that ENSO events vary significantly, and on should not assume that all the responses shown will occur during every event or even simultaneously. In particular, the coincidence between events in the Pacific and Atlantic is far from secure.

rapidly associated with the "phenomenon known as El Niño" (to use the jargon of the United Nations agencies). Recent research has shown that it casts a far wider shadow and has effects that are global in extent: it is, in truth, the dominant interannual climate signal on the global scale as Philander and Rasmusson pointed out in 1985. If, for the moment, we assume that the relaxed or normal state of the tropical weather patterns is the existence of a well-developed trade wind belt in both hemispheres, then a strong zonal pressure gradient across the tropical Indo-Pacific (high in the east and low in the west), and thus a high value of the SOI, is also the norm. An ENSO event is foreshadowed by a period of anomalous southwesterly winds in the Tasman Sea and initiated by a weakening of the trades in the western Indo-Pacific. The event develops by the progressive extension eastwards of outbursts of rainy, westerly winds initiated between a succession of pairs of tropical cyclonic cells, north and south of the equator in the western Pacific.

These bursts of eastward wind stress transport surface water to the east and induce the propagation of Kelvin waves along the equatorial waveguide; consequently, the thermocline tilts across the entire Pacific basin, deepening in the east as it shallows in the west. There are also smaller scale changes in thermocline depth along the equator associated with the passage of the Kelvin waves: deepening equatorward and shoaling poleward. A fully developed ENSO event is characterized by weakening of coastal upwelling on the western coasts of the Americas and along the equator so that the surface waters of the eastern tropical ocean are warm, oligotrophic, and overlie a deep thermocline. Even if coastal upwelling continues in these circumstances, nitrate-rich water may not be brought to the surface. The biological consequences of all this are well described by Arntz (1986). In contrast, in the midocean portion of the subtropical gyres (at 20–25°N), an increase in overall productivity of about

65% occurs and there are significant changes in the functioning of the pelagic ecosystem (Karl *et al.*, 1995); reduced wind stress leads to decreased upper ocean mixing and allows the development of nitrogen-fixing, phosphate-limited *Trichodesmium* blooms.

Anomalous conditions may also occur in the trade wind zone of the Atlantic during some ENSO events, as noted previously. This was observed in both 1968 and 1984 (Hisard, 1980; Tomczak and Godfrey, 1994); in both instances, the expected coastal upwelling did not occur during boreal summer in the Gulf of Guinea, heavy coastal rainfall occurred in the winter dry season, and the whole of the eastern part of the ocean had an anomalously deep and warm mixed layer. In the same year, cold anomalies occurred in the northwest Atlantic near Newfoundland. Recently, the response of the Atlantic to ENSO events has been more satisfactorily analyzed (Bakun, 1996; Binet, 1997). During a Pacific ENSO event, the anomalous low-pressure region (warm, wet, rising air) over central America strengthens the Atlantic trade wind system, and the westward slope of the sea surface is enhanced as is thermocline uplift in the east, leading to anomalously cool SST in the eastern tropical Atlantic. When the ENSO event relaxes, high pressure (cool, dry, sinking air) over central America causes a relaxation of the Atlantic trades, a slumping back of the sea-surface slope, and a deepening of the mixed layer in the eastern part of the ocean. The seasonal oceanic-scale tilt of the thermocline [see Chapter 7, Western Tropical Atlantic Province (WTRA)] in response to seasonally varying trade wind strength is then reestablished. The ENSO situation is thus marked by the following observable processes: The NECC (and the Guinea Current) are strengthened, unusually high saline subsurface water surfaces in the Gulf of Guinea, and there is anomalous southward flow along the Namibian coast, leading to the "Benguela Niño" that was observed in 1934, 1950, and 1963 and perhaps subsequently.

More generally beyond the trade wind zone and the monsoon areas, where the most important shifts in weather patterns occur, ENSO events are associated with a general strengthening of the midlatitude westerly winds of the Northern Hemisphere, though there is much variability between ENSO events in the actual sites at which the effects occur, and a stronger Aleutian atmospheric low pressure over the North Pacific. Winter weather systems over northern Canada are modified so that the spring breakup of ice in Hudson's Bay occurs up to 3 weeks earlier in strong ENSO years. Associated with these far-field effects, a warming of the whole northeast Pacific and a northward shift of the subarctic boundary, defined as the zoogeographic transition between subarctic and central water masses and biota, may also occur in ENSO years. This effect is especially marked at the eastward margin of the ocean along the coasts of Oregon, British Columbia, and Alaska and is caused partly by poleward Kelvin waves passing along the continental margin and partly by changed atmospheric forcing at the sea surface in the northeast Pacific. In the northwest Pacific, the seas to the east of Japan (25–45°N) experience a negative temperature anomaly of about 1°C largely caused by the changed weather patterns. Anomalous poleward extension of the ranges of individual species of pelagic fish and nekton (Longhurst, 1966) and changes in bulk biomass of plankton both in the California Current and the oceanic North Pacific in ENSO years (Chelton *et al.*, 1982) are related to these oceanwide changes in near-surface circulation.

Binet (1997) has reviewed comparable events in the Canary Current. Here, two periods (in the early 1970s and the late 1980s) when coastwise wind stress was anomalously strong resulted in strengthening of the upwelling cells and a very significant extension southwards of the temperate sardine *Sardina pilchardus,* and threefold increase in its abundance, at the expense of the home range of the tropical sardines *S. aurita* and *S. maderensis.* During these wind-stress anomalies, *S. pilchardus* extended its range even south of Senegal, onto the Arguin Bank (see GUIN below). We may reasonably suppose that changes in organisms at such relatively high trophic levels as sardines must reflect very profound changes in the planktonic ecosystem on which their growth depends.

Especially strong ENSO events, like that during 1982–1983, may generate sufficiently energetic, poleward Kelvin waves along the continental margin that these establish conditions for the radiation westwards of long Rossby waves at midlatitudes (35–40°N) which progress very slowly across the North Pacific and eventually (a decade after the original ENSO event) modify the latitude at which the axial flow of the Kuroshio extension occurs. The consequent advection of anomalously warm water to the northeast Pacific has the same amplitude and extent as that which occurs by atmospheric forcing during an ENSO event (Jacobs *et al.,* 1994). In such circumstances, we may suppose that the effective coordinates of the Kuroshio Current Province will be modified, a fact which should be readily detectable in satellite imagery of SST.

It is the provinces of the Indo-Pacific trade wind biome that will show the greatest modification, and especially the western Pacific Warm Pool Province (WARM) whose eastern boundary will extend far to the east of its normal position at about the date line. In extremely strong events, the conditions across the whole Pacific basin, from the Indo-Pacific archipelago to the coastal boundary of the Americas, may have sufficiently uniform and oligotrophic conditions as to be considered a single province (WARM-ENSO). In this case, though weather systems, biota, and seasonal production cycles will be strongly modified within them, the coastal boundary provinces (California Current, Central American Coastal, and Humboldt Current Coastal Provinces) will probably retain their identity and the characteristics of the coastal boundary zone of the ocean. There may be some change in the effective coordinates of the eastern part of the zonal province representing the transition between subarctic and subtropical regimes [North Pacific Transition Zone Province (NPPF)]; apparently, this province will curve polewards at its eastern extremity under such conditions.

In the Atlantic Ocean, the distinction between the two zonal trade wind provinces (WTRA and ETRA) may become so slight that we should recognize them as a single entity (TRAT-ENSO) having the usual poleward boundaries. It is probably only in the trade wind biome and perhaps the equatorward parts of the westerlies biome that the boundaries of provinces are effectively modified during a strong ENSO event; elsewhere, though the normal seasonal production cycle within provinces may be modified, the coordinates of the provinces themselves are unlikely to change. Though the date of onset and the intensity of the monsoon over India are a function of the SOI (Shukla, 1987) and have important ecological consequences in the Indian Ocean (Longhurst and Wooster, 1990), there seems to be no reason to think that the provinces will lose their identity.

LONGER SCALE TRENDS AND CHANGES

The pulsed, periodic occurrences of ENSO events occur against a background of decadal and secular-scale changes such as those which have dominated recent times and were probably initiated by conditions in winter 1976–1977 when the whole North Pacific weather pattern changed state as the Aleutian low deepened, shifting storm tracks further south than normal. From then until at least the early 1990s, the Southern Oscillation remained in the state of a weak ENSO event so that the tropical ocean unexpectedly remained in a quasipermanent warm mode. Full development of ENSO events has been more frequent during this period: In the period from winter 1976–1977 to 1995 the mean return interval has been less than 2 years, and only a single cold year (1988–1989) with full development of the trade winds has occurred (Graham, 1994; Miller *et al.*, 1994). Currently, the 1997–1998 ENSO event is developing and promises to become yet another very strong warm event. During this whole period, ENSO-like anomalies have frequently dominated the North Pacific, with significant ecological effects; upwelling in the California Current has been constrained by a cap of light, warm water, and winter zooplankton biomass there progressively declined by about 80% from the 1970s to the 1990's (Roemmich and McGowan, 1994), whereas in the central North Pacific the carrying capacity of the ecosystem has significantly increased, with integrated chlorophyll almost doubling from 1975 to 1985 (Venrick *et al.*, 1994). It is worth noting that the step-shift in 1976–1977 that Miller *et al.* demonstrated by integration of 40 environmental variables is reflected in open ocean chlorophyll, whereas the zooplankton biomass data for California show a steady declining trend throughout the whole period, as if buffering the effect.

In higher latitudes, a long-term oscillation between two apparently stable states has been uncovered in the well-studied North Atlantic: the North Atlantic Oscillation (NAO). The meteorological index in this case is the departure from the mean pressure difference between the Azores high-pressure cell and the Iceland low-pressure cell during winter. Low values of the NAO occur when the Azores high lies north of its mean position and western Europe experiences cool northerly winds (and eastern North America experiences the opposite); high values of NAO occur when the Iceland low is very strong and western Europe experiences southwesterly weather from the Atlantic (and again the opposite occurs for eastern Canada).

Using sea level and meteorological records, it has been possible to reconstruct a record for the Gulf Stream transport and the strength of the midlatitude westerly winds extending back to 1850 within which long-term trends could be identified. The most striking change was a strong increase in Gulf Stream transport and westerly winds from 1920 to about 1950 (high NAO) and then a decline to original levels by about 1970; sea-surface temperature increased off western Europe to maxima in about 1950 (Colebrook, 1976). This is the "warming of the 1930s and 1940s" which led to major shifts of species distributions in the North Atlantic, most important the penetration of the southern parts of the boreal arctic province (ARCT) by cod (*Gadus atlantica*), which built up stocks progressively further north along the coast of western Greenland (Cushing and Dickson, 1966). From 1950 to the 1970s the Azores–Bermuda high-pressure cell increased in intensity and northerly winds increased progressively along the seaboard of western Europe (Colebrook, 1986; Dickson *et al.,* 1988) causing significant changes in the regional oceanography (low NAO). Upwelling on the Portuguese coast was unusually strong, and the strong northerly winds in winters also created a high-salinity anomaly east of Greenland which could be traced for the following 10 years during its advection around the North Atlantic circulation.

Striking trends in plankton abundance in each of 12 compartments around Britain (corresponding approximately to Northeast Atlantic Shelves Province and parts of North Atlantic Drift and Altantic Subarctic Provinces) occurred during this period and reversed in the 1980s. All showed a progressive reduction of both phytoplankton and zooplankton biomass from 1950 to 1970, with some recovery thereafter. This is consistent with a delay and weakening of the spring bloom due to the deepened mixed layer anticipated from the increased strength of wind stress during these years.

However, there is no reason to think that either the seasonal production cycles characteristic of the North Atlantic provinces or the effective location of the boundaries between them will have changed significantly during this multi-decade excursion of atmospheric forcing (which, incidentally, is the wider expression of the "Russell Cycle" of zooplankton biomass and clupeid fish species of the western English Channel), characterized most simply by the frequency of westerly weather at the British Isles, from the norm of the first decades of this century (Russell *et al.,* 1971). We have to bear in mind that the ecology of the North Atlantic provinces—probably in common with all provinces of the westerlies and polar biomes—can exist in a range of states between two extreme conditions.

In the trade wind biome, where reversing monsoon winds and monsoon currents dominate the regional environment, we are used to strong interdecadal variability in the strength and reliability of the monsoon regime. These are some of the factors which force drought, famine, and warfare. For the eastern Arabian Sea, we have excellent records for most of the twentieth century not only of the variability of monsoon winds but also of the oceanographic and biological consequences (reviewed by Longhurst and Wooster, 1990). As shall be discussed in Chapter 8 (see INDW), the southwest monsoon forces upwelling on the west coast of India and this process, in turn, produces diatom blooms and supports a large population of *S. longiceps,* the oil sardine which feeds on them. Because *S. longiceps* is a very short-lived species whose abundance responds rapidly and sensitively to year-to-year changes in ocean conditions, fishery records can tell us much about long-term changes in circulation and upwelling.

However, we also have direct information for both the Indian Ocean and the Atlantic. Used with appropriate caution for steric effects, sea level is an indicator of upwelling: Sea level is highest, indicating no upwelling, during the northeast monsoon and lowest, because of strong upwelling, during the southwest monsoon. Long time series can be constructed for several stations, and those from Cochin and Mangalore are particularly useful. Average seasonal variation of the sea level at Cochin was calculated for the period 1939–1987.

Using the Cochin data for 1939–1987 we can compare sea level for the monsoon period of locally forced Ekman divergence (May–September) with sea level for months when remotely forced, baroclinic upwelling occurs (March–April). The two kinds of upwelling seem to be somewhat independent because the strongest decadal trends occur in the April data, which show a trend of increasing sea levels beginning about 1942 and reaching its maximum around 1960, after which a slightly lower level was sustained until the end of the data set. On the contrary, for the monsoon months (exemplified by June) decadal trends are less conspicuous but there is an important change of state from weak to strong upwelling over a period of 2 or 3 years culminating in 1960.

Somewhat similar trends occur in a longer record (1878–1982) for Bombay, which shows rising sea levels from about 1930 until the mid-1950s, with a subsequent minor retreat which was initiated earlier than at Cochin (Fig. 6.2). It is unlikely that the occurrence of highest sea levels along the west coast in the mid- to late 1950s at two such distant stations could be a matter of chance. Comparison of the trends in Cochin sea level and oil sardine catches confirms the anticipated relationship (Fig. 6.3). A long period of relatively low but variable abundance from 1900 to the late 1950s was transformed by a massive population increase in 1960, coincident with the onset of a period of strong upwelling. Both upwelling intensity and sardine population were sustained, with appropriate variability, until at least the late 1980s. These statistical data thus confirm the more anecdotal information concerning the relationship between diatom blooms and oil sardine response which are noted in Chapter 8 and are a powerful illustration of the ecological effect of decade-scale climatic changes in the wind patterns over the ocean.

In the Atlantic, along the coast of tropical West Africa and especially off Ghana and the Ivory Coast, the sustained threefold increase in sardine abundance that was initiated in the late 1970s and apparently continued until at least the mid-1990s must be explained. This biological event is matched by increased zonal wind stress but not, as might be expected on this upwelling

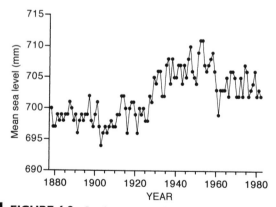

FIGURE 6.2 Bombay mean sea level, 1978–1982. A long record of sea level on the west coast of India showing the secular-scale variability in the annual mean from which may be inferred the relative strength of the southwest monsoon winds in each year and hence in the strength of the upwelling (and biological enhancement; see Fig 6.3) along the western coast of the Arabian Sea if not also along the coasts of Somalia and Arabia.

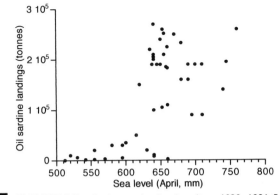

FIGURE 6.3   Cochin sea level and sardines, 1939–1986. Relationship between sea level in April on the west coast of India and the landings of oil sardines in the inshore fishery. It is reasonable to assume (i) that the inshore fishery yield reflects relative abundance of oil sardines, (ii) that abundance of this rapidly growing fish reflects differential growth of the diatom stocks on which it depends directly for food, and hence (iii) that the strength of the monsoon (see Fig. 6.2) determines the abundance of these important components of the pelagic ecosystem.

coast (see GUIN), in enhanced upwelling strength, which is obviously the first kind of the biological enhancement we should seek to explain such a massive increase in sardine abundance. In fact, the explanation is quite unexpected and is surely explicable in terms of the larval retention area/adult stock size hypothesis of Sinclair (1988) as argued convincingly by Binet (1997). Since the late 1970s, the Pacific Ocean has not returned to a full relaxation from ENSO conditions and hence the Atlantic Ocean has, during the same period, experienced strengthened flux in the eastward zonal currents which, at Cape Palmas (Ivory Coast) and Cape Three Points (Ghana), have supported much larger than usual turbulent gyres downstream of each cape. In this case "downstream" refers both to the very shallow Guinea Current and to the contrary Guinea Undercurrent, which frequently surfaces on this coast. These are the retention areas of the larvae of *S. aurita,* the sardine of this upwelling region, and their enlargement is sufficient reason for the observed increased stock size.

These two case studies are directed at two sardines having quite different ecologies: *S. longiceps* is one of the few species of clupeids to specialize in direct grazing on blooms of large diatoms, often *Coscinodiscus* spp., whereas *S. aurita* is a copepod specialist, in this case dependent on the population of *Calanoides carinatus.* Both provide evidence, in their time-dependent variability, of the extent to which the whole pelagic ecosystem undergoes variability on the same scale in the tropical seas.

CONCLUSION: A MAP OF PROVINCES FOR A STRONG ENSO EVENT?

Integrating the information reviewed in this chapter, it would be simple to assemble a map of the provinces as we might expect them to appear during a strong ENSO event. This would accommodate the more uniform conditions across the tropical Pacific and Atlantic Oceans that we shall expect to characterize a strong ENSO event: The map would show a single province across the trade wind zone of each of the two oceans. The eastern boundary of NPPF would also have to be adjusted slightly, as discussed previously.

Recalling that the basic climatological map of province boundaries was formulated with the composite CZCS seasonal images as one of the inputs, the reality of the ENSO provinces map might be validated by recourse to 90-day composite sea-surface color images from the CZCS sensor leading to an ENSO event (1982 spring, summer, and fall; 1982–1983 winter; and 1983 spring) and a the relaxed state (1979 spring, summer, and fall; 1979–1980 winter). Such images are the only ones available at the time of writing and give only rather incomplete coverage, although they are sufficient to enable the assertion to be made that the chlorophyll field for this particular event did not fully support the conclusions reached in this chapter concerning the reorganization of tropical Pacific provinces during ENSO events. Though the region of enhanced chlorophyll in the eastern tropical Pacific, along the equator, and lining the countercurrents is weaker in the fall to winter period of 1982–1983 than in 1979, the normal pattern is still existent and the Pacific Equatorial Divergence Province is a region–as it is defined in the normal map–of enhanced surface chlorophyll. On the other hand, the ENSO situation is indeed marked by the expected increase in surface chlorophyll values in the north subtropical gyre reported by Venrick.

Should we conclude from this lack of support for rearranged provinces that a single, climatological arrangement will satisfactorily represent the actual ecological compartments of the tropical ocean? Probably not, because even 90-day composites of the CZCS data for the open ocean give very incomplete coverage. The images are an unsatisfactory basis on which to reject the logic of the arguments made in this chapter, and the only conclusion that seems sensible is that at least they do not support the opposite case: that the ecological geography of the ocean describes such an ephemeral and variable field that its formal description is impossible if numerical coordinates are to be assigned to its compartments.

7

THE ATLANTIC OCEAN

In the following chapters, each devoted to a single ocean basin, you should bear in mind the basic geography of each ocean because it is often the simple dimensions and spatial relationships of topography and coastlines that determine the puzzling details of circulation patterns. The symmetrical planetary forces that drive physical circulation and mixing can be modeled mathematically with great predictability but operate within irregular basins shaped by the asymmetric drift of continental masses during geological time.

The major geographical characteristics of each ocean are clearly to be seen in the good atlas which readers will undoubtedly have at their elbow. We shall deal with each ocean in turn and include with each its marginal seas. In the case of the well-endowed Atlantic basin, we shall include the Arctic Ocean and the Caribbean, Mediterranean, Black, and Baltic Seas. As Tomczak and Godfrey (1994) point out, this gives it by far the longest latitudinal extent of any ocean basin— 21,000 km from the Bering Straits over the pole and down to Antarctica.

It is a remarkable fact how little emphasis is placed on the individual geography of each ocean basin in most texts on oceanographic processes, whether these are physical, chemical, or biological. An intimate knowledge of the geography of each basin, and a critical understanding of how the current arrangement of continents determines the rates of processes (physical, chemical, and biological) in each ocean, must be one of the most important items in any oceanographer's tool kit. Unfortunately for most, it is something they just have to pick up on the job. Each of the chapters to follow will start by listing how the relative positions of the continents, and the alignment of their coasts, determine their regional circulation and stability patterns and hence forces their mean ecological distributions.

For the Atlantic Ocean, the following are among the significant relations. South America extends much farther poleward into the Southern Ocean than does Africa, and so creates a unique asymmetry in the surface temperature fields of the South Atlantic that has no homolog in either Indian or Pacific Oceans. Interruption of the strong westerly flow of the Circumpolar Current induces a northwards loop along the continental edge east of Tierra del Fuego and the Falklands plateau which continues until it meets the southward flow of the warm western boundary current off Mar del Plata. If the Falkland Islands arose from deep water, instead of from the shoal Falkland plateau, the circulation pattern of the whole southern Atlantic would be different. Again, if Cape Hatteras did not exist, the Gulf Steam would separate from the American continent at a different latitude and the ecology of the whole North Atlantic would be different from what it is today. Perhaps most important, if the triangular protuberance of Brazil and the Guianas lay farther north, the flow of the westward trade wind currents would be more evenly divided between North and South Atlantic. Equatorial surface water from both hemispheres would not, as it does now, flow almost entirely into the Gulf Stream of the North Atlantic and, consequently, 10°C water would not penetrate to 60°N off Iceland.

In winter, wind stress (a quadratic function of wind speed, stronger in gusty than in steady winds) and heat loss at the sea surface combine to force deep mixing against the remaining summer stratification of the water column. Between the Iceland low and the Azores high, wind stress ($>250 \times 10^{-2}$ dyn cm^{-2}) and heat flux (-40 to -60 W m^{-2} year^{-1}) both take very high values over the North Atlantic (Hellerman, 1967) and drive winter mixing to 750–900 m, especially in the 50–60°N zone from Ireland to Newfoundland. One of the revelations of the Coastal Zone Colour Scanner (CZCS) monthly chlorophyll fields was to show graphically the result of solar heating and stabilization of the water column in spring. These images provided confirmation that the North Atlantic supports by far the most extensive and strongest spring bloom anywhere in the oceans, judged on the basis of chlorophyll accumulation. A consequence of deep mixing and a spring bloom that takes nitrate to very low levels is the "nutrient hole" of the North Atlantic, which is a striking anomaly on the global nutrient field. This deficit, which extends to 1000 m, is presumably maintained because export of organic material to the sea floor is faster than horizontal advection of nutrients into the region of deep mixing. This process results in a deep nutrient front at 1000 m across the ocean at about 30°N for nitrate, phosphate, and silicate that has no counterpart in other ocean basins. Why we do not see the same phenomenon in the Southern Ocean, where wind stress is even greater in winter, or in the relatively windy North Pacific are questions I shall address in the following chapters.

In the North Atlantic, the bifurcation of the Gulf Stream over the Newfoundland Basin adds complexity to the gyral system, which is reflected in its dynamic biogeography, and requires that we recognize compartments within the gyral circulation, though they may not be readily identifiable in surface features. From about 40 to 45°W the Gulf Stream feeds a southeasterly flow (the Azores Current) directly into the gyral circulation and a northeasterly flow (the North Atlantic Current), of which part recirculates into the gyre only when it encounters the European continent. The midocean ridge topography restricts the main gyral recirculation to the western basin: There are significant biological differences between the eastern and western parts of the gyre.

ATLANTIC POLAR BIOME

Because the coastal geography of the polar regions of the Atlantic is so complex, especially in the Canadian archipelago, and because the marginal ice zone which migrates through these regions during the summer has many of the characteristics of a coastal zone, it is not useful to recognize a coastal boundary zone poleward of the North Atlantic continental shelf provinces. In the Boreal Polar Province (BPLR) and the Atlantic Subarctic Province (SARC), therefore, we include the narrow continental shelf regions enclosed by each.

Boreal Polar Province (BPLR)

Extent of the Province

BPLR comprises the Arctic Ocean, between North America, Greenland, and Asia, and all but one of its marginal seas: the Chukchi, East Siberian, Laptev, Kara, and Beaufort Seas. Only the northern Barents Sea is considered to be part of BPLR; the southern part is assigned to the SARC because it is strongly influenced by Atlantic Current water and does not freeze in winter. Outflows of polar surface water through the Fram Strait and the Canadian arctic archipelago extend conditions characteristic of the province to the Greenland and Labrador coastal currents and to the inter-island passages of the archipelago and Baffin Bay. Because of the complex geography of this province, and its characteristic surface brackish layer and extensive ice cover, it is convenient not to recognize a coastal boundary zone. Consequently, Hudson's Bay is included in this province.

Continental Shelf Topography and Tidal and Shelf-Edge Fronts

The Arctic Ocean has wide continental shelves: Off Asia, the Eastern Siberia, Lapta, and Kara Seas are extensive shallow-shelf seas into which many large rivers open (Lena, Yenisei, and Ob), and the shelf width reaches 700–800 km along much of the north coast of Asia. Off northern Alaska the shelf reaches only 250 km and narrows even more to the east across the Beaufort Sea toward the Mackenzie delta. The shelves of the north coasts of Greenland and the Canadian archipelago are much narrower, though the archipelago itself extends over 15° of latitude. The continental shelf of Labrador is remarkable for its anomalous depth, with the shelf break occurring at about 400 m.

Arctic continental shelves show a complex density circulation, behaving as salt-wedge estuaries in summer and, by the deep export of salt-rejection brine produced as the sea freezes, negative estuaries in winter. The surface water of continental shelves is more dilute and shows greater seasonal salinity variation than the central part of the Arctic Ocean (Carmack, 1990).

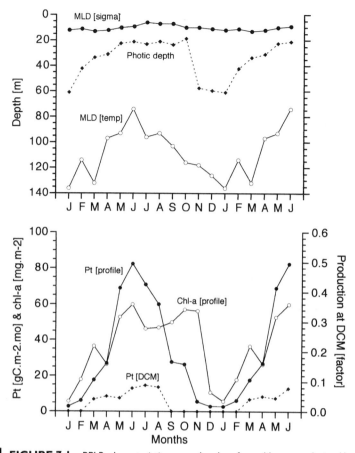

FIGURE 7.1 BPLR: characteristic seasonal cycles of monthly averaged mixed layer and photic depths, chlorophyll at the surface, and rate of primary production both depth-integrated and at the DCM. Data sources are discussed in Chapter 1.

Though little eddy activity occurs below the permanent ice cover of the Arctic Ocean, where mesoscale features are thought to be very long lived (Muench, 1990), ice-edge fronts and shelf-edge fronts are significant elsewhere. The position of the summer ice edge of the Beaufort and Chukchi Seas is constrained by the position of the shelf break. Warm, northward currents are steered by troughs on the Chukchi shelf, and the ice-edge frontal system therefore follows this topography (Muench, 1990). The Chukchi–Beaufort Sea marginal ice zone (MIZ) is typical, with strong input of meltwater along the ice-edge front.

Defining Characteristics of Regional Oceanography

The Arctic Ocean itself is almost completely enclosed. Surface water enters the ocean through the shallow (45 m) Bering Straits and through the eastern part of the deep Fram Strait, with some entrainment from the Barents Sea north around Novaya Zemyla. These source waters are modified by river runoff and meltwater in summer and by salt rejection during freezing in winter (Jones *et al.*, 1990).

Though some polar surface water leaves the Arctic Ocean through the Canadian archipelago, by far the greatest outflow is through the Fram Strait, which is the only deep-water connection with other oceans. The strongest flow is on the western side of the Fram Strait, entraining some Atlantic surface water around the eddy over the Molloy Deep (>5 km) west of Spitzbergen (Muench, 1990). This flow occurs at all seasons and is the source of the narrow, buoyancy-forced East Greenland Current, whose seawards boundary is the strongly marked East Greenland Polar Front, which defines the outer boundary of the province. The East Greenland Current is 150–200 km wide, conforms to the continental edge, and is continuous along the coast as far as Cape Farewell, which turns the current. At the East Greenland Polar Front the stream of polar surface water (always <0.0°C) meets warmer, saltier Atlantic water, and at the front itself there is significant eddying associated with a narrow wind-driven jet current aligned along the ice edge.

The province is not so clearly defined on the western side of Greenland because the surface water of the West Greenland Current, still landwards of the Polar Front, has by this time been significantly modified by Atlantic water entrained across the front. In the Davis Straits, surface water from the Arctic Ocean enters the western side of the Labrador Sea. It is convenient to include the Labrador Current to about 52°N in the polar province. During winter, the front between the Labrador Current flow and the warmer Atlantic water corresponds, as in the East Greenland Polar Front, with the ice edge. Mesoscale eddying in the front is reflected in the pattern of drifting pack ice (Muench, 1990).

Polar surface water of the Arctic Ocean extends to about 200 m and is cold (<0.0°C) and fresh (<34.4°C). Dilution of surface water within the polar basin by meltwater and river runoff from northern Russia results in a shallow brackish layer above a 40 to 50-m deep halocline. The same occurs in Baffin Bay during summer, where a shallow (10–25 m) brackish layer occurs within the "Arctic water" (actually polar surface water transported by and modified in the Greenland Currents) that extends down to 150–500 m (Muench, 1970). Similar conditions occur in ice-free periods in the sounds and passages of the Canadian archipelago. Surface water having salinity characteristics of Pacific and Atlantic surface water, respectively, occurs in the Chukchi Sea (approx 30%) and north of Spitzbergen (approx 34%).

The polar province is dominated by permanent ice cover which opens significantly during summer only in the coastal seas to the north of Asia, Alaska, and the Canadian Northwest Territories. North of Russia, the Kara Sea becomes almost ice free to about 1200 km from the coast, the Lapta Sea partially so to about 200 km from the coast, and the Eastern Siberian Sea very incompletely ice free. Summer landfast ice associated with small island groups at 100 and 140°E between the three Siberian seas may be continuous with Arctic Ocean ice. The summer ice-free zone of the Chukchi and Beaufort Seas is about 150–200 km wide, the position of the ice edge being determined by northward streams of warm Pacific water through the Bering Straits (Muench, 1990). The Chukchi and Beaufort ice-edge fronts are strong and active and are associated with wind-driven jet currents (Muench, 1990). Narrow ice-free passages open in summer through the Canadian archipelago and thence into Baffin Bay, which becomes almost totally ice free in late summer. Open-water polynyas occur during winter throughout the Canadian archipelago at locations predictable from topography and currents, especially the well-known "North Water" at the northern end of Baffin Bay. The East Greenland Current transports broken pack ice south to about 70°N, the ice being restricted to the west of the Polar Front; during winter, more continuous sea–ice cover extends south to Cape Farewell.

Biological Response and Regional Ecology

Pelagic ecology in the BPLR has local characteristics determined by the varying period of sea–ice coverage and by the effects of coastal topography. It will be convenient, therefore, to discuss three ecological zones: (i) those areas which are permanently ice covered, (ii) the marginal ice zone during retreat in spring or formation in autumn, and (iii) the extensive open-water areas that occur in some regions in mid- to late summer.

The biota characteristic of these three ecological phases are similar and now reasonably well explored (see, for example, the review by Longhurst *et al.*, 1989), though they occur in different abundance and their characteristic seasonality is modified to match local conditions. To some extent, biota perform as if at the extreme end of their range of possible thermal adaptation: Arctic heterotrophic bacteria perform as a uniform psychrophilic assemblage and metabolize glucose and amino acids actively down to freezing temperatures but also contain organisms capable of opportunistic growth at mesobiotic temperatures of approximately 30°C (Li and Dickie, 1984). Diatoms and coccolithophores are generally assumed to dominate polar phytoplankton, but recent work has revealed that a substantial proportion of chlorophyll, cell numbers, and RuBPC activity are actually contributed by the picoplankton fraction (cells <3.0 μm diameter) and these perform 10–25% of all carbon fixation compared with >50% in the tropical ocean (Trotte, 1985). Generally, 60% of algal biomass is contributed by >35 μm cells, whereas >50% of respiration is contributed by <1 μm cells (bacteria and other microheterotrophs). Limited by low temperatures are the prokaryotic cyanobacteria and prochlorophytes, which occur only in relatively low abundance and serve as a biological marker for the transport of southern water in summer. No novel physiological mechanisms need be invoked to explain the success of arctic phytoplankton in extremely cold water. It would, in fact, have been possible to predict their performance by extrapolation from what is known of the physiology of temperate-zone organisms (Li and Dickie, 1984). Growth is, generally, limited by light and temperature rather than by conventional nutrient levels.

The general characteristics of arctic zooplankton, especially copepods, have been well understood for many years. Growth rates are very slow, individuals are large compared with congeners in warm seas, seasonal ontogenetic migration dominates vertical distribution, and copepods dominate the total biomass. Recently, it has been found that protist microplankton comprise only about 10–20% of total biomass. Lipid storage is very important for copepods, especially in diapausing copepodites in winter, when ammonium excretion is reduced to very low levels (Head and Harris, 1985). It is especially in the copepodite 5 stage that lipid is accumulated in preparation for the adult molt and egg production at depth during the last winter of each multi-year generation cycle (Conover, 1988). These generalizations may be extended to other arctic biota: Chaetognaths (e.g., *Sagitta elegans*) also have a 2-year life cycle and reach an unusually large size in BPLR.

Returning now to the three ecological zones of BPLR: the Arctic Ocean, by far the largest part of the province (3×10^6 km²), remains covered year-round with pack ice whose drift follows the general circulation of the basin; open water occurs only in parts of its marginal seas during summer. This central region of the province, at the top of the world, will serve as a model for all parts of BPLR which remain icefast in summer, with the caution that in ice-covered passages within the archipelago strong currents must advect biota from adjacent open-water areas. In this case, biota found below the ice may not have developed there.

The dilute surface water mass of the Arctic Ocean originates in both Pacific and Atlantic Oceans and lies everywhere above a halocline in the upper part of which a strong nutrient maximum occurs that is not mediated by local biology. Rather, it is due to the fact that the halocline water originates on the surrounding continental shelves and carries the nutrient signatures of Pacific and Atlantic ocean water—high in both cases, but highest in Pacific water—so that the Chukchi Sea nutrient maximum is stronger than that of the Kara Sea. Though biological uptake of nitrogen does occur in the surface water, levels remain relatively high compared with those of other oceans; 4 or 5 μM is normal in the central Arctic Ocean (Jones *et al.*, 1990). For silicate and phosphate, concentrations above and below the halocline are similar.

Though the permanent ice cover of the central Arctic Ocean is only 2 or 3 m thick, it reduces incident radiation by 80–100%. This is sufficient, however, for ice algae to be a feature of the ice-covered portions of BPLR during summer, and phytoplankton blooms may even occur at times and places where the sea ice becomes snow free, albeit at rates of about one-third of the water column over the surrounding continental shelves (Legendre *et al.*, 1992; English, 1961). Interstitial growth of *Melosira arctica* and loosely attached mats of both centric and pennate diatoms are the usual ice algae organisms.

The water column below the Arctic Ocean ice is inhabited by a sparse but permanent zooplankton community, its biomass (0–200 m) dominated, as elsewhere in the Arctic, by calanoid copepods: *Calanus glacialis, C. hyperboreus,* and *Metridia longa* and larger numbers, but smaller biomass, of *Pseudocalanus, Oithona, Microcalanus,* and *Oncaea*. In the Nansen Basin, *Oncaea borealis* forms 40–80% of the individuals. It may be noted that the geographical distribution of *C. glacialis* very closely matches the boundaries of the BPLR province (Conover, 1988). *Calanus hyperboreus* at these latitudes requires 3 years to complete its life cycle: in Year 1 it achieves growth to copepodite 2, in Year 2 it reaches copepodite 4 or 5, and in the third year it reaches the adult stage and reproduces (Conover, 1988; Smith and Schnack-Schiel, 1990). It is characteristic of the Arctic Ocean that many of the most abundant copepods extend from the surface to at least 1000–1500m, and there appear to be few if any deep-water specialists (Groendahl and Hernroth, 1986), at least above 2500 m.

Diel migrations are generally thought to be largely absent in summer, though Groendahl and Hernroth (1986) did find both *O. borealis* and *Metridia longa* performing diel migrations in the Nansen Basin, to the north of Spitzbergen, over a depth range of several hundred meters when the daily irradiance cycle was sufficiently strong to induce such behavior. Nevertheless, ontogenetic migrations are the principal determinant of herbivore distribution in the vertical plane, and Conover (1988) described the seasonal succession of copepods below fast ice as an "ontogenetic escalator" on which waves of the F_0, F_1, and F_2 generations (successively and in that sequence) of *C. hyperboreus* and *C. glacialis* rise toward the surface, whereas *Pseudocalanus* remains close below the ice surface. With the erosion of sheets of diatoms from the undersurface of the ice as summer advances, all copepods come to be concentrated in the upper 20 m of the water column. At this season, depth partitioning of the near-surface layer occurs between species and their growth stages, which tend to be differentially distributed with depth.

As Walker Smith (1987) emphasized, ice edges or MIZs occur in a variety of geographic locations in this province and are influenced differently by Pacific and Atlantic conditions and biota, both over deep and shallow water, and form either an abrupt edge or a zone as much as several 100 km wide. For these reasons, it is not easy to generalize their ecology. Nevertheless, we must try, and it may help if we concentrate our attention on the properties of a receding ice edge in spring.

It is principally the effects of the freeze–thaw cycle of seawater that set the ecological stage in the MIZ by determining the strength of near-surface stratification. Thawing sea ice in spring releases fresh water to create strong density gradients, whereas freezing in the autumn partitions seawater between freshwater ice and strong brine which is rejected and sinks, causing instability and deep mixing. In late winter (March and April), as daylight returns, the surface water column is deeply mixed, nutrient levels are high, and net carbon fixation is not yet established (Smith and Brightman, 1991). It is precisely the effect of a thawing, receding ice edge in spring which leads to strongest stability and conditions most likely to lead to phytoplankton bloom, especially where (as expected) cell growth will be principally light limited. Where a spring bloom in a MIZ reduces nutrient levels below optimal, a further enhancement of algal growth may occur by ice-edge upwelling forced by wind-driven transport of surface water away from the ice and its replacement from below. For all these reasons, we may expect that MIZs are a focus for biological activity. Indeed, this is seen in the CZCS imagery for the Greenland and the Labrador Currents, both characterized by a receding ice edge in summer and each of which tends to be defined in satellite chlorophyll fields as an area of enhanced chlorophyll. Very high surface chlorophyll values also occur in open-water regions of the Chukchi Sea and eastwards along the Beaufort Sea, where the ice edge recedes very rapidly during summer. However, the bloom in the Chukchi Sea appears to be partly the result of algal biomass transported northwards through the Bering Straits in the Anadyr Current along the western coast (see Chapter 9, North Pacific Epicontinental Sea Province). The Kara Sea, east of Novaya Zemyla, has higher surface chlorophyll in summer than the Barents Sea to the west, which carries little winter ice (see Chapter 7, Atlantic Subarctic Province).

MIZs develop from west Spitzbergen through the northern and eastern Barents Sea with the vernal retreat of the ice edge. These induce upwelling, which may be brief, wind-induced events, or longer density-induced events (Johanessen, 1986). The low-salinity surface layer is on the order of 10–30 m deep at its shoalest position near the ice edge when density-driven or wind-forced upwelling occurs. The MIZ of the Chukchi Sea appears only in June, but only a small part of the sea becomes ice free; extremely strong stratification occurs at the ice edge, and reported rates of primary production seem improbably high (150–7000 mg C m^{-2} day^{-1}). At any rate, a strong chlorophyll maximum occurs at the base of the pycnocline and nitrate levels are reduced to undetectable. In the Barents Sea, the MIZ retreats very rapidly northwards across shallow water in spring, whereas in the deep Fram Strait it remains close the boundary between the northward flow of Atlantic water and the southward flow of polar water. In both regions, nitrate is reduced to low values after the summer bloom occurs. In high summer, at the deep chlorophyll maximum in the northeast Spitzbergen MIZ, C. hyperboreus and C. glacialis consume from 65 to 90% of the daily primary production while supporting about 35% of it by the excretion of ammonium (Eilertsen et al., 1989a).

Where extensive open water develops during summer, as in Baffin Bay and the Chukchi Sea, nutrients may be stripped from the surface brackish layer by algal, principally diatom, blooms. Strong algal blooms sustained by topographic upwelling occur in many places in the Canadian archipelago, notably in Lancaster Sound; a spring bloom occurs in the open water of Baffin Bay, though it is not long sustained after nutrient limitation occurs. In Baffin Bay, a summer chlorophyll maximum occurs consistently between the 1 and 10% light levels, usually near the base of the pycnocline at which depth a nutricline for nitrate also occurs. The production/chlorophyll ratio (P^B, the photosynthetic index) is usually 4–7 g C (g chl)$^{-1}$ day^{-1} compared with about 0.2 g C (g chl)$^{-1}$ day^{-1} for the ice-fast regions of the Arctic Ocean (see Longhurst et al., 1989). Values of the photosynthetic quotient (PQ = molar ratio of oxygen evolved to carbon fixed) of 1.3–1.8 suggest that some nitrate metabolism continues even at very low nitrate levels.

In this open-water summer situation there is a strong differentiation in depth selection by zooplankton species. A clear separation occurs between those occupying the photic zone (Calanus and Pseudocalanus) and those occurring deeper (Metridia, Euchaeta, and Microcalanus). As in warmer seas, detailed depth distributions (obtained with multiple nets or continuous sampling systems) indicate that herbivores tend to select either the depth of maximum production rate (C. glacialis) or the slightly deeper chlorophyll maximum (C. hyperboreus). There is thus a deep biomass peak across the pycnocline and also a smaller biomass peak 2–5 m below the surface, associated with the near-surface feeding often reported for arctic copepods (Herman, 1983; Longhurst et al., 1984; Sameoto et al., 1986). Copepods exhibit strong diel feeding rhythms, even in the absence of diel migration.

Though the relationship is variable from station to station, the primary production/herbivore consumption ratio often takes a positive value in the photic zone (Longhurst and Head, 1989). Moreover, the standing stock of algal cells in surface waters in late summer appears to be unlimiting to herbivore grazing: Less than 1% is removed per day in Baffin Bay at this time and the fate of most algal cells must be to sink through the pycnocline when this is destroyed by winter mixing. Rapid, mass sinking of diatom-dominated algal floc is suspected to be a feature of the final phase of blooms in this province.

The flow of Arctic water in the Greenland coastal currents and through the Canadian archipelago receives only a single significant input of riverine water—from the shallow water of Hudson Bay, which itself receives a large part of the runoff from the great Canadian Shield. This eventually passes through the Hudson Strait, to the north of Labrador, and is the source of anomalously high nitrate concentrations in summer on the Labrador shelf, downstream of the Hudson Strait. It has been suggested that this enriches the whole of the Labrador shelf ecosystem relative to shelf areas north of the Hudson Strait.

Synopsis

Photic depth as shown is strongly affected by ice cover and winter darkness which occurs over much of the province (Fig. 7.1). A layer is always present because of the brackish surface layer, whereas winter cooling extends much deeper. The primary production rate conforms to irradiance and sun angle, peaking at the date of maximum day length. Accumulation of chlorophyll tracks primary production, except (i) from July to September, when grazing pressure is released by the descent of *Calanus* to overwintering depths so that accumulation occurs though primary production rate falling; and (ii) in April, before *Calanus* rise again to the photic zone and before grazing commences seriously.

Atlantic Arctic Province (ARCT)

Extent of the Province

This province lies between the edge of the Greenland coastal currents and the Oceanic Polar Front (Dietrich, 1964) or Subarctic Front (Meincke, 1984) that crosses the ocean diagonally from Flemish Cap to the Faeroes, and therefore includes the central part of the Labrador Sea, a broad zone trending northeast toward Iceland from the Flemish Cap, and the central part of the Nordic Sea from Iceland to the Fram Strait. The limits of the province are rather variable, and its southern boundary probably cannot often be traced in satellite imagery.

Continental Shelf Topography and Tidal and Shelf-Edge Fronts

Except for the narrow shelf of Iceland, this is a purely oceanic province.

Defining Characteristics of Regional Oceanography

The Labrador Sea, the Atlantic Ocean south of Greenland, and the Nordic Sea all have special characteristics and the grouping of these regions into a single province is partly a matter of convenience, recognizing the related but different nature of the surface water masses across this boreal zone of the Atlantic Ocean, and also because the surface circulation north of the Polar Front is dominated by cyclonicity in comparison with the anticyclonic, subtropical gyre to the south. Because of entrainment across the Polar Front, this province has some of the characteristics of both the North Atlantic Drift and the polar water masses.

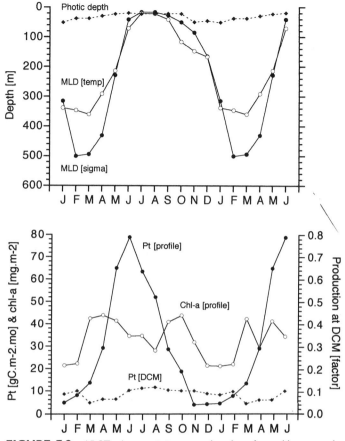

FIGURE 7.2 ARCT: characteristic seasonal cycles of monthly averaged mixed layer and photic depths, chlorophyll at the surface, and rate of primary production both depth-integrated and at the DCM. Data sources are discussed in Chapter 1.

James Swift (1986) defines the Arctic waters in the Nordic Seas as a "hydrographic middle ground lying between the domains of the polar and Atlantic waters . . . relatively cold (0–4°C) and saline (34.6–34.9)." This entity, which corresponds closely to what is proposed here as the Atlantic Arctic Province, is bounded to the northwest by the front at the edge of the Greenland coastal current and to the south and east by the frontal region at the edge of the flow of Atlantic water toward the Norwegian coast. There is confusion over how these fronts are named, and I do not follow Swift's suggestion, but use "Greenland Coastal Front" for the edge of the polar water flowing south from Fram Strait (see BPLR) and "Oceanic Polar Front" (OPF), in the classical sense, for the Atlantic water front.

Though the definition of the Oceanic Polar Front is simple, its location at the sea surface is complex. It is the conjunction between warm, salty, subtropical gyral water and cold, less saline subpolar water and is thus the extension of the "North Wall" of the Gulf Stream (sometimes termed the Polar Front of the

Atlantic). The North Wall retains coherence eastwards as the OPF as far as the mid-Atlantic Ridge at about 30°W (Dietrich, 1964). Thence, the OPF is more difficult to locate unequivocally but passes to the east of Iceland, where it becomes highly meandering, continues east along the Faeroe–Iceland Ridge, and then northwards along the Jan Mayen section of the Mid-Atlantic Ridge (Dietrich, 1964; Johanessen, 1986; van Aken *et al.*, 1991; Wassman *et al.*, 1991). Most of the Atlantic water which passes north of Iceland to form the surface water in the arctic parts of the Nordic Sea does so in a relatively narrow stream around the western coast of Iceland from the Irminger Current. The line of the OPF is the boundary between the Atlantic Arctic Province (ARCT) and the adjacent SARC as far as the latitude of Bear Island, where it is convenient to cut eastwards to the line of winter ice cover at about 15°E. Atlantic water transported north through this line toward Fram Strait will then be considered part of ARCT.

Ice cover of the polar water of the BPLR along the coasts of Greenland extends patchily and irregularly into the Arctic Province during some winters, and broken pack ice is characteristic of the regions bordering on BPLR. In the south, along the cold side of the OPF there is strong flow of relatively cold, low-salinity water from the west, recirculating within the Arctic Province as the "northern cyclonic gyre" (Krause, 1986) between the southern Labrador Sea and Iceland. The OPF is associated with a wide, very active eddy field and exchange of Atlantic surface water occurs across it from the south. An arm of the North Atlantic Current penetrates into the Arctic Province as the Irminger Current; some of this flow then passes through the eastern Denmark Strait into the Nordic Sea, and the rest is entrained westwards along the East Greenland Polar Front and eventually into the eastern Labrador Sea. The eastern, poleward limb of the Labrador Sea gyre has more Atlantic qualities than the western equatorward limb which entrains water from the Canadian archipelago.

The surface waters of the Greenland and Labrador Seas have a relatively lower stability than the adjacent areas so that winter cooling and wind mixing cause unusually strong deep convection. In fact, these seas are two of only three global centers of deep convection of surface water in winter; the Weddell Sea off Antarctica is the third. Consequently, at the end of winter a very deep trough (<900 m) in the topography of the mixed layer extends in an arc eastwards from the southern Labrador Sea to the continental edge of northwest Europe; it occupies a somewhat different area according to the data and criteria used to display it (e.g., Levitus, 1982; Robinson *et al.*, 1979; Glover and Brewer, 1988; Woods, 1984), but apparently it always lies to the north of the Oceanic Polar Front. This deep trough is associated with high winter nitrate values in the mixed layer, reaching 16 μM.

North of Iceland, mixed layer depths are deepest in the central Nordic Sea (approx 500 m), whereas a thermocline dome (<100 m) appears in the mixed layer data (Levitus, 1982) associated with the northward-turning flow of Arctic water east of Iceland. Not only is winter mixing deepest at 50–60°N in the Atlantic but also, because the subsurface nitrate field slopes with the baroclinicity of the subtropical gyre, initial mixed layer nitrate concentration in spring are progressively higher toward the northern edge of the gyre (Glover and Brewer, 1988; Yentsch, 1990).

Biological Response and Regional Ecology

The spring bloom of algae (as indicated in CZCS images) occurs earliest (in March) to the north and east of Iceland, and subsequently develops throughout the province simultaneously but patchily, rather than by progressive polewards shoaling of the seasonal pycnocline as occurs further to the south. The discontinuity between the Arctic open-ocean spring bloom and that of the Atlantic further south is important and largely unrecognized. If the CZCS image indications are correct (and they appear to be supported by observations of very early, transient blooms in the central Labrador Sea), the arctic/subarctic bloom at 60–65°N is initiated at about the same time as the start of the bloom at about 35°N in the subtropical gyre (see, for instance, Fig. 4 of Campbell and Aarup, 1992). In the Arctic province, blooms may be determined to an important extent by salinity-driven density stratification due to the freeze–melt cycle of sea ice and the modification of Atlantic water by lower salinity polar surface waters. This is a most important distinction between the algal dynamics of the Arctic and subarctic provinces and the West Wind Drift and subtropical gyre provinces to the south. The colonial prymnesiophyte *Phaeocystis pouchetti* appears often to be the dominant spring bloom organism, and such blooms are capable of supporting extremely high rates of primary production (Smith *et al.*, 1991) and may occur either after a diatom bloom, when silicate becomes limiting, or as the dominant organism from the onset. *Phaeocystis* blooms may extend simultaneously across the whole of the province in the Greenland Sea in April and early May (when the Atlantic spring bloom in the North Atlantic Drift Province (NADR) is being initiated by thermal stratification approx 30° of latitude to the south) with new (i.e., nitrate-based) primary production rates of about 2 g C m^{-2} day^{-1} or about the same as occurs at polar ice edges.

Herbivore ecology has the same characteristics as in BPLR, and the same organisms are dominant. The general outlines of mesozooplankton ecology have been explored in the central Labrador Sea (Keilhorn, 1952) and the Greenland Sea (Richter, 1994). Because the Greenland Sea region of the ARCT province is much influenced by mesoscale eddying at the Polar Front, inclusions of Atlantic water are rather frequent; within these, expatriate populations of *C. finmarchicus* are common, whereas where Arctic water dominates, *C. hyperboreus* is the dominant copepod. *Calanus glacialis*, the characteristic copepod of BPLR, is rare.

The ARCT province is clearly distinguished from NADR to the southeast by the "mean community body size" for copepods computed by Hays (1996); a line from Iceland to the Flemish Cap, closely coinciding with the boundary proposed here between these two provinces, divides small copepods (mostly about 0.5 mg wet weight) to the south in NADR from large copepods to the north in ARCT (mostly about 2.0 mg wet weight). The same line divides two characteristic patterns of diel vertical migration (DVM). Mean community DVM values computed by Hays (DVM = [night biomass − day biomass]/night biomass) from about 130,000 Hardy Continuous Plankton Recorder samples archived for this region show clearly that in ARCT this index takes values of about 0.8, whereas in NADR values of 0.1–0.5 are more usual. DVM is strongest in summer, weakest in winter in both provinces, and not only intensity but also timing differs on either side of the line defined above: in ARCT, maximum values for DVM occur at midsummer, while in NADR there are spring and autumn peaks.

Ontogenetic, seasonal vertical migration follows the same seasonal pattern as I have noted for BPLR and shall describe in more detail for SARC in the following section, and *C. hyperboreus* has overwintering depths of 1000–1500 m in the Greenland Sea (Hirsche, 1991). Though some populations of this copepod may require only a single year to complete their life cycle, the probable duration for most populations of this province is 2 years. Deep convection events do not occur with great regularity every winter but must, when they do occur, disrupt the arrangement of the deep water masses within which overwintering populations of copepods reside. Richter (1994) proposes that it is this process that causes the sporadic but massive recruitment failures of copepods known to occur in this region. That deep convection events are a highly variable forcing on the environment of this region has become clear now that this process is regarded as a potentially critical link in the global ocean carbon cycle; atmospheric forcing has changed so that for most of the past 40 years, deep convection has carried surface water to little more than 1000 m, instead of all the way to the bottom at 4000 m as previously.

Synopsis

Mixed layer deepens significantly in winter, with a rapid near-surface stabilization in April and May which is not captured by archived data (Fig. 7.2). Photic depth is always shallow and pycnocline is illuminated only briefly in summer. Primary production rate conforms to the irradiance cycle and is tracked by chlorophyll accumulation except in March, prior to the rise of *Calanus* from depth, and in September and October after they descend to overwintering depths. Balance between production and consumption is closer than in BPLR to the north.

Atlantic Subarctic Province (SARC)

Extent of the Province

From about the middle of the Rejkanes Ridge, south of Iceland, the SARC province includes both the coastal and Atlantic streams of the Norwegian Current and also the southeastern Barents Sea, including some areas which would be considered as coastal boundary regions in more southerly provinces. The boundary between BPLR and SARC in the Barents Sea is the well-marked thermal front between Arctic and Atlantic water which passes close around Bear Island and meanders eastwards from Hope Island at about 74–78°N.

Continental Shelf Topography and Tidal and Shelf-Edge Fronts

Except for the Lofoten Bank (68°N), the central bank of the Barents Sea, and the bank connecting Hope Island, Bear Island, and Spitzbergen, the whole province comprises ocean depths.

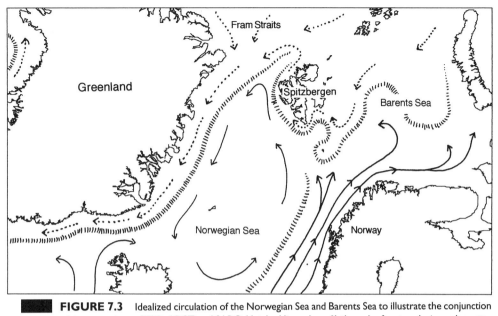

FIGURE 7.3 Idealized circulation of the Norwegian Sea and Barents Sea to illustrate the conjunction of three provinces: BPLR, ARCT, and SARC. Hatched lines show (i) the polar front enclosing polar water (BPLR), which follows the shelf edge to the south and east of Spitzbergen and continues around the coast of Greenland into the Labrador Sea, and (ii) the front between the rim current of warm Atlantic water along the Norwegian coast (SARC) and the more variable flow of mixed water in the Norwegian Sea and to the south of Iceland (ARCT).

Defining Characteristics of Regional Oceanography

This is a complex and difficult oceanographic region (Fig. 7.3). Where the North Atlantic Current turns eastward toward the Norwegian coast, at about 55°N south of Iceland, is a convenient point to set its southwest corner. Here, Atlantic subtropical water may be said to begin its odyssey toward the Arctic Ocean. The North Atlantic Current exists here both as a frontal jet (the Oceanic Polar Front) and as a broader west wind drift. The OPF, whose location was described in the previous section, forms the northwest boundary of the province separating Atlantic water from Arctic water, in the sense of James Swift.

There are two sources of surface water for the SARC province, but its principal characteristic is that it carries the influence of warm Atlantic water far to the north, even to the Kara Sea (Dickey *et al.*, 1994). The eastwards zonal flow to the south of the OPF passes across the Iceland–Faeroes Ridge south of the "Iceland Gap Front" (Johanessen, 1986) and feeds into the Norwegian Atlantic Current. Water from the Baltic and northern North Sea passes along the Norwegian coast as the Norwegian Coastal Current.

The Norwegian Coastal Current is bounded by the salinity-dominated Norwegian Coastal Front, which maintains a highly unstable eddy field (Johanessen, 1986), modifying both its Atlantic and coastal components as they pass northwards. Off North Cape, most water from the combined flow passes polewards into the southern Barents Sea, while some continues northwards

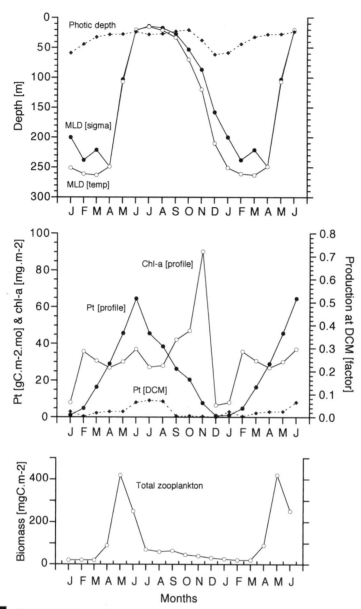

FIGURE 7.4 SARC: characteristic seasonal cycles of monthly averaged mixed layer and photic depths, chlorophyll at the surface, and rate of primary production, both depth-integrated and at the DCM. Zooplankton data at OWS M from Østvedt (1955); other data sources are discussed in Chapter 1.

above the continental slope west of Spitzbergen to Fram Strait as the West Spitzbergen Current. The water passing into the southern Barents Sea feeds the gyral circulation, which returns westwards to the south of Spitzbergen. A meandering convergence system extends toward Bear Island from the central Barents Sea, separating the westward return flow from the eastward flow in the southern part of that sea and from the northward flowing West Spitzbergen Current. It is this front which is at once the southern limit of winter ice cover and the boundary between SARC and the adjacent BPLR region. The edge of winter ice passes south at about 45°E so that the extreme eastern end of the Barents Sea, from the White Sea to Franz Josef Land, is considered part of BPLR. South and west of this line, an ice-free zone stretches discontinuously toward the west, even in winter, and it is induced by Atlantic water passing east from the Barents Sea.

In the oceanic parts of this province, winter mixing is deep, especially along the OPF and to the south of Iceland, though not so extreme as in the southern-most parts of the ARCT province.

Biological Response and Regional Ecology

We have good information on the cycles of pelagic production and consumption for several regions: in the Barents Sea at Ocean Weather Station (OWS) "I," at 59°N just south of Iceland, and at OWS "M" at 64°N in the Norwegian Sea. It was, of course, studies at OWS "M" of the interaction between the mixed layer depth and the critical depth, D_{cr} (briefly, mixing to this depth causes integrated algal photosynthesis just to match integrated respiration), that enabled Sverdrup (1953) to formulate his classical model of a spring bloom. His findings, dating from 1949, were closely matched by observations at OWS I from 1971 to 1974, during which very frequent multidepth profiles for density, chlorophyll, nutrients, and zooplankton were made.

Perhaps because of the presence of low-salinity surface water of Arctic origin, the spring bloom at OWS "I" is earlier than would be predicted by the models of mixed layer evolution and initiation of the algal bloom for the North Atlantic of Wolf and Woods (1988) and Strass and Woods (1988). Chlorophyll and ^{14}C data for 1972 and 1974 show that phytoplankton growth rapidly follows the reestablishment of a shallow mixed layer (e.g., Williams and Robinson, 1973). These earlier observations from the weather ships have been confirmed recently by daily data obtained from a moored optical profiling array (Dickey et al., 1994) at 59°N, near OWS "I". Ephemeral blooms associated with temporary near-surface pycnoclines after periods of calm weather occur as early as mid-March, and continuous spring bloom conditions begin around April 15, with mixed layer chlorophyll reaching 3 or 4 mg m^{-3} by about May 10. In 1972, nitrate was reduced from >10.0 to <2.0 μM in about 7 days in April, whereas in 1973 the same uptake occurred only during June. In the early stages of the bloom, during April and May, chlorophyll concentrations of 0.25 mg m^{-3} extend well beneath the mixed layer, to 150–200 m, suggesting that un-grazed cells were sinking through the weakly established pycnocline.

Chlorophyll values remain relatively high throughout the summer, usually >2.0 mg chl m^{-3} in the core of the near-surface, high-chlorophyll layer; some nitrate remains apparently unutilized in the photic zone throughout the summer, consistent with rather active vertical flux across the nutricline induced by frequent wind events. Recent incubation experiments (Martin et al., 1993) with Joint Global Flux Study (JGOFS) samples from the boreal North Atlantic did not suggest that Fe limitation of nitrate uptake might be occurring. Late summer blooms of coccoliths, probably *Emiliania huxleyi,* are consistently observed as backscattered light in surface color images in the southern part of SARC (Brown, 1995).

The seasonal life cycle of *C. finmarchicus,* the dominant calanoid copepod, is well described by a series of multidepth zooplankton profiles which were obtained at OWS "I" almost weekly from late March to mid-October in 1971–1974. These 111 profiles comprise 4700 individual, metered plankton samples, each representing a mean depth interval of only 12 m from the surface to 500 m. Five copepodite stages and adults of four species of copepods (*C. finmarchicus, Pareuchaeta norvegica, Metridia lucens,* and *Pleuromamma robusta*) comprised >80% of the biomass of more than 200 sorted copepod species. Each passes through several generations each summer.

The following four distinct phases can be identified in the *C. finmarchicus* seasonal cycle:

1. *Rise of the overwintered population:* From the middle of March until the end of April, the overwintered population of fifth-stage copepodites (C5) and adults dominates the biomass. During this period, C5 individuals and adults rise from overwintering levels deeper than 500 m. By about April 20, essentially the whole population is shallower than 100 m. From about April 10, the early copepodites of the first generation begin to appear.

2. *Production of first generation:* Throughout May and during the first 10 days of June, population biomass increases very fast with the growth of copepodites, and in the second half of this period adults of the first generation appear. The population remains almost entirely within the upper 150 m, with many profiles showing crowding into the upper 50 m. Diel migration, if it occurs, is shallow.

3. *Multigeneration period, some individuals descending:* From mid-June until mid-September, population biomass in the upper 500 m declines progressively, even as the late copepodites and adults of the G_2 generation appear. While there must be some continual loss of biomass to predation, the decline in biomass during the late summer is caused by the progressive migration of cohorts to depths $\geq$500 m. After mid-June, most profiles show bimodal populations with layers of high abundance in the upper 50–100 m and also deeper than 250–350 m. The upper population contains early and late copepodites, whereas the deep populations are almost entirely C4 and C5.

4. *Main population at overwintering depths:* Between mid-September and mid-October, there are few early copepodites (and no C1 at all), and the bimodal distribution is progressively replaced by profiles in which most of the biomass is layered deep, usually below 300–350 m, and in such a way as to show that only the upper parts of the deep layers are being sampled.

At OWS M in the Norwegian Sea, the same seasonality occurs: *Calanus finmarchicus* biomass is also concentrated in mid-winter at 600–1000 m. In May, the surviving population aggregates shallower than 100 m. Østvedt observed that the numbers of *C. finmarchicus* declined progressively during winter so that only about 15% survived until spring, and the same thing was observed at OWS I, where 70–80% of the biomass fails to return to the surface in spring (Longhurst and Williams, 1992).

Pareuchaeta norvegica, the second largest component of copepod biomass, also rises from overwintering depths below 500 m from about March 15 to April 10. Reproduction occurs at depth in late winter so the rising population is quite different from that of *C. finmarchicus.* Early (C1–C3) copepodites of the G_1 generation already numerically dominate the population. During the summer the biomass of *P. norvegica* is maintained at a more consistent level than that of *C. finmarchicus,* and there is less evidence of the progressive establishment of deep layers and little indication that the population descends to overwintering depths before mid-October. Thus, *P. norvegica* spends a shorter period at depth during the winter than *C. finmarchicus.*

The next ranking species, *M. lucens* and *P. robusta,* appear not to undertake seasonal migrations unless they descend after sampling terminated in October. *Pleuromamma* and *Metridia* are, in any event, strong diel migrant genera wherever they occur. Among the protists, *Globigerina bulloides* and *G. quinqueloba* are useful indicator species of the foraminiferan species assemblage 1 (see Chapter 1).

Turning now to the southern Barents Sea, which is not more than 500 m deep and over most of its area it is significantly shoaler than this, I distinguish three areas with somewhat differing ecology: (i) the region adjacent to the MIZ bordering the BPLR province to the north and east; (ii) the area adjacent to the northern coast of Scandinavia, with its narrow shelf region; and (iii) the main central area. There is significant between-year variability in the location of the ice edge, though this is much more consistent in the western than in the eastern Barents Sea. Down the west coast of Spitzbergen and close around the south of Bear Island, variation is on the order of only a few tens of kilometers. In the eastern end of the Barents Sea, this distance increases to about 500 km. In 1984, the ice front lay zonally along 75°N to encounter the northwest coast of Novaya Zemyla; in 1979, it lay southeasterly from Hope Island to the White Sea, a full 5° of latitude further south at 45°E.

Processes affecting stability of the water column are not identical and, hence, the timing of the spring bloom differs in each (Loeng, 1991). The spring phytoplankton bloom in both Atlantic (SARC province) and Arctic (BPLR province) surface water masses in the Barents Sea appear to comprise the same groups of species, usually *Chaetoceros socialis* and the colonial prymnesiophyte *Phaeocystis pouchetii,* though the two dominant calanoids of the Barents Sea are separated by this boundary: *Calanus glacialis* to the north and *C. finmarchicus* to the south. There is some evidence that the latter has poor reproductive success in the colder, eastern parts of the Barents Sea portion of the SARC province.

Adjacent to the MIZ of the polar province, whose location differs significantly between years, a surface layer of low-salinity water floods south and west over the more saline Atlantic water and creates sufficient near-surface stratification to sustain a very early spring bloom as soon as sun angle increases sufficiently, perhaps as early as mid-March. Where the influence of surface melt-water does not reach, thermal heating of the central regions which are mixed to the sea bottom in winter may not produce sufficient stability for a bloom to be initiated until much later in the summer—as late as June.

Statistical variability of the parameters of the photosynthesis/irradiance (P/I) relationship support the view that in this, as in other arctic areas, photosynthesis is mainly controlled by physical variables, though consumption by herbivores in summer and ammonium recycling modify the relationship (Rey, 1990).

Coastal water near and above the continental shelf retains some stability even during winter and here also a spring bloom begins as soon as light intensity is sufficient. Though blooms in ice-melt surface water toward the boundary with BPLR rapidly form a very strong deep chlorophyll maximum at close to 50 m depth, the bloom in coastal water often has uniform chlorophyll throughout the mixed layer (Mitchell et al., 1991a), as does the bloom in the central regions later in the season. These characteristic zonal features of the Barents Sea bloom have led to some difficulties in the interpretation of the CZCS images, which did not sense the subsurface chlorophyll of the northern border regions of the southern Barents Sea.

The Barents Sea bloom is well described by Sherman and Alexander (1989). It may be initiated by diatoms or by *P. pouchetti*, but small flagellates characterize postbloom conditions in the classical manner. Winter nitrate in the photic zone is lower than that in other polar seas (12–14 μM) and may be reduced to low or even undetectable levels after the bloom. A deep chlorophyll maximum (DCM) develops by mid-summer within the upper 50 m, though there is great between-year variability in both nitrate uptake and DCM development, forced by changes in stability and pycnocline formation in the surface layer. In June 1982 there was no DCM and residual nitrate was about 1.5 μM, whereas in 1980 at the same station and in the same month, nitrate was undetectable and there was a strong DCM at 30 m. In 1981, the situation was intermediate. Associated with these changes, there is also strong between-year variability in the primary herbivore population; in 1980–1982 there were 200,000–500,000 individuals m^{-2} of *C. finmarchicus* in the central Barents Sea. In 1983 and 1984, there were just 10,000.

Progressively during the summer, the DCM deepens to come to lie on the upper slope of the nitracline. Very high sedimentation rates of algal cells may occur during and just after the spring bloom, especially if this is dominated by *Phaeocystis*. Rates of up to 1 g C m^{-2} day^{-1} have been recorded, though the average rate in spring is about one-third this value (Wassman et al., 1991). Part of this pulse of sedimentation will be the result of zooplankton grazing and the release of rapidly sinking fecal pellets, and there is some evidence that interannual variability extends also to the percentage of total seasonal primary production consumed by primary herbivores, mainly *C. finmarchicus*. In summer 1980, the calanoid herbivores consumed about 70% of the total primary production, though they took a much smaller percentage in 1983 and 1984, when massive sedimentation occurred. Eilertsen et al. (1989b), on the other hand, suggest that the greater part (80–95%) of the phytoplankton biomass produced in spring settles unconsumed to the sea floor, though later in the season they find much closer coupling between production and consumption. In any event, this rain of organic material to the sediments, whether directly as aggregates or indirectly as fecal pellets, is surely the key to the rich benthic fauna and demersal fish stocks of the Barents Sea. Currently, this is not absolutely true since we have now learned, to the great cost to the livelihood of fishermen and fish-plant workers, that high-latitude demersal stocks are dangerously vulnerable to competitive industrial fishing. Which organisms now consume the benthos, still presumably fattening on the rain of algal cells, is altogether another story.

The species of herbivorous plankton in the SARC part of the Barents Sea and their life cycles are similar to those of the Norwegian Sea. The principal difference must be (though this seems not to have been investigated) that since their normal overwintering depths are not available in the Barents Sea, the population probably aggregates into the few deep basins, such as the 500-m-deep Bear Island Channel. This suggestion is made because this is how the same species manages its affairs in the Gulf of Maine, where it encounters the same problem. I shall discuss this under Northwest Atlantic Shelves Province (NWCS).

Finally, I must note that the effect of Baltic outflow may create considerable uncertainty in interpreting the distribution and performance of planktonic algae in this province. The Norwegian Coastal Current carries high concentrations of "gelbstoffe" originating in the Baltic Sea, which is perhaps equivalent to 5–10 mg chl m^{-3} in satellite imagery (Sakshaug, personal communication) which must be discounted when evaluating the spring bloom which occurs along the Norwegian coast.

Synopsis

Moderately deep winter mixing, with a rapid near-surface stabilization in April and May, may not be captured by archived data (Fig. 7.4). Photic depth illuminates pycnocline only in summer as algal biomass declines. Primary production rate conforms to the irradiance cycle and is tracked by chlorophyll accumulation except in March, prior to the rise of *Calanus* from depth, and in August–October, after they descend again to overwintering depths, an effect which is more strongly marked than in the adjacent ARCT.

ATLANTIC WESTERLY WINDS BIOME

The central fact to be explained in any synthesis of North Atlantic pelagic ecology is the seasonal anomaly in the chlorophyll field that has the greatest areal dimension and greatest excursion of chlorophyll values of any region in the oceans. Deep winter mixing by the westerly winds and by thermal convection sets up conditions throughout the poleward portion of the anticyclonic gyral of the North Atlantic basin for a spring bloom which is anomalous at the global dimension.

The Subtropical Convergence, which lies across the gyre at 25–30°N (between the trades and the westerlies), forms a surface frontal system and a rational southern limit to the winter-mixing regions of the westerly winds biome.

North Atlantic Drift Province (NADR)

Extent of the Province

NADR comprises part of the westwind drift region of the North Atlantic in the sense of Dietrich (1964). To the north it is bounded by the OPF (see

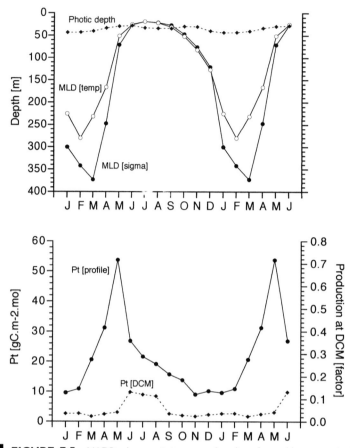

FIGURE 7.5 NADR: characteristic seasonal cycles of monthly averaged mixed layer and photic depths, chlorophyll at the surface, and rate of primary production, both depth-integrated and at the DCM. Data sources are discussed in Chapter 1.

Atlantic Arctic Province) and the Subarctic Front above the Iceland–Faeroe Ridge: These fronts lie zonally with an average latitude of 55–56°N. To the south, a separation lies across the ocean at about 42°N between the northeasterly flow of the North Atlantic Current and the associated westwind drift and the southeasterly flow of the Azores Current into the northern limb of the anticyclonic subtropical gyre (Krause, 1986). For convenience, this line is here taken to lie zonally along 42°N, though operationally it should be possible to trace the line dividing the flow into the southeasterly Azores Current and northeasterly North Atlantic Current. In the east, the edge of the European continental shelf is taken as the boundary of the province.

Continental Shelf Topography and Tidal and Shelf-Edge Fronts

Off northwest Europe, thermal fronts occur along the break of the continental shelf, especially over the relatively steep shelf break from 45 to 49°N (Armorican to Celtic shelves) and especially from May to October. They are expressed at the surface as meandering linear zones of cool water, observable in satellite thermal imagery especially during quiet sea conditions. The dynamic processes which produce the European shelf break front at 45–49°N also induce a continuous supply of nutrients, leading to high levels of algal biomass at the front (see pp. 28–30).

Defining Characteristics of Regional Oceanography

The bifurcation of the flow of subtropical water of the Gulf Stream extension occurs in the vicinity of the mid-Atlantic Ridge at about 30°W. Here, the gyral flow becomes less coherent and emerges as two streams, the North Atlantic Current continuing toward the northeast and the Azores Current toward the southeast (Krause, 1986). This province is characterized by slow eastward drift of surface water, except along the axis of the North Atlantic Current (NAC), whose flow is an order of magnitude faster and forms the frontal jet of the OPF along the northern boundary of this province. The slow eastwards drift that occurs over most of the province contributes only minor southwesterly flow into the eastern limb of the subtropical gyre.

Meridional sections obtained along 30°W close to the alignment of the mid-Atlantic Ridge with an undulating, towed instrument package (Sea Rover) indicate that the axes of both the Azores Current (at about 47°N on this occasion) and the NAC (about 49°N) are associated with markedly increased values of potential vorticity that coincide spatially with the surface temperature signals of these two flows (Woods, 1988).

High wind stress in autumn, combined with rapid heat loss, induces significant deepening of the mixed layer, so that by mid-winter the pycnocline slopes down to the east from about 300 m in the central ocean to more than 500 m along the European continental edge, extending the deep mixed layer trough of the ARCT province. Development of a vernal shoal pycnocline is associated with increasing sun angles and relaxation of wind stress.

Biological Response and Regional Ecology

The biological regime is dominated by the seasonal succession of deep winter mixing followed by vernal stratification; the details of the sea-surface chlorophyll field are determined by the distribution of eddy kinetic energy. The seasonal succession is forced entirely by local wind stress and surface radiation, as in all provinces where winter wind stress and cooling significantly force an increase in the depth of the mixed layer.

It has generally been assumed that there is a simple northward progression of the spring bloom in the NADR, and that stratification that is shoaler than the critical depth for net algal growth is also induced progressively northwards. However (as is so often the case), reality is rather more complex. This is clear from a review of the monthly CZCS images and also by the University of Kiel Sea Rover transects along 30°W that were repeated almost weekly though spring and summer of 1984–1986 (Strass and Woods, 1988, 1991).

Though the onset of the spring bloom can indeed be traced in seasonal CZCS images as a northward progression between March and June of the 0.2 mg chl m^{-3} signature, a seasonal shift of the higher chlorophyll values (0.8–1.0 mg chl m^{-3}) toward the north is not clear in the climatological images or in those for individual years. The regularity of the progressive northward shoaling of the mixed layer has also been overstated. Rather, some regions have consistently lower values throughout the spring and summer: e.g., the eastern part of the west wind drift at about 20°W and over the full meridional extent of the province. Conversely, in the western part of the province, there tends to be a region of high surface chlorophyll during the summer compared with relatively low values in the eastern part of the province.

This zonal difference is consistently observed in CZCS chlorophyll imagery but does not appear to have been noticed previously. The most probable mechanism would seem to be a conjunction between the baroclinic upsloping of nitrate isopleths toward the edges of the anticyclonic gyre and the eddy fields associated with separation of the Gulf Stream/NAC flow from the continental edge. Maximum variability of the sea-surface elevation in the northwest Atlantic indicated by Geosat and Topex-Poseidon images corresponds well with the general area of consistently enhanced surface chlorophyll.

The Sea Rover sections along 30°W for 1985, which will serve as an excellent model for seasonality in this province (Strass and Woods, 1988, 1991), show three seasonal phases: (i) the start of a spring bloom (mid- to late April), (ii) the transition to summer oligotrophy (end of June to beginning of July), and (iii) a late summer situation (end of August to beginning of September).

The Start of a Spring Bloom

The first indications (April 18–24) of a bloom occurred where meandering flow of the NAC was strongest, across 44–47°N, with a small southerly outlying patch at about 40°N to the south of this province, in the North Atlantic Subtropical Gyral Province (NAST). Elsewhere, the bloom had not started. By April 24–29, consistent with rapid onset of density stratification, a near-surface bloom with values >1.25 mg chl m^{-3} extended clear from 39 to 50°N, a distance of 1500 km. It seems likely, then, that initial stratification (and the first patches of surface bloom) is related to the effect of rising sun angle on cells brought near the surface in mesoscale eddy dynamics rather than on density stratification which occurs almost simultaneously across a rather wide zonal swath of ocean. Finally, it should be remembered that it is possible for a spring bloom to be initiated under conditions which apparently defy Sverdrup's model—in the absence of stratification; this has been reported for the Gulf of Maine (see Northwest Atlantic Shelves Province).

Transition to Summer Oligotrophy

By the end of June, the transition to oligotrophic conditions had occurred, though patchily, poleward to 46°N. There was residual nitrate in the photic zone only in small areas having exceptionally shoal pycnoclines, suggesting an ephemeral event. Maximum chlorophyll values occurred near the Oceanic Polar Front and from there to 46°N the DCM lay at, or shallower than, the mixed layer depth. To the south of 46°N, nitrate was depleted and the chlorophyll maximum was significantly deeper than the bottom of the mixed layer. Oligotrophy propagates poleward during the summer at about 3° of latitude a month, whereas the DCM deepens, once established, at about 10 m per month.

Late Summer Situation

By the end of August, everywhere south of the Oceanic Polar Front, which on this occasion lay at 52–54°N, a chlorophyll maximum occurred within the seasonal pycnocline, deepening toward the south at about 3.5 m per degree of latitude. Only to the north of the Oceanic Polar Front (in ARCT) was the chlorophyll maximum within the mixed layer, and here it had maximal values. Late summer blooms of coccolithophores occur consistently in the northern part of NADR (as part of a much larger area including much of the southern part of SARC) and are especially frequent bordering the coastal boundary biome, such as over the Rockall Channel.

This province lies centrally in the area of the North Atlantic spring bloom, which has generally been believed to be an ideal case of the diatom–copepod food chain. However, we now know, thanks to the results of the 1989 North Atlantic Bloom Study (NABE; Ducklow and Harris, 1993) that nano- and picoplankton contribute heavily to total primary production in the Atlantic spring bloom and that diatoms are not consistently the dominant large cells. Silicate was reduced to limiting values before nitrate limitation occurred at both of the NABE time series stations (18 and 40°W); these represent the zonal extent of the southern part of the province at about 45°N, so this unexpected result must be typical of the NADR. Before silicate limitation, the dominant large cells were diatoms (mostly *Rhizosolenia, Fragillariopsis, Thalassema, Thalassiosira,* and *Nitzschia*) at 18°W and dinoflagellates, with relatively few diatoms, at 40°W. Dominance shifted rapidly to an abundance (104 cells ml^{-1}) of small (2–5 μm) flagellates as soon as silicate limitation occurred, apparently without reduction of the overall chlorophyll concentration. The unconsumed, silicon-depleted diatoms sank out, leaving behind an abundant mucopolysaccharide residue in the photic zone. Prior to silica limitation, large cells (5 μm) contributed 50% of primary production and nanoplankton (1–5 μm) contributed about 37%, but after limitation these fractions contributed equally. Throughout the bloom, picoplankton contributed about 13–15%.

During the spring bloom, protistan microheterotrophs contributed 11% of total living organic carbon compared with 4% contributed by zooplankton. Protists comprise the principal sink for primary production, consuming 25% of the standing stock of plant cells per day and 90% of primary production (phytoplankton growth rates being about 0.4–0.7 doublings per day), whereas grazing by zooplankton herbivores took less than 10% of daily production. Of the copepods, small forms (e.g., *Oithona*) took as much chlorophyll as the medium- (*Metridia*) and large-size classes (*Calanus* and *Pleuromamma*) combined.

This province is the southern limit of the boreo-arctic *C. finmarchicus,* the central range of *C. helgolandicus,* and the northern limit of the ranges of *C. tenuicornis, Neocalanus gracilis, Nannocalanus minor,* and *Calanoides carinatus.* In NADR, therefore, a range of depth strategies occurs (Williams and Conway, 1988): Single annual generations of both *C. finmarchicus* and *C. helgolandicus* perform seasonal ontogenetic migrations, wintering at 400–900 m as C5 and rising to the mixed layer in spring and summer, with the former remaining a little deeper than the latter. *Neocalanus gracilis* remains within the upper 200 m throughout the year as does *C. tenuicornis,* which consistently avoids the near-surface layer and occurs at 20–200 m in all months. These two, and the other southern species, reproduce year-round. For a comparison between patterns of diel vertical migration in this and the adjacent province to the northwest see SARC, pp. 115–116. Foraminiferan assemblages 2 and 3 are characteristic of the province: *Globigerina sacculifer, G. aequilateralis,* and *G. inflata* typically dominate these useful indicator organisms.

Synopsis

Very deep winter mixing and an early spring near-surface stratification which is not captured by archived data occur and pycnocline lies within the euphotic zone from May to September (Fig. 7.5). Spring pulse of production rate occurs in March to May but is rapidly nutrient limited. Chlorophyll biomass accumulation is more constrained by consumption than in SARC, though there is some indication that this constraint is removed in September, when *Calanus* descends and chlorophyll accumulates for 60 days although productivity continues to decline.

North Atlantic Subtropical Gyral Province (East and West) (NAST)

Extent of the Province

This province is bounded to the west and northwest by the eddy field of the Gulf Stream and to the northeast by the bifurcation of flow between the Azores Current and the North Atlantic Drift; that is, at about 40–42°N. To the south, the boundary is the Subtropical Convergence, weakly defined along about 25–30°N as a series of largely subsurface thermal fronts, reaching >1°C/10 km and consistently identifiable, especially in the west, in all months except July–September. In the east, the equatorward limb of the gyre is formed by the offshore Canary Current. The division into eastern and western components follows two cues: both the southerly edge of the Azores Current and, further south, the mid-Atlantic Ridge.

Continental Shelf Topography and Tidal and Shelf-Edge Fronts

There is no continental shelf in this province save for the island platforms of Bermuda and the Azores and the Canary Islands.

Defining Characteristics of Regional Oceanography

This province represents that part of the anticyclonic midlatitude gyre that lies below the influence of the westerly winds, relatively weak at these latitudes. By this definition, some winter mixing occurs, though this is usually weaker than in the provinces further to the north because not only are winds slighter here but also, as noted in Chapter 3, the ability of wind stress at the sea surface to deepen the mixed layer diminishes equatorwards.

The southern limit of the province is the Subtropical Convergence (STC) which, in the North Atlantic, is a relatively weak front, so we should take care in examining the evidence for its existence. Although the front is expressed as a subsurface thermal gradient, the thermal signature at the surface may be rather

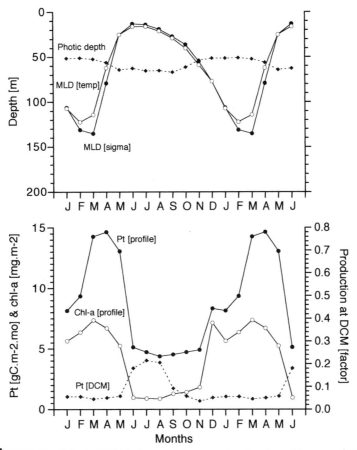

FIGURE 7.6 NAST(W): characteristic seasonal cycles of monthly averaged mixed layer and photic depths, chlorophyll at the surface, and rate of primary production, both depth-integrated and at the DCM. Note that graph for NAST(E) is almost identical. Data sources are discussed in Chapter 1.

slight, and because the STC is also associated rather frequently with cloud cover, it may not be observable in satellite imagery on a routine basis.

To the south and east of Bermuda, the transition between midlatitude westerlies and the trade winds from the east drives convergence of surface water and geostrophic flow eastwards (Iselin, 1936; Voorhuis, 1969) along the STC. The thermal fronts comprising the STC thus separate the anticyclonic gyre (or the Sargasso Sea, if we apply this term to the gyre west of the mid-Atlantic Ridge) into a northern subtropical and a southern tropical portion at about 25–30°N. To the south of Bermuda, a transition has long been known to occur across which subtropical, winter-mixing conditions are replaced by tropical conditions in which surface water in winter never falls below 20°C. Schroeder (1965) traced the thermal front of the STC from 70°W eastwards to 40°W, always between 20 and 31°N; farther east at 30°W, the front was investigated to the south of the Azores by Gould (1985). The STC is rich in eddy activity at all scales; shear instability and random superposition of internal waves produce a small-scale structure along the front (Voorhuis and Bruce, 1982; Toole and Schmidt, 1987), whereas wave-like baroclinic eddies (800 km, 200 days) induce persistent westward-propagating sea-surface temperature (SST) anomalies at the front (Halliwell *et al.*, 1991). These jets and eddies of cooler/warmer water across the frontal zone induce mesoscale frontogenesis, whose movement along the front is spaced 3–5 days apart (Leetmaa and Voorhuis, 1978).

The western part of the province includes the Northern Sargasso Sea, which is typified by the presence and formation of 18°C mode water (Worthington, 1986); in winter this breaks the surface from 30°N almost to 40°N, and its meridional extent narrows toward the east (Schroeder, 1965). The mixed layer deepens toward the center of the Sargasso Sea, in the vicinity of Bermuda, from shoaler regions in the eastern and western parts of the province.

Errant cold-core eddies originating in Gulf Stream meanders may propagate beyond the average eddy field, which I suggest is best considered to be part of that province. The sequence of biological processes within them has been discussed in the previous section. In addition to this eddy population, isolated seamounts support Taylor columns and may also spawn cyclonic, warm-core eddies observable in sea-surface elevation imagery as depressions in the sea surface. Eddies occur, for instance, downcurrent from the Corner Rise seamounts (Richardson, 1980). Since the pycnocline is deepened in these anticyclonic eddies, it is unlikely that biological enhancement will occur within them. It is, however, downstream of the Canaries that we have the best information of island eddies and their biological effects (Aristegui *et al.*, 1997). Here, filaments from the Canary Current coastal upwelling and mesoscale eddies (which are generated by perturbation of the southward flow of the offshore Canary Current by the islands) dominate the chlorophyll field. Both cyclonic and anticyclonic eddies are generated in the wake downstream of the islands at intervals of several days to a few weeks at all seasons, and both types of eddies often have elliptical or irregular form, suggesting that they are not yet in geostrophic balance. The pycnocline of cold-core cyclonic eddies takes the form of a central dome, which requires an outward flux of surface water. This pattern may be modified by wind stress and a consequent secondary vertical circulation in the Ekman layer. Warm-core anticyclonic eddies have a bowl-shaped pycnocline with some shoaling around the periphery of the eddy. Vertical nutrient flux in both types of eddy is determined by the appropriate physical circulation.

The weakly defined velocity maximum of the Azores Current lies a little to the south of the Azores and along a thermal front at about 35 or 36°N, between approximately 17 and 18°C, and also includes a modest salinity difference (about 36.5% to the north and 36.5% to the south). This front is deformed over the mid-Atlantic Ridge and is weaker into the eastern component of this province and becomes lost in the mesoscale eddies associated with the eastern boundary current and the Canaries. Here, the STC is called a "subtropical front" by Fernandez and Pingree (1996).

The topography of the mid-Atlantic Ridge constrains the main recirculation of western boundary current water within the western basin (e.g., Richardson, 1985), and the biological properties of the eastern and western basins differ significantly. Therefore, for some purposes, it may be useful to consider treating the western and eastern basins as distinct subprovinces (NAST-E and NAST-W).

In this case, we should draw the boundary along the topography of the mid-Atlantic Ridge, running southwest from the Azores.

Biological Response and Regional Ecology

Because the geographical Sargasso Sea corresponds with the subprovince NAST-W, there is a very appropriate marker to demonstrate that the line between the eastern and the western subprovinces is biologically significant: floating masses of gulf weed (*Sargassum natans*) are abundant to the west but curiously absent to the east of this line. There are also significant morphological and ecological differences between *Sargassum* in this province and in the NATR to the south of the STC (Niermann, 1986; Butler *et al.*, 1983, and references therein). Other east–west biological differences may be expected and some will be discussed later.

Though Gordon Riley (1957) obtained 2 years of data at OWS E at 35°N 48°W, well to the northeast of Bermuda, the most comprehensive description of the biological seasons in the pelagic ecosystem in this province is that derived from the recent Bermuda Atlantic Time Series (BATS) obtained at 31°N, just south of the island over deep water (Michaels and Knap, 1996). The station was occupied for several days on 111 occasions from 1989 to 1994 and this extraordinary data set allows almost weekly precision over this 6-year period: Seasonal and between-year signals are unambiguous in the data. Because the BATS station is very close to OWS S, studied since the 1950s, decadal variability in pelagic production is also observable for this region.

Simply put, when not complicated by the entrainment of Gulf Stream meanders and eddies, winter mixing and deep convective events associated with the formation of 18°C-mode water carry the pycnocline down to 200–300 m. Deepening occurs progressively from late summer to February, and apparent shoaling to 20–40 m occurs very rapidly during March. Nitrate is available in the mixed layer briefly (usually 0.5–1.0 μM and for <30 days) when mixing is at its deepest. A short winter bloom of varying duration occurs between February and April, and the period when chlorophyll is enhanced is usually shorter than the period of enhanced rate of primary production. Because the NO_3/PO_4 ratio may exceed Redfield values in the Sargasso Sea generally, and specifically at the BATS station, it is thought that nitrogen fixation occurs deep in the photic zone.

The seasonal biological cycle at BATS is weaker than that at the more dynamic BIOWATT site (closer to the Gulf Stream eddy field and influenced more by short-scale, local wind mixing by gales that pass north along the coast) and appears to be similar to the weak spring bloom found at the 32°N, 20°W JGOFS 1989 station in the NAST-E (Jochem and Zeitzschel, 1993). However, we must be very careful when comparing data sets not obtained simultaneously. The 1955–1994 data at Bermuda show strong decadal and between-year differences, principally forced by the relative length of the winter mixing period and the depth of mixing. These factors determine the nutrient levels in the mixed layer when stratification is reestablished. Depth of winter mixing at Bermuda varies from 150 to >250 m and the duration of the bloom (as indicated by enhanced chlorophyll) varies from 40 to 120 days; winter mixing was especially deep and of long duration in 1991–1993, apparently associated with the concurrent ENSO events of those years. Also during the period 1958–1971, winter mixing was carried unusually deep (often to 350–450 m) and mixed layer nitrates in the range of 2–5 μmol kg^{-1} have been hindcast for that period. At OSW E, for the years 1950–1952, Gordon Riley found a rather weak seasonal cycle for nanoplankton but rather strong pulses of diatoms (mostly *Chaetoceros*) in April, which developed very rapidly after surface stratification was established.

A DCM develops everywhere when nitrate is reduced to limiting levels. It lies at about 70–80 m across the province, as usual somewhat deeper than the depth of maximum photosynthetic rate and the biomass maximum for photosynthetic cells, which must follow from the generalization that the C/chl ratio decreases with increasing depth of the DCM. Li (1995) shows that the DCM for cyanobacteria and prochlorophytes can be traced oceanwide across NAST. In late spring the composition of the planktonic biota of the mixed layer is different from that of the NADR province to the north: Small phytoplankton (<10 μm) form 45% and large cells (>10 μm) only 4% of total living biomass. In NADR, these contribute much more equally to the total phytoplankton biomass.

McGillicuddy and Robinson (1997) have suggested that the observed rate of new production in the mixed layer of the Sargasso Sea is consistent with an independently-derived net flux of 0.5 mol N m^{-2} year into the mixed layer, most of which is delivered during the formation and intensification of cyclonic eddies in a manner consistent with the observed eddy field. About 30% of the flux may result from the interaction between mesoscale flow and wind-driven surface currents. They point out that it is important to consider the interaction between the light field in the oligotrophic ocean and vertical motions, of both signs, induced within eddy fields. "Cold" cyclonic eddies import nutrients up into the euphotic zone, whereas "warm" anticyclonic eddies carry phytoplankton-rich, nutrient-depleted water down below the lighted layer.

The "subtropical front" (or STC) of Fernandez and Pingree (1996) supports local enhancement of primary production and biomass accumulation, as is generally the case for oceanic fronts. Winter productivity is in the range 0.8–0.9 g C m^{-2} day^{-1} and is about twice the rate for those parts of this province not influenced by frontal or mesoscale eddy processes. These authors suggest that because of the great spatial extent of the frontal signature in the Azores Current system, primary production within the subtropical front is of major significance for regional carbon budgets. There is also some evidence that the phytoplankton species composition within this frontal region has affinities with the Sargasso Sea, as befits the Azores Current which has, of course, its root in the Gulf Stream.

The DCM is a site of especially intense trophic activity in this province as elsewhere, and it is a preferred depth zone for herbivorous mesozooplankton. Consumption follows a strong diel cycle not only for diel migrant micronekton and copepods but also for those resident in the photic zone. For the diel vertical migration of micronekton and large copepods, consumption is lowest in late afternoon and peaks around midnight. It is in this province that the role of small heterotrophic nanoflagellates in "the grazing and nurturing of planktonic bacteria,", as John Sieburth (1984) put it, was worked out.

NAST-E includes a sector of the offshore Canary Current, clear of the field of eddies and filaments generated within the coastal boundary zone. Here, mixed layer depths are shoaler, and enhanced chlorophyll biomass is indicated by the CZCS images, particularly south of the Canaries, though it will be important to separate the effects of eddies from the filaments of high chlorophyll induced by tidal mixing (and other coastal processes) at the islands. The features of the chlorophyll field are explicable by the presence of island-induced mesoscale eddies and filaments of chlorophyll-enhanced upwelled water. Vertical nutrient flux in cyclonic eddies occurs centrally by isopycnal transport, and therefore it is strongest where the pycnocline dome is shoalest and some diapycnal mixing may occur across the shoaled pycnocline. In many upwelling situations chlorophyll enhancement occurs most strongly a little downstream of the surface nutrient maximum and so, in this case, chlorophyll is highest around the edge of the cyclonic eddies. Anticyclonic eddies frequently interact with filaments of water with enhanced chlorophyll, causing the chlorophyll signal to spiral inwards toward the center of the eddy as also occurs in Gulf Stream rings.

Throughout the province, within which there are >300 individual seamounts (not all associated with the mid-Atlantic Ridge) of sufficient topography to sustain Taylor columns, it may be expected that some areas of enhanced surface chlorophyll will be topographically controlled, even in midocean. Their downstream eddy field may, like the Canaries, introduce features into the surface chlorophyll field. Saltzman and Wishner (1997a) briefly review the ecological effects that may be anticipated where seamounts populate a deep ocean region. Uplifting of isotherms and upwelling of nutrients, interaction with diel vertical migrant zooplankton and consequent patch formation, and induction of relatively high biomass of pelagic fish are among the more frequently noted effects.

Synopsis

Moderate winter mixing occurs so that pycnocline lies within the photic zone from April to November (Fig. 7.6). Primary production rate is minimal in late summer (but up to 20% occurs in the DCM) but begins to increase as soon as mixed layer deepens in the autumn, reaching a nutrient-limited peak in spring with only 5% in the DCM. Chlorophyll accumulation closely tracks productivity, indicative of tight coupling between production and consumption. Seasonal cycles in eastern and western parts of the province are very similar.

Gulf Stream Province (GFST)

Extent of the Province

The Gulf Stream Province (GFST)comprises the Gulf Stream from Cape Hatteras to the Newfoundland Basin, where flow of the NAC across the poleward limb of the subtropical gyre begins (Mann, 1967). The landward margin of the province is the North Wall of the Gulf Steam. The field of cold-core eddies within the warm water of the jet current is included in the province, whose seaward limit is most usefully defined by the distribution of these eddies.

Continental Shelf Topography and Tidal and Shelf-Edge Fronts

From the Straits of Florida to Cape Hatteras (though we are considering that region to be part of the NWCS province), the Gulf Steam conforms closely to the continental slope outside the Blake Plateau. The shelf narrows and the continental edge veers north and then west where the Gulf Stream diverges from the coast.

Defining Characteristics of Regional Oceanography

Western boundary currents owe their relative strength to westward propagation of Rossby waves, accumulating energy on the western margin of oceans, and its subsequent dissipation within a western boundary current. The Gulf Stream is the western boundary current of the North Atlantic and is prominent not only in hydrographic sections (e.g., those of Schroeder, 1965) but also in the surface thermal field as a group of closely spaced isotherms.

Originating in the Florida Straits, the velocity maximum of the current is topographically locked to the edge of the continental shelf as far north as Cape Hatteras, where the coastline veers westwards. Here, the Gulf Stream separates from the continental edge and proceeds across deep water toward the northeast. This departure from the coastal boundary biome is a convenient southern limit to the GFST province. The region of separation is dynamic, and episodes of surfacing of nutrient-rich North Atlantic central water occur in response to

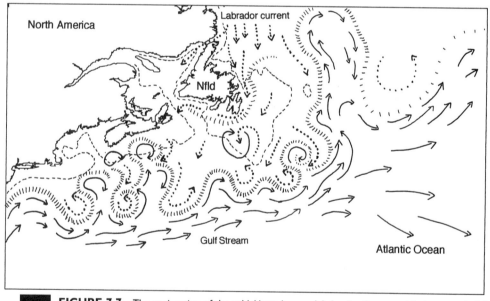

FIGURE 7.7   The conjunction of the cold (dotted arrows) Labrador Current with the warm (solid arrows) Gulf Stream. The dotted line is the 200-m isobath and the hatched line represents fronts between warm and cold water, the most prominent being the North Wall of the Gulf Stream. Between Labrador water, which dominates the shelf, and Gulf Stream lies the mixed body of slope water. Irruption of a warm-core eddy onto the shelf entrains slope water seawards and expatriates shallow-water pelagic organisms. Note the indication of a bifurcation of flow into the Gulf Stream extension and North Atlantic Current on the east side of the map.

offshore meanders of the frontal jet. Once separated from the coast, the Gulf Stream then behaves as a free inertial jet until it encounters the Newfoundland Rise, around which it continues as the Gulf Stream Extension, progressively dissipating into (i) water which drains continuously southeastwards into the western recirculation gyre of the North Atlantic, whereas the main flow begins its bifurcation into (ii) the Azores Current and (iii) the NAC (Fig. 7.7). A semipermanent loop of Gulf Stream water regularly passes around the Newfoundland Basin before continuing eastwards as the NAC: This is a prominent feature in surface fields of temperature and several biological variables. This loop, a "ring-meander," occasionally pinches off a mesoscale eddy which carries warm-water organisms far to the north in the Labrador Sea.

Slope water of the coastal boundary biome, containing a contribution originating in the Labrador Current, forms the cold wall to the west or north of the stream which is usually prominent in the thermal field, with the warm side of the stream being less so. During winter, the isotherms representing the Gulf Stream are usually 10–18°C and the boundary of the 18°C water of the Sargasso Sea indicates the warm edge of the flow. During summer, the formation of a warmed, shallow mixed layer right across the slope water, Gulf Steam, and Sargasso Sea changes the thermal gradient to about 13–21°C across the cold wall of the current. However, in winter, isothermal water in the Gulf Stream extends deeper than in the 18°C-mode water of the subtropical gyral province beyond the seaward boundary of Gulf Stream. In winter, this region has what must be the strongest horizontal temperature gradient of SST anywhere in the sea: 18°C Sargasso water and the pack ice of the Gulf of St. Lawrence are only 300–400 km apart. No wonder, then, that this is a region of rapid weather change and strong mesoscale variability in SST.

The free inertial jet of the Gulf Stream forms both cyclonic and anticyclonic meanders; when these turn acutely enough to become pinched off, mesoscale eddies are formed. Each is either a ring of isolated Gulf Stream water enclosing an area of cold slope water (cold-core, sea-surface depressed cyclonic eddy on the seaward side of the current) or a ring of warm, Sargasso water (warm-core, sea-surface elevated, anticyclonic eddy on the landward side). These rings, both cyclonic and anticyclonic, are deep (1000–3000 m), large (100–300 km), long-lived (6 month to 1 year), and move slowly during their lifetime contrary to the flow of the current. At any one time, 2–10 rings may be observed in satellite imagery (thermal, surface elevation, or chlorophyll) on each side of the current. The water contained in a ring is progressively modified to the characteristics of the surrounding water mass.

Though similar eddy fields occur in many other areas—notably in the Pacific homolog of the Gulf Stream (see Chapter 9, Kuroshio Current Province)—it is in the GFST province that these rings have been most closely investigated (e.g., Richardson, 1976; Ring Group, 1981; Wiebe and McDougall, 1986; Joyce and Wiebe, 1992). Some will quarrel with the logic, but this province is best defined as comprising not only the frontal jet of the Gulf Steam, from Cape Hatteras to the Newfoundland Basin, but also the field of cold-core eddies associated with its flow. Small numbers of eddies will travel beyond the boundaries of the province (Richardson, 1983): it may be claimed that this violates the logic, though not the convenience, of the definition. The smaller number of

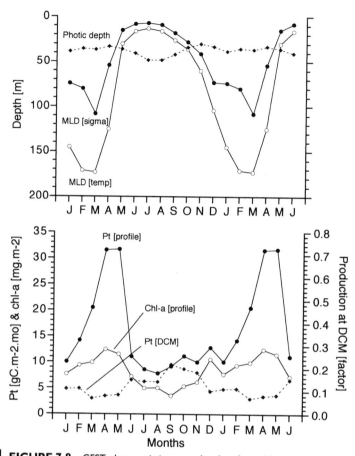

FIGURE 7.8 GFST: characteristic seasonal cycles of monthly averaged mixed layer and photic depths, chlorophyll at the surface, and rate of primary production, both depth-integrated and at the DCM. Data sources are discussed in Chapter 1.

warm-core eddies that become inserted in slope and even shelf water are treated as a special characteristic of the coastal boundary province NWCS.

Biological Response and Regional Ecology

The ecology of the warm water of the Gulf Stream resembles that of the adjacent Sargasso Sea (see North Atlantic Subtropical Gyral Province). It is differentiated by the consequences of the meandering of the frontal jet and by the associated population of cold-core eddies. We shall therefore concentrate our attention on these eddies and on some other special features associated with the meandering frontal jet current. In fact, a glance at seasonal CZCS images will be enough to demonstrate that the surface chlorophyll field of the northwest Atlantic is dominated by algal blooms occurring within the eddy field of the Gulf Stream as far east as the mid-Atlantic Ridge. In this eddy field the spring bloom is continuous from spring to autumn compared with the eastern

Atlantic at similar latitudes, where mesoscale eddies are less numerous and energetic. It has commonly been thought that the passage of the Gulf Stream across the New England seamount chain induces the formation of particularly energetic meanders and eddies, but satellite thermal imagery shows that this is not the case (Cornillon, 1986).

John Woods (1988) has explored and modeled the biological consequences of the dynamic changes in pycnocline topography that are associated with instability of mesoscale jets, such as the Gulf Stream. He points out that mesoscale jets are inherently unstable, with high isopycnal potential vorticity, and soon develop meanders with along-stream wavelengths of 10–100 km. Vortex contraction on the flanks of the anticyclonic warm-core meanders should cause upwelling to occur, with downwelling within the cyclonic meanders. This vertical motion is reflected in the sea-surface temperature field, and primary production rates are higher in the anticyclonic meanders where both pigment maximum and the nitracline rise along the upward-sloping isopycnals. This should result in a patchy distribution of mixed layer chlorophyll of the same scale as the meanders. Such along-stream variations in chlorophyll and production rates, related to the physical effects of meandering, have in fact now been observed (Lohrenz et al., 1993). A more general cross-stream effect on production rate and standing stock was suggested by Yentsch (1974) as a result of Rossby's theory of the Gulf Stream, by which water is drawn into the right side of the jet and discharged toward the left along the upwards-sloping isopycnals. Together with cross-frontal mixing of relatively pigment-rich slope water, this should result in the cold north wall of the Gulf Stream being observable—at the right season—as a chlorophyll front.

Woods (1988) has generalized his analysis to suggest that regions with seasonally high isopycnal potential vorticity (Q, linearly related to density stratification N, the Brunt–Väisälä frequency) should exhibit abundant mesoscale enhancement of the surface chlorophyll field. As he shows from monthly maps of Q for the North Atlantic, there is a very good fit with areas of anomalously high chlorophyll in the seasonal CZCS images. It should, therefore, be no surprise that the area included in the GFST province should correspond well with an area of consistently higher than background chlorophyll values in these images. There are major between-year differences in the CZCS images, but the 7-year climatology for spring (AMJ) and summer (JJA) trimesters shows the Gulf Stream to be populated by many chlorophyll "hot spots" prior to the bloom over the adjacent continental shelves. In some years this is very strongly the case (1980), whereas in others (1982) the GFST bloom is continuous and simultaneous with the bloom over the shelf.

A seasonal series of mesoplankton samples at 36°N reveals the general annual cycle of vertical migrations in the Gulf Stream system (Allison and Wishner, 1986). In winter, biomass in the upper 200 m is low both day and night in the slope water and in the adjacent north wall of the stream relative to biomass in spring and early summer. There is about an order of magnitude difference in the winter/summer ratio. In the Gulf Stream itself and the adjacent Sargasso water, there is little seasonality in near-surface biomass. In May, biomass is concentrated at 0–100 m both by day and night, progressively deepening offshore, and little diel migration occurs. By September, diel migration is strongly established from the slope to the Sargasso water, with a subsurface maximum at night at about 80–90 m in all areas, near the DCM that develops during the summer months. Such strong diel migration to 400–600 m in a region with such active horizontal advection must cause very significant horizontal redistribution of biomass.

Cold-core eddies were well studied in the late 1970s by the Ring Group (1981), mostly out of Woods Hole. Young rings contain water having the ecological characteristics and biota of slope water, and the central doming brings the 15°C isotherm close to the surface. The rate of primary production within a cold-core eddy is, at least initially, higher than the surrounding oceanic water by a factor of about 1.7; this is similar to the general ratio between slope water and the open Sargasso Sea. If shed at the end of winter, a spring bloom may occur within the young ring, and the chlorophyll maximum layer weakens (by a factor of 5–10 in maximum chlorophyll values) and descends to about 100 m as the summer progresses, coming to lie on the upper part of the nutricline within the ring. This evolution is due to both seasonal processes and, as the ring ages, progressive replacement of the slope water flora and fauna by species typical of the Sargasso Sea. Though it will not concern us here, the reaction of the entrained mesozooplankters in the warming ring is to descend progressively into cooler water: Thus, mature cold-core rings have warm-water species above and shelf species below. Vertically integrated biomass of cold-core rings may thus exceed that of surrounding water because of the addition in such data of a fully developed immigrant Sargasso biota above and the remains of a refuged slope water community below.

Synopsis

Moderate winter mixing occurs so that the pycnocline are within the photic zone only from May to October (Fig. 7.8). Primary production rate is minimal in late summer, increases when mixed layer deepens in autumn, and reaches a sharp nutrient-limited peak in spring (in April and May). Chlorophyll accumulation and seasonality is weak and variable, though it tracks productivity and indicates close coupling between production and consumption. Production at DCM is relatively higher (20% of total) in the very clear water of late summer.

Mediterranean Sea, Black Sea Province (MEDI)

Extent of the Province

The Mediterranean Sea, Black Sea Province (MEDI) includes the whole of the Mediterranean basin, distinguished by the Ancients from the "ocean stream" lying beyond the Pillars of Hercules at Gibraltar. I include the marginal Adriatic Sea, the Aegean Sea, and also the Black Sea. The landlocked seas of Asia (the Caspian Sea and the fast-disappearing Aral Sea) are regarded as saline lakes no longer part of the ocean.

Continental Shelf Topography and Tidal and Shelf-Edge Fronts

This province comprises mostly deep basins. In the Mediterranean, extensive shelf depths occur only in the northern half of the Adriatic and the Gulf of

Gabes on the African coast, whereas the coasts of the Black Sea are steep-to, with the exception of the Crimean peninsula, which is bordered by extensive shelves reaching as far as 200 km wide off Odessa. The Sea of Azov is a very shallow embayment which is largely isolated from the main basin behind the Crimea.

Defining Characteristics of Regional Oceanography

It will be convenient to discuss the Mediterranean and Black Seas separately because of their very different characteristics. The Mediterranean is an evaporative basin constrained by a shallow sill at the Straits of Gibraltar, whereas in the Black Sea the salt balance is approximately in equilibrium despite its shallow sill. The fast and continuous surface flow from the Black Sea into the Mediterranean through the narrow passages of the Bosphorus and Dardanelles is a unique feature in the circulation of the world ocean. The Danube contributes two-thirds of the 304 km^3 of fresh water which enters the Black Sea annually. One of the most useful and comprehensive accounts of Black Sea ecology is still that to be found in Hedgepeth (1957), to which I owe much.

Mediterranean Sea

It is only in winter, when the zonal band of the westerlies lies farthest equatorward, that the planetary wind field is effective in the Mediterranean; therefore, wind stress at the surface is strongest and most uniformly distributed from December to March. For most of the year, the dominant wind fields are local and orographic, though even in winter there are local irruptions of especially strong northerly winds, cold and dry. In the western Mediterranean, these are the adiabatic "mistral" winds channeled down the Rhone valley, and in the Adriatic the similar "bora" is not appreciated by the residents of Trieste. In the absence of large-scale geostrophic circulation, the physical oceanography of the Mediterranean is strongly influenced by these and other local wind fields. The circulation of surface water is characterized especially in the eastern basin by a number of semipermanent gyres resembling in scale the large mesoscale eddies of the open ocean. However, these are not errant eddies and their location, together with their associated fronts and jet currents, is generally predictable. Their existence and location have been revealed by satellite imagery in recent decades.

The Mediterranean Sea comprises two partially isolated basins within each of which a cyclonic surface circulation occurs (see Robinson and Malanotte-Rozzoli, 1993; Minas and Nival, 1988) and the details of coastline alignment impose many smaller, semipermanent gyres (Fig. 7.9). The narrowness of the Sicilian Channel (140 km) partially isolates the gyral circulations of the eastern and western basins, which Millot (1992) regards as two separate Mediterranean seas. The Tyrrhenian Sea, partially enclosed by Sicily and Sardinia-Corsica, and the Adriatic Sea, behind the narrow (70 km) Strait of Otranto, each have a partially enclosed cyclonic gyral circulation.

Because the two cyclonic gyral circulations of the western and eastern Mediterranean are only partially isolated, there is a general cyclonic flow around the whole basin, with the surface water becoming progressively saline and the return flow progressively deeper. This process preconditions water entering the

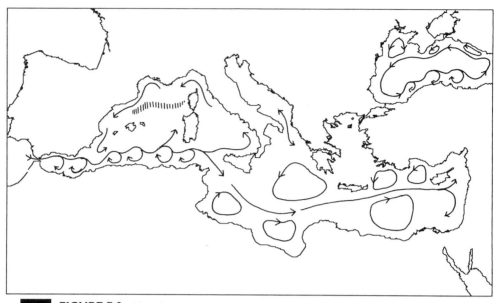

FIGURE 7.9   Map of the generalized circulation of the Mediterranean Sea to illustrate the effects of coast forms on persistent eddies and fronts. Note the rim currents of the Black Sea and western Mediterranean and the eastern Mediterranean central jet. Some coastal eddies, such as that occupying the Alboran Sea immediately east of Gibralter, are effectively permanent, whereas others, such as those to the east along the Algerian coast, are locked more loosely to prominent capes and are more variable. Comparison with CZCS images will show the striking extent to which coastal eddies and dipoles are locations of chlorophyll enhancement.

Ligurian Sea and the Gulf of Lions so that mistral wind episodes in winter which strongly cool the surface water mass (a process which produces prominent cool plumes in the surface thermal field) induce persistent deep convection and the formation of Mediterranean deep water near the coast. In this case, a cool anomaly in the surface thermal field does not necessarily denote upwelling as it does almost everywhere else. It is this mechanism which is responsible for the relative vertical uniformity of Mediterranean water masses.

At the Straits of Gibraltar, Atlantic water enters the Mediterranean at the surface and passes eastwards along the African coast as the density-driven, topographically locked Algerian Current, which continues into the eastern basin through the Sicilian Channel (Millot, 1992) as the Ionian–Atlantic stream.

The flow of the Algerian Current along the southern coast of the western basin (that is, from Morocco to Tunisia) generates vorticity of both signs and flux within a field of mesoscale eddies that occupy the whole basin to the latitude of Corsica (Millot, 1987), being separated from westwards flow along the European coast at the divergent North Balearic Front. The Western Alboran Sea, the first basin entered by Atlantic surface water, is occupied by a persistent anticyclonic gyre forced by the orientation and topography of the Straits of Gibraltar. In the eastern part of the Alboran Sea, circulation is more variable. A persistent feature, the Almeria–Oran Front, is associated with the Alboran Sea gyres. East of the Alboran Sea, along the Algerian–Tunisian coasts, the flow forms unstable meanders around a series of eddies which propagate eastward.

In the eastern basin, the already meandering Ionian–Atlantic stream at the Sicilian Channel becomes a mid-Mediterranean jet that can be traced far into the Levantine Basin passing between flanking cyclonic eddies (Cretan, Rhodian, and West Cyprus) to the north and anticyclonic eddies to the south (Shikmona and Mersa Matruh) whose quasi-permanent position is determined by topography (POEM Group, 1992). A coastal anticyclonic loop current around the Ligurian Sea between Sicily and Greece creates a series of (mostly) cyclonic eddies.

Though tidal streams in the Mediterranean are generally weak, this is not the case everywhere; the Venturi effect over the shallow sill of the Straits of Messina produces tidal currents that are unusually strong, and violent, local upwelling occurs. This is, of course, the whirlpool of Greek antiquity lying between six-headed, dog-barking Scylla, and Charybdis, who swallowed the sea and vomited it back again thrice daily.

Black Sea

With its anoxic interior, the Black Sea is the most extreme case of a meromictic basin in the present-day ocean (see Murray, 1991, to which I owe much of what follows). Excess precipitation together with runoff from the rivers Danube, Dniester, and Don creates a surface low-salinity layer overlying a halocline at about 100 m which is sufficiently strong to prevent ventilation of the interior of the sea. Below the halocline, the basin lacks oxygen and has high concentrations (increasing with depth) of dissolved H_2S. A fast, permanent flow pours southward through the Bosphorus—making a ferry crossing at Istanbul a memorable affair. Some Mediterranean water enters the Black Sea as a bottom flow along the Bosphorus (whose bottom slopes down toward the north), though much of this undercurrent water is entrained back into the fast outflowing surface current (Caspers, 1957; Sorokin, 1983).

The basin-scale circulation is cyclonic and appears as the Rim Current, strongest near the coasts, though how this is forced is not yet fully understood (Sür et al., 1994). Although the positive curl of the wind stress has long been the accepted explanation, it may be more complex. From the mouth of the Bosphorus a strong current runs eastwards along the Paphlagonian coast, forming the strongest flow of the Rim Current (40–80 km wide) that circles the whole Black Sea at the 200-m depth contour. The Rim Current is associated with two principal cyclonic gyres which occupy the eastern and western basins (divided south of the Crimea). Especially along the southern and eastern coasts, there is strong mesoscale vorticity in the meandering flow (Oguz et al., 1992), whose features propagate eastwards at 10–15 km a day.

The two major gyres are constrained by the shelf edge that runs zonally across the basin at the latitude of southern Crimea, and flow on the northern shelves themselves is more variable. Smaller anticyclonic gyres lie between these and to the west of the Crimea, downstream of the main coastal flow. During winter, the two-gyre circulation may break down, to be replaced with a single, more elongated cyclonic gyre.

This circulation pattern explains the topography of the halocline and the oxic/anoxic interface which lies at about 150 m near the centers of circulation of the two main cyclonic gyres and deepens to >200 m around the coastal margins. Because the chemistry of the oxic/anoxic interface is so intimately connected with biological processes, we shall defer discussion of it to the following section.

Biological Response and Regional Ecology

The ecological characteristics of the two seas are sufficiently different that to place them in a single province is largely a matter of convenience. Both, however, have been subjected to significant modification during this century, not only from land-based sources of contamination but also perhaps more important by reduced runoff from the major rivers entering the basins. Nitrate values in the mixed layer of the Black Sea have increased significantly in the past 25 years. However, the most fundamental of these changes has been the loss of the annual Nile flood, which has been held in recent decades behind the Aswan High Dam; the result of this loss has been a very significant modification of the ecology of the eastern Mediterranean. The artificial Lessepsian connection between the Mediterranean basin and the Red Sea is of great significance for taxonomic biogeography because of expatriation of Indo-Pacific species through the Suez Canal but less so for the basic ecology of the eastern Mediterranean. Introduced species have, on the other hand, modified the ecology of the Black Sea.

Mediterranean Sea

The seasonal cycle of primary production and consumption resembles that of the subtropical Atlantic. Winter mixing causes nitrate to become available in the photic zone and a relatively weak late-winter bloom ensues, which is followed by a long period in which the typical profile includes a DCM.

Major upwelling events occur in the Sicilian Channel which appear to be wind induced; they occur mostly with a lag of 3 days after a westerly gale (often originating in a Mistral event in Provence) passes through the area. They are usually based on the southwestern tip of Sicily (near Cape Granitola), where they often involve the entire southern coast of the island. They may extend across the entire Sicilian Channel to the African coast (Piccioni *et al.*, 1988).

In the eastern Mediterranean, the most dominant surface enrichment was induced by the annual discharge prior to the closure of the Aswan High Dam and the control of the annual Nile floods. This feature is now almost absent and current conditions in the Levant Basin (which has, in its center, water of clarity equal to that of the central oceanic gyres) there is a only subsurface algal bloom in bottom water over the shelf, perhaps induced by the sporadic upwelling of nutrients across the shelf break (Townsend *et al.*, 1988).

The CZCS images show a general increase in chlorophyll accumulation in winter, especially in the western Mediterranean, and this confirms observations that it is only in winter months that positive net community production exceeds respiration and that a winter production pulse (mid-January to mid-February) occurs off Southern Spain (Rodriguez *et al.*, 1987). Both near-surface and subsurface chlorophyll maxima were dominated by small autotrophic cells: 80–100% of the biomass passed a 10-μm mesh and 20–60% passed even a 1-μm nuclepore filter. In the Adriatic in summer, Revelante and Gilmartin

(1994) found a twofold higher biomass of larger cells compared with the rest of the water column, even though picoplankton formed 50% of the biomass.

By late spring, a DCM is established essentially in both the eastern and western Mediterranean basins; typical profiles show the chlorophyll maximum and nutricline occurring at the base of the thermocline at 75–80 m, with the topography of this surface sloping according to geostrophic flow. In anticyclonic features, the DCM may be associated with the depth of the photic zone rather than with the nitracline, which may be carried very deep (to about 250 m). Except in such situations, the depth of the DCM generally coincides with the nitracline rather than with density surfaces, suggesting that primary production is limited primarily by nutrient supply and only secondarily by light and other factors. The depth of maximum primary production rate occurs a little above the DCM. At, or close to, the DCM is the expected layer of abundant zooplankton.

These features generally deepen through the summer, at a rate of about 15–20 m per month (Estrada et al., 1993). Because deep and intermediate water masses are formed within the Mediterranean basin by the modification of surface water, subpycnocline nutrient levels are significantly lower than those in the open ocean, with maximum values in middepths of 9.5 μM at 250 m in the western basin (Coste et al., 1988). In fact, at the Straits of Gibraltar, the balance between nutrients transported in the incoming and outflowing water masses translates into a net gain of nutrients for the Mediterranean basin.

During the postbloom period, the processes which bring relatively nutrient-rich water to the surface in the Mediterranean are as diverse as the topography and as variable as the wind regimes. For example, the thermal field at the sea surface for August 5, 1987, shows cool anomalies in the Alboran Sea (clearly gyral in this case), to the east of Majorca, to the east of the Strait of Bonifacio, along the whole west coast of Sardinia, south and east of Sicily (again, gyral in form), on the west coast of Greece, and southwest of Crete. The largest cool anomaly extends from the Gulf of Lions as far south as Corsica, and those with the weakest thermal gradients appear on the coast of Libya and western Egypt.

There are persistent surface chlorophyll features in the Alboran Sea related to upwelling, both along the east coast of Spain and around the Alboran eddies. A jet of Atlantic water along the Spanish coast induces geostrophic upwelling and a very shallow (<50 m) chlorophyll maximum. Upwelling is forced around the anticyclonic gyre which occupies the western Alboran Sea (Minas et al., 1991). Farther east, a divergent front between Spain and the Balearic islands between southward coastal and northward offshore flow is associated with upwelling and chlorophyll enhancement (Estrada and Margalef, 1988). Cyclonic eddies generated at the coast within the Algerian Current are not wind induced but force upwelling over sufficient spatial and temporal scales to induce algal growth and surface enhancement of the chlorophyll field (Millot, 1987).

The copepods of the Mediterranean basin resemble those of the subtropical Atlantic, and since the sill depth excludes all abyssal biota that do not perform vertical migrations, the vertical ranges of many species extend much deeper here. The copepods of mass occurrence in the photic zone are typically *Oithona spp.*, *Clausocalanus spp.*, *Neocalanus gracilis*, and *Centropages typicus*, whereas *C. helgolandicus* is restricted almost completely to much greater depths. *Centropages typicus* is an important herbivore which occurs at the DCM by day; part

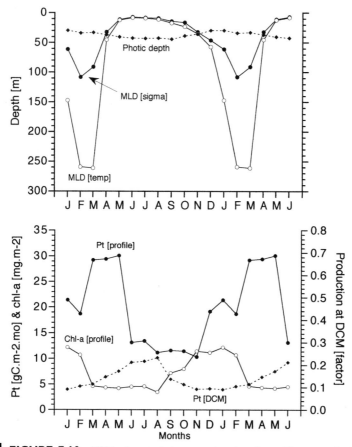

FIGURE 7.10 MEDI: characteristic seasonal cycles of monthly averaged mixed layer and photic depths, chlorophyll at the surface, and rate of primary production, both depth-integrated and at the DCM of the Mediterranean Sea. Data sources are discussed in Chapter 1.

of the population passes up through the thermocline into the surface mixed layer at night.

Black Sea

Though we shall be mostly concerned with processes occurring at shallower depths, any account of the ecology of the Black Sea must start with the extraordinary fact that from some depth between 80 and 200 m (shoaler in midgyre, deeper near the coast, and apparently some tens of meters shoaler in recent years than previously, according to information offered in Murray, 1991), oxygen is absent and hydrogen sulfide concentrations increase progressively downwards. At this redox interface, there is a null zone tens of meters thick in which both gases are present at low concentrations. In the interface, photosynthetic oxidation of H_2S by sulfur-oxidizing bacteria (*Thiobacillus*) occurs, resulting in high concentrations of elemental sulfur. Other phototrophic, green (bacteriochlorophyll-β) sulfur bacteria, probably *Chlorobium phaeobacterio-ides,* are also active in the redox interface.

In the oxygenated water above the redox layer, a thermocline develops in summer that has associated with it a DCM due to normal autotrophic algae. However, in the oxygenated surface layer nitrate has a very unusual profile. In the redox zone the nitrogen cycle is extremely active, and this layer is a sink for nitrate, nitrite, and ammonium. Consequently, rather than a nitracline, we find a subsurface nitrate maximum in the Black Sea. Below this maximum, nitrate is utilized by the activity at the redox layer and, above, by autotrophic cells of the euphotic zone (Sorokin, 1983; Murray and Izdar, 1989; Murray, 1991). This nitrate maximum consistently had values of about 7.0 μM in 1988 compared to about 3 μM in 1970. A similar increase in the integrated May to December rate of primary production (100 to 300 mg C m^{-2} day^{-1}) and chlorophyll concentration (0.2 to 0.4 mg chl m^{-3}) has also been observed between 1970 and 1990.

A winter bloom occurs rather generally between November and March in the Black Sea, with highest chlorophyll values and growth rates in the latter month (Vedernikov and Demidov, 1991; Krupatkina *et al.,* 1991), replacing the summer community of coccolithophores and dinoflagellates with a surge of diatoms. On the Anatolian coast, *Chaetoceros* sp. comprised >90% of large algal cells. A twofold increase in the rate of primary production and an order of magnitude increase in standing stock of chlorophyll are associated with the bloom, which continues even after the mixed layer has deepened to about twice the depth of the photic zone, enabling nitrate from the nitrate maximum to be mixed upwards. This appears to be a case in which a bloom occurs without a "floor" of density stratification so that individual cells must be competent to survive temporary mixing excursions below the 1% illumination depth.

The Black Sea plankton is overall much less diverse than that of the Mediterranean: 15 species of copepods compared with 304, for example. Coccoliths, radiolarians, siphonophores, salps, and pteropods are essentially absent. The distribution of mesozooplankton in the Black Sea is unique (Vinogradov *et al.,* 1985). By day, the dominant species form a layer of high abundance (2.5–38 g m^{-3} wet weight) over a depth interval of only 5–20 m, coinciding exactly with the isopleth for 0.4–0.5 ml O$_2$ liter^{-1} which occurs in the null zone (or redox interface) discussed previously. Following this isopleth, the actual depth of the zooplankton layer varies from 50 to 150 m. The distribution of organisms within this narrow layer is precise: The upper part is occupied by the tentaculate ctenophore *Pleurobrachia pileus,* the middle portion by late copepodites and adults of *C. helgolandicus,* and the lower part by the chaetognath *Sagitta setosa.* The small *Pseudocalanus elongatus* may also be present. At night, diel migration alters the pattern: The copepods and chaetognaths rise to the near-surface layers (0–20 m), whereas the ctenophores adjust their depth only slightly, rising by about 20 m. Thus, during each diel cycle, the copepods must run the minefield of ctenophores passively awaiting their passage while avoiding the hunting chaetognaths, which track their migration. Seasonal vertical migration also occurs but is rather unusual. The cold-water forms, such as *C. helgolandicus, P. elongatus,* and *Oithona similis,* occur throughout the oxygenated layer in winter but descend to depths $\geq$50 m during summer. In general, their depth distribution is constrained between the oxycline and the isotherm of their maximal temperature, usually 10–14°C.

During the spring bloom the CZCS images reveal a response of the chlorophyll field to the dynamic instabilities of meanders within the Rim Current, and during the summer patches of upwelling occur along the Anatolian coast, probably influenced by interaction between topography and the flow of the Rim Current rather than by wind stress. Here the images reveal a rich population of jets, dipoles, and eddies (Sür *et al.*, 1994), within which surface divergences and upwelling occurs followed by chlorophyll enhancement. Discharge of nutrients from rivers, especially the Danube, are associated with massive algal blooms, probably the coccolith *Emiliania huxleyi*. In this region, the contributing organisms and the magnitude of blooms seem to have been progressively changing in recent years toward blooms dominated by nanoplankton, including coccoliths (Mihnea, 1997). It is in regions such as the Black Sea, which is rich in topographic detail and ephemeral physical and biological events that a real revolution in understanding is already at hand as a result of satellite imagery.

There have been recent and quite major perturbations in the ecological balance of the pelagial in the Black Sea. The stocks of pelagic clupeid fish have almost completely collapsed, and declared landings of *Engraulis* dropped by an order of magnitude between 1979 and 1990. To what extent this was due to overfishing in the classical sense or to a concomitant collapse in zooplankton biomass has not been established. A predatory ctenophore (*Mnemiopsis leidyi*) from the estuarine habitat of North America was introduced in the 1980s— presumably in the ballast water of tankers returning empty to the oil ports— and underwent a population explosion to the detriment of its food: mesozooplankton, including fish eggs and larvae. This foreign organism continues to survive in the Black Sea.

Synopsis (Mediterranean Sea only)

Winter cooling, which extends deeper than the density-criterion mixed layer, is very brief so the pycnocline lies within the photic zone from March to November (Fig. 7.10). Primary production rate is minimal in late summer, and increases when mixed layer deepens in autumn. Chlorophyll accumulation is rapidly overtaken by loss in spring as herbivore consumption builds up to balance production.

ATLANTIC TRADE WIND BIOME

Here we encounter for the first time in our descriptions of biogeochemical provinces the effects of distant physical forcing on algal dynamics that characterizes the tropical ocean. The intensification of the western jet current along the northern coast of Brazil and the Guianas by the trade winds over the western Atlantic during boreal summer requires that geostrophic balance be maintained by the tilting of the oceanwide thermocline about a meridional hinge line at 20–25°W (Houghton, 1983; Hastenrath and Merle, 1987). Mixed-layer depths increase in the west and decrease in the east, where a seasonal algal bloom is then induced by the local effect of moderate seasonal intensification of meridi-

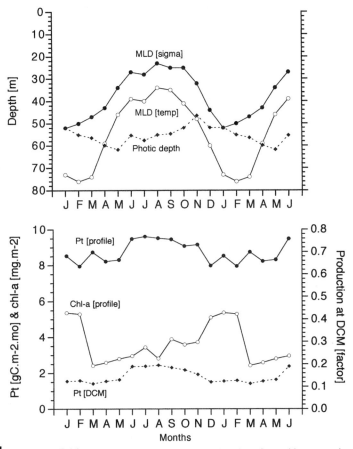

FIGURE 7.11 NATR: characteristic seasonal cycles of monthly averaged mixed layer and photic depths, chlorophyll at the surface, and rate of primary production, both depth-integrated and at the DCM. Data sources are discussed in Chapter 1.

onal winds (Longhurst, 1993). Local biological responses to these phenomena are thus primarily distantly forced, not locally forced as in the westerlies biome.

North Atlantic Tropical Gyral Province (NATR)

Extent of the Province

The North Atlantic Tropical Gyral Province (NATR) comprises the North Atlantic gyre south of the STC which runs zonally across the ocean at about 30°N (see North Atlantic Subtropical Gyral Province). The southern boundary is the limit of westerly flow along the thermocline ridge at the North Equatorial Current (NEC)/ North Equatorial Countercurrent (NECC) conjunction, or about 10–12°N. The western boundary is taken to be the edge of the coastal boundary biome seawards of the Antilles and the Bahama islands. It includes the continuation of the offshore Canary Current south of the Canaries and the flow into the NEC and then to the western limb of the gyre.

Continental Shelf Topography and Tidal and Shelf-Edge Fronts

The Cape Verde Islands (17–19°N) lie within the eastern limb of the anti-cyclonic gyre and their downstream eddies modify the thermocline topography of the outer Canary Current as it passes to the southeast, away from the African continent (see North Atlantic Subtropical Gyral Province for a discussion of this process at the Canaries).

Defining Characteristics of Regional Oceanography

For this province, we lack the wealth of information—particularly from time-series observations—that is available for the provinces to the north and for the equatorial ocean to the south. Perhaps because it corresponds rather well with the zonal region, having consistently the lowest surface chlorophyll of the North Atlantic—well seen as a big blue hole in the seasonal CZCS images—it has attracted little attention from oceanographers. It was, for instance, ignored by McClain *et al.* (1990) in their CZCS-based models of phytoplankton growth for 11 boxes intended to characterize the main ecologies of the North Atlantic. Like many other provinces which follow in this and later chapters, we shall just have to do the best with what we have got.

The NATR lies below the trades of the North Atlantic, separated from the westerlies by the Azores High, which is often expressed as a series of high-pressure cells lying southwest–northeast across the ocean at about 30°N and is best developed in boreal summer: maximum westward wind stress occurs at about 20°N in the Canary Basin. The NATR province comprises the oceanic flow around the southern half of the anticyclonic gyre of the North Atlantic; much of this flow is constrained to the northwest quadrant and within the Gulf Stream recirculation gyre. NATR thus includes the ageostrophic, 200-m-deep, wind-driven flow (Fiekas *et al.*, 1992) of the southern part of the offshore Canary Current, the NEC, and the Antilles Current along the western boundary.

To the south, from May through December, a thermocline ridge lies along 10°N, trending somewhat northeast–southwest and deepening to the west (Hastenrath and Merle, 1987). Divergence along this ridge occurs between flow of the NECC (along the southern flank of the ridge) and flow of the NEC (along the northern flank of the ridge). During the boreal winter, when the NECC become discontinuous west of 20°W, separation is difficult between westward flow of the NEC and of the equator-crossing South Equatorial Current (SEC), though the conjunction still occurs at about 10°N.

Shallow, winter wind-driven overturn occurs to about 25°N (Schoeder, 1965; Levitus, 1982). A warm (>23°C) mixed layer may reach 100 m compared with summer depths of 25–50 m. Further south, mixed layer depth is dominated by responses to geostrophic forcing; pycnocline topography reflects current transport and vector. Thus, a deep thermocline trough lies southwest–northeast across the province at about 20–25°N in most months, with greatest mixed layer depths at 80–120 m in winter (December–May) and rather shoaler (60–80 m) during the rest of the year; these greatest depths occur more often in the western than in the eastern part of the gyre. These facts may be held to suggest that NATR is a heterogeneous unit and could with advantage be divided between NAST, where winter mixing occurs, and Western Tropical Atlantic Province (WTRA) fully in the trade wind zone. I hope that a glance at the figures illustrating the seasonal cycles in the three provinces will convince the skeptic that the current proposals are reasonable. Furthermore, the seasonal thermocline tilt of WTRA does not extend to NATR: winter deepening occurs for a different reason in the two provinces.

Biological Response and Regional Ecology

There is little comprehensive information on the mixed layer nitrate field, but values appear to be uniformly low (0.2 μM) at all seasons (Wroblewski, 1989; Glover and Brewer, 1988, ICITA Atlas). Although higher values in wintertime (<2.0 μM) appear to occur along the southern boundary of the province at about 10°N, these are probably better attributed to distant-field effects of the zone of Ekman suction in the NECC of the adjacent WTRA (see below).

This region of the North Atlantic has a consistently low and uniform surface chlorophyll field with a seasonal cycle of very small magnitude, though even here there is a significant seasonal change in solar irradiance (500–800 W m^{-2} in February and >1000 W m^{-2} in June). Only in the Canary Basin do we find patches of enhanced values, caused by the island effect downstream of the individual Canary Islands and in the few detached eddies and jets of upwelled water from the inshore Canary Current (Hernandez-Guerra et al., 1993).

It is here, and in WTRA to the south, that Herbland's concept of the typical tropical profile was developed, and although there are few seasonal studies of phytoplankton ecology in this province, Jochem and Zeitzschel (1993) interpret the algal and nutrient situation at 18°N, 20°W in March and April as showing no evidence that a spring bloom had occurred. They suggest that winter convective mixing is sufficiently weak that it mixes only the nutrient-depleted water above the depth of the nutricline, normally at about 60 m. It is during the second half of the year, when the mixed layer is shoaler than the depth of 1% irradiance by about 20 m, that the model of Platt and Sathyendranath computes highest vertically integrated production rates. The CZCS data support this view because there is evidence of a progressive accumulation of chlorophyll occupying most of the area of the province during this period. Though the distant-field effect of the high chlorophyll in the NECC peaks in winter along the southern edge of NATR, and dominates the NATR-integrated chlorophyll signal, this conclusion is not invalidated. Apparently, we know almost nothing of the dynamics of herbivore consumption and its seasonality in this province and can only speculate what might lead to the weak accumulation of chlorophyll in winter when no increase in primary production rate is indicated.

Synopsis

Late winter mixing is sufficiently weak that the pycnocline is within photic zone for 9 months (April–December) (Fig. 7.11). The primary production rate is slightly higher when pycnocline is illuminated in summer and there is very weak winter accumulation of chlorophyll, perhaps indicating an effect of winter mixing to constrain herbivores.

Caribbean Province (CARB)

Extent of the Province

The Caribbean Province (CARB) comprises the enclosed Caribbean Sea and Gulf of Mexico and is bordered by the American coastline from Florida to Venezuela and by the island arc from the Bahamas to Trinidad. Sometimes known as the American Mediterranean Sea, it is divided by the constriction of the Yucatan Channel into two basins. Because these are rather dissimilar in their production processes, it would be quite logical to consider the Gulf of Mexico and the Caribbean to be two different provinces. Here, it is simply a matter of convenience to unify them.

Continental Shelf Topography and Tidal and Shelf-Edge Fronts

The whole coast of the Gulf of Mexico has a significant shelf width, widest to the west of Florida (250 km) and continuous along the Mexican coast to the equally wide Campeche Bank off the Yucatan peninsula. There are also several other major areas of shallow carbonate platform of continental shelf dimension: the Bahamas Bank and the Mesquite Bank on the east coast of Honduras. The low-lying coastline of the Gulf of Mexico (Florida to southern Yucatan) is largely free of coral formations because of limiting winter temperatures and excessive runoff of fresh water and river sediments, but fringing reefs are abundant in the Caribbean and an offshore barrier reef occurs along the eastern coast of Yucatan and south onto the Mesquite Bank. The islands of the Antilles arcs are typically endowed with leeward reefs on their Caribbean coasts. Reefs in the Atlantic are taxonomically much less diverse and lack many of the characteristic biota of Indo-Pacific reefs. Living reef formations here are rarely exposed at low water.

Warm filaments of Loop Current water (see below) may intrude onto the shelf of the Northern Gulf of Mexico, especially along western Florida (Paluskiewicz, 1983), forming a frontal eddy feature on the shelf. River discharge fronts have also been observed in the Gulf of Campeche.

Defining Characteristics of Regional Oceanography

The major source of energy for motion in the currents in this province is wind. The northeast trades blow almost constantly over the Caribbean, though the Gulf of Mexico has a more seasonally changeable wind regime. The principal surface water masses entering this province are the flow of the North Brazil Current (NBC) and that of the NEC. The NBC passes around Trinidad, turning west along the continental slope into the southern Caribbean. There is a seasonal switch in the amount of NBC water entering the province in this way. During February to May, when the NECC surface flow is discontinuous (see Western Tropical Atlantic Province) or does not extend west of midocean, a broad (150–200 km) shallow flow from the NBC (importantly modified by Amazon and Orinoco water) enters the province between Trinidad and Barbados. This water is brackish, turbid, and has relatively high chlorophyll. During the remainder of the year, much of this water is diverted eastwards, and the

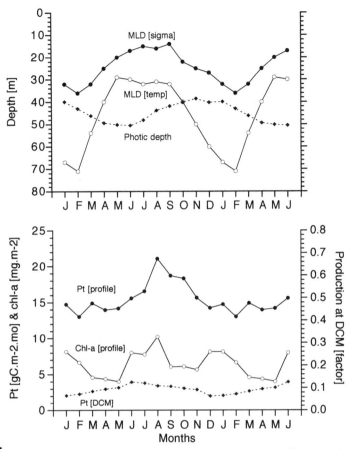

FIGURE 7.12 CARB: characteristic seasonal cycles of monthly averaged mixed layer and photic depths, chlorophyll at the surface, and rate of primary production, both depth-integrated and at the DCM. Data sources are discussed in Chapter 1.

water entering the Caribbean between Trinidad and Barbados is of NEC origin, of high clarity, and with little vertical structure in the upper 100 m (Borstad, 1982; Farmer *et al.*, 1993).

NEC water also passes into the southeast Caribbean through the passages on the Lesser Antilles arc (Roemmich, 1981; Kinder, 1983), where it forms about half of the total influx, and into the northeast Caribbean through the Windward Passage (Cuba/Haiti) and Mona Passage (Puerto Rico/San Domingo). These multiple-source flows of surface water result in a rather complex pattern of circulation within the Caribbean which is detectable in satellite imagery. Flow through the passages induces eddies and vortices, initially of about 100 km diameter but increasing in dimension downstream (Kinder, 1983).

The major flow of the Caribbean Current passes around the southern part of the Caribbean toward the Yucatan Channel, through which water leaves the Caribbean and enters the Gulf of Mexico. A minor countercurrent flows eastwards between Puerto Rico and Venezuela carrying flow from the Windward and Mona passages into the westwards flow around the south of the basin (Duncan et al., 1982).

Flow through the Yucatan Channel along the northeast edge of the Campeche Bank forms a true western boundary current: the Loop Current that passes northwards across the Gulf of Mexico over deep water. Here, it becomes unstable and the position of its average flow represents an anticyclonic loop around the eastern Gulf, passing from the Campeche Bank to the western edge of the Florida shelf, which it then follows around and out through the Florida Straits. The Loop Current penetrates furthest into the western Gulf when it separates from the Campeche Bank at a relatively western position (Molinari and Morrison, 1988).

The dynamic shifts of the Loop Current induce the pinching off of cold- and warm-core eddies that are a major feature of the circulation and nutrient budget of the western part of the Gulf. Warm, anticyclonic rings separate from the Loop Current at about 6- to 17-month intervals (mean = 11), mostly in winter (Vukovich, 1988), and move into the western Gulf. These rings are 100–300 km in diameter, have a lifetime of 1 year, and may interact dynamically with the continental shelf edge (Vukovich and Waddell, 1991). Cold-dome cyclonic eddies (80–120 km) occur on the eastern/southern side of the Loop Current, evolving into tongues or quasi-permanent meanders of cool water (Vukovich and Maul, 1985) that are a characteristic feature of the eastern Gulf.

The seasonal mixed layer depth cycles of the northern and southern basins are rather different because of their different current patterns which the topography of the pycnocline reflects. Each deepens in winter and shoals in summer with about the same seasonal difference as in the adjacent open ocean of the NATR province. The Gulf of Mexico, with its lower surface salinity, mixes down to about 50 m in the Loop Current, whereas in the Caribbean the mixed layer goes down to 100 m around Trinidad but only to about 60–70 m along the coast in the Caribbean Current jet. In summer the Gulf of Mexico has a very shoal mixed layer (10–20 m), whereas in the Caribbean it is 50 m deep in the east and slopes upwards to about 30 m near the American coastline.

Biological Response and Regional Ecology

A deep oligotrophic profile is characteristic of the area, maxima of chlorophyll, phytoplankton carbon, and microzooplankton carbon occurring within the pycnocline and above a nitracline (Hobson and Lorenzen, 1972). When the pycnocline is deeper than about 100 m, the DCM is found to be extremely weak.

Associated with the oligotrophic profiles, the vertical distribution of mesoplankton follows the usual pattern: The >150-μm mesozooplankton is dominated by copepods (comprising 90% in numbers and >50% of biomass) and concentrated in the mixed layer, with a typical suite of warm-water genera (Clausocalanus, Euchaeta, Scolecithrix, and Nannocalanus) including diel migrant genera (especially Pleuromamma and Sergestes), which shuttle between 300–400 m and the photic zone (Hopkins, 1982). Though these studies are not explicit, we may assume that the usual relationship exists between residence depths of mesoplankton and the features of the phytoplankton profile.

The CZCS images suggest that regional-scale changes in mixed layer chlorophyll are relatively minor and that the chlorophyll field is dominated by singular processes associated with features of the complex topography and circulation characteristic of the province. It is thought that the provincial seasonal cycle obtained from the CZCS data is, in fact, a spatial integration of the sum of these individual processes. Thus, it is to the singular processes that we should direct our attention.

A constant feature of the chlorophyll field in the Gulf of Mexico is the offshore entrainment of shelf water and the upwelling of nutrients induced by the anticyclonicity of the Loop Current in the eastern Gulf (Paluskiewicz *et al.*, 1983), whereas anticyclonic eddies significantly modify the nutrient budget as they propagate into the western Gulf. The anticyclonic eddies and the Loop have, of course, bowl-shaped pycnocline topography so that nitrate isopleths are brought up into the lighted zone around their perimeters (Walsh *et al.*, 1989). Both families of rings exhibit chlorophyll enrichment which is observable in satellite images (Yentsch, 1982; Salas de Leon and Monreal-Gomez, 1986; Trees and El-Sayed, 1986). Nitracline doming in cyclonic eddies shed from the Loop Current into the eastern Gulf also produces features in the chlorophyll field.

Coastal upwelling also occurs along the southern coast of the Caribbean. The presence of a western boundary and the pattern of wind stress curl sets up divergence south of 12°S and convergence north to about 15 or 16°N (McClain and Firestone, 1993). Ekman suction brings cooler water to the surface especially along the eastern coast of Venezuela near Margarita Island, where inshore temperatures can fall to about 22°C. While it is known that these surface coolings occur in August and September, their seasonality is unclear. It has been suggested that this process is observable in the CZCS images as areas of enhanced chlorophyll, but confusion with the pigment plume from the Orinoco outflow is likely. This influx of turbid, brackish, green water is carried by the Guiana Current into the southeastern corner of the Caribbean between Trinidad and Grenada and is probably responsible for the strong signal seen in seasonal CZCS images; however, once the plume rounds Trinidad, the rate of primary production within it falls rapidly to oceanic background levels, though the water is still discolored in CZCS images (Bonilla *et al.*, 1993).

The Mesquite Bank lies in a region of extremely high rainfall, and river discharge of sediments results in a brackish, turbid, organized flow along the coast. Beyond this stream, offshore water rides up onto the bank (as it also does on the Campeche Bank) and provides an ideal environment for abiotic carbonate precipitation, potentially observable as white turbidity in satellite images.

Not to be confused in CZCS images with a phytoplankton bloom, the Bahamas Bank, overlain with extremely clear oceanic water of the NEC, appears as a high pigment region that exactly matches its outline; this is surely due to symbiotic and benthic algal chlorophyll showing through a few meters of the clearest ocean water. Apparently, high chlorophyll signals, perhaps for the same reasons, are also associated with the Florida Keys and the bank south of Cuba on which stands the Isle of Pines.

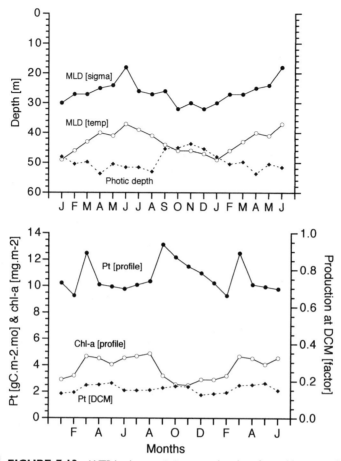

FIGURE 7.13 WTRA: characteristic seasonal cycles of monthly averaged mixed layer and photic depths, chlorophyll at the surface, and rate of primary production, both depth-integrated and at the DCM. Data sources are discussed in Chapter 1.

Synopsis

Though extending over a longer period than in the equivalent NATR province, winter mixing is sufficiently weak and transparency sufficiently high so that pycnocline lies within the photic zone at all times (Fig. 7.12). Primary production rate is highest in boreal mid-summer. The accumulation of chlorophyll appears anomalous and is probably forced by the late summer events in the southeastern Caribbean.

Western Tropical Atlantic Province (WTRA)

Extent of the Province

The WTRA comprises the tropical Atlantic west of 20°W and south of the thermocline ridge marking the flank of the seasonally varying NECC lying across the Atlantic at about 8–10°N. In the south, the northern edge of the seasonally varying flow of the SEC toward the northwest marks the limit of the province. The SEC itself is referred to the South Atlantic Province (SATL).

Continental Shelf Topography and Tidal and Shelf-Edge Fronts

The NECC interacts with the edge of a continental shelf only at its termination, west of Senegal at about 20°W. It is in this area, and in consequence of the topographically guided northward swing of part of the NECC flow, that the Guinea Dome is formed seasonally (see below).

Defining Characteristics of Regional Oceanography

The pelagic ecology of this province is dominated by two consequences of the seasonally varying strength and position of the hemispheric trade winds and of the intertropical convergence zone (ITCZ) between them: (i) The thermocline of the whole tropical Atlantic tilts seasonally about two axes—zonal and meridional, and (ii) the zonal extent of eastward flow in the NECC is seasonally variable. The base of information on physical processes in the WTRA is rich: The tropical Atlantic has been the site of several recent, international studies (GATE, the Atlantic Tropical Experiment of Global Atmosphere Research Programme; SEQUAL, the Seasonal Response of the Equatorial Atlantic; and FOCAL, the French Ocean and Climate in the Equatorial Atlantic) of ocean–atmosphere dynamics. Unfortunately, we have few comparable ecological studies: The biological component of EQUALANT of the early 1960s provided few data useful for our purpose, though it included a voyage of the Nigerian research ship *Kiara* that included my personal encounter (45° wire angles and long parallel glassy slicks at the surface) with the equatorial undercurrent, which I had not then read about!

The seasonal changes in depth of the thermocline in the western ocean are related to the migration of the ITCZ so that the mixed layer is deepest in boreal summer when the ITCZ is in its most northerly location; during this season of increased westward wind stress, the thermocline of the whole tropical Atlantic tilts to the west. Because this phenomenon has even more ecological importance in the east of the ocean, we shall discuss it in detail in the section devoted to the Eastern Tropical Atlantic Province (ETRA).

Sverdrup (1947) showed that it was the curl of wind stress rather than wind stress itself which drives a generic NECC (Philander, 1985) and this has, as will be shown, important biological consequences: Ekman suction (Isemer and Hasse, 1987) causes divergence to occur along the axis of the NECC.

In boreal winter, when the ITCZ is at its most southerly position, the NECC lies under the northeast trades, whereas in the boreal summer the stronger southeast trades curve toward the north over the NECC as they approach the ITCZ. Eastward flow of the NECC occurs along the southern flank of the geostrophic ridge which extends right across the ocean at about 10°N (Garzoli and Katz, 1983; Dietrich *et al.,* 1970; Hastenrath and Merle, 1987). This ridge deepens progressively weswards to 60–80 m in the western Atlantic from 20 to 40 m in the east, following the general downsloping of the tropical Atlantic thermocline in response to westwards stress of the trade winds.

During the boreal summer when zonal trade wind stress is greatest in the western half of the tropical ocean, the North Brazil Current (NBC) flows as a western jet which is retroflected eastwards at about 5°N, topographically locked to the Demarera Rise. This flow thus contributes to the NECC at the season of its greatest westwards extent (Boisvert, 1967; Bruce *et al.*, 1985; Anderson *et al.*, 1981; Borstad, 1982; Müller-Karger *et al.*, 1988). Mesoscale low-salinity features near the origin of the NECC during this season represent anticyclonic eddies shed at the retroflection with a periodicity of about 50 days (Anderson *et al.*, 1981). The similarity of the Amazon (4°N) and Demerara (8°N) eddies of the NBC to the Great Whorl of the Somali Current, with respect to latitude, season, and mechanism of formation, is striking (Bruce *et al.*, 1985). Differences in steadiness of separation may be related to the different orientation of the coasts (Johns *et al.*, 1990).

The rather complex seasonal changes in the NECC may conveniently be summarized as follows. In August and September, flow is continuous across the ocean at about 10°N. In October, a discontinuity in eastward flow appears at about 35°W, and the two now-separated areas of eastward flow regress toward Africa and America. By March, the origin of eastwards flow lies at about 18°W (not far from the African coast), and only a small area of eastward flow remains off America at about 50°W (Boisvert, 1967; Bruce *et al.*, 1985; Garzoli and Richardson, 1989). Surface flow is now westwards, is indistinguishable from the SEC, and is in approximately the same direction as the trade winds.

Throughout the year, when it reaches the African coast at about 15–20°W, the NECC flow continues to the east, as the Guinea Current, though some water passes north along the coast of West Africa toward Senegal and into the bight south of Cape Verde. Deeper eastwards flow in the underlying North Equatorial Undercurrent circulates around a permanent cyclonic feature at about 10°N, 22°W which has been compared with a mountain sitting on the end of the thermocline ridge between the NEC and NECC. This is the Guinea Dome, a permanent feature associated with the flow fields of the NEC and the NECC, which is most strongly developed in summer. Its relative development appears not to be related to variance in the local Ekman vertical velocity fields responding to local wind, and it is more likely another distantly forced feature of the regional oceanography of this province (Siedler *et al.*, 1992).

Biological Response and Regional Ecology

The chlorophyll field of the WTRA, as seen in seasonal CZCS climatology, responds to the westward tilt of the thermocline in summer by showing a clear demarcation near the 15–20°W pivot line: To the west, near-surface chlorophyll values are consistently lower than those to the east. Otherwise, the chlorophyll field of the WTRA is dominated by three singular processes which force algal blooms in the NECC (Longhurst, 1993): (i) In the west, upwelling appears to occur around very large anticyclonic eddies shed from the divergent flow in the eastward retroflection of the NBC; (ii) across the open ocean, there is positive Ekman suction and divergence along the ridge topography denoting the northern flank of the NECC; and (iii) off the African coast, vertical motion in the cyclonic Guinea Dome brings the nitracline into the photic zone. In addition, the strongly meandering flow of the NECC itself induces vertical motion within cyclonic eddies and due to eddy/eddy interactions, this motion is thought to be a nonnegligible source of nutrient flux to surface waters (Dadou *et al.*, 1996).

Though it has been suggested that high-nutrient, high-turbidity water discharged from the Amazon accounts for the observed "chlorophyll" signal in CZCS images (e.g., Müller-Karger *et al.*, 1988; Johns *et al.*, 1990), it is not clear how either nutrients or turbidity could be conserved so far from the river mouth over oceanic depths. In fact, as Amazon water moves north along the shelf it undergoes predictable transformation, sequentially, through (i) an inner zone, in which algal growth is light limited by suspended sediments; (ii) a zone of active growth after sufficient suspensoids have sunk out; and (iii) an outer zone, in which algal growth is nutrient limited (Curtin and Legeckis, 1986; DeMaster *et al.*, 1991). The nutrient-limited zone lies inshore of the 100-m isobath, so it is likely (but not certain) that the high chlorophyll values associated with the Amazon Eddy are at least partly algal blooms occasioned by upwelling around the anticyclonic feature, as in the Somali and Agulhas retroflection eddies. The pigment signal in the Amazon Eddy may also be enhanced to an unknown extent by fluorescence of refractory humic substances which are probably conserved far from the river mouth, though it is to be noted that most of the water advected into the eddy is slope water from the NBC, originating far offshore away from any influence by the Amazon. However they are formed, the series of adjacent eddies off this coast form one of the most remarkable phenomena in the global CZCS images.

Eastwards from the eddy field, across the middle of the ocean, a sufficient explanation for the zone of high chlorophyll values associated with the NECC lies in strong vertical Ekman flux from 20 to 40°W from June to October, creating divergence along the crest of the thermal ridge that lies between the NECC and the NEC (Isemer and Hasse, 1987). Furthermore, as pointed out by Yentsch (1990), underlying the NECC is a baroclinic ridge in the subsurface nitrate field which contains a value of $<16.0 \ \mu M$ at 150 m, similar to concentrations at the same depth south of Greenland. This will render any physical mechanism that tends to draw subsurface water toward the surface more effective in supplying nutrients to the photic zone.

Seasonal changes in vertical Ekman velocity support this model: The greatest vertical flux occurs when chlorophyll values are highest. In January, vertical Ekman velocity along the NECC is weak but variable in sign, and by April, a broad band of zero vertical transport corresponds with the area to be occupied later by the NECC. When the ITCZ is in its northernmost position from July to September and the southeast trades dominate, a zone of exceptionally strong positive curl of the wind stress lies along 10–12°N from 20–40°W (Isemer and Hasse, 1987; Hastenrath and Lamb, 1977); this translates into Ekman suction (upwelling) rates of 2 or 3×10^{-5} m sec (roughly 2 m day^{-1}) along the whole length of the NECC during this season (McLain and Firestone, 1993). Toward the end of this trimester, the locus of highest values comes to lie nearer the African coast. Averaged over the year, the band of positive Ekman suction remains a prominent feature from northern Brazil to Senegal and is the most prominent feature of Ekman vertical velocity in the North Atlantic Ocean at any latitude (Isemer and Hasse, 1987).

A large triangular area, apparently of enhanced chlorophyll, from southern Mauritania to Guinea–Bissau and extending southwest to 20–25°W is prominent in CZCS images for the boreal winter at the eastern termination of the

NECC. This feature cannot readily be reconciled with the anticipated consequences of upwelling in the Guinea Dome (Voituriez and Herbland, 1982) because the boreal winter chlorophyll field is much more extensive than can
reasonably be attributed to that source alone. During the remainder of the year,
high chlorophyll values are restricted to the known coastal upwelling area of
Senegal–Mauritania and to the turbid water of the Bissagos shelf to the south
in the coastal Guinea Current Coastal Province (GUIN).

Significantly, it is during the boreal winter that Harmattan winds (Hastenrath, 1985) carry eolian dust to the ocean over an area corresponding generally
to the enhanced pigment signal (Duce and Tindale, 1991; Duce *et al.,* 1991;
Donaghey *et al.,* 1991). This dust, carried to the coast of Florida, introduces
errors into the CZCS atmospheric radiance algorithms designed to correct for
single-component aerosols (Cardon *et al.,* 1991). It is likely, therefore, that the
available CZCS images are incapable of detailing the chlorophyll field below the
Harmattan haze extending southwest from the coast of Senegal–Mauritania.
Therefore, it may not be easy to identify the location of the Guinea Dome in
CZCS images. Although subthermocline, high-nitrate water does not surface in
the Guinea Dome, as it does in the Costa Rica Dome (Voituriez and Herbland,
1982), we may expect that this feature will be observable in the SeaWiFS imagery based on more critical sensors.

The ecological profile everywhere in the WTRA is characteristic of an oligotrophic situation. Depending on the depth of the mixed layer—from almost
100 m deep in the west to 20–30 m in the Guinea Dome—the chlorophyll and
productivity maxima and the subjacent nitracline lie either at the top of the
pycnocline or deep within it. We have little information on the vertical distribution of biota, but there is no reason to believe that the eastern tropical Pacific
profiles (see Chapter 9, North Pacific Equatorial Countercurrent and North
Pacific Tropical Gyre Provinces) will not serve as competent models for the
tropical Atlantic. We shall have reason to refer again to the Harmattan dust veil
leaving Africa westwards on the northeast trade winds when we discuss those
regions in the eastern Pacific where nitrate is not the limiting nutrient but rather
Fe from eolian sources. In the tropical Atlantic, we have no reason to believe
that the limiting inorganic nutrient for plant growth is not nitrate.

Synopsis

Seasonality in mixed layer depth is distantly forced, deepening in the second half of the year (Fig. 7.13). Pycnocline is shoaler than photic depth at all
times. Seasonality in primary production rate and chlorophyll accumulation is
sufficiently weak as to be insignificant and perhaps forced in the data archives
by unusual events.

Eastern Tropical Atlantic Province (ETRA)

Extent of the Province

This province extends eastwards across the ocean from a meridional
boundary at the 20°W thermocline hinge line (see below). Its northern limit is
the convergence between the Guinea Current and the westward flow of the
northern limb of the SEC. The southern boundary is set at about 10°S at the

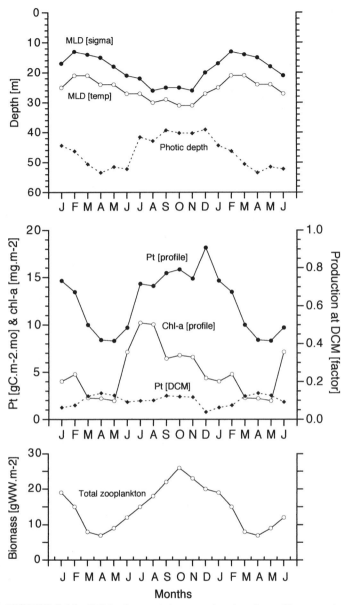

FIGURE 7.14 ETRA: characteristic seasonal cycles of monthly averaged mixed layer and photic depths, chlorophyll at the surface, and rate of primary production, both depth-integrated and at the DCM. Zooplankton data from Soviet surveys, quoted by Blackburn (1982); other data sources are discussed in Chapter 1.

convergence between the weak, narrow, and variable South Equatorial Countercurrent (SECC) and the southern limb of the SEC, which belongs to the SATL.

Continental Shelf Topography and Tidal and Shelf-Edge Fronts

There are none in this province.

Defining Characteristics of Regional Oceanography

The currents in this province, and hence the topography of the thermocline, are complex and rendered the more so by the change of sign of the Coriolis force at the equator, which of course the province straddles. The northern limb of the anticyclonic (counterclockwise in the Southern Hemisphere) gyre of the South Atlantic Ocean is the SEC, which passes westwards across the ocean as two streams north and south of the narrow, weak, and variable SECC that lies at about 5°S. The two streams diverge in their flow toward the western boundary currents of the North and South Atlantic Oceans, respectively. The northern stream of the SEC carries warm, low-salinity water originating in the eastern Gulf of Guinea and is separated from the cooler, more saline Benguela water of the southern stream of the SEC across a thermal front at about 10°S in the west and 15°S in the east at the Benguela divergence (Shannon, 1985; Voituriez, 1982).

Symmetrically below the equator is the 250- to 300-km-wide eastwards flow of the Atlantic Equatorial Undercurrent (EUC), originating north of Brazil. This has a vertical thickness of 150–250 m, sloping upwards progressively to the east. Depth of the EUC core is 50–75 m, and inversion with westward flow of the SEC lies at 15–35 m (Neumann, 1969; McCreary et al., 1984; Voituriez, 1981). The topography of the thermocline below the SECC, which theoretically should carry a ridge along the equator, is disturbed by the presence of the EUC that deepens the isotherms for water <20°C and by a narrow band of wind-driven divergence at the equator, aligned according to the predictions of Cromwell (1953). This divergence shoals the isotherms for >20°C toward the surface. The SECC runs along the equatorward slope of a ridge (or very elongate zonal cyclonic feature; see below) at about 5–10°S which separates it from the westward flow of the southern stream of the SEC.

The southeast and northeast trade winds are separated by calms (resultant winds, 1 or 2 m sec^{-1}) at the atmospheric ITCZ, which from November to April is stationary at 6°N on the African coast, lying southwest across the ocean. In the next 3 months, the ITCZ moves rapidly northwards and by August reaches 15°N. This reflects the rapid extension of the strong (5–8 m sec^{-1}) southeast trade winds across the equator which surge along the north coast of Brazil into the western basin of the tropical Atlantic, strongly intensifying the zonal wind component there. Meanwhile, wind stress strengthens only marginally in the eastern Atlantic and principally in its meridional component (Hastenrath and Lamb, 1977). Almost the entire province lies under negative wind curl during this season, with the southern limit of the SEC approximating to the change of sign in vertical motion (Gordon and Bosley, 1991; Ding et al., 1980).

From May to October, as a direct consequence of the impulsive intensification of zonal wind stress, the thermocline shoals in the ETRA and deepens to the north of Brazil. To the east of a pivot line at 15–20°W, the base of the thermocline (defined as the depth of the 15°C isotherm) shoals from 145 to 90 m, whereas the upper thermocline (the 20°C isotherm) shoals by about 20 m; thus, both the thermocline and the mixed layer above are vertically compressed. Understanding this oceanwide response to seasonal intensification of zonal wind stress in the Atlantic has been a major recent accomplishment of tropical oceanography (e.g., Moore et al., 1978; Philander, 1979; Picaut; 1981; Servain *et al.,* 1982; Houghton, 1983, 1989, 1991; Merle, 1983; Merle and Arnault, 1985; Hastenrath and Merle, 1987; Duing *et al.,* 1980).

The shoaling of the thermocline is accompanied by a cooling of the mixed layer over the whole of the ETRA province, leading to minimum surface temperatures somewhat south of the equator. The observation that the shoaling of the thermocline is a distantly forced change in the subsurface thermal structure is supported by the fact that the associated cooling of the mixed layer in the ETRA occurs while net heat gain across the sea surface is positive. Elsewhere in the tropical Atlantic, sea-surface temperature changes are phase-linked to net heat gain across the sea surface (Houghton, 1991). The cooling appears to occur because intensification of local meridional winds during the same season is sufficient to erode the upper isotherms of the uplifted thermocline so that water entrained from it will further cool the mixed layer, resulting in the observed regional SST cooling. Response of the mixed layer to the anomalous, rapid collapse of the trade winds in February 1984 is an additional confirmation that tilting is indeed distantly forced and is not a response to local wind stress (Verstraete, 1992). Imposed on the effects of the thermocline tilt are the consequences of divergence and instability of the EUC along, and slightly south of, the equator as seen (i) in isolated linear patches of cooling SST early in the season, (ii) in the location throughout the season of the coolest SST, and (iii) in the linear extension of enhanced values west of the wider region of chlorophyll enhancement.

The relationship and extent of the various entities described as cyclonic gyres in the eastern tropical Atlantic are unclear, as are the details of circulation in this undersampled region (Shannon, 1985). A large active cyclonic thermal dome or gyre lies in the eastern ocean at about 11°S, 6°E that has the SECC on its northern slope and the offshore flow of the Benguela Current (Peterson and Stramma, 1991) on its southern slope. It may be of some biological significance because it is most strongly developed above the area of lowest subsurface oxygen levels in the Atlantic, which is a feature related to surface productivity (Bubnov, 1972; Chapman and Shannon, 1985). In this Angola Dome, the pycnocline reaches closer to the surface than it does in the larger gyre (and indeed than it does in the Guinea Dome), perhaps with significance for algal growth processes (Voituriez, 1981). For convenience, I include the Angola Dome within the ETRA province, placing the eastern boundary with GUIN along an arbitrary longitude.

Biological Response and Regional Ecology

A prominent feature of the CZCS images for the Atlantic Ocean is the relatively high chlorophyll (0.3–0.5 mg m^{-3}) that occurs almost uniformly over the eastern tropical Atlantic in June–August east of a meridional boundary at about 20–25°W. In the following 3 months, this retreats slightly eastwards, but chlorophyll enhancement at the equatorial divergence becomes more evident and extends farther to the west. Through boreal summer and fall, the CZCS images show a chlorophyll hot spot (about 2 mg m^{-3}), varying in its location but apparently corresponding to the Angola Dome in the extreme southeast of the province.

Though the evolution of the seasonal algal bloom is best followed in the 1978–1979 CZCS images, the incomplete coverage of subsequent years is sufficient to affirm that the 1979 pattern is typical even though some differences between years were noted. In 1980, 1984, and 1985, the bloom in the last trimester was less strong than in 1979. In the third trimesters of 1982 and 1983, the bloom extended farther north, closer to the African coast, whereas in 1981 it covered a larger area than in any other year, extending continuously from the Angola coast to north of Brazil at 35–40°W. A filament of high chlorophyll frequently occurred at 10–20°W along and just south of the equator; this can be seen in the level 3 composites for October–December 1978, 1979, 1982, 1984, and 1985 and corresponds, as discussed later, to the location of the consistently highest mixed layer nitrate concentration during the season of thermocline uplift.

This seasonal pelagic bloom occurs (i) where seasonal variability in insolation is slight, (ii) where surface wind stress varies only modestly between seasons, and (iii) where curl of the wind stress is weak. The factors which force the bloom have been interpreted as follows (Longhurst, 1993): The bloom occurs after distant forcing by intensified zonal wind stress in the western ocean causes uplift of the thermocline in the eastern ocean and during the season when local meridional wind stress is maximal. Increased local wind mixing erodes the upper thermocline and increases vertical transport of cool (high-nutrient) water across the greatest thermal gradient. Combined with the dynamic uplift of the nutricline in relation to the light field, an algal bloom is initiated that utilizes the nitrate advected into the mixed layer and then erodes the nutricline down to the depth at which light becomes limiting. The bloom subsequently continues because dynamic uplift of the thermocline is maintained against the positive heat flux and mixing due to meridional wind stress, both of which would tend to increase the depth of the mixed layer as the cool season progresses. This will sustain vertical flux through the nutricline.

How far can this model be supported by observation? Primary production rates double in the boreal summer in the eastern tropical Atlantic, whereas the CIPREA data of 1978–1979 at 4°W, integrated over the euphotic zone, show no seasonal change at the equator itself but show an increase in summer production both to the south and to the north (Herbland *et al.*, 1983). During the FOCAL voyages, meridional chlorophyll sections were made across the equator at 4, 23, and 35°W (Herbland *et al.*, 1985). These sections support the inferences from CZCS data: Not only does the DCM at 4°W shoal from 50 m to about 25 m but also chlorophyll values within it double. The same occurs, but less clearly, at 23°W. At 35°W, in the WTRA, the DCM remains at about 60–70 m in both seasons with essentially unchanged values.

Recent meridional sections for temperature, nitrate, and chlorophyll show an oligotrophic profile for these variables across the tropical Atlantic from 6°E to at least 23°W (Herbland *et al.*, 1987; Oudot and Morin, 1987). The nutricline and the DCM co-occur in the upper few meters of the thermocline in both seasons, and nitrate is at a limiting concentration in the mixed layer throughout the year, except immediately in the equatorial divergence zone during the cool season. The depths of the DCM and nutricline respond predictably to the basinwide tilting of the thermocline. Herbland *et al.* comment that these phenomena imply a simple upward displacement of the thermocline rather than the equatorial upwelling favored by Oudot and Morin.

Thus, in February, the DCM lies at 50–55 m near the 25°C isotherm in the upper thermocline with maximum values of only 0.05 mg chl m^{-3}, and the nutricline lies in the upper thermocline near the 20°C isotherm. As this isotherm shoals in the cooling period, the nutricline shoals with it but comes to lie deeper in the thermal gradient, along the 17 or 18°C surface. Similarly, by July, the DCM lies relatively deeper in the thermocline and close to the 20°C isotherm, though its absolute depth rises to 25 m at 1 or 2°S, and it now has maximal chlorophyll values of 0.8–1.0 mg chl m^{-3}. During both seasons, there is a coastwise zone at 4 or 5°N where the DCM is shoaler and contains higher chlorophyll values (see Guinea Current Coastal Province). Consequently and predictably, the DCM always lies close to the depth at which NO$_3$ reaches 1.0 mol liter^{-1}.

The cool water which lies along and just south of the equator is associated with enhanced (<5 μM) nitrate levels at the surface because the shoaling of the thermocline above the EUC coincides with the occurrence of stronger turbulence there than to north or south (Hebert *et al.*, 1991), though surface cooling caused by shear and vertical turbulent mixing in the EUC can account for only one-third of the total cooling within 1.5° of the equator (Verstraete, 1992). There is greater short-term variability of integrated primary production along the equator than there is only >100 km to north and south, as shown by the CIPREA sections at 4°W (Voituriez, 1982). This is consistent with what we may expect from equatorial divergence. Transient patches of cool water, associated with elevated nitrate levels, are seen as isolated zonal filaments of chlorophyll in the CZCS images

Unlike the eastern tropical Pacific, in the tropical Atlantic there appears to be no general constraint on the utilization of nitrate by lack of the trace nutrient Fe. Residual nitrate, reported by Oudot and Morin (1987), along the equatorial divergence during boreal summer represents the normal situation in which rate of supply exceeds utilization rate or may suggest that another element, perhaps silicate, may temporarily limit the increase in chlorophyll biomass.

The Angola Dome is most strongly developed in the boreal winter, when the doming of the pycnocline brings the nutricline closer to the surface and into higher light levels (Voituriez and Herbland, 1982). Chlorophyll enhancement occurs as a result of this uplift (see above), though in neither the Angola Dome nor in the Guinea Dome of the WTRA province does the nutricline actually break the surface.

It has been known since the early 1980s that the seasonal algal bloom is dominated by small cells, though the data reported then did not reflect the revelation, derived from seagoing flow spectrometry, that cyanobacteria and prochlorophytes (including submicrometer cells) are probably relatively dominant in the chlorophyll biomass at the bottom of the photic zone. Be that as it may, in 1985 Herbland reported that in the nitrate-depleted mixed layer, cells <1 μm dominate the chlorophyll biomass (mean $= 71\%$), whereas at the nitracline such cells comprise only 50%. In the nitrate-replete deep pycnocline, they comprise only a very small proportion, and here larger cells predominated. In the oceanic algal bloom, the larger cell fractions comprise $<15\%$ of chlorophyll biomass. It is only in the coastal upwelling off Ghana and the Ivory Coast (in the GUIN coastal province) that larger cells dominate the bloom, as appears to also occur in the coastal upwelling off northwest Africa in the eastern coastal boundary province (Blasco *et al.*, 1981). For this reason, Herbland suggests that the whole of the open tropical Atlantic should be considered a single ecosystem. This relative importance of different size fractions in open ocean and coastal blooms is repeated in the warm waters of the Pacific (Chavez, 1989) and suggests another reason to distinguish the chlorophyll signature during the boreal summer along the Guinea coast from that of the open ocean.

There are few modern studies of higher trophic levels in the ETRA, but Le Borgne (1981) correlated standing stocks (expressed as integrated mixed layer dry weight) of mesozooplankton with chlorophyll and found a positive correlation ($DW_{zoo} = 74.5\ Chl^{1.05}$) with a slope not different from 1. He also found that the mesozooplankton biomass is aggregated at the DCM and has a relationship with the nitracline and the pycnocline that depends on the absolute depth of this suite of features. When detailed mesozooplankton profiles are obtained in the eastern tropical Atlantic, it is predictable that they will resemble the many that are already available for the eastern tropical Pacific. At even higher trophic levels, there is some evidence in catch-rate maps for tuna that ETRA produces more of the two shallower swimming tropical species (yellowfin and skipjack) than does the deeper and more oligotrophic mixed layer of the WTRA. Conversely, the deeper-swimming bigeye (*Thunnus obesus*) shows less east–west difference in abundance in the tropical Atlantic. As with all such fishery-dependent evidence, we must note that perhaps this observation simply reflects the fact that purse seiners and bait boats will obviously find their work easier above the relatively shoal thermocline of the ETRA.

Synopsis

Distantly forced seasonal changes in the depth of pycnocline are significant and greater than those in the west of the ocean, but pycnocline nevertheless remains within the photic zone at all times (Fig. 7.14). Primary production rate is inverse to the depth of pycnocline, and chlorophyll accumulation closely tracks seasonal changes in primary production rate. Herbivore biomass is known to respond in a predictable manner to chlorophyll accumulation, confirming the assumption that close coupling occurs between production and consumption.

South Atlantic Gyral Province (SATL)

Extent of the Province

The SATL comprises the anticyclonic circulation of the South Atlantic, excluding the coastal boundary currents (Brazil and Benguela), which are treated as separate provinces [see Brazil Current Coastal Province (BRAZ) and Benguela Current Coastal Province (BENG)]. The east and west boundaries of the province are therefore within the eddy fields seawards of these two currents. The northern boundary is taken as the northern limit of the southern limb of westward flow of the SEC from Africa to the coast of South America, and the southern boundary is the limit of the biological enhancement associated with the Subtropical Convergence Province (see Chapter 10).

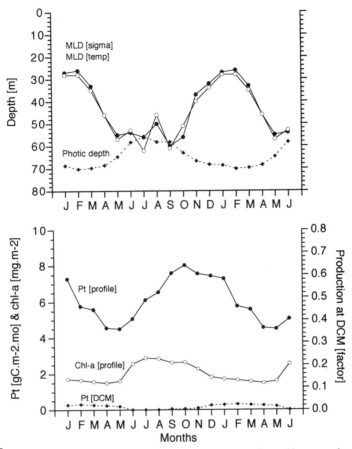

FIGURE 7.15 SATL: characteristic seasonal cycles of monthly averaged mixed layer and photic depths, chlorophyll at the surface, and rate of primary production, both depth-integrated and at the DCM. Data sources are discussed in Chapter 1.

Continental Shelf Topography and Tidal and Shelf-Edge Fronts

There are none in this province.

Defining Characteristics of Regional Oceanography

As will become clear, this is one of the least well-researched regions of the oceans, far from the major oceanographic research institutes and the troubles of the world. The Southern Atlantic Accelerated Research Initiative of the U.S. Office of Naval Research during the 1980s worked mostly at the edges of the ocean in the Falklands Conjunction and the Benguela Current. It is hoped that for more general information will be provided from the current World Ocean Circulation Experiment (WOCE), but data from WOCE cruises for this region currently have not yet been processed or analyzed usefully.

The subtropical gyre of the South Atlantic is strongly displaced toward the west and is not entirely the homolog of the North Atlantic gyre because of the geographical differences between the two basins—principally, (i) the heat equator, and the ITCZ between the northern and southern trade wind systems, lies north of the true equator; (ii) the South Atlantic is not fully enclosed on its eastern margin since Africa extends south only to 35°S; and (iii) the Andes form a more complete block to the planetary westerlies and extend further both poleward and equatorward than the Rockies of North America. Consequently, the Greenland winter low-pressure cell (60°N) associated with westerlies from 30 to 55°N has no counterpart over the South Atlantic, where full development of westerlies occurs only further poleward as the circumpolar winds pass around the southern tip of the continent (Peterson and Stramma, 1991).

In both winter and summer, a high-pressure cell lies over the South Atlantic at about 20–30°S and consequently the effective confluence between the westerlies and the trades, below which we shall expect to find the oceanic Subtropical Convergence Front, lies across the southern margin of the ocean from just north of the Falklands plateau to just south of Cape Town (see Subtropical Convergence Province) in the general vicinity of 35–45°S or even further poleward in austral winter. It must be noted that the North Atlantic equivalent passes south of Bermuda and close to the Azores and is thus generally between 30 and 35°N.

Thus, the subtropical gyre of the South Atlantic is more persistently under the influence of the trade winds than that of the North Atlantic, yet a significant seasonal variation in its mixed layer depth is nevertheless forced by seasonal variation in wind stress and surface heat flux. Winter heat loss at the sea surface is remarkably uniform (ranging between only -100 and $-125°$W m^{-2}) in July over the whole region from 10°S in the SEC right down to 45°S. It is no surprise that even at relatively low latitudes a deepening of the mixed layer occurs during the austral winter (Levitus, 1982), when the ITCZ is at its most northerly position and highest wind speeds and wave heights occur from the equator to 25°S. The greatest mixed layer depths (<100 m) occur along a zonal trough at about 20°S in July, across the central part of the gyre.

Flow of the Benguela Current begins to turn northwestwards (to leave the region designated as the BENG eastern boundary current province) and detaches from the coast as a broad stream at about 15–30°S, passing through a gap in the topography of the Walvis Ridge and broadening as it enters the subtropical gyre to feed into the southern stream of the SEC (see Eastern Tropical Atlantic Province), where some eddying occurs. The axis of this flow lies across the ocean in a northwest/southeast direction into the intensified western boundary current which flows southwards along the coast of Brazil (see Brazil Current Coastal Province for definition of the seawards boundary of that province) as the Brazil Current.

Return flow into the southern limb of the gyre is fed from the Brazil Current, part of which departs from the shelf at about 35–37°S, forming a southerly loop to about 40–42°S, before turning northeast around the Rio Grande Rise. This flow then departs to the east as the South Atlantic Current and proceeds along the northern side of the Subtropical Convergence Zone. Seasonal changes in the latitude of the Brazil/Falkland Current confluence reflect a general seasonal shift in the whole subtropical gyre.

Because the retroflection loop of the Agulhas Current southwest of the Cape of Good Hope (see Chapter 8, East Africa Coastal Province) carries more eddy kinetic energy than anywhere else in the Southern Hemisphere, large warm-core eddies are shed when an Agulhas intrusion into the Atlantic occurs, usually 5–10 times each year. These unusually energetic, large (300-km) eddies have very long lifetimes and some survive to reach the Brazilian coastal boundary, where they have been resolved in GEOSAT altimeter data. They are therefore of potential significance in the structure of the whole subtropical gyre (Peterson and Stramma, 1991; Shannon, 1985).

In describing the flows around the periphery of the SATL province, we should not lose sight of the fact that the whole comprises an anticyclonic subtropical gyre, excluding only its coastal boundary zone. Consequently, the bowl-shaped isopleths for nutrients slope upwards toward the edges, an effect which is clearly evident in the chlorophyll field.

Biological Response and Regional Ecology

Modern information on the ecology of the open South Atlantic is not available. In the open subtropical gyre, surface chlorophyll values are low throughout the year over most of the area, though the CZCS images suggest that some increase in chlorophyll biomass occurs in late austral fall and winter. The general pattern is consistent with the baroclinicity of the nitrate field, with higher chlorophyll values consistently around the periphery and minimal values in the central regions where the mixed layer is very deep. CZCS images also indicate higher chlorophyll values on the eastern side of the gyre, especially in the period from October to December, downstream from the departure of the Benguela Current from the coastline. These patches of enhanced chlorophyll extend well offshore, and in some images form a field extending almost halfway across the ocean (NASA CZCS browse files). The eddying at the Walvis Ridge also appears to induce higher surface chlorophyll in some images.

We can, however, predict that when this region is properly investigated, the vertical profiles of nutrients, light, autotrophy, and consumption will have similar relationships to density profiles as those that are typical in the Sargasso Sea.

Synopsis

Shallow winter mixing so that the pycnocline remains within the photic zone except when water clarity is reduced by weak accumulation of cells occurs briefly in May–July of the austral winter (Fig. 7.15). Accumulation fails to track longer than usual periods of relatively high primary production, and losses reduce cell population after only 2 months of accumulation. Close coupling between production and consumption must be assumed, as in all other subtropical gyres.

ATLANTIC COASTAL BIOME

In consulting the descriptions of the individual coastal provinces in this and the following chapters, you will encounter descriptions of the local expression of some phenomena discussed more generally in Chapter 2. Fronts at the shelf edge or in shallow shelf sea areas have common mechanisms which will not be repeated each time a fresh example is encountered. Similarly, the general mechanism of wind-driven coastal divergence and upwelling exemplified by the eastern boundary currents (but by no means restricted to them) will not be reviewed each time it is met afresh.

Northeast Atlantic Shelves Province (NECS)

Extent of the Province

The Northeast Atlantic Shelves Province (NECS) comprises the continental shelf of western Europe, from northern Spain to Denmark and then into the Baltic Sea. The edge of the deep Faeroe–Shetland Channel and the Norwegian Trench forms the separation between this and the SARC province.

Continental Shelf Topography and Tidal and Shelf-Edge Fronts

This is one of the larger continental shelf regions (1.63×10^6 km^{-2}), of which the Baltic Sea represents about a quarter and the North Sea about one-third. Like other coastal regions, this province may be rationally subdivided almost infinitely, and a first-level subdivision of the province may be useful for some purposes. If so, the following entities would probably be the classical candidates for the primary divisions: (i) the North Sea, from the straits of Dover to the Shetlands; (ii) the English Channel from Dover west to Ushant; (iii) the southern outer shelf from northern Spain to Ushant, including the Aquitaine and Armorican shelves off western France; (iv) the northern outer shelf, including the Celtic Sea and the Irish, Malin, and Hebrides shelves off Britain; (v) the Irish Sea; (vi) the central Baltic (Gottland) Sea; and (vii) the Gulfs of Bothnia and Finland. However, there is another way of subdividing the region which is more sensitive to ecological reality.

The North Sea has a shallow central region, the Dogger Bank (about 20 m deep; it was dry land during the last glaciation), but deep basins like those in the NWCS province do not exist, unless we equate them with the Gottland Deep of the central Baltic (about 200 m) which, because of the shallow sill and weak ventilation of the Baltic Sea, is anoxic; the extent of the anoxic area is rather variable and depends on the relative degree of ventilation through the Kattegat each year. In these shallow seas, wind mixing and tidal stirring are sufficiently strong that sediments are almost everywhere sand or gravel; soft muddy sediments occur only in restricted regions of the North Sea and in coastal embayments, especially on the British coasts and, of course, in the deeper basins of the central Baltic Sea. Tidal amplitude varies from almost zero at the amphidromic points of the M2 tide to 10 m in cone-shaped embayments such as the Bristol Channel or the Golfe de St. Malo.

During winter, when the open North Atlantic is mixed to a depth of several hundred meters, the water column over the shelf is mixed to the bottom. Once stratification is reimposed in the spring, tidal fronts at the boundaries between mixed and thermally stratified shelf water are prominent in satellite infrared imagery of sea-surface temperature and are quite predictable in location. In summer, they occur principally (i) across the western entrance of the English Channel, (ii) across the northern Celtic Sea at the mouth of the Irish Sea, (iii) within the northern Irish Sea, and (iv) along a line from northeast England to

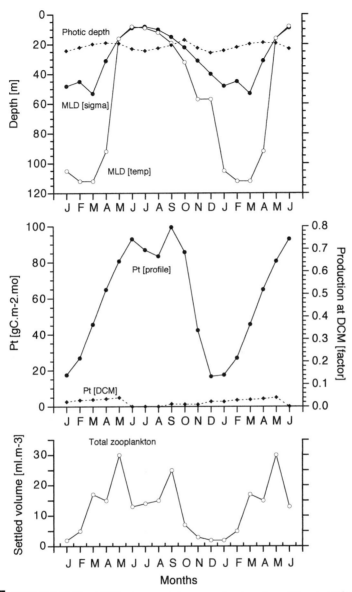

FIGURE 7.16 NECS: characteristic seasonal cycles of monthly averaged mixed layer and photic depths, chlorophyll at the surface, and rate of primary production, both depth-integrated and at the DCM. Zooplankton data from western channel from Harvey (1953); other data sources are discussed in Chapter 1.

the Friesian coast. Other similar fronts, on a smaller scale, occur throughout the islands to the north and west of Scotland and on the western Danish coast. The seasonal development of stratification on the shelf causes the these fronts to migrate seasonally: As the thermocline shoals and deepens, this movement is onshore in the spring and offshore in the autumn, covering a distance of >200 km in the western channel (Pingree, 1975; Pingree and Griffiths, 1978; Pingree *et al.*, 1975, 1982). The location of these fronts is sufficiently predictable that they afford a different regional subdivision of the open sea shelf of this region, which is sensitive to ecological differences.

Thus, in summer, we would identify

1. A central area of vertically mixed water occupying most of the English Channel and the southern North Sea as well as the central Irish Sea
2. Stratified areas occupying the northern part of the North Sea above a line from Denmark to Yorkshire and the whole of the outer Atlantic-facing shelf from Shetland to Spain, interrupted off Ushant by an extension of No. 3
3. A transitional zone between Nos. 1 and 2 across which the shelf sea fronts migrate with the tidal cycle. This occupies the western English Channel with an extension to the shelf edge off Ushant, the northern Irish Sea, and an arc across the southern North Sea from Yorkshire to the Dutch coast to Denmark.

In addition to these tidal fronts, a variety of other shallow-water, coastal, and estuarine fronts exist in this and other coastal provinces and must be recognized. Some of these are examples of tidal fronts in which no major salinity gradient exists and are characterized by the critical h/u^3 parameter; for example, the "marginal front" of Lyme Bay on the south coast of England. Estuarine fronts, characterized by turbidity and salinity gradients, occur in many places along all coastlines, though not many have been studied in detail. Among those which have are the Severn estuary/Bristol Channel region and the Loire estuary. Because the hydrography of the Baltic Sea is dominated by large freshwater inputs (see below) and has a narrow entrance to the North Sea through the Kattegat, there is a very strong salinity gradient, having the characteristics of a frontal zone, within this channel to the east of Denmark.

Once stratification has been imposed in spring, a shelf-break front exists as a meandering belt of relatively cool surface water (a 1 or 2°C temperature anomaly) just beyond the shelf edge and has been best studied off the Celtic Sea and the Armorican–Aquitanian shelf. How it may be formed has been discussed in Chapter 2. In brief, the most widely accepted explanation now involves the generation of internal standing waves on the thermocline at the shelf edge where the tidal stream encounters the rough topography of the seabed. It is considered the outer boundary of this province.

Defining Characteristics of Regional Oceanography

This large area of continental shelf lies beside a region of weak oceanic drift, so transport over the shelf is dominated by local winds and major geostrophic flow along the continental slope is lacking. The outer shelf is largely occupied by water which is homohaline but thermally stratified in summer, except from northern Denmark to the west of Scandinavia, where low-salinity Baltic outflow induces a haline stratification. Though tidal streams dominate perceived flows, residual circulation in the North Sea is cyclonic, concentrated into a rim current along the coastline. None of this flow is as strong as that over the deep water of the Skagerrak, reinforced by the Baltic discharge of brackish water and feeding directly into the Norwegian Coastal Current of the SARC province.

On the outer shelf between Ireland and the Bay of Biscay, nutrient dynamics in summer are strongly influenced by a linear zone of remnant "winter water," which is colder and denser than surrounding water yet trapped below the zone where summer stratification is most strongly developed. These "bourrelets froids" (Le Fèvre, 1986) lie along the shelf and are dome shaped in section and have the potential to act as a nutrient reserve and even as an accumulator of nutrients during the summer months.

The Baltic Sea is a relatively small Mediterranean sea (mean depth, 57 m), having a shallow sill depth (<20 m) via the Kattegat. Water balance depends on river discharge and episodic ventilation of the deeper water mass by irruptions of saline water eastwards over the sill. There is thus permanent salinity stratification (halocline at 30–40 m, depending on sill depths between the different basins) at its head (Bothnian Sea and Gulf of Finland). The inner Baltic has a very low surface salinity (1–3 ppt) and this allows ice cover to develop over the whole of the northern Baltic in winter. As over the open shelf, thermal stratification develops in summer, with a thermocline at 30–40 m.

Biological Response and Regional Ecology

This province is distinguished in the regional CZCS chlorophyll data as a region with many seasonally and locally enhanced chlorophyll features. We shall have to explain both the seasonality and the localization in these images. We shall recognize four ecological seasons: (i) mixed conditions and light limitation in winter, (ii) a nutrient-limited spring bloom, (iii) stratified conditions during summer with localized zones of high chlorophyll along fronts, and (iv) a second general bloom when autumn gales break down the summer stratified condition. This is, of course, a shorthand version of the classical plankton calendar for the European continental shelf that was worked out many decades ago at the principal European marine biological stations.

However, it is not, of course, really so simple; for instance, in the area of permanently mixed water where tidal streams are too strong and water is too shallow for summer stratification to develop. Here, because tidal friction is constantly supplying nitrogen to the water column from benthic regeneration processes, and because the limiting tidal fronts are constantly transporting nitrogen into the mixed areas, we should not expect nitrogen limitation to limit a bloom and we should not expect the bloom to begin as early in the spring as it does offshore because of greater light limitation due to suspensoids. In such areas there is a midsummer bloom in which the rate of primary production is a simple function of irradiance. In the mixed area of the western English Channel, in March the rate of primary production starts to increase from low values in winter and this increase is maintained steadily until a maximum rate is achieved in July, which is then followed by an isotonic decrease until the winter minimum is reached again in November. During the whole period, diatoms dominate the large algal cells and dinoflagellates are negligible. The same seasonal cycle occurs in the permanently mixed, highly turbid Severn estuary and the macrotidal Bristol Channel, with a progressive seasonal bias toward a spring bloom in the outer, less turbid region. It must be noted that despite the great quantities of agricultural nitrogen transported to the European continental shelf in river effluents, there is little evidence that this has yet affected the general pattern of algal nutrient limitation in summer in stratified water.

The typical phytoplankton cycle with large spring diatom and weaker autumn dinoflagellate blooms seems to occur most typically in the central North Sea (Colebrook and Robinson, 1965). In the southern North Sea the spring and autumn peaks may be of about the same magnitude, except in the mixed region off the Dutch coast where a single summer bloom is the rule. In the northern North Sea and the western English Channel, the spring bloom is stronger, relative to the autumn bloom, than elsewhere. The Baltic Sea has a double-bloom cycle, and the autumn bloom may reach even higher chlorophyll values than those during spring (Kullenberg, 1983). In the Irish Sea, and probably elsewhere, the spring bloom starts in shallow embayments (in the case of a recent study, in Dundalk Bay) and occurs progressively later offshore and farther to the north.

The seasonal succession has been followed most closely in the western English Channel, where Holligan and Harbour (1977) clearly distinguished a near-surface spring bloom (<4 mg chl m^{-3}, 0–15 m, April) from a summer subsurface bloom in the thermocline (2–4 mg chl m^{-3}, 20–25 m, May–September, and fueled importantly by regenerated NH$_4$) and an autumn near-surface bloom (<2 mg chl m^{-3}, 0–15 m, late September to October). The spring bloom was dominated by diatoms, which occurred abundantly in the subsurface bloom until May, when they were progressively replaced by dinoflagellates and flagellates; this replacement was completed by midsummer. In the autumn bloom, diatoms again became important. This series can be used for a model for most of the stratified regions of this province and also has been observed in the northern North Sea during the Fladen Ground Experiment (FLEX). The spring bloom of diatoms usually develops faster than herbivores can increase their consumption rate by population buildup; consequently, much of the plant biomass sediments to the sea floor to provide at least a part of the regenerated nitrogen utilized by the microalgae of the summer phytoplankton. Such an imbalance of copepods and diatoms has been observed in several locations in this province.

Recent research on the components of autotrophic production reveals that the modern generalization that autotrophic pico- and nanoplankton are a vital and important component is also valid here in shelf waters: Joint and Williams (1985) compute that 36% of primary production is by the 0.2- to 1.0-μm cell fraction and that 77% is by the 0.2- to 5.0-μm fraction. Even more recent work suggests that production by autotrophic picoplankton accounted for 50% of production prior to the spring bloom, but that thereafter its absolute production rate changed little: The seasonal increases are almost entirely due to cells >2 μm.

The summer profile on the shelf may be further typified by that of the Celtic Sea, which is a two-layered system in which the upper water is at summer temperatures and the lower is at about 8 or 9°C, which is close to winter values. At the interface, in the thermocline, the DCM is associated with maxima of microflagellates and ciliates. Bacterial biomass is uniformly high in the upper layer and uniformly low in the lower layer. Mesozooplankton partition this vertical space: The Atlantic-boreal *C. finmarchicus* is restricted to the lower layer, the temperate *C. helgolandicus* is restricted to the upper layer, and the euphausiids (*Nyctiphanes couchi* and *Meganyctiphanes norvegica*) make diel migrations between the two layers. Joint and Williams (1985) compute that the demands of herbivores could only be met on the assumption that they can take particles in the 1- to 5-μm range. Microplankton consumption in the DCM in the Celtic Sea has independently been computed as removing 13–42% of the standing stock of autotrophs daily. In the western English Channel two chaetognaths, like the two species of *Calanus*, partition the water column: *Sagitta elegans* occupies the cold, lower layer and *S. setosa* the upper layer. None of these findings are at variance with observations of vertical profiles in summer in the central North Sea or in the FLEX box in the northern North Sea, although here *C. finmarchicus* is dominant.

Tidal and shelf-break fronts are observable throughout this region as linear enhancements of the chlorophyll field (Holligan, 1981), though it is not simple to separate chlorophyll and turbidity signatures in the CZCS images. The central and northern North Sea have clearer water during the summer and winter trimesters than in the spring and autumn, suggesting that the chlorophyll signal is at least an important part of the whole since this pattern is compatible with the occurrence of spring and autumn blooms. Additionally, in this region it has been shown that the neritic zone within a coastal front (that occurs almost universally along coastlines) may, in fact, represent areas of greatly enhanced chlorophyll. This is the case along the southern coast of Ireland where a neritic zone of cooler water, confined within a coastal front, is enriched (8–10 mg chl m^{-3} compared with <1 mg chl m^{-3} offshore).

The pelagic ecosystem of the Baltic is very complex (Segerstråle, 1957). In the same water body there is a gradient from estuarine organisms in the west (Kattegat: salinity, 10–15 ppt) to lacustrine organisms in the east (Gulf of Bothnia: 2 or 3 ppt). The glacial relict copepod *Limnocalanus grimaldii* occurs in >97% of plankton tows in the Gulf of Bothnia and in <10% in the Belt Sea east of Denmark where the salinity gradient is sharpest. The marine medusa *Aurelia aurita* is abundant in the western Baltic but occurs rarely and does not reproduce in the Gulf of Finland. The glacial relict mysid (*Mysis relicta*) occurs in deep water. Brackish water copepods (*Eurytemora hirundoides* and *Acartia bifilosa*) and cladocera (*Bosmina maritima*) are an important component of the fauna, with the latter organism at times being the most abundant planktonic crustacean in summer and an important food for herring. Other cladocera (*Podon* and *Evadne*) and rotifers are especially important in the gulfs in summer. Large cells of the phytoplankton are dominated by diatoms and Cyanophycae, especially *Nodularia*, *Anabaena*, and *Aphanizomeno*.

In the Landsort Deep, in the main basin of the Baltic Sea, the mesozooplankton exhibit some resource partitioning in a water column having a pycnocline at 70–80 m, associated with a nutricline and (presumably) a DCM (Ackefors, 1966). Small copepods sort themselves by depth horizons, with the abundant species having their maximum biomass at the following horizons: *Acartia* sp., 15 m; *Temora longicornis*, 20–25 m; *P. elongatus*, 50–100 m.

In the Belt Sea, a seasonal study (Smetacek *et al.*, 1984) shows a very typical cycle. A spring bloom occurs in shallow water, utilizing nitrate that has accumulated in winter, during which large-scale sinking of plant cells occurs. This bloom is followed by an early summer population maximum of herbivorous zooplankton and consequently very little sedimentation of plant cells. By mid-summer a complex food web has developed which is based largely on regenerated ammonium. Finally, in the autumn there is a bloom closely resembling the spring bloom, but in this case it uses nitrate which has accumulated in stagnant subthermocline water during the summer.

Synopsis

Winter mixing carries the pycnocline to the bottom from midshelf shorewards and to moderate depths seawards of this line (Fig. 7.16). Primary production rate has an extended summer maximum, matched by herbivore abundance, whereas CZCS data on chlorophyll accumulation are unreliable, showing a late winter maximum.

Eastern (Canary) Coastal Province (CNRY)

Extent of the Province

The Eastern (Canary) Coastal Province comprises the southerly coastal flow of the eastern boundary current of the North Atlantic from Portugal to Senegal, its seaward boundary being the convergent front at the outer limit of the zone of anticyclonic curl of wind stress approximately 200–400 km offshore (see Chapter 4 for a general discussion of the wind field in eastern boundary currents and its effects on circulation). The province includes the field of eddies associated with this convergence.

Continental Shelf Topography and Tidal and Shelf-Edge Fronts

The continental shelf of this province is the most extensive of any eastern boundary current, with the widest areas occurring in the Cape Ghir–Cape Jubi and the Cape Bojador–Cape Blanc bights. Here, the 200-m contour lies 100–120 km offshore, whereas off the capes themselves it narrows to 20–30 km. In the extreme south, the shelf widens again progressively southwards from Cape Verde to the Bissagos archipelago. The Arguin Bank just south of Cape Blanc at 20°N requires special attention. The persistent hydrographic features of the Canary Current are locked to the topography of the continental shelf, and localities where strong upwelling occurs are determined by the alignment of the coastline.

Defining Characteristics of Regional Oceanography

Broad equatorward flow along the eastern limb of the North Atlantic subtropical gyre is seasonally continuous from Portugal to Senegal but is usually termed the Canary Current only along the African coast, where velocities are higher and flow is more consistent; it is convenient to use the term Canary Current System to refer to the flow from Cape Finisterre (43°N) to where it separates definitively from the coast near Cape Blanc (21°N). Further south, it is also convenient to include the region of more complex circulation off

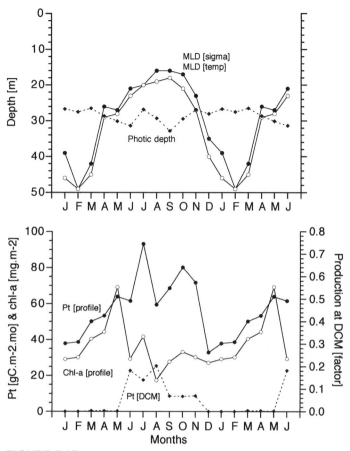

FIGURE 7.17 CNRY: characteristic seasonal cycles of monthly averaged mixed layer and photic depths, chlorophyll at the surface, and rate of primary production, both depth-integrated and at the DCM. Data sources are discussed in Chapter 1.

Mauretania and Senegal, including Cape Verde at 15°N and extending to the Bissagos Islands (11°N).

The Canary Current System is synonymous with the coastal boundary of the temperate subtropical region of the eastern North Atlantic and is the area over which interaction between oceanic flow and continental topography occurs. This coastal boundary zone comprises offshore and inshore parts (Mittelstaedt, 1983, 1991); the inshore zone, within which upwelling occurs, is restricted to the continental shelf, whereas the offshore zone, within which interaction occurs between upwelling and oceanic conditions, occupies the area above the continental slope. This, of course, represents the characteristic situation for all eastern boundary currents.

South of Cape Blanc, water originating in the NEC recirculates through the poleward bifurcation of the NECC into a poleward compensation flow along the coast (Mittelstaedt, 1991), and this circulation requires a large offshore semipermanent anticyclonic gyre. Poleward flow also occurs at the coast in semipermanent eddies in the bight south of Cape Ghir (31°N), at Cape Jubi (27°N), and wherever mesoscale cyclonic eddies (100–300 km) form on the landward side of the main flow of the offshore current (Fedoseev, 1970), even as far north as the coast of Morocco.

The wind field over the Canary Current System is typical of those over eastern boundary currents (Bakun and Nelson, 1991). Zero wind-stress curl coincides with the offshore zone of maximum equatorward wind velocity and with the main flow of the offshore equatorward current. Landwards, cyclonic curl strengthens upwelling, whereas offshore, anticyclonic curl forces a convergent front. This wind field interacts with the intertropical convergence zone in boreal summer and autumn at about 15°N.

Cyclonic curl maxima are associated at the coast with the major upwelling centers discussed below, whereas some anticyclonic curl in the coastal winds may occur south of Cape Finisterre in northwest Spain as a local response to the topography of that coast. A persistent lobe of anticyclonic curl also interrupts the coastal cyclonic curl field between Cape Bojador and Cape Sim in autumn and winter, though strong upwelling occurs here during other seasons (Bakun and Nelson, 1991). The wind stress at the sea surface is more intense here than in some of the other major upwelling systems. The mean annual alongshore component of the wind stress off northwest Africa is -1.5 dyn cm^{-2} compared to -0.5 dyn cm^{-2} off Oregon or $+0.8$ dyn cm^{-2} off Peru (Barber and Smith, 1981). Recall that positive wind stress in the Southern Hemisphere represents equatorward stress, equivalent to negative values in the Northern Hemisphere; recall also that alongshore wind stress divided by the Coriolis force is an index of coastal upwelling, and that wind stress itself is proportional to the square of the wind speed.

It is the seasonal meridional migration of the trade wind system, and hence of the velocity and curl maxima in the coastal regime, that drives the basic seasonal cycle of upwelling, which consequently is most intense in boreal summer (June–October) off Portugal and in boreal winter (January–May) off Senegal. In the central region from 20 to 25°N, upwelling is more continuous (Wooster et al., 1976; Mittelstaedt, 1991). Imposed on this seasonal cycle are local event-scale changes of wind strength and direction, which for a few days may produce or suppress upwelling (Van Camp et al., 1991) and thus radically change the surface thermal field within a short period.

The climatic seasonal mean fields of temperature and density (e.g., Mittelstaedt, 1991) indicate that upwelling occurs continuously along sections of the coast (e.g., the persistent elongated cell of cool water along the Moroccan coast), but with small topographically locked cells of greater intensity. The instantaneous situation, revealed in satellite-derived fields of temperature and pigments, is quite different: Cool filaments exhibiting strong vorticity, mostly rooted in the persistent coastal upwelling cells (Wooster et al., 1976), extend over and beyond the shelf, especially in the central region around 30–35°N and in the vicinity of Cape Blanc at 20–22°N (Van Camp et al., 1991).

Where the Canary Current detaches from the coast, the convergence zone between it and poleward flow from the south is characterized by the persistent giant filament often observed seawards from Cape Blanc at about 21°. This is independent of local winds and appears to be forced by convergence between the Canary Current after its separation from the slope and the poleward countercurrent, which is forced to loop to the southwest on encountering the Canary Current flow and to form a quasi-permanent cyclonic eddy. Furthermore, as suggested by Gabric *et al.* (1993), when upwelling-favorable winds are sustained on this shelf, the focus of upwelling at the base of the giant filament may progress offshore until it is situated above the shelf break so that oceanic water is entrained to the surface.

Wind stress off northwest Africa forces deeper mixing than off Peru or in the California Current because of relatively higher wind stress and also because of the different density structure of this eastern boundary current which lacks, for example, a light surface layer of low-salinity water. In addition, the winds are upwelling favorable for almost all the year off Africa but are favorable for only half the year off Oregon. Off northwest Africa, mixing frequently reaches to the bottom and this has consequences for the accumulation of the organic material formed in the euphotic zone during upwelling blooms.

The dimensions of the continental shelf define the inshore part of the coastal boundary province, whereas the seaward edge of the upwelled water mass takes a complex form in the surface temperature field (and as seen in satellite imagery). In hydrographic sections, this edge appears as a prograde front at which upwelled water may move downward on the inshore side, whereas upwelling of water occurs on the seaward side. A thermal discontinuity in the region of 17 or 18°C, apparently defining aged upwelled water, can be identified in many thermal images (e.g., Mittelstaedt, 1991).

The Canary Islands lie about 100–300 km offshore and may induce a counterclockwise meander of the offshore part of the coastal boundary zone around the archipelago at about the 2000-m isobath, a feature that then influences the surface temperature field.

Between Cape Blanc and Cape Verde, the surface countercurrent represents the inshore limbs of mesoscale eddies. In this situation, a divergence occurs on the seawards side of the poleward flow, with weak upwelling, whereas convergence occurs on its inshore side. Therefore, we may also expect coastal and offshore loci of upwelling, and this appears to be confirmed by observation. Tidal currents, and tidal mixing on the shelf, are relatively weak throughout the region, so continental shelf tidal fronts are not to be expected. However, domes or ridges (bourrelets froids) of well-mixed bottom water occur in upwelling foci, as they do on the Armorican shelf off western Europe (see Northeast Atlantic Shelves Province). Especially between Cape Bojador and Cape Barbas, upwelled NACW water forms cells of shelf bottom water, within which nutrient levels may be unusually high (Hughes and Fiuza, 1982; Minas *et al.*, 1982).

Actual locations of stronger (or more frequent) than average upwelling, forced by cyclonicity in the coastal wind field, depend principally on the orientation of the coast and on bottom topography, both where shallow banks occur off capes and where submarine canyons cross the shelf. The principal foci for upwelling are the following:

• Cape Ortegal/Cape Finisterra (42 to 43°N): Though upwelling occurs in summer along the whole Galician coast, and within the mouths of the rias, it is most intense where the coast turns southerly between these two capes (Fraga,

1981), and here intensification occurs because of the conjunction of differing subsurface water masses. Some confusion is possible in color images between algal production due to upwelling south of Cape Finisterre and production within the adjacent coastal rias due to river discharge of nutrients.

• Cape St. Vincent (37°N): Upwelling behind this southwest-jutting cape is topographically reinforced, strong, and extends along the coast of the Algarve, where red tide blooms are frequent accessories to upwelling events. Off Huelva, a front often extends in a southwesterly direction and marks the limit of the upwelling process.

• Cape Bedouzza (33°N), Cape Ghir (31°N), and Cape Jubi (28°N): All produce intense local upwelling, peaking in frequency and intensity during summer. Upwelling is strong because of favorable angles between the coastline and the direction of the trade winds (Mittelstaedt, 1991). Discharge from rivers draining from the Atlas Mountains may cause some difficulty of interpretation along this sector of the coast.

• Cape Bojador (26°N) in Western Sahara: An upwelling focus in both spring and summer. From this cape south to Cape Blanc, the upwelling season is longest. Perhaps onshore flow may restrict the spread of the upwelled water and restrict it to a rather narrow zone along the coast (Mittelstaedt, 1991).

• Cape Blanc (21°N): A major upwelling focus where greatest intensity and frequency of upwelling is during the first half of the year; its effects frequently extend sufficiently southward that this and the next two sites to the south become a single feature.

• Arguin Bank (20°N): Occupies the bight to the south of Cape Blanc and is a special case. There is a strong excess of evaporation over precipitation in the shallow water around this biologically rich area of shallows and offshore banks. Episodic cascades of the dense water formed on the bank down the continental slope are replaced almost instantly with cool recently upwelled water, bringing nutrients to the shallows (Van Camp *et al.*, 1991).

• Cape Timiris (19°N): This is a special case because a shelf-crossing submarine canyon focuses the upwelling flow here (Herbland and Voituriez, 1982; Roy, 199), and an immensely productive, plume-like ecosystem develops like that off Peru at 15°S (see Chapter 9, Humboldt Current Coastal Province). The persistent giant cold filament along the convergent front at the Canary Current separation is a feature of this region.

• Cape Verde (15°N): The most southerly focus of upwelling, which extends to about 13°N off southern Senegal. Upwelling is restricted to winter months and the region is entirely under the influence of the NECC and of tropical water in the second half of the year. Farther to the south, the coast bends eastwards into the Gulf of Guinea and a nonupwelling, tropical coastal regime is met (see Guinea Current Coastal Province).

Biological Response and Regional Ecology

Although the dominant algal blooms in this coastal province are forced by the upwelling processes described previously, a spring bloom must also occur in the northern part of the province, from northern Spain to northern Morocco, some months prior to the onset of the summer coastal upwelling season. Otherwise, mixed layer nutrients are controlled by wind-driven variations in vertical transport, whereas the dimension of upwelling cells is smaller than the area over which appropriate wind stress is applied. Water depth restricts upwelling to a band only 10–20 km wide (Barber and Smith, 1981).

There is a meridional gradient in the consequences of upwelling for nutrient availability: South Atlantic Central water that is upwelled south of Cape Blanc is relatively nutrient rich ($NO_3 = 14$–20 μM), whereas from Cape Blanc to the Iberian peninsula relatively nutrient-poor North Atlantic Central water and Mediterranean water upwells ($NO_3 = 6$–9 μM). A complex front between the two regimes occurs off Cape Blanc so that the South Atlantic Central water (SACW) may at times be available as source water for upwelling as far north as Cape Barbas (Minas *et al.*, 1982). Though different levels of nutrients occur in the two water masses, their N/Si ratios are similar. High nutrient levels at the surface to the southwest of Cape Blanc are probably the effect of the mesoscale cyclonic dome that occurs persistently. Barber and Smith (1981) remind us that we should think not only about new production fueled by upwelled nitrate in eastern boundary currents but also about nitrogen regenerated there as NH^4. Because of the episodic nature of upwelling pulses, and blooms, much of the organic matter produced sinks unconsumed to the sediments and here, in the Canary Current, the relatively high wind stress and strong equatorward and cross-shelf currents prevent accumulation and ensure that benthic utilization and remineralization rates are high. A further consequence of the relatively high wind stress and deep wind mixing in the northwest African upwelling cells is weak near-surface stratification during the immediate postupwelling production phase of each episode; compared to the other eastern boundary currents, accumulating cells will more frequently be mixed down, even below the euphotic zone, and their concentration diluted. One of the findings of the CUEA project in which Oregon, Peru, and northwestern Africa were compared was that because of high wind stress, primary production rates are relatively lower in this province than off the western coasts of America (Barber and Smith, 1981).

The giant filament off Mauretania, reaching as much as 450 km offshore, may appear as a persistent chlorophyll feature or it may be seen principally in the temperature field. To explain the appearance of a chlorophyll feature so far offshore, Gabric *et al.* (1993) proposed a mechanism for the conservation of nutrients in the upwelled water as it advected to the west. The algal cells seeding the upwelling must be entrained from below the photic layer when upwelling occurs over the shelf break: In this event, a delay occurs before a bloom can be initiated, as the cells undergo physiological conditioning to a photic environment.

An intuitive and simple negative relationship between chlorophyll and temperature is apparent in the few satellite images for which both fields are available (Van Camp *et al.*, 1991), similar to the relationship established in the South African ocean color and upwelling experiment (Shannon, 1985; Lutjeharms *et al.*, 1985). The chlorophyll values derived from ship data and CZCS images (e.g., Minas *et al.*, 1982) are in good agreement, though larger areas with surface values >20 mg m^{-3} appear in satellite images than might have been expected from ship data. It should be noted that in some cases these are where river discharge is significant, for instance, to the north of Senegal and Gambia where river plumes are transported in the poleward counterflows

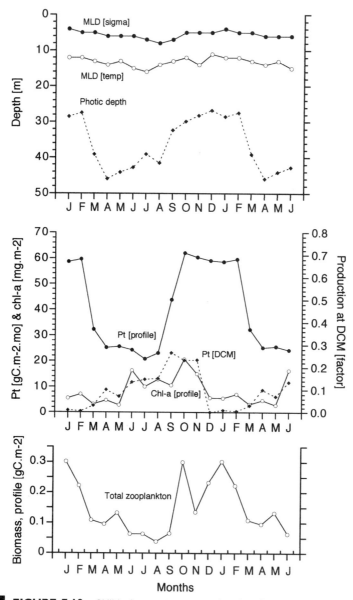

FIGURE 7.18   GUIN: characteristic seasonal cycles of monthly averaged mixed layer and photic depths, chlorophyll at the surface, and rate of primary production, both depth-integrated and at the DCM. Zooplankton data from Longhurst (1983); other data sources are discussed in Chapter 1.

close to the coast south of Cape Verde and Cape Blanc (Fig. 9 of Van Camp *et al.*, 1991).

Most of the current information about transfers within the pelagic ecosystem for the Canary Current is at the diatom–copepod level. Copepods are distributed differentially according to water depth. At 20°N, over the shallow Arguin Bank, the dominant species are *Acartia clausi*, *Temora turbinata*, *On-*

caea spp., and other small forms, whereas along the outer edge of the bank, over deep water, there are concentrations of larger forms, especially *Calanoides carinatus, Centropages chierchiae, Rhincalanus nasutus, Pleuromamma borealis,* and *Metridia lucens*. We should especially note the occurrence of *Calanoides,* a specialist of low-latitude upwelling systems. Although it does not appear to have been described, we may expect by analogy with what occurs elsewhere that during nonupwelling periods, the population of *Calanoides carinatus* will descend here into deep water off-shelf as a population dominated by stage C5. The other species are typical warm-water, open-ocean forms of which the last two are strong diel migrants. There is also a spatial separation between copepod and salp dominance: In, and inshore of, the coastal countercurrent at 17–19°N, copepods dominate, whereas seawards of the divergence at the outer edge of the countercurrent, salps dominate.

Over deep water, the mesozooplankton profile closely resembles those for the eastern tropical Pacific. Biomass is highest in the upper 100 m (not well specified in profiles from this region) and a secondary maximum at 200–400 m represents diel migrants at their daytime residence depths. Over shallow water, in the same region, a sharp chlorophyll maximum at about 10–15 m containing many diatoms was associated with maximum abundance of a wide range of genera: *Corycaeus, Oithona, Oncaea, Euterpina, Muggaeia, Oikopleura,* and others.

Because this province lies between the Atlantic and the Sahara desert, a special characteristic of the wind regime is the dust-laden nature of the northeast trades wherever their back trajectory lies over the desert (Hastenrath, 1985); these Harmattan winds bear a heavy burden of mineral dust over the ocean from Morocco to the Gulf of Guinea and may deposit as much as 25 g m^{-2} $year^{-1}$ at the sea surface off Senegal. This deposition is among the heaviest anywhere in the oceans and is similar to deposition of loess clays in the northwestern Pacific; such eolian deposits may have regional biological significance and may be a major source of turbidity in inshore waters.

Synopsis

This integration does not represent a single upwelling cell (Fig. 7.17). Rather, all upwelling cells have shorter upwelling seasons; this integration best fits the central region from 20 to 25°N. Upwelling and wind mixing modify the shoal pycnocline, and the relationship between productivity and chlorophyll accumulation suggests rapid buildup of herbivore consumption after the production rate increases.

Guinea Current Coastal Province (GUIN)

Extent of the Province

GUIN extends from the mouth of the Gambia River eastwards along the coastline of tropical West Africa to the vicinity of Cape Frio (17°S) at the Angola/Namibia border. Along the Guinea coast the seawards boundary is simply the convergence between the Guinea Current and the northern stream of the SEC (at 2–3°N 4°W); along the coast from Cameroon to the Angola Bight, *faute de mieux* (see below) we should have to take the line of the Angola/

Benguela Front, though this (enclosing the large offshore Angola cyclonic gyre) really carries us too far seawards from the coastal boundary zone. For convenience, then, an arbitrary line along 6–8°W will suffice.

Continental Shelf Topography and Tidal and Shelf-Edge Fronts

The continental shelf of tropical West Africa is narrow (about 25 km) except from the Bissagos Islands to Sherbro Island (a distance of about 1000 km), where a width of about 150 km is reached. Other minor extensions occur off Ghana and at the large island of Fernando Po. The shelf is generally flat and the coastline straight; inshore, muddy deposits dominate at depths shoaler than the permanent thermocline, below which the bottom is sandy with linear fossil coral banks. Two major rivers open onto the shelf: the Niger, through its wide deltaic region into the Bight of Biafra, and the Congo, through a narrow mouth over the Congo canyon which cuts across the shelf. Rainfall is exceptionally heavy in the Bight of Biafra (associated with the terrain of the Cameroon mountain and the high island of Fernando Po) and also from Liberia to Guinea-Bissau, where many small rivers and creeks open onto the western part of the shelf. Reef corals exist only on Annobon, the most seawards of the Bight of Biafra islands; otherwise, the coastline is fringed with sandy beaches and mangrove vegetation where there is protection from surf. Surface effects of biological enhancement at shelf-edge fronts have been observed along the Guinea coast (Longhurst and Pauly, 1987).

The demarcation between mixed and stratified regions on the shelf is relatively simple. The pycnocline meets the shelf everywhere at about 25–45 m, except during upwelling episodes (see below), so that inshore of this line the water column is mixed, and seawards it is stratified. This demarcation has important biological consequences for benthos and demersal fish because it clearly divides inshore, warm-water communities from offshore, cool-water forms.

Defining Characteristics of Regional Oceanography

This coast must be dealt with in two sections: (i) the Guinea coast, from Gambia east to Cameroon in the corner of the Bight of Biafra, and (ii) the Central African coast, from Cameroon to southern Angola at about 15°S. We have a great deal more information about the Guinea coast than the Central African coast, much of it from the colonial period.

This province lies between Cap Blanc (Senegal) and Cape Frio (Angola) where the Canary and Benguela Currents diverge from the coastline. The Guinea Current continues the flow of the NECC along the whole of the Guinea coast with some seasonal changes in strength and constancy; flow along the Central African coast is much more variable, though tropical surface water (27–28°C) normally occurs to about 15°S along the coast of Angola (Shannon *et al.*, 1987).

Guinea Coast

The very shallow flow of the Guinea Current streams eastward above a more constant, westward countercurrent. Should the flow of the Guinea Current weaken sufficiently, the countercurrent may shoal so that surface flow is reversed. During the boreal summer when the thermocline of the whole eastern tropical Atlantic shoals (see Eastern Tropical Atlantic Province) and when, moreover, it slopes upwards toward the coast, the westward countercurrent is most likely to surface (Verstraete, 1992). A sharp pycnocline lies below a shallow (30 m) mixed layer (27°C) which in boreal summer is almost eroded by seasonal wind stress so that surface temperatures fall to 24 or 25°C and the isotherms for 19–23°C are widely spaced. Seasonal cloud cover modifies the cycle of solar irradiance, which varies from about 400 g cal cm^{-1} day^{-1} in boreal winter to about 250 g cal cm^{-1} day^{-1} in boreal summer. The seasonal sea-surface temperature minimum of June and July is thus a product both of thermocline erosion and of reduced irradiance.

In addition, the surface water masses are under the influence of the alternation of dry and wet seasons. During the June–September rainy season, surface salinity is reduced, especially in the Bight of Biafra, by the extremely heavy rainfall on the adjacent Cameroon coastal massif. Lutjeharms and Meeuwis (1987) described as an "upwelling cell" the region of cool surface water seen in satellite thermal imagery between Fernando Po and the coast, mostly in March to August, during the period of greatest rainfall on the Cameroons. This interpretation must be incorrect, however, since these "cells" are supposed to occur over an area of unusually shallow and muddy continental shelf. River discharges are often visible in CZCS images as chlorophyll-like plumes, especially in the Guinea–Sierra Leone and Nigeria–Cameroon sections of the coast. During the dry, Harmattan season (January–March), significant falls of eolian dust occur at sea in the western part of the region (see Chapter 4) and a secondary, less evaporative cooling of sea-surface temperature occurs.

From eastern Liberia to western Nigeria, the shoaling of the thermocline is sufficiently strong that it constitutes coastal upwelling. This process is highly variable between years and is strongest along the coastline of Ghana and the Ivory Coast, contiguous with the "Dahomey Gap" in the forests ashore, where coastal upwelling occurs especially frequently off Cape Palmas (7°W) and Cape Three Points (2°W). Though the general cooling associated with the shoaling and eroding of the thermocline occurs each year, significant upwelling occurs only sporadically except off this central region. Upwelling, even when it occurs, is not a continuous process but a series of isolated events, each lasting days to weeks in duration (Berrit, 1961; Longhurst, 1964; Philander, 1979; Servain *et al.*, 1982; Houghton and Colin, 1986; Verstraete, 1992).

There are two components to these upwelling events. Impulsively strengthened zonal wind stress during boreal summer in the western tropical Atlantic not only tilts the thermocline so as to shoal in the east (see Eastern Tropical Atlantic Province) but also induces equatorially trapped, eastward-propagating Kelvin waves which, on reaching the African coast, are diverted along the coast as Rossby waves contrary to the Guinea Current. These waves travel westward along the Guinea coast, doming the thermocline and nutricline as they pass (especially where steep or protruding topography is encountered). Their passage forces local upwelling events that may cause the westward undercurrent to surface, and the events can be tracked westward along the coast using individual coastal station data. Though local wind stress is usually of the correct sign to generate coastal Ekman divergence, it is of quite insufficient strength to induce the observed vertical movement that occurs down to 500 m at the start of the cooling season.

Central African Coast

Here the coastal oceanography has not been fully investigated, and the current streams are rather variable. Centered at about 13°S, 5°E, the large cyclonic gyre of the Angola Dome forces southward subsurface transport along the coast on its eastern side into the shallow subsurface Angola coastal current. This stream joins the coastal flow to the south, originating near Cape Lopez (1°S), which carries EUC water. Surface wind drift over the whole of the Congo–Angola Bight is toward the northwest, especially during the boreal summer between Cape Lopez and Luanda. This suggests that if the surface wind drift weakens sufficiently, the coastal current will appear to be reversed at the coast, as it is off Nigeria. South of Cape Lopez, coastal upwelling cells occur from Cape Lopez to near Lobito (11°S). These are not as well documented as those along the Guinea coast, but they occur during the same season and are presumed to be similarly forced—by Rossby waves propagating southward along the shelf edge (Lutjeharms and Meeuwis, 1987; Verstraete, 1992; Voituriez and Herbland, 1982).

Biological Response and Regional Ecology

The shoal thermocline typical of this province is very sharp and overlies a nitracline. The depth of 1% irradiance is usually between 15 and 45 m off Nigeria; this province is characterized by a vertical arrangement of properties typical of oligotrophic situations. During boreal summer, studies at coastal stations from Sierra Leone to Nigeria recorded an increase in mixed layer chlorophyll and in phytoplankton cell numbers and a decrease in water clarity consistent with a bloom during this season (Bainbridge, 1960a; Longhurst, 1964). The nitracline off the Central African coast shoals into the upper 20 m during the same season, when a seasonal chlorophyll maximum also occurs (Binet, 1983). The FOCAL meridional section at 4°W (Oudot, 1987) also demonstrates the upsloping of thermocline, nutricline, and DCM toward the coast in July, with a shoreward increase in chlorophyll values.

The CZCS images confirm that the boreal summer bloom is a reality, that it is very variable between years, and that it is differentiated from the equatorial bloom by a zone of blue water. The same images also confirm the existence, especially in the central Guinea coast, of persistent and localized upwelling cells whose biological effects have been observed since the early years of oceanographic research off West Africa, and interest in them continues today (e.g., Bainbridge, 1960b; Mensah, 1974; Houghton and Mensah, 1978; Sevrin-Reyssac, 1993).

In some of the larger embayments, such as the wide Sierra Leone estuary where tidal streams are strong, the seasonal production cycle is the reverse of that of the open sea, and algal cells are probably limited by light rather than by nutrients. Phytoplankton growth (about 850 mg C m^{-2} day^{-1}) is maximal in the boreal winter dry season, when estuarine water is relatively clear, and it is greatly reduced (<75 mg C m^{-2} day^{-1}) in the summer wet season when inshore water carries a heavy load of silt. Furthermore, during the dry season, blooms occur preferentially at neap rather than spring tides and thus when the inshore water is clearest. The growth of large diatoms in inshore blooms in this province supports a large population of phytoplankton-feeding clupeids (*Ethmalosa dorsalis*) in the inshore waters.

Planktonic consumption of inshore blooms is small relative to their production. While in the wet season, at Sierra Leone, the apparent mesozooplankton demand is computed at about 91% of daily primary production, in the dry season only 5% of the bloom is apparently required to sustain the zooplankton. Because in this case the demand of the benthic community is even less (about 1%), much of the dry season bloom must be exported from the system or buried in the coastal mud banks as organic matter.

We have some information on offshore plankton profiles which suggests that the epiplankton is quite typical of tropical seas; at night, plankton biomass in the 0–25 m zone is 5–10 times greater than below 25 m, whereas by day biomass is equally distributed from 0 to 50 m. Genera performing these minor diel migrations are chiefly *Nanocalanus, Paracalanus,* and *Neocalanus.* Except over water >500 m deep, the major diel migrants *Metridia* and *Pleuromamma* are absent.

Much attention has been given to the ecology of *Calanoides carinatus,* a large species that also appears in low-latitude upwelling situations. Here it has been recorded over the outer shelf during boreal summer, and preferentially during upwelling episodes, from Sierra Leone to Angola (Petit and Courties, 1976). Its fidelity to this ecology is shown by the fact that I observed it abundantly on the shelf off Nigeria only during an upwelling episode, in a region where upwelling episodes are very rare. Off Ghana, Mensah (1974) showed that only C5's survived the end of the upwelling season, during which three or four generations appeared, with the development from egg to adult taking only 14–18 days. During the nonupwelling season, the population is maintained as C5's estivating beyond the shelf below 500 m. Because this period of 7 or 8 months could not be sustained on the oil sac accumulated during feeding on diatoms in the upwelling period, Mensah investigated feeding at depth and found this species could also capture bacterioplankton. In this instar, maxillary setules have an interval of 1–4 μm. The same cycle has been observed on the Central African shelf off Zaire by Binet (1976), who found six to eight generations during the upwelling period but only C5's at 800 m off the shelf during the remainder of the year.

These large calanoid copepods are a principal food of the West African sardine (*Sardinella aurita*) whose appearance off Ghana and Ivory Coast (Cury and Roy, 1987) coincides with the rise of *Calanoides* to the surface waters.

Synopsis

Seasonally invariant very shoal pycnocline, always significantly shallower than photic depth, is a negative correlate of chlorophyll biomass (Fig. 7.18). Primary production rate is light limited and is enhanced during the dry, cloudless boreal winter, when herbivore biomass also accumulates. Chlorophyll accumulation responds both to seasonal changes in primary production rate and to the seasonal increase in herbivores.

Benguela Current Coastal Province (BENG)

Extent of the Province

BENG extends along the coastline of Africa from the Cape of Good Hope north to Cape Frio (18°S). Excluded is the retroflection region of the Agulhas Current south and southwest of the Cape which is included in the EAFR coastal province. Included is the offshore eddy field of the Benguela Current.

Continental Shelf Topography and Tidal and Shelf-Edge Fronts

The continental shelf off southwestern Africa is wider than that off Angola, especially south of 20°S, where it reaches 150 km off the Orange River. It also includes anomalously deep zones; double shelf breaks are common at about 150–200 and 300–500, m respectively. Shelf topography plays a significant role in determining the locations of surface features observable in the circulation and the width of the shelf is usually <10 km at salient capes. The coastline

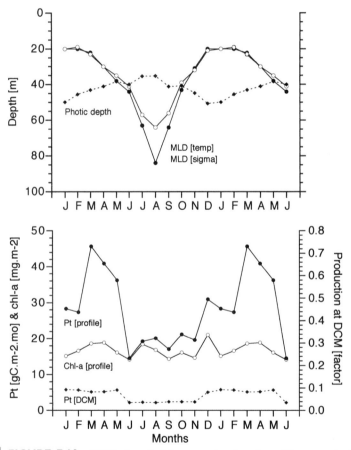

FIGURE 7.19   BENG: characteristic seasonal cycles of monthly averaged mixed layer and photic depths, chlorophyll at the surface, and rate of primary production, both depth-integrated and at the DCM. Data sources are discussed in Chapter 1.

is low lying and very dry, with sand dunes a dominant feature and rocky out-crops at prominent capes. A shelf-edge frontal feature is discussed as part of the cross-shelf circulation that results from upwelling at the coast.

Defining Characteristics of Regional Oceanography

The Benguela Current is the dominant feature of the oceanography off the west coast of southern Africa and forms the eastern limb of flow around the subtropical gyre of the South Atlantic. Flow into the Benguela originates in the southern limb of the subtropical gyre (as the South Atlantic Current), to which are added frequent, but irregular, intrusions of Indian Ocean water from the Agulhas retroflection feature (see Chapter 8 and Fig. 8.8).

This current system is complex but well explored and recently reviewed (e.g., Hart and Currie, 1960; Shannon, 1985; Chapman and Shannon, 1985). Equatorwards flow is carried in a series of jet currents (strongest in summer) and eddies associated with coastal and continental shelf topography, together with a series of wind-forced upwelling cells (Shannon, 1985; Lutjeharms and Meeuwis, 1987). The broad equatorward flow, parallel to the coast, has two components that diverge off Cape Columbine at 33°S: (i) Aligned with the shelf edge is a meandering offshore core of maximum velocity, the shelf edge jet, and (ii) the coastal Columbine jet that is transformed progressively northwards into variable coastal flow. Flow of the shelf edge jet is topographically steered and lies consistently over the continental slope from the cape as far north as Luderitz (27°S), where separation from the slope first occurs, and seawards of which place there is pronounced divergence of flow.

The separation of the Benguela Current westward into the northern limb of the southern subtropical gyral circulation is completed in the vicinity of Cape Frio at 18°S, where the warm water of the seasonally shifting Angola–Benguela Front is encountered (Shannon *et al.*, 1987). The offshore flow passes through a gap in the deep topography of the Walvis Ridge at 20°S. The Angola–Benguela Front represents the convergence zone between the poleward surface flow of tropical surface water in the Angola Current and the equatorward flow of cool Benguela water and is marked at the surface by a temperature gradient reaching 4°C per 1° latitude. Tongue-like mesoscale intrusions of warm water occur across the front, especially in austral summer, and the front also marks the transition between a strongly stratified tropical water mass to the north and weak stratification to the south; here also there is a strong discontinuity in equatorward wind stress.

The wind stress field forcing the surface flow of the Benguela Current is typical of that overlying eastern boundary currents and is forced by interaction between the atmospheric anticyclone of the southeast Atlantic and the desert coast of southwest Africa. Day–night alternation of the land–sea breeze system is strongly developed (as it is off all coasts backed by major deserts), with the sea breeze having a fetch of 100–150 km. As Shannon (1985) comments, this can be expected to modulate diel processes in the upwelling system.

As also occurs in the Canary Current, there is significant deposition at the sea surface of aerosol particulates; off southwestern Africa eolian dust events are associated with offshore katabatic ("berg") winds from the deserts of south-

western Africa, which blow strongly for periods of several days, often simultaneously over as much as 1500 km of the coastline between 20- and 30°S
(Shannon, 1985). Such offshore wind events result in a succession of small
upwelling cells, restricted to <10 km from the coast.

Winds are perennially favorable to upwelling south of about 15°S. During
austral summer the ITCZ loops far south over the continent and seasonality in
upwelling at the coast is associated with the seasonal march of the wind system.
Over the Benguela, cyclonic wind stress curl lies inshore of the wind velocity
maximum, itself lying above the velocity maximum of the offshore current.

The coastal zone of cyclonic wind stress originates between Cape Point and
Cape Columbine (33–35°S), where it is narrow, and widens to the north until
at about 20°S the field becomes very patchy, especially in austral winter and
spring. In austral summer and fall, cyclonic curl dominates as far as 700 km
from the coast at this latitude. Cyclonic curl maxima and upwelling-favorable
winds occur near Luderitz and Cape Frio, though in this northern region the
seasonal difference in upwelling winds is less than that in the southern region
of the Benguela and restricted to slight enhancement of upwelling in the northern region in April and May. In the northern region, upwelling-favorable winds
are persistent not only seasonally but also on shorter time scales and this persistence distinguishes the northern from the southern Benguela region, where
upwelling is modulated at periods of about 1 week by relaxation or reversal of
coastal winds forced by the passage of cyclones to the south of the Cape
peninsula.

Coastal upwelling in the Benguela Current, as in other eastern boundary
currents (Mooers *et al.,* 1978), results in cross-shelf circulation and a set of
fronts related to rotary motion within upwelling cells. Near the shelf break,
there is frequently a prominent front at which upwelled water sinks on the
shoreward side and offshore water diverges on the seaward side. Where upwelling is especially vigorous, it may be circumscribed by a surface front closer
inshore than the shelf-break convergence/divergence front. The shelf-break feature appears at the sea surface as a strong thermal front (20–30 km wide) that
can be traced north to Cape Frio, though the frontal zone becomes more diffuse
toward the north. The outer part of the divergence zone is associated with
surface slicks above internal waves (Shannon, 1985). Particularly pronounced
divergence zones, and jet currents, occur off Cape Columbine and the Cape
peninsula. Beyond the shelf-break front, an eddy field marks the edge of the
clear waters of the subtropical gyre of the South Atlantic. Cold-core eddies
appear as isolated areas of chlorophyll enhancement along this frontal zone.

The upwelling centers identified by Shannon (1985) and Lutjeharms and
Meeuwis (1987) are not always observable as entities in the CZCS images—
isolated from the filaments and eddies resulting from general regional upwelling. Coincidentally, one of the most striking of the early SeaWiFS images is of
an upwelling center below cloudless skies representing an atmospheric high-
pressure cell. Nelson and Hutchings (1983) suggest that upwelling occurs preferentially where the continental shelf is narrowest, off salient capes, and this is
confirmed by the satellite imagery. For practical purposes, the upwelling centers
can be grouped as follows:

• Cape peninsula (34°S) and Cape Columbine (32°S): Upwelling is largely restricted to the austral summer trimester (December–February), and the average temperature of upwelled water in both cells is about 17°C. The upwelling cells off Cape Columbine may generate cold filaments that extend 500–700 km offshore and have a longshore spacing of 200–300 km, whereas those off the Cape peninsula are about half that dimension. Each of these upwelling cells is frequently bounded seawards by a jet current, and off Cape Columbine this jet is isolated from the shelf edge jet by a clear divergence zone (Shannon, 1985). In summer, upwelling may extend around the Cape peninsula, as far east as Cape Agulhas (35°S 20°E).

• Hondeklip Bay, Namaqualand (29°S): A regional maximum of cyclonic wind stress curl and upwelling occurs persistently in the bight south of the Orange River; the shelf here is narrow and deep, and the source of water brought to the surface in coastal upwelling is also deep. A persistent mesoscale cold filament or tongue originates at this upwelling center. Upwelling response to wind stress is slower than that in the regions to the south; maximum upwelling occurs in October–December, and minimum upwelling occurs in May–July.

• Luderitz to Walvis Bay (24–22°S): This is the most intense upwelling area, especially off Luderitz itself, and forms a major environmental barrier in the Benguela Current system. Upwelling occurs throughout the year, being seen in >80% of images scanned by Lutjeharms and Meeuwis (1987) over a 3-year period. Upwelled water is quite cool (mean temperature is 16.5°C at Luderitz and 17°C in the Namaqua cell) because the shelf break is exceptionally deep and the source of upwelled water lies at 200–300 m. The seaward extent of upwelling is greatest here, reaching almost 300 km off Luderitz, and the upwelling event scale is long compared with that of regions to the south. Shannon (1985) comments that here the event scale resembles Peru and northwest Africa rather than Oregon or the southern Benguela.

• Central (21–23°) and northern Namibia, especially Cape Frio (19–17°S): Upwelling is continuous throughout the year but strongest in the austral winter trimester (June–August), with some extension into spring (September–November). During summer the water column is sufficiently stratified so as to constrain upwelling somewhat. Upwelled water is 16°C (mean of upwelling cells is 19.5°C at Cunene and 18.5°C in the Namibia cell). The continental shelf off central Namibia is narrower and shoaler than to the south, with a shelf break at 140 m about 100 km offshore. For this reason, very cool water is unavailable as the source of upwelling. The upwelling cells in this sector extend about 150 km offshore. There may be a persistent convergence between these upwelling cells and those off Luderitz to the south.

Biological Response and Regional Ecology

While the upwelling centers described above are the foci of enhanced algal growth and surface chlorophyll, the Benguela Current is sufficiently dynamic that upwelling occurs seasonally essentially the whole length of the coast. Three zones can be recognized in each upwelling cell: (i) an inshore zone where nitrate and silicate are supplied and utilized by an algal bloom; (ii) a middle zone where one of these nutrients becomes limiting; and (iii) an outer zone, associated with the meandering equatorward jet current, where algal growth is limited. Sinking of aggregates and particles occurs principally in the middle zone, and if this is over shelf depths, remineralization will be important. In general, the chlorophyll-enhanced water follows the outline of the shelf edge except where it narrows to the south of Luderitz, and during upwelling episodes there is often a near-coastal band of relatively clear, recently upwelled water, with the maximum chlorophyll concentration of the inshore zone in midshelf about 15–25 km offshore. Chlorophyll biomass after upwelling forms a near-surface maximum (Shannon and Pillar, 1986). In the quiet periods between upwelling events, red tides are common all along the Benguela.

The filaments and eddies (chlorophyll values of 3–10 mg chl m^{-3}) of up-welled water may merge with the next upwelling center so that it is frequently impossible to separate the production from each individual center in satellite color images. In this event, the whole of the Benguela Current appears as a coastal band of high chlorophyll. An offshore enhancement of chlorophyll which occurs persistently over the Orange Banks (150 km offshore) may be associated with a shelf-break divergence zone, but it has not clearly been demonstrated that shelf-break upwelling occurs in the Benguela (Shannon, 1985).

Oxygen-deficient (<2.0 ml liter^{-1}) and anoxic water may be brought to the surface by coastal upwelling. The main southeast Atlantic subsurface oxygen minimum layer lies north of the Benguela–Angola front between Cape Frio and Luanda associated with the cyclonic gyre of the Angola Dome. However, off Namibia, especially Walvis Bay, and off Luderitz, as well as to the north of Cape Columbine, oxygen deficiency may be induced *in situ* in shelf waters (Chapman and Shannon, 1985). Especially at Walvis Bay, anoxia of upwelled water (which may even contain H_2S) has significant consequences for shelf biota, including fish kills. The formation of oxygen-deficient water over the shelf is associated with the production and remineralization of organic material from phytoplankton and fecal pellets formed after the advection of nutrients into the photic layer in upwelling cells (Chapman and Shannon, 1985).

The biological response to upwelling differs significantly between the northern and southern Benguelan regions. Comparing Namibia with the Cape Province, Chapman and Shannon (1985) suggest that in the south upwelling occurs mostly as a series of short pulses, whereas in the north it is more continuous during the upwelling season. That the southern upwelling should be more dominated by diatom growth than the north is probably a consequence of the short time scale of the pulses of upwelling which occur. In the north, coupling between herbivore biomass and plant growth is close and this results in rapid silica depletion, as diatom frustules are incorporated in copepod fecal pellets and contribute to the large deposits of biogenic silica that occur in the shelf deposits off Namibia. In the south, more silica is recycled in the water column. In these circumstances, it is probable that nitrate is the limiting nutrient in the south, off the Cape peninsula, whereas silica limits plant growth in the north, off Namibia, as is reported for the Peru–Chile Current (see Chapter 9, Humboldt Current Coastal Province).

There is little endemism among the biota of the Benguela system: Mesozoo-plankton diversity is low relative to the Agulhas Current to the east and the herbivore species could be predicted: *Calanoides carinatus* is the dominant copepod, associated with species of *Paracalanus, Clausocalanus, Centropages,* and the rest. There is a different emphasis in the dominant large algal cells between neritic (diatoms) and offshore (dinoflagellates) and between northern (dinoflagellates) and southern (diatoms) regions. Similarly, for the mesozoo-plankton, offshore and to the north, there is high diversity and low biomass with many gelatinous organisms, whereas inshore and to the south, low diversity and high biomass are dominated by crustaceans.

The Benguela current is the habitat—as are all the other eastern boundary currents—of important pelagic clupeid stocks which consume calanoid copepods and diatoms. Clupeids often exist as species pairs (one large and one small) in eastern boundary currents and here *Sardinops ocellatus* and *Engraulis capensis* share the habitat. Feeding concentrations of sardines regularly occur along the upwelling front over the outer continental shelf.

Synopsis

The mean condition does not correspond with any of the upwelling cells, but shows the generally deeper austral winter pycnocline and the effects of austral summer upwelling on primary production rate (Fig. 7.19). Chlorophyll accumulation is generally very weak when averaged over the whole province monthly.

Northwest Atlantic Shelves Province (NWCS)

Extent of the Province

The NWCS province comprises the continental shelf and slope water from Florida to the Grand Banks of Newfoundland. Its offshore boundary is the north wall (the inshore wall) of the Gulf Stream from Florida to Newfoundland; arbitrary lines are drawn around the east and north sides of the Grand Banks.

Continental Shelf Topography and Tidal and Shelf-Edge Fronts

The continental shelf of eastern North America shoals progressively southward from the anomalously deep shelf off Labrador until, beyond Cape Hatteras, it is represented by the coastal plain ashore. It is of great width off Newfoundland and encloses two semi-enclosed shallow seas, the Gulf of St. Lawrence and Gulf of Maine. If it was desirable to subdivide this province, it would be logical to do so as follows: (i) shelf from Newfoundland to southwest Nova Scotia, (ii) Gulf of St. Lawrence, (iii) Gulf of Maine and Bay of Fundy, (iv) shelf from Georges Bank to Long Island, (v) Middle Atlantic Bight, and (vi) Florida shallows south of Cape Hatteras. It will be useful in this account to recognize this subdivision for some of the material.

The topography of this shelf includes the following characteristic features:

1. There are several unique deep (<200 m) basins at midshelf off Nova Scotia and in the Gulf of Maine which have consequences for pelagic ecology.
2. Below Saguenay, the St. Lawrence river follows a deep, ice-cut trough which continues across the Gulf and passes out across the shelf through the Cabot Strait (45–47°N) as the Laurentian Channel, which is both deep (<400 m) and wide (<100 km); this has consequences for circulation and upwelling.

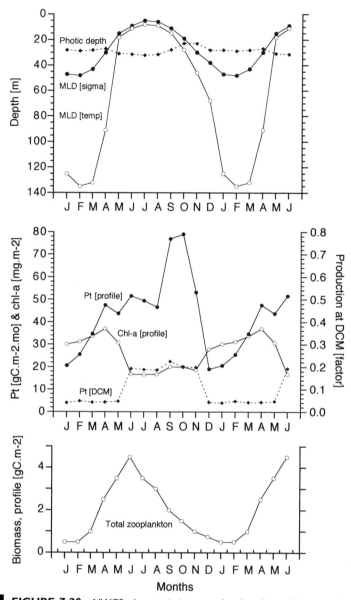

FIGURE 7.20 NWCS: characteristic seasonal cycles of monthly averaged mixed layer and photic depths, chlorophyll at the surface, and rate of primary production, both depth-integrated and at the DCM. Zooplankton data are from Davis (1987); other data sources are discussed in Chapter 1.

3. Georges Bank is a large, shallow offshore feature surrounded by deeper water and bearing parallel sand ridges that are sufficiently shoal as to exhibit features at the sea surface. The Grand Banks of Newfoundland are also largely separated from the coast by the deeper water of the Avalon Channel.

Though the water column over the shelf is thoroughly mixed in winter, and in places it is ice covered, stratification is established in summer except where tidal stirring in shallow water maintains a mixed water column. In winter, only the thermal front between shelf and slope water is not broken down. It is here and in the NECS province that these phenomena have been most closely studied (see Chapter 2).

Tidally mixed areas with peripheral tidal fronts occur off southwest Nova Scotia, and the central well-mixed region of Georges Bank is separated from surrounding stratified water in summer across strong tidal fronts around the slope of the bank (Loder and Greenburg, 1986). Other tidally mixed areas occur in the Gulf of Maine, across the mouth of the Bay of Fundy, around Grand Manan Island, and also on the Nantucket Shoals south of Cape Cod (Townsend, 1992). Within the Gulf of St. Lawrence, tidal mixing and tidal fronts occur mainly in the channels between Anticosti, Prince Edward Island, and the mainland and also around the Magdalen Islands (Pingree and Griffiths, 1980). Further south, in the Middle Atlantic Bight (Long Island to Cape Hatteras) tidal fronts will parallel the coast separating an inner neritic and an outer open shelf zone. South of Cape Hatteras, the water column over the shelf is permanently stratified.

A shelf-break front reflects the instability between the shelf and slope water masses and is a consistent feature of this province (Smith and Petrie, 1982). This is periodically modified as warm-core eddies which have been shed from the meandering Gulf Stream flood over the outer shelf, creating instabilities and jet currents in the shelf water. At the same time, bursts of topographic Rossby waves propagate across the shelf as internal bores or solitary waves (Smith and Sandstrom, 1986); this mechanism is responsible for significant transport of nutrient-replete slope water up onto the shelf. Where the Gulf Stream lies over the slope in the southern part of the province, energetic frontal eddies more frequently extend into shallow water.

Defining Characteristics of Regional Oceanography

Circulation is extremely dynamic and variable in this province, which is not a surprising observation in the light of the extraordinary conjunction of polar and subtropical conditions. In winter, the pack ice of the Gulf of St. Lawrence is only about 300 km from the north wall of the Gulf Steam and hence of the 18°C surface water of the Sargasso Sea. I believe this is a unique condition. The pelagic ecology of this province is determined by three factors: the conjunction and partial mixing of water masses of very different characteristics, the partitioning of the area between stratified and mixed situations in summer, and the formation of sea ice in winter.

The influence on the circulation of the province of several different water masses, whose boundaries and properties are observable in satellite SST images, requires some initial description. These surface water types are particularly important to distinguish from Cape Cod northwards. The Labrador Current transports low-salinity, Arctic water south along the continental shelf and slope of Labrador, and separation into an inshore and offshore component occurs. The inshore component passes along the Avalon Channel off the east coast of Newfoundland and contributes to shelf water formation on the Grand Banks;

some also enters the Gulf of St. Lawrence to the west of Newfoundland, through the Straits of Belle Isle. The offshore component of the Labrador Current passes around the continental slope of the outer Banks and contributes to the slope water lying between the Gulf Steam and the waters of the continental shelf. The slope water which is formed by this conjunction lies as a distinguishable zone between the Gulf Stream and shelf water bodies.

Circulation in the Gulf of St. Lawrence passes around a complex eddy field, which is induced by the several large islands and the complex coastline, and is influenced by the discharge of the St. Lawrence River, which reaches 30×10^{-3} m^{-3} sec^{-1} at times of peak flow in summer. This flow, together with Labrador Current water that has entered directly from the north through the Straits of Belle Isle, drives an outward stream at the surface through the Cabot Strait at all times. Ice cover forms during winter in the Labrador Current and over much of the Gulf of St. Lawrence, and during the period of spring breakup, fractured pack ice is transported south out of the Gulf and along the eastern coast of Nova Scotia. The low-salinity, cold water which passes out onto the Scotia shelf from the Gulf of St. Lawrence moves southwards along the coast with the shelf water mass, circulates around the Gulf of Maine, rounds Georges Bank, and so passes southwards past Cape Cod to form the shelf water of the Middle Atlantic Bight. South of Cape Hatteras, the Gulf Stream lies along the break of slope quite close to the coast, and inshore there is modified subtropical surface water.

The outer shelf is characterized by meandering southerly flow throughout the year, though this is highly sensitive to variability in sustained wind direction. With sustained longshore wind from the southwest, local upwelling cells form at many points along the coast. Coastal upwelling occurs off southwest Nova Scotia persistently, and also intermittently along the southeast-facing coast when wind forcing is from a favorable direction.

Biological Response and Regional Ecology

In terms of integrated seasonal primary production and (at the other end of the trophic series) the potential production of demersal and pelagic fish, this is one of the most productive regions of the ocean.

Analysis of CZCS imagery in this province is confounded by much suspended sediment in shallow water, but a spring bloom clearly occurs as a component of the general North Atlantic bloom. The Continuous Plankton Recorder (CPR) database (Colebrook, 1979) differentiates seasonal cycles for (i) the Grand Banks (strong spring blooms, weak autumn blooms, and copepods peak in late summer), (ii) the Nova Scotia shelf (weaker and more similar spring and autumn blooms and copepods without seasonal peak), and (iii) the Cape Cod region (algal abundance barely higher in summer than winter and few copepods in winter and many during the rest of year). It must be remembered that the CPR algal data are based only on large cells, but even so these data suggest that the physical processes discussed previously do determine the regional pattern of algal growth dynamics in the province and that the areas of chlorophyll enhancement observed in the CZCS images during summer can mostly be related to frontal dynamics and to coastal upwelling.

However, in the Gulf of St. Lawrence, the Bay of Fundy, the offshore Georges Bank and along the neritic zone from Maine southwards, it is currently impracticable to separate algal blooms from coastal turbidity in the CZCS images: We must await images from the SeaWiFS sensors. The Bay of Fundy is a special case of a macrotidal embayment, similar to the Bristol Channel–Severn Estuary system and recalling the tropical macrotidal estuary of the Sierra Leone river (see Guinea Current Coastal Province), in that the regional cycle of algal blooms is sustained only in the outer, less turbid region. Further in, where turbidity at spring tides results in a photic zone <1 m deep, the spring–neaps tidal cycle dominates the ecological succession.

Since the entire Gulf of St. Lawrence has an estuarine circulation, the discharge of brackish water at the surface toward the open ocean induces a landward flow of deep water through the Cabot Straits, resulting in the upwelling of nutrient-rich, high-salinity water at the head of the Laurentian Channel north of the Gaspé peninsula (Dickie and Trites, 1983). A spring bloom occurs over the whole western Gulf as soon as ice cover is cleared and algal growth is no longer light limited, having particularly high values (<200 mg C m^{-2} hr^{-1}) in the region of upwelling and also further down the Gulf where it is reinforced by strong tidal mixing over the Magdalen shallows, and in the region of the tidal front between Anticosti Island and the Gaspé peninsula.

In the Gulf of Maine (see Townsend, 1992) upwelling occurs at many small estuarine fronts but has been held to be a major factor in overall productivity, with the water in the deep basins being constantly renewed by episodic pumping from deep slope water sources. However, the effects of tidal mixing dominate the distribution and strength of algal blooms: The principles discussed already for the northeast Atlantic continental shelf (see Northeast Atlantic Shelves Province) apply here. Blooms are initially light limited in mixed areas, then persist longer because of constant recharging of nutrients by benthic regeneration. In stratified areas, an oligotrophic profile rapidly forms as the initial nutrient charge is utilized. In winter, very dense aggregations of C. finmarchicus and the euphausiid M. norvegica occur in the deep troughs and basins both in the Gulf of Maine and on the eastern Canadian shelf. As in the northeastern Atlantic, these are essentially deep-water expatriates over continental shelves— though they may still be the dominant organisms of their trophic group there— and the shelf basins are to be regarded as deep-water refuges. In winter, when seasonal migration would have carried them to depths of 1000 m in the open ocean, extremely dense aggregations of oceanic copepods and euphausiids occur close to the bottom in these basins, whose depths exceed 200 m.

Most areas south of Cape Cod have a rather variable seasonal cycle, with highest values tending to occur toward the end of summer, and only the slope water has a classical seasonal cycle of productivity with a single spring peak. The summer DCM is progressively deeper seawards, following the sloping thermocline. There are seasonal changes in the vertical distribution of mesoplankton herbivores relative to the chlorophyll feature: During summer, there is a coincidence in depth between chlorophyll and larger herbivore abundance, whereas at the end of summer the copepods begin to descend into the lower layer below the pycnocline. On, and south of, Georges Bank C. finmarchicus is scarce, and the dominant copepods, both numerically and by biomass, are Oithona spp. which are distributed throughout the water column even in stratified water.

A spring bloom over the deep water of the Gulf of Maine can precede water column stratification (Townsend *et al.*, 1992), and this may be a general but overlooked process wherever spring blooms occur over deeper water. The principle involved appears to be that deep penetration of light in a clear, winter-mixed water column may support cell growth rates that overcome the vertical excursion rates induced by wind stress at the surface, especially in calm weather. This process is assisted by the nature of the spring-bloom cells in some cases, including gelatinous colonies and diatom chains with very low sinking rates. In such situations, it is also possible—as has been demonstrated for the Arabian Sea—that the presence of a layer of light and heat-absorbing chlorophyll may be a contributing factor in the eventual establishment of stratification. This phenomenon should be watched for in the open North Atlantic.

Cross-frontal transfer of nutrients plays a significant role in supplying nitrate to support the summer bloom in the tidally mixed parts of Georges Bank (Loder and Platt, 1985; Horne *et al.*, 1989). A leading candidate for cross-frontal transport is the residual circulation associated with tidal current interactions. The computed rate of this transfer is sufficient to support observed new production at the front itself (67% of total production is nitrate based) and within the enclosed mixed area on the bank, where only 27% of production is fueled by nitrate in summer.

In the Mid-Atlantic Bight, once again, the same principles apply, though the mixed region is linear and neritic so that chlorophyll increases regularly seawards: At the same time, the percentage of total production contributed by nanoplankton also increases seawards. Consistently higher rates of primary production occur in the New York Bight, associated with discharge of urban sewage into an area of strong tidal fronts and mixing. The shallower, northern part of Georges Bank has only slightly lower productivity, and primary production in these two areas is 25% higher than in any other part of the Mid-Atlantic Bight (O'Reilly and Busch, 1984). Below the warm, clear surface water over the shelf of the Mid-Atlantic Bight in summer there is a cold, well-lit, chlorophyll-rich layer of bottom water; this originates in the Gulf of Maine and on the Nova Scotia shelf when shallow summer stratification isolates the cooler bottom water. Advected off-shelf at Cape Hatteras and vertically mixed in the Gulf Stream flow, this water body contributes significantly to the nutrient budget of the Gulf Stream (Wood, 1996).

In the South Atlantic Bight (Carolina Capes region) the western, cyclonic edge of the Gulf Stream generates wave-like perturbations that propagate northwards along the continental edge to induce upwelling of cool nutrient-rich Atlantic Central water at topographic features, especially during summer (Blanton *et al.*, 1981). Because the upwelled water mass has great clarity, subsurface algal blooms may occur across the shelf in the bottom water, and these upwelled water bodies support dense concentrations of small copepods. Thus, abundance of most mesoplankton species on the Florida shelf increases progressively toward deeper water which has high particulate load (Paffenhöfer, 1983).

The shelf-break front off eastern Canada is a site of bioaccumulation as well as injection of nutrients onto the shelf. In spring, both copepod abundance and chlorophyll are three or four times higher than anywhere over the shelf. Vertical transport mechanisms at the front are sufficient to account for the supply of nutrients from below, whereas convergence at the front between shelf and slope water is sufficient to account for the lateral aggregation of mesozoo-plankton. This increase of productivity at the shelf break, at least along the Grand Banks and Scotia shelf, is evidently associated with benthic enrichment. It is here that the most consistent populations of demersal fish, especially large gadoids, occur.

Warm-core rings, though carrying Sargasso Sea and Gulf Stream water masses from the slope water regime onto the continental shelf between Cape Cod and Newfoundland, may be associated with algal blooms and surface chlorophyll enhancement if deep convection occurs in them at the end of winter. Some upwelling may also occur during the final decay phase of the ring (McCarthy and Nevins, 1986). However, probably more important are the biological consequences of an interaction between a warm-core ring and the continental edge; a filament of cool shelf or slope water may be entrained around the warm-core ring and then carry cool-water biota from the New England or Nova Scotia shelf far offshore.

Synopsis

Winter mixing carries the pycnocline to the bottom from midshelf shore-wards and to moderate depths seawards of this line (Fig. 7.20). Primary pro-duction rate begins to increase in late winter and is matched for the first few months by chlorophyll accumulation, which subsequently is reversed by con-sumption as the herbivore populations increase.

Guianas Coastal Province (GUIA)

Extent of the Province

Guianas Coastal Province (GUIA) extends from Cape de Sao Roque (5°S) in Brazil to Trinidad (10°N) within the offshore limits of the North Brazil (or Guiana) Current along the northern coast of Brazil.

Continental Shelf Topography and Tidal and Shelf-Edge Fronts

There is a significant width of continental shelf along the whole province, widest in the region of the Amazon mouth where it reaches about 200 km. Many rivers open into this province, draining the rain forest regions of conti-nental South America: the Para, Amazon, Essequibo, and Orinoco rivers are the largest. The coastal regime is consequently highly turbid, especially north of the Amazon mouth, and migratory mud banks occur and migrate along the coast (Wells, 1983).

The Amazon is a special case (well described by Geyer et $al.$, 1996) and discharges, as is well-known, more water than any other river and lies directly on the equator. At peak flow (May) this discharge is about 220,000 m^{-3} sec^{-1} and at minimum flow (November) about 100,000 m^{-3} sec^{-1}. Such a strong flow prevents seawater from entering the river mouth so that a strong salinity front occurs about 100 km offshore, coincident with a turbidity front and above the transverse shoals formed by deposition of sedimentary material. Even at this distance offshore, water depth is only about 15 m. Although tidal streams are

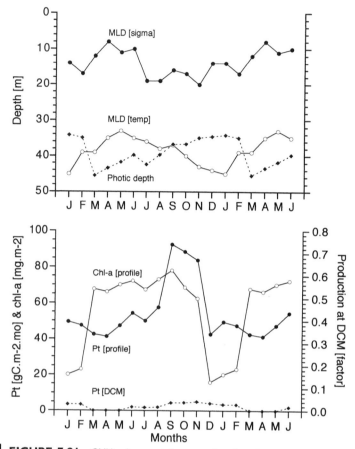

FIGURE 7.21 GUIA: characteristic seasonal cycles of monthly averaged mixed layer and photic depths, chlorophyll at the surface, and rate of primary production, both depth-integrated and at the DCM. Data sources are discussed in Chapter 1.

amplified by the wide shelf so that cross-shelf velocities reach 200 cm sec^{-1} in the frontal zone, the suspended load in the bottom water is so great as to dampen vertical mixing by tidally induced bottom stress. The shallow, near-surface, low-salinity plume passes northwest along the coast, the form taken by its bounding salinity front being wind-dependent. Southeast winds induce a "fast" plume of brackish water to head along the coast just offshore of the transverse shoal, whereas northeast wind produces a "slow" plume that balloons out across the shelf, with very little northward transport from the river mouths.

In the dynamics of the plume, there is another interesting example of the consequence of the low value of the Coriolis acceleration at low latitudes (see Chapter 3). In mid- to high latitudes, the Coriolis force would induce the plume to turn anticyclonically across the shelf (clockwise, that is, in the Northern Hemisphere), but here the dynamics of the plume respond directly to the generally landward wind stress, and this force is sufficient to maintain a general flow along the coast to the northwest.

Defining Characteristics of Regional Oceanography

The North Brazil Current (NBC) transports water into the North Atlantic from the South Atlantic subtropical gyre delivered by the transequatorial flow of the SEC. The flow of the NBC northwards along the coastline of South America is augmented by confluence with the weaker NEC north of the equator (Boisvert, 1967). The NBC is a warm, shallow (<200 m) stream and is one of the sources of water masses for the equatorial undercurrent (Metcalf and Stalcup, 1967); the coastwise flow, though always to the north, alternates seasonally between a coastal jet (boreal summer) and a series of coastal eddies (boreal winter). Flow in the NBC is modified from the equator northwards by the Amazon discharge of fresh water (at 6×10^{12} m^{-3} year^{-1}, this is the greatest discharge from any single river) and also from the Orinoco (1.0×10^{12} m^{-3} year^{-1}) at 8°N.

During boreal summer, the surge of the southeast trades into the western Atlantic transforms the NBC, north of Recife, into a western jet (now transporting 35 Sv compared with 10 Sv at the end of boreal winter) with high current speeds so that, although the flow is topographically locked to the continental edge, it becomes unstable between the mouths of the Amazon and Orinoco with a double periodicity (25–40 and 60–90 days). Part of the flow which overlies the continental slope is retroflected to the east into the open Atlantic in a series of very large, persistent eddies at 4°N as the Amazon eddy, and at 8°N as the Demerara eddy (Bruce et al., 1985), forming at the 60- to 90-day periodicity. It is convenient to consider these eddies, which form a prominent feature extending so far seaward, as part of the western tropical Atlantic province (see Western Tropical Atlantic Province). It has been suggested that the eddies represent offshore transport of the Amazon discharge plume, but it seems more likely that the Amazon discharge is mostly transported coastwise and, as originally suggested by Ryther et al. (1967), in a series of lenses or eddies (300–500 km diameter) which migrate northwards along the coast and that in this case it must be mostly slope water which is retroflected seawards.

Drogued, satellite-tracked drifters have been deployed in the NBC just offshore of the Amazon shelf and near the river mouths. Most of these buoys subsequently track northwest along the shelf, with those nearest the shelf edge describing a series of tight anticyclonic eddies as they approached and entered the Caribbean near Trinidad. Several passed at a shallow angle toward the edge of the shelf and entered the retroflection eddies before passing into the North Atlantic Equatorial Current and heading off toward the Caribbean or Gulf of Mexico. One, placed in the NBC over the slope off the Amazon, entered a retroflection eddy and passed from there into the NECC before reaching the coast of West Africa.

Coastal upwelling occurs from northern Brazil to Surinam, as indicated by the upslope of isopycnals toward the coast and the occurrence of relatively cool water in the upper 10 m close inshore. In fact, this is a predictable consequence of the development of a coastal wind-driven jet current along the western boundary of an ocean and the prevailing vector of coastal wind stress during the boreal summer, when southeast trades are strongest (Gibbs, 1980).

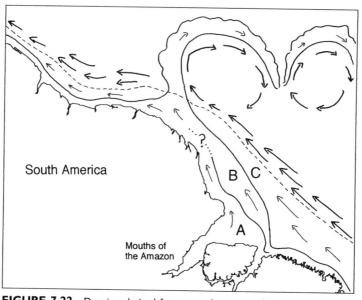

FIGURE 7.22 Drawing, derived from several sources, of the plume of the Amazon River (more properly, of the three major rivers draining the Amazon basin) in boreal summer when the jet current along the coast is most strongly developed. Represented is the inshore zone of high turbidity (A, >100 mg/liter of suspended material) and the outer zone of maximum chlorophyll (B, 2–4 mg m^{-3}); on the outer shelf of the Amazon (C), these two values are around 1.0 and 0.1, respectively. The two retroflecting eddies are persistent features at this season, carrying apparently high chlorophyll only on their northern sides, and are of controversial origin

Biological Response and Regional Ecology

This region is one of the more difficult for the separation of chlorophyll from turbidity in satellite images, especially where the turbidity field is influenced by the Amazon and Orinoco discharges. At the Amazon mouth, turbidity inshore is sufficiently high that algal growth is light limited and the fresh water discharged at the mouth has a nitrate concentration of about 15 μM, which is small compared with the discharge of other great rivers: The Yangtse, Yellow, and Mississippi rivers have freshwater end members in the range 50–100 μm nitrate. Perhaps when the Amazon rain forest is entirely cleared, and the basin reduced largely to agriculture as in the other examples, its freshwater drainage will come to have similarly high nitrate values.

Three ecological zones (Fig. 7.22) are recognized in recent studies of the ecology of the Amazon shelf (Smith and DeMaster, 1996): zone A, a coastal zone inshore of the transverse shoals, where turbidity is sufficiently high as to inhibit photosynthesis and nitrate remains high, averaging 7 μM at the surface in all seasons; zone B, along and just seawards of the turbidity front, where primary production ceases to be light limited and chlorophyll accumulates (2–4 mg chl m^{-3}) in a surface bloom while nitrate is taken up; and zone C, seawards of the chlorophyll plume, still only in about 40–50 m depth, where nitrate becomes limiting and chlorophyll values are reduced to 0.2–0.6 mg chl m^{-3} and nitrate is always <0.5 μM. It is possible that the chlorophyll accumulation in the very narrow, linear zone B is enhanced by upwelling of nitrate-replete shelf water along the salinity front (as in an estuarine circulation). Currently, there appears to be very little evidence for the availability to algal cells in zones 2 or 3 of nitrogen regenerated by benthic microbial or water column metabolism as ammonium.

The coastally trapped discharge from the Amazon, and the subsequent input from the Orinoco discharge, combine significantly to modify the ecology of the eastern shelf of Venezuela and of the Gulf of Paria (Bonilla *et al.*, 1993). Most of the nutrients discharged by the Orinoco remain in this area, where benthic regeneration is very active, so their contribution to the eastern Caribbean is problematical, though the low-salinity signal of Amazon and Orinoco water can be detected as far away as Puerto Rico. Significant amounts of nitrogen probably do not survive the passage through the coastal ecosystems and the Gulf of Paria (unless recycled nitrogen is entrained in the regional flow) so that the CZCS pigment feature so prominent in the eastern Caribbean is probably not caused by riverborne nutrients, as suggested by Müller-Karger *et al.* (1989), but rather is either a turbidity signal or is related to the coastal upwelling which occurs west of Trinidad and Margarita Island (see Carribean Province).

Synopsis

Mixed-layer depth is seasonally invariant as is CZCS-indicated chlorophyll except for a brief period of lower values in boreal winter (Fig. 7.21). The effect of freshwater discharges is seen in the permanent shallow pycnocline, whereas the deeper themocline lies at about the photic depth in this province of low water clarity.

Brazil Current Coastal Province (BRAZ)

Extent of the Province

The BRAZ province extends from the bifurcation of the flow of the SEC east of Recife at 10°S between this province and GUIA, south along the east coast of South America to about 35°S, off Argentina, where the northward flow of the Falkland Current is encountered in the important confluence region, which is discussed in the next section. The outer boundary of this rather narrow province seldom occurs beyond the 2000-m contour.

Continental Shelf Topography and Tidal and Shelf-Edge Fronts

The coastline is remarkably straight and the only topographic features of importance occur at 25°S, just north of Rio de Janeiro, where there are two prominent capes: Cape Sao Thomé and Cape Frio (not to be confused with the similarly named cape in the Benguela Province), seawards of which the shelf has a complex topography of valleys and submarine canyons. South of this latitude the continental shelf is wider and circulation more complex.

Defining Characteristics of Regional Oceanography

The Brazil Current is the weakest of the western boundary currents because 75% of the 16 Sv transported westwards across the equatorial southern Atlantic by the SEC enters the incipient NBC (headed toward the Caribbean) by transequatorial flow rather than remaining in the Southern Hemisphere subtropical gyre (Peterson and Stramma, 1991; Gordon and Greengrove, 1986).

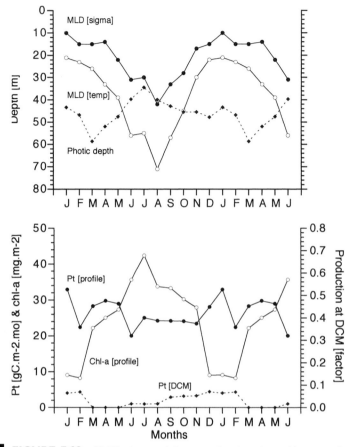

FIGURE 7.23   BRAZ: characteristic seasonal cycles of monthly averaged mixed layer and photic depths, chlorophyll at the surface, and rate of primary production, both depth-integrated and at the DCM. Data sources are discussed in Chapter 1.

South of 30°S, the Brazil Current strengthens (at about 5% per 100 km between 20 and 35°S) by entrainment from a recirculation cell lying between the continent and the Rio Grande Rise. This gyre originates at the confluence of the Brazil Current with the northward flow of the Falkland Current.

As the Brazil Current proceeds southwards, its maximum flow lies over the shelf edge but much of its transport remains across the shelf. Seasonal variation in mass transport (maximal in austral summer and minimal in winter) is directly related to the seasonal change in wind stress curl. The strongest (weakest) transport occurs when the confluence region is furthest south (north), according to Matano *et al.* (1993). The latitude of the confluence, which determines where separation of the Brazil Current from the continent occurs, is farther to the north during austral winter and spring; this may be related to the general seasonal shift of wind systems and to a seasonal meridional shift of the subtropical gyre (Peterson and Stramma, 1991). Imposed on this seasonality, there is also some short-term variability in the southward extent of the Brazil Current. When a loop which it has pushed unusually far south retreats to the north, it may be shed as a series of warm-core eddies that pass into the Antarctic Circumpolar Current (Partos and Piccolo, 1988).

Though intrusions of SACW across the shelf and shelf-edge upwelling also occur off southeastern Brazil, and on the inner shelf of San Paulo State, they have been most intensively studied at Cape Frio (22°S), where there is a large quasi-permanent offshore meander (Peterson and Stramma, 1991) so that up-welling regularly occurs east and north of the cape. This is a complex process, combining wind stress and the domination of local bottom topography by a major canyon which cuts across the shelf (Mascarenas *et al.*, 1971; Moreira da Silva, 1971). Winds from the northeast force water to upwell across the plateau that lies east of the cape and to surface either at Cape Frio itself or somewhat to the north. To the south of Cape Frio, a quasi-permanent coastal cyclonic eddy forms an eastward compensation current (Mascarenas *et al.*, 1971). Up-welling may be inhibited during periods when the Brazil Current remains for a period topographically locked to the coastline.

Upwelling–downwelling episodes occur very quickly in response to changes in wind direction, which themselves are very variable and episodic, responding to the cyclical development and collapse of the atmospheric anticy-clone east of southern Brazil. A pulse of northeast winds will very rapidly draw SACW, of 18°C and 10 μM NO_3, to the surface from about 300 m. Such episodes occur more frequently in austral summer than in winter, during which the frequent passage of cold fronts and southwest winds maintains an almost continuous downwelling, oligotrophic situation along the outer shelf.

Biological Response and Regional Ecology

The frequency and mechanism of upwelling episodes at Cape Frio are known, but the ecological consequences are only inferred (Moreira da Silva, 1971), though the physiological parameters of phytoplankton involved in up-welling blooms respond predictably to light, nutrients, and temperature (Gonzalez-Rodriguez *et al.*, 1992; Gonzalez-Rodriguez, 1994). The upwelling of SACW is episodic, and the biological response is rapid, though the upwelled water does require conditioning by superficial heating and stratification before primary production rate increases significantly.

The anticipated physiological conditioning of algal cells has also been con-firmed in this upwelling, where it has been found that the physiological status of phytoplankton cells responds to upwelling and downwelling conditions with appropriate values of the light-saturation curve. A typical upwelling event lasts from 5 to 10 days, during which successive reductions in wind speed induce surface heating so that the rate of primary production starts to increase by about Day 5 and introduces the "production phase" of the event. During this phase, the mixed layer (typically 10 m deep, 15–20°C at the surface, and 15–20 mmol NO_3) accumulates about only about 5 or 6 mg chl m^{-3} from a production rate of about 10 mg C m^{-3} hr^{-1}. Subsequently, as the system returns to a downwell-ing state, chlorophyll biomass rapidly falls by an order of magnitude, though productivity remains at about 2 or 3 mg C m^{-3} hr^{-1} for some time. This suggests that the upwelling bloom is rapidly and heavily grazed, a suggestion for which there is some experimental support. There is also some indication that the mag-

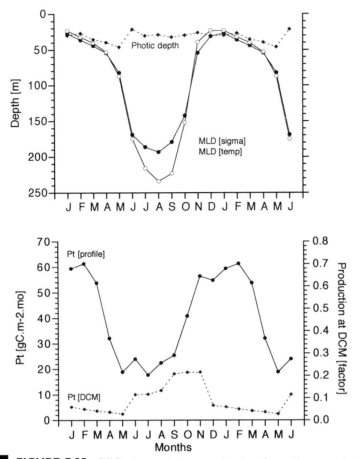

FIGURE 7.25   FKLD: characteristic seasonal cycles of monthly averaged mixed layer and photic depths, chlorophyll at the surface, and rate of primary production, both depth-integrated and at the DCM. Data sources are discussed in Chapter 1.

nitude of the production cycle may respond to the initial inoculum of cells in the upwelling parcel of water rather than to its nutrient load.

Farther polewards, at 25–30°S (or from Santos to Cabo de St. Marta), recent studies during austral summer over and beyond the shelf have confirmed the oligotrophic nature of the tropical water (0–200 m) borne by the Brazil Current and the ecological effects of intrusions of SACW across the shelf (Metzler *et al.*, 1997). These surveys (of the Brazilian COROAS expeditions) found highly variable rates of nitrate uptake by phytoplankton in this hydrographically dynamic region. Differences between inshore and oceanic profiles of production rate and chlorophyll biomass, and the concomitant physiological constants, were predictable.

The CZCS images for this coast are not very helpful and seasonal images do not appear to record the upwelling episodes at Cape Frio. The principal feature observed is the wide shelf area between about 28°S and the confluence region, probably representing both shelf processes and the effluent from the

River Plate which discharges Parana River water of very high turbidity, though constrained by a sharp turbidity front, visible in NOAA-AVHRR imagery. This feature, probably a typical shelf sea front indicating the limit of tidal stirring near the mouth of the River Plate, migrates seasonally: upstream and westwards in summer (minimum discharge) and downstream and eastwards in winter (maximum discharge).

From distributional studies of copepods, we may infer something of the regional ecology in the northern part of the current. Species lists, and the vertical distribution of species, suggest a typical warm-water assemblage with a epiplankton extending to about 75–100 m, dominated numerically by *Clausocalanus, Corycella, Calocalanus,* and *Euchaeta,* with diel migrants (*Pleuromamma*) lying from 100 to 800 m by day.

Synopsis

The data used do not capture the coastal processes described previously (Fig. 7.23). Offshore, weak winter mixing carries mixed layer depth below the photic depth for several months, yet primary production rate is almost invariant seasonally and chlorophyll accumulation is inverse to depth of mixing. This observation suggests forcing of the chlorophyll cycle by variable consumption rather than variable production rate.

Southwest Atlantic Shelves Province (FKLD)

Extent of the Province

The Southwest Atlantic Shelves Province (FKLD) comprises the Argentine shelf and Falklands plateau from Mar del Plata (38°S) to Tierra del Fuego (55°S), including the oceanographic confluence region between the Falklands and Brazil currents. The seawards boundary is perhaps best approximated to the 2000-m contour around the Falklands plateau and northwards along the Argentinian continental shelf to about 35°S, or wherever in real time the southerly flow of the Brazil Current impinges on equatorward flow.

Continental Shelf Topography and Tidal and Shelf-Edge Fronts

From the River Plate south to Tierra del Fuego, this is one of the widest and flattest continental shelves anywhere in the oceans, and because of this fact alone it is useful to separate it from the rest of the South Atlantic coastal zone. This province comprises both the Argentine–Patagonia shelf and the triangular Falklands plateau, where (at 50°S) the continental shelf widens to about 800 km and there are several major embayments on the coast (Bahia Blanca and the Gulfs of San Matias and San Jorge). South of the Falklands, deep water occurs in the bight to the east of the Straits of Magellan. A shelf-break front occurs at least along the whole north–south section of the Patagonian shelf edge from 37 to 47°S (Glorioso, 1987), separating the cool subantarctic water offshore from the warmer shelf water. Tidal dissipation rates are high and an inshore tidally mixed zone lies shorewards of the 40-m isobath, where a slight escarpment on the shelf topography may occur, and is especially well developed in relation to headlands (e.g., at the Valdes peninsula at 42°S). Seawards of the mixed area, and separated from it across typical tidal fronts, summer stratification occurs.

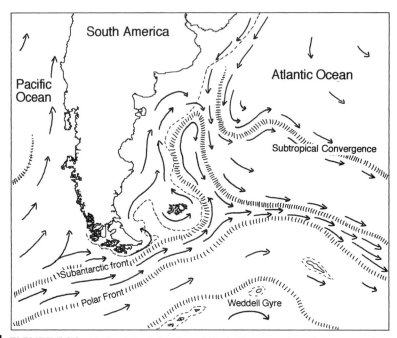

FIGURE 7.24   Idealized circulation through Drake Passage to show the interaction between circulation and topography in the conjuction zone between the Brazil and Falkland currents, emphasizing the retroflection of the subantarctic front around the Falkland plateau. The CZCS images (Fig. 1.4) frequently suggest chlorophyll enhancement along the shelf edge north of the Falklands. The 200-m isobath is shown as a dashed line, and hatched lines represent ocean frontal zones, including that at the edge of the Weddell Gyre. Note that an instantaneous satellite image of sea-surface temperature or elevation would show that each front and its associated flow is a strongly eddying feature.

Defining Characteristics of Regional Oceanography

This province has a very complex circulation that accommodates several singularities (Fig. 7.24):

1. The Subantarctic Front of the southeast Pacific Ocean, part of the flow of the Antarctic Circumpolar Current, is entrained into lower latitudes on rounding Cape Horn and flows as the subpolar (western boundary) Falkland Current.

2. The confluence of the southward Brazil Current and the northward Falkland Current at about the latitude of the River Plate requires a combined offshore flux whose location is seasonally variable between 38.5 (July–December) and 36.5°S (January–March), and it is determined by the relative strength of flux in the two currents.

3. Though it will perhaps not influence pelagic ecology of the region, I note that the water column here is as complex as anywhere in the ocean: There are seven identifiable subsurface water masses originating in the North Atlantic, the South Pacific, and the Southern Ocean.

The Falkland Current carries subantarctic water northwards along the continental slope from the latitude of Cape Horn, first around Birdwood Bank and then across the Falkland Plateau along the 1000-m isobath. Flow continues north along the continental slope to the confluence with the Brazil Current at about 35–40°N (Peterson and Whitworth, 1989). Cold-core eddies are shed, particularly across the Falkland plateau, and the more saline shelf water originating in the Brazil Current is thereby modified. Eddy generation is also especially active at the confluence of the Brazil and Falkland currents, in the general latitude of Buenos Aires, because here the Falkland Current makes an abrupt cyclonic loop and returns toward the southeast, parallel with the seawards flow of the Brazil Current. This, the Brazil/Falklands Confluence, continues eastwards across the ocean as the Subtropical Convergence and stands out in satellite imagery as a region rich in eddies.

Here, there occurs the same interaction between tidal mixing with the seasonal cycle of winter mixing and summer stratification that has been investigated principally in the NECS province. The neritic region is mixed year-round by tidal stress, whereas deeper water over the shelf is mixed in winter and stratified rapidly in spring as seasonal heat flux is accumulated within the upper 30 m of the water column. Because the shelf has so little topography, the tidal front separating the mixed from stratified water in summer runs parallel to the coast along the 40-m isobath and clearly separates coastal from outer shelf water. The latter is limited by the shelf-edge front that also runs almost uniformly parallel to the coast.

Biological Response and Regional Ecology

Though we have little direct information, there is no reason to believe that the description of processes associated with midlatitude continental shelves having active tidal fronts (e.g. NECS and NWCS provinces) should not also be generally relevant to this province. The CZCS seasonal climatology images show areas of enhanced chlorophyll concentration in spring and summer and clear water in autumn and winter. The spring bloom begins simultaneously both offshore (north of the Falklands) and in the neritic zone; the deeper water of the bight east of the Straits of Magellan appears to bloom later. In both austral spring and summer, high-chlorophyll water is advected around the recirculation gyre of the Brazil Current.

Dynamic eddying at the persistent and prominent shelf-break front that lies north of the Falkland Islands appears as a consistent linear chlorophyll feature in a high percentage of daily images, especially during the austral summer. The tidally mixed area along the inner (mixed) half of the shelf also appears as a consistent chlorophyll feature (sometimes, off the Gulf of San Jorge, with clearer water inshore) as does the shoal water around the Falkland Islands, where tidal mixing presumably also occurs.

Synopsis

The FKLD province is a typical high-latitude shelf regime—deep winter mixing to bottom over shoal water (and down to 250 m off the shelf edge)—with pycnocline illuminated only very briefly in summer (Fig. 7.25). Primary production rate tracks irradiance in spring and fall, with a broad (4-month) summer peak, corresponding to a period of high chlorophyll over the whole Falkland plateau in CZCS images.

8

THE INDIAN OCEAN

The Indian Ocean, north of the Subtropical Convergence zone, is the smallest ocean basin (about 50×106 km²) and has some special characteristics which must be accommodated when it is partitioned into ecological provinces. It is, unfortunately, rather poorly known except in one corner. This is especially unfortunate because of the relative complexity of the circulation of this ocean, forced by the seasonal reversal of the dominant wind systems.

The International Indian Ocean Expedition (IIOE) of 1959–1965 was an early version of the international multiship investigations that have since come to dominate the progress of oceanography: unfortunately, though much biology was done, very little ecology emerged and because the wide grid of stations was sampled quite irregularly there is no uniform field of values for any property. We can therefore infer only very superficially the ecological functions of the pelagic systems of the open ocean. In fact, this was expressly not the intention of the organizers; rather, the intention was to permit complete freedom of action to the scientists aboard the 11 participating ships. A centrally planned grid of stations and agreed procedures to wring answers from critical problems would have been "ideal but unrealistic," according to the principal coordinator.

Fortunately, we have come a long way since those days, and the intense and disciplined exploration of the ecological consequences of monsoon reversal off Somalia was one of the most successful investigations of the Joint Global Ocean Flux Study (JGOFS: see Smith *et al.,* 1991; Burkill *et al.,* 1993a). As will be shown, having agreed on the scientific issues, the multinational participants of the JGOFS Arabian Sea Process Study systematically examined the physical processes induced by monsoon reversal and their consequences for biological production. Therefore, paradoxically, although we remain in almost as great ignorance as before the IIOE about the ecology of the Indian Ocean as a whole, the ecology of its marginal Arabian Sea is probably as well understood as any other region.

The most important singularity of the Indian Ocean is the seasonally reversing monsoon wind system which dominates the ocean climate north of about 25°S and carried the dhows from Arabia to Zanzibar and back again "laden with sandalwood, apes and ivory" and other merchandise that we would rather forget. Reversal of monsoon winds rapidly induces reversal in both shallow and deep currents because (as noted in Chapter 3) wind stress in low latitudes induces momentum rather than mixing. Indeed, the reversal of the Somali Current and the rapid spin-up of a deep boundary jet in response to the onset of the southwest monsoon is the classical model for this phenomenon in low-latitude oceans. Any partitioning of the Indian Ocean must accommodate two seasonal circulation states.

The scheme of ecological provinces proposed here owes much to the analysis of Colborn (1975), who partitioned the Indian Ocean by reference to the monthly evolution of the mixed layer, and the gradient of the thermocline, in each of 274 "subareas," together aggregated into 40 "primary areas having distinct thermal characteristics" and these again were gathered into a smaller number of "major geographic provinces." Banse (1987) and Brock et al. (1991) both interpreted their studies of thermal structure and algal blooms in terms of Colborn's classification of the northwest Arabian Sea. For convenience of comparison, the equivalence of the provinces proposed here with Colborn's areas is noted in each section.

The monthly thermocline topography for the Indian Ocean compiled by Rao et al. (1989) follows Colborn in defining mixed layer depth solely by thermal criteria; where heavy rainfall occurs at sea (as in some parts of the monsoon provinces) or where river runoff lowers the salinity of the surface layers, a purely thermal definition of mixed layer depth is not entirely satisfactory as was pointed out by Banse (1987). Global mixed layer topography based on density criteria has been computed by Levitus (1982), but his maps have only a limited information content for the Arabian Sea. Despite all these difficulties, changes in the topography of the thermocline and in sea-surface elevation do reflect the major seasonal changes in circulation during northeast and southwest monsoons (Hastenrath, 1989).

The northeast monsoon forces weak westward flow across the whole ocean between India and the equator so that some southerly flow must occur at the coast of Africa. The Arabian Sea and Bay of Bengal are occupied by variable currents during this season. The strong, equator-crossing southwest monsoon of boreal summer reverses these flows, spinning up anticyclonic gyres in the Arabian Sea and Bay of Bengal, establishing eastward flow across the ocean north of the equator, and, most dramatically, establishing a deep, fast jet current northwards along the Somali coast. Though the response of the ocean is swift, often occurring within 1 month, there is some lag so that the extrema for ocean circulation occur in February and August.

The Coastal Zone Color Scanner (CZCS) images of the Indian Ocean confirm that the equatorial divergence (characteristic of Atlantic and Pacific Oceans) is here relatively weak and ephemeral: instead, convergent eastward flow in the intense Indian Equatorial Jet (600 km wide) occurs at the transition between southwest and northeast monsoons. This flow does not support the same biological effects as equatorial divergence in the other oceans.

The geographic scheme for six provinces proposed here is sensitive to these current reversals and also to the unusually wide range of regional conditions that exist in the Indian Ocean and its adjacent seas: the evaporative basins of the Red Sea and Persian Gulf, the dilution basin of the estuarized Bay of Bengal, the effects of the monsoon flow reversal through the Indonesian archipelago, the dynamic seasonal eutrophication of the northwest Arabian Sea, and finally, the extremely oligotrophic low-latitude, open-ocean region.

There are several possible ways in which the Arabian Sea, north of the Indian Monsoon Gyres Province (MONS) and East Africa Coastal Province (EAFR) provinces, could be partitioned using the principles proposed here. None is entirely objective, but the reasons for the choices are as follows:

• The coastal boundary along the west coast of the Indian continent from 8 to 25°N is relatively narrow and has special characteristics that make it sensible and simple to recognize this as a separate province [Western India Coastal Province (INDW)].

• From Somalia to the Gulf of Oman the southwest monsoon forces a wide zone of coastal divergence; this feature is continuous with an offshore region of Ekman suction forced by the same wind system. Both are areas of strong biological enhancement. It is unreasonable to separate these two related processes, so it is best to group part of the deep northwestern Arabian Sea together with the Somalia–Arabian coastal boundary in a single province: Arabian Sea Upwelling Region (ARAB).

• The two adjacent evaporative basins, the Red Sea and the Persian Gulf, are taken together simply as a matter of convenience and because together they make an interesting and instructive comparison [as the Red Sea, Persian Gulf Province (REDS)].

INDIAN OCEAN TRADE WIND BIOME

Indian Monsoon Gyres Province (MONS)

Extent of the Province

The MONS province extends from the hydrochemical front at 10°S (see below) north to the offshore limits of the coastal provinces. The province also includes the central Bay of Bengal and the southern part of the Arabian Sea. Thus, it comprises Colborn areas 2 and 12–21.

Continental Shelf Topography and Tidal and Shelf-Edge Fronts

Though continental shelf topography is not relevant to this province, the shallow water of the Maldive Islands aligned along 73°E, from the equator to

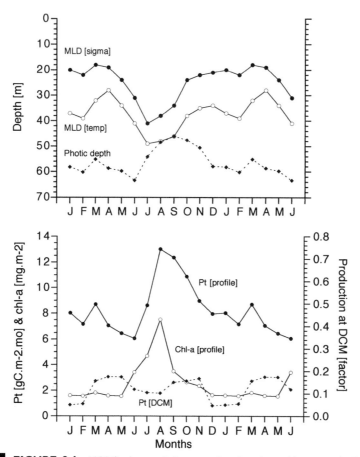

FIGURE 8.1 MONS: characteristic seasonal cycles of monthly averaged mixed layer and photic depths, chlorophyll at the surface, and rate of primary production, both depth-integrated and at the DCM.

the Indian continental shelf at 12°N, has consequences for circulation and biological processes.

Defining Characteristics of Regional Oceanography

This province is synonymous with the reversing monsoon gyre of the Indian Ocean which has high surface temperatures and relatively low salinity, as befits an extension westwards of the warm water pool of the Pacific Ocean (see Chapter 9, Western Pacific Warm Pool Province).

The monsoon gyre occupies the whole northern Indian Ocean south to the "hydrochemical front" of Wyrtki (1973a); this front lies above a zonal thermocline ridge at 10–15°S and separates high-oxygen, high-salinity regions to the south from low-oxygen, low-salinity regions to the north. This is a convergent front and lies on the equatorward flank of the perpetual westwards flow of the South Equatorial Current (SEC). In the eastern part of the province it is less well marked and I place the tropical–subtropical distinction (based on subsurface nitrate) at 20–22°S along the 110°E IIOE section of Tranter (1979) as the boundary.

When the northeast monsoon becomes established in boreal winter, and as the Arabian Sea anticyclonic gyre weakens, westwards flow across the ocean [as the Northeast Monsoon Drift or the North Equatorial Current (NEC)] is established between about 2°N and the Indian subcontinent. This flow originates in the Bay of Bengal and in water that has been transported westwards through the Indonesian archipelago; accordingly, surface salinities are low in the eastern part of the province during this season. Westward divergent flow at the equator is weaker than that in other oceans and is best developed in February and March. The southward coastal current that develops along East Africa during the northeast monsoon returns eastwards across the ocean in the South Equatorial Countercurrent (SECC) along the northern slope of the thermal ridge at 5–10°S (see Fig. 1 in Reverdin and Fieux, 1987). Convergence thus occurs at about 2–5°S between the NEC and the SECC (Wyrtki, 1979).

In boreal summer, the establishment of the southwest monsoon collapses the winter circulation and, in April and May, forces the reversal of the NEC eastwards as the Southwest Monsoon Current (SMC) from about 5°S to 10°N. The SMC is associated with a thermocline trough lying close to the equator. At the same season an anticyclonic gyre is spun up in the Arabian Sea and flow from the now-northward Somali jet passes around the eastern limb of the gyre and eastwards as the SMC along the same latitude that the NEC occupies in winter. The equatorial undercurrent of the Indian Ocean is singular in that it is not an oceanwide, permanent feature of the circulation: There is no persistent westward wind stress and thus no persistent westward wind drift to be compensated by subsurface return flow. We need concern ourselves only with the surface equatorial jet (600 km wide) that passes eastwards along the equator briefly at each transition between monsoon seasons: in April and May and in September and October.

The entire province has a permanent thermocline, usually at 30–50 m, except from Sumatra to the south of Sri Lanka where it which deepens in the southwest monsoon to 100 m (Colborn, 1975). In the area of the oceanwide zonal monsoon currents, the thermocline lies shallower (also thinner and steeper) in the west and deeper (also thicker and weaker) in the east, forced by eastwards wind stress in the equatorial region and resultant transport in the equatorial countercurrent (Hastenrath, 1989).

An important low-oxygen layer underlies the thermocline at intermediate depths through much of the eastern Arabian Sea, weakening progressively southwards toward the hydrochemical front discussed previously; it is maintained by the slow passage through these depths of southern water, already oxygen deficient, combined with very high rates of oxygen consumption resulting from high sinking rates of organic material from near-surface algal blooms (Olson *et al.*, 1993; Kamykowski and Zentara, 1990; Vinogradov and Voronina, 1961).

Because of their unique characteristics, the two northern embayments are discussed separately.

Arabian Sea Gyre

In order to understand the oceanography of the southern part of the Arabian Sea the whole gyral circulation must be noted briefly. This account should be read alongside the account of the dynamic upwelling processes that occur with monsoon reversal in the ARAB province that occupies the western and northern parts of the Arabian Sea.

At the onset of the southwest monsoon winds, the pycnocline is shallow and the profile typically oligotrophic. The mixed layer of the central part of the gyre deepens by 40–50 m as an anticyclonic gyre is spun up, with the center of the bowl lying rather centrally in the basin at about 10°N 63°E. The rate of deepening is greatest just after the onset and is associated with anticyclonic wind-stress curl. Cooling in the central Arabian Sea is caused by downward transfer of heat and evaporative heat flux. In the southeastern Arabian Sea, the mixed layer deepens by 25–35 m and cools by 2°C (Colborn, 1975; Rao, 1986, 1990; Hastenrath, 1989; Bauer et al., 1991). During the northeast monsoon, the wind maximum is a weaker, broader jet which forces only insignificant Ekman dynamics, though some downwelling is predicted (Bauer et al., 1991).

Bay of Bengal Gyre

Compared with the Arabian Sea, circulation in the Bay of Bengal is weaker and less predictable, and its response to monsoon reversal is more complex. Circulation and mixed layer depth is determined by winter cooling in the north of the Bay of Bengal and by river water (principally from the mouths of the Ganges and Irawaddy) as well as by the reversing monsoon winds and the influence of the zonal currents lying across the south of the embayment. The northern area has low surface salinities year-round so that the 33.4 isohaline lies zonally at the sea surface across the Bay of Bengal at 10–15°N. Though a gyral circulation is induced by the northeast monsoon in winter, several mesoscale cyclonic features occur persistently off the Indian and Burmese coasts and in the Andaman Sea (Rao and Sastry, 1981).

The topography of the mixed layer of the central Bay of Bengal is less predictable than in the Arabian Sea, perhaps because of undersampling. Thermal stratification is strongest in April and May because of the northern excursion of the sun and cloudless skies, though since salinity distribution largely determines mixed layer depth, the pycnocline remains rather shallow and abrupt. With the onset of the stronger winds of the southwest summer monsoon, this density structure is broken down by mixing, overturn, and cooling due to heavy cloud cover. During the same period, the anticyclonic gyral circulation breaks down, and the Bay of Bengal then contains two minor cyclonic eddies, which must include pycnocline uplift at their centers. One is in the northern part, in July–September, centered at 20°N 90°E. The other appears in October off Madras at 85°E, moves southerly, and dissipates off Sri Lanka in December (Banse, 1990a).

During the northeast monsoon, an anticyclonic gyre forms in the Bay of Bengal (Wyrtki, 1973b). This forms a shallow western jet current along the east coast of India from February to June which can be traced in the surface thermal field as a warm feature, rooted in the poleward boundary flow up the east coast of Sri Lanka and then extending, with anticyclonic curl, into the cooler water of the northeast sector of the Bay of Bengal (Legeckis, 1987). Mixed layer depths in the north of the Bay of Bengal (15–17°N) deepen by 50–75 m as the northeast monsoon season progresses (Colborn, 1975) and a thermal ridge develops, lying southwest–northeast from Madras to Burma (minimum mixed layer depths along this ridge are 30–40 m) and thus separating the Bay of Bengal circulation from an area of deeper mixing to the northwest of Sumatra: there, mixed layer depths reach 90 m during the northeast monsoon.

Biological Response and Regional Ecology

Both CZCS images and the IIOE data fields show that low surface chlorophyll values of <0.05–0.10 mg chl m^{-3} are almost totally unrelieved, except in the Arabian Sea and coastal regions. During the northeast monsoon, when chlorophyll values are at their lowest (e.g., Krey and Babenard, 1976), nitrate is absent at the surface except in the coastal boundaries, where there are some scattered patches <0.5 μM (McGill, 1973), particularly at the head of the Bay of Bengal, caused either by river discharge or by doming in cyclonic eddies, and in the east of the province where there is flow through the Indo-Pacific archipelago.

Equatorial divergence and enhanced production in the equatorial zone is weaker than in the other oceans, though it has been reported to occur (see references in Zeitzschel, 1979); however, careful examination of all available CZCS level 3 images ($n = 64$) of the Indian Ocean revealed no bloom along the equator during any season (Longhurst, 1993). Only in the western basin, where weak southward flow along the African coast in boreal winter retroflects eastwards, is there a bloom aligned along the equator, in some years extending as far east as the Maldives; the coastal part of this feature is due to eddy upwelling in the retroflection area (Kabanova, 1968).

In winter, cooling and some deepening of the mixed layer of the southern Arabian Sea induces a weak algal bloom reaching only about 0.2 mg chl m^{-3}, varying in strength and extent between years (Banse and English, 1993). When this process is active (as in winter 1979–1980) the enhanced surface chlorophyll signal, as observed in CZCS images, lies in an arc across the whole of the monsoon gyre poleward of about 15°N. The meridional line of the Maldive Islands along 73°E forms a feature in CZCS surface chlorophyll images: In boreal winter with westwards flow across this line, chlorophyll enhancement occurs to the west, downstream. This situation is reversed in boreal summer, when flow is in the opposite direction.

The boreal spring intermonsoon (in May) is the ultimate oligotrophic season and values >0.10 mg chl m^{-3} remain only in very small coastal patches in the Arabian Sea. With the onset of the southwest monsoon in succeeding months, and the rapid spin-up of the ocean response, surface nitrate values respond; beyond the upwelling region (the ARAB province) biological uptake is sufficiently rapid that surface nitrate is undetectable or occurs as temporary patches of <0.5 μM. Chlorophyll values at the surface are enhanced (to >0.3 mg chl m^{-3}) over much of the western and northern part of monsoon gyre, whereas lower values occur in the eastern half of the Arabian Sea basin. The production/loss balance computed from the CZCS images suggests that these processes are closely coupled during this period. However, in this region, outside the Arabian Sea upwelling bloom province, biological profiles very rapidly settle into the typical oligotrophic pattern with a deep chlorophyll maximum (DCM) at 30–50 m. In fact, some of the earliest observations and analysis of the DCM were made during the IIOE, and the relation between DCM and the nutricline had already been identified. Steepening and weakening southwards, it was observed that the feature was the result of active algal growth rather than passive accumulation at density discontinuities. High chlorophyll values also occur during the southwest monsoon in the northern half of the Bay of Bengal; even in this estuarized embayment a DCM is a normal feature in the profiles (Sharma and Aswanikumar, 1991).

The deep oxygen minimum (at about 200–500 m) of the Arabian Sea extends (progressively attenuated) over much of the northern Indian Ocean though it apparently has little impact on the vertical distribution of mesoplankton beyond the Arabian Sea (Madhupratap and Haridas, 1990). The vertical distribution of typical tropical genera here follows closely the pattern we shall encounter in the eastern Pacific, for which there are many and detailed accounts.

Further information on ecological cycles in the extreme eastern part of this province can be inferred from the Australian IIOE sections to the south of Java (see Indian South Subtropical Gyre Province).

Synopsis

Pycnocline depth responds to flow in Somali Current, lying deepest in eastward flow, but always remains shoaler than photic depth (Fig. 8.1). There is a sharp increase in primary production rate at onset of southwest monsoon and coastal upwelling and perhaps some response (March) to winter northeast wind mixing. Productivity at the DCM increases to around 20% during two seasons of declining chlorophyll. The sign of chlorophyll accumulation tracks that of change in primary production rate, indicating high consumption or loss rate.

Indian South Subtropical Gyre Province (ISSG)

Extent of the Province

The Indian South Subtropical Gyre Province (ISSG) extends from the hydrochemical front at 10–15°S to the Subtropical Convergence at about 30°S. The eastern margin is the Australian coastal boundary at the outer edge of the Leuwin Current, and to the west the coastal boundaries of the EAFR province at the outer edge of the Agulhas and East Madagascar Currents and thus, Colborn areas 22, 23, 25–29, and 30–32 (in part).

Continental Shelf Topography and Tidal and Shelf-Edge Fronts

Major areas of flat shallow banks, <200 m deep, lie along the Mauritius–Seychelles Ridge (Nazareth, Saya do Malha, and Seychelles Banks).

Defining Characteristics of Regional Oceanography

This is the subtropical gyre of the southern Indian Ocean for which we have very little organized knowledge; some of the IIOE investigations did survey this

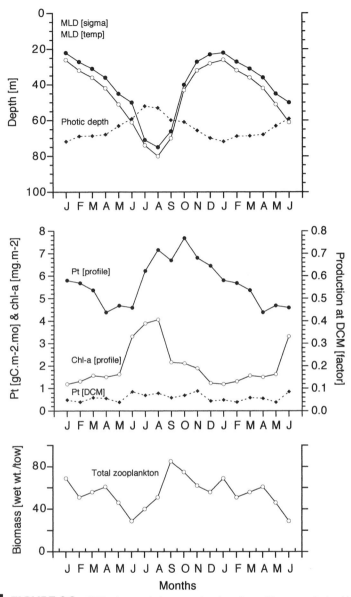

FIGURE 8.2 ISSG: characteristic seasonal cycles of monthly averaged mixed layer and photic depths, chlorophyll at the surface, and rate of primary production, both depth-integrated and at the DCM. Data sources are discussed in Chapter 1. Zooplankton data are from Tranter and Kent (1969); other data sources are discussed in Chapter 1.

province, especially by means of a few long meridional sections by the then-USSR ships, which generated some data on primary and secondary production. Only in the southeastern part did the well-planned Australian sections produce a body of organized ecological data season by season.

Circulation is more variable than in the Atlantic and Pacific subtropical gyres in response to the seasonal march of the monsoon winds over the Indian

Ocean. During the boreal summer, the southeast trades cross the equator north-wards with the intertropical convergence zone (ITCZ), which comes to lie across the Indian subcontinent along the southern flank of the Himalaya in August; the wind-driven anticyclonic circulation south of the equator is then fully established. Anticyclonic wind stress curl extends from about 10–15°S to the southern limb of the gyre, associated with deepening of the mixed layer centrally in the gyre; this effect is strongest in austral summer and autumn, reaching maximum effect at 30°S. In boreal winter, when the ITCZ lies south of the equator, winds and wind-driven circulation are weaker and more vari-able, and the central thermocline trough shoals and becomes subdivided into a series of separate basins separated by shallower ridges.

The zonal thermocline ridge at about 10°S is the limit of westward flow of the low-salinity SEC on its southern slope and of the eastward flow of the SECC along its northern slope. Below the southern slope of the ridge, and thus below the salinity minimum, Wyrtki's hydrochemical front slopes down southwards and separates the high-nutrient/low-oxygen water of the monsoon gyre from the low-nutrient/high-oxygen water of the subtropical gyre. The location of the ridge changes seasonally; as the SEC strengthens during the southwest mon-soon, the thermal ridge moves progressively northwards (to 5°S) and shoals [mixed layer depth (MLD) = <30 m] in the southwest monsoon, whereas during the northeast monsoon it is at 10°S and deeper (MLD = <40 m).

The low salinity of the water mass transported in the SEC originates partly from the southerly, eastern limb of the Bay of Bengal gyre, partly from the low-salinity water of the Southeast Asian archipelago, and partly from the effects of a belt of very heavy rainfall (>200 cm/year) from 3°N to 10°S, east of 60°E, which dilutes surface water in the northern part of the SEC. Finally, there is input to SEC water from rainfall and rivers as it passes north of Madagascar. Mixed layer depth may be influenced more by salinity distribution than by temperature.

Though the flow along the Subtropical Convergence Zone (SCZ) is placed in a separate province, it is worth noting that return eastward flow in the poleward limb of the subtropical gyre occurs along the SCZ at about 40°S. The core of this flow is the current band (or oceanic jet) of the South Indian Ocean Current (Stramma, 1992), continuous from the Aghulas Return Current to the western coast of Australia. Bulk transport in this oceanic jet diminishes progres-sively eastwards as eddies and meanders pass water into the interior of the gyre. It is the equatorward edge of this field of eddies and meandering flow that is proposed as the limit of the ISSG province.

Across the whole subtropical province in boreal winter, the thermocline is shallower in the west, deeper in the east, and driven by eastwards wind stress along the ITCZ. Polewards of 15–20°S, austral winter deepening of the mixed layer to 95–150 m from summer values of 33–46 m occurs along the thermo-cline trough between Madagascar and northwest Australia, around which flows the gyral circulation (Colborn, 1975). The whole forms yet another subtropical anticyclonic gyre having nitrate isopleths deepest in the central regions and shallowest around the periphery. Once again (see Chapter 7, South Atlantic Province; Chapter 9, South Pacific Subtropical Gyre Province), this peripheral shoaling of nitrate is clearly and generally reflected in the surface chlorophyll field.

Biological Response and Regional Ecology

This is a region of persistently high water clarity, and surface chlorophyll is usually <0.05 mg chl m^{-3}. Divergent flow and potential upwelling occurs on the 10°S thermocline ridge and should be reflected in biological enhancement at the surface. Indeed, a maximum of surface chlorophyll occurs there in austral winter (July–October), with surface chlorophyll increasing from a background of 0.05 to a maximum of about 0.2 mg chl m^{-3}. This occurs when the mixed layer deepens through the photic zone, and nutrient entrainment may be presumed to occur. The production/loss balance computed from the CZCS surface chlorophyll suggests that coupling between herbivores and phytoplankton is extremely close: accumulation tracks the rise in primary production rate only very briefly, then declines well before productivity slackens.

Examination of all available CZCS level 3 images and also the seasonal global images revealed discrete seasonal blooms clearly associated with the shallow banks on the Mauritius–Seychelles Ridge (5–20°S) and the Chagos archipelago during austral winter. This appears to confirm previous suggestions that the Mauritius–Seychelles Ridge causes divergence in the SEC during this season, leading to nutrient enrichment of surface water on its western side; similar effects occur at the Seychelles themselves, which lie athwart the thermal ridge at 10°S (Ragoonaden *et al.,* 1987; Vethamony *et al.,* 1987).

The best source of seasonal ecology is the Australian IIOE meridional transect at 110°E, south of Java, and although this transect was close to the eastern margin of the province, maps of (for example) 100 m nitrate for the whole ocean suggest that extrapolation to the central part of the province is not unreasonable. Integrated primary production has similar values (40–60 mg C m^{-2} hr^{-1}) in June through October from 10°S down to 35°S during the austral winter seasonal bloom, though both this seasonal field and that for integrated chlorophyll (<20–30 mg chl m^{-2}) are rather patchy. Frequently, an ephemeral maximum occurs in the far north of the transect (in the MONS province). The seasonal field for integrated 200-m night zooplankton biomass shows higher values occurring between 10–12°N and 30°S.

Interpreting the results of the IIOE transect, Tranter (1979) summarized the seasonal cycle for the eastern MONS and ISSG as follows:

- Austral spring, early summer—meridional transport of subtropical water into MONS
- Late summer—impoverishment of ISSG by thermal stratification
- Autumn—meridional transport of tropical water into ISSG
- Austral winter—enrichment of tropical water in MONS by dynamic uplift of SEC

Coupling between production and consumption is close in these regions: for MONS and ISSG, Tranter (1979) shows that nitrate, primary production rate, and zooplankton biomass all vary seasonally by about 40–50% of the annual mean value, whereas phytoplankton biomass (on the evidence of chlorophyll) varies only by about 20%. This conclusion agrees with what can be deduced from the CZCS data (see above) and from general assumptions about pelagic ecology in warm seas. However, looking more closely at evidence for trophic succession, Tranter found the following pattern: in MONS (tropical water) the timing of zooplankton biomass correlates well with productivity and nitrate, but in ISSG (subtropical water) chlorophyll correlates positively with productivity but negatively with zooplankton. He infers that consumption response is instantaneous in MONS but is lagged by one cruise interval (several weeks) in ISSG.

Synopsis

There is moderate austral winter mixing, when pycnocline lies briefly deeper than photic zone (June–August) (Fig. 8.2). Primary production rate responds to nutrient input from deepening mixed layer, rising to a broad summer–autumn peak, initially tracked by accumulation of biomass, but consumption reduces biomass after only 2 or 3 months accumulation as herbivore population responds. Chlorophyll biomass peak is therefore brief and early.

INDIAN OCEAN COASTAL BIOME

Red Sea, Persian Gulf Province (REDS)

Extent of the province

This province comprises the Red Sea north of the Straits of Bab-el-Mandeb and the Persian Gulf within the Straits of Hormuz. The Gulfs of Aden and Oman are considered to be part of the adjacent coastal boundary province (ARAB).

Continental Shelf Topography and Tidal and Shelf-Edge Fronts

The Red Sea comprises a shallow coastal shelf surrounding a very narrow deep basin geologically continuing the East African rift valley. Depths of <100 m occupy 41% of the total area of the basin and the shallow area is wider south of about 20°N, reaching 120 km at Massawa (Head, 1987). Over much of the Red Sea, the shallow shelf is occupied by active coral reefs. It is these reefs, and the effect of the heavy brine filling the deep rift, which mostly have occupied marine ecologists who have had the chance to work in the Red Sea.

The Persian Gulf comprises a series of deeper basins, reaching to 80 m, along the north coast, and a series of shallower banks (often <20 m) along the eastern and southern coasts. The south coast is an area of active coastal accretion by biotic carbonate sedimentation in shallow "sabka" lagoons and offshore organic reefs (Shinn, 1987).

Defining Characteristics of Regional Oceanography

These two evaporative basins, one deep and the other shallow, have little in common other than their proximity, so their circulations are discussed independently. As already noted, in an entirely logical system it would be well to separate these two entities as individual provinces; they are maintained together here purely as a matter of convenience.

Red Sea

The Red Sea is a model evaporative basin so that deep circulation in the long, narrow rift valley occupied by the sea is largely driven by evaporative

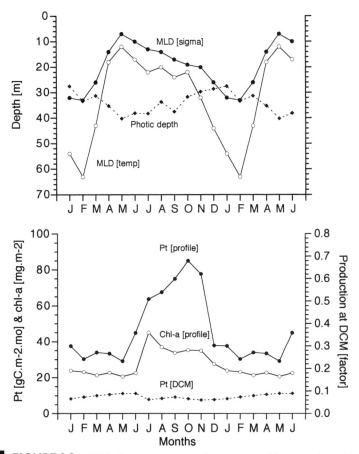

FIGURE 8.3   REDS: characteristic seasonal cycles of monthly averaged mixed layer and photic depths, chlorophyll at the surface, and rate of primary production, both depth-integrated and at the DCM. Data sources are discussed in Chapter 1.

processes and surface flow may be contrary to the prevailing wind stress. The circulation pattern is strongly influenced by the existence of a shallow sill, in this case of only 110 m, at the Straits of Bab-el-Mandeb. Excess of evaporation over precipitation in the basin of about 2 m year^{-1} forces a constant influx of surface water through these straits from the Arabian Sea. Winter cooling to about 18°C at the northern end of the Red Sea of water already having a salinity of 42% creates strong deep convection cells in which the deep water (anomalously warm and saline at 21.5°C and 40.6) of the Red Sea basin is formed (Dietrich *et al.*, 1970). At the straits of Bab-el-Mendab deep water flows out below the surface influx at a rate of about half that of the Mediterranean outflow over the sill of the Straits of Gibralter. Surface wind-driven streams are generally weak and eddying, but flow occurs toward the south along the whole length of the western coast of the Red Sea during the season of northerly, summer winds, but at other times the density-driven flow will cause surface drift contrary to these winds, especially up the eastern coast. Thus, the general surface circulation is cyclonic.

The depth of the mixed layer is approximately the inverse of surface temperature, except where it is destroyed by winter mixing or tidal effects in shallow water. In boreal spring, thermocline depths decrease, until the summer situation is reached, with minimum depths of about 30 m on the eastern side of the rift. In the southern Red Sea, the summer thermocline is sustained above the uplift of the permanent weak thermocline forced by the very active flow of a cool-water core northwards through the straits of Bab-el-Mandeb under the influence of the southwest monsoon. This shallow permanent thermocline persists to about 22°N in the Red Sea. In the northern Red Sea, the summer thermocline, at about 25 m, lies above an almost isothermal deep water mass.

The monsoon reversal has only minor influence over the Red Sea, where the prevailing winds are along the axis of the rift valley and from the north; during winter the wind is reversed over the southern part so that a wind convergence occurs at 18–22°N (Edwards, 1987). Surface inflow at Bab-el-Mandeb is therefore stronger in winter than in summer. At times of stronger than usual northerlies, the water column at these straits may comprise three layers: a thin (40 m) surface, wind-driven outflow; a deep density-driven outflow; and an intermediate low-salinity inflow. Interannual variability in inflow through Bab-el-Mandeb is also strong and can be detected by differential spreading of Arabian Sea water (Ganssen and Kroon, 1991).

Persian Gulf

The shallow (<80 m in the deepest parts) Persian Gulf is oceanographically an extension of the surface water of the Arabian Sea, and there is no shallow sill between it and the Gulf of Oman as there is between the Red Sea and the Gulf of Aden. Evaporation greatly exceeds both precipitation and the input of river water from the Euphrates and Tigris, and salinity reaches 50% in shallow water on the Arabian coast. A slow cyclonic circulation is maintained.

In winter, a weak thermocline is established in the outer part of the Persian Gulf at 30–40 m, but during the intense heating of the boreal summer an isothermal layer effectively ceases to exist, with surface temperatures reaching 32°C above a thermal gradient to bottom water of 22–24°C. Salinity reaches 40 ppt along the Iranian coast and off Arabia, where the outflow of dense bottom water occurs, especially in the deeper channels.

Surface drift occurs in response to the strong wind events that sweep the Gulf, especially northeast "Shamal" winds along the axis of the gulf that break down the density stratification to at least 60 m and temporarily destroy the mean cyclonic circulation.

Biological Response and Regional Ecology

The single significant source of new dissolved nutrients in the Red Sea is the surface water flowing in from the Gulf of Aden, though tidal mixing may transport regenerated nutrients into the photic zone. In the two northern gulfs (Suez and Eilat-Aqaba) local inputs of industrial effluents have induced local blooms. Over most of the Red Sea, the strong pycnocline is an effective and permanent barrier to the vertical mixing of nutrients from below. Consequently, the overall productivity of the mixed layer decreases progressively northwards.

Offshore Red Sea water has great clarity and primary production is generally extremely low (about 100 mg C m^{-2} day^{-1}), though locally, in coastal situations, values may be much higher. A seasonal cycle has been described for the eastern coast at the latitude of Jeddah with maxima in January and July (Dowidar, 1983). It is only in the southern region that there appear to be significant offshore algal blooms; the CZCS level 3 images show that the entire southern area of wide shallow shelf (that is, south of about Massawa) has relatively high surface chlorophyll during the southwest monsoon period. The area of enhanced chlorophyll appears to be defined by the edge of the shelf between Massawa (15°N) and Port Sudan (20°N). To the north of this region, the CZCS level 3 images show only a narrow, intermittent near-coastal bloom.

Though the Red Sea has an anomalous density profile, the profiles of algal biomass and nutrient distribution closely resemble the typical tropical situation with a significant DCM (Weikert, 1987). Picoplankton (0.2–2.0 μm) are the dominant-size fraction of phytoplankton and contribute about 75% of both autotroph biomass and primary production (Gradinger *et al.*, 1992). At the other end of the algal size spectrum, prominent blooms of *Oscillatoria erythraeum* are frequent in the open parts of the Red Sea and are perhaps the reason for its name; however, the significance of these blooms for basin-scale carbon fixation and for the general pelagic ecosystem has not apparently been assessed.

The Red Sea is a net importer of zooplankton by the density-driven influx in the south though many species of expatriated Indian Ocean species do not long survive the extreme conditions of the Red Sea. It has been suggested that this nitrogen flux may partly balance the negative nitrate budget associated with the two-layered pattern of exchange at the entrance to the Red Sea. Obviously, zooplankton diversity is attenuated northwards, and in the extreme northern Red Sea relatively few oceanic species survive.

The effects on vertical profiles of plankton of the strong subsurface oxygen minimum (O_2 <1.3 ml liter^{-1}) in the Red Sea between 100 and 700 m have been investigated. While epiplankton avoid the layer so that biomass rapidly declines from 50 to 100 m above the sharp pycnocline, the subsurface zooplankton biomass maximum lies within the O_2 minimum layer. Here, diel migrants (e.g., *Pleuromamma indica*) have their daytime residence depths, as this genus does elsewhere, along with apparently resting populations of *Rhincalanus nasutus* (Weikert, 1984).

CZCS images suggest that the Persian Gulf is relatively turbid throughout the year with maximum clarity in the winter quarter (January–March) and we may expect this to be due, in part, to suspended carbonate particulates. Surface chlorophyll values in the range of 1 or 2 mg chl m^{-3} occur throughout the central part of the Gulf at the end of summer, with even higher local maxima along the Trucial Coast (Dorgham and Moftah, 1989). These chlorophyll values are almost an order of magnitude higher than those in the adjacent Gulf of Oman at the same season. Of the larger (net) size fraction of algae, *Trichodesmium* and *Anabaena* were the major fraction of total biomass, especially in areas where surface nitrate was undetectable.

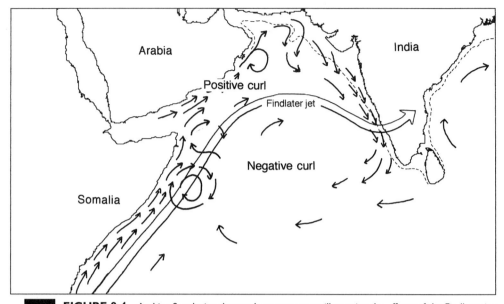

FIGURE 8.4   Arabian Sea during the southwest monsoon illustrating the effects of the Findlater jet of the low-level atmospheric circulation which crosses the region (solid line). To the south of the jet, negative curl of the wind stress deepens the mixed layer, to the north positive curl causes the mixed layer shoal and it is an important factor in coastal upwelling. Note the anticyclonic features of the Great Whirl and the Socotra Eddy off Somalia and the similar eddy east of Arabia. These are persistent and observable in CZCS images.

Synopsis

Moderate winter mixing occurs to well below photic depth so that pycnocline is illuminated only from April to November (Fig. 8.3). Primary production rate follows the effect of the southwest monsoon in the Arabian Sea but does not rise to so sharp a peak as in MONS or ARAB. Sign of chlorophyll accumulation approximately matches the cycle of productivity, again suggesting close coupling between production and consumption.

Northwest Arabian Upwelling Province (ARAB)

Extent of the Province

The ARAB province includes the coastal areas of the northern Arabian Sea from Somalia to Pakistan. The upwelling region extends from the coastline to an offshore convergence zone having local characteristics (Colborn: 1, 3, 6, and 20 in part).

Continental Shelf Topography and Tidal and Shelf-Edge Fronts

The African and Arabian coasts have a very narrow (<5–10 km) shelf right through the upwelling areas, except for a small shallow area of banks southwest of Socotra Island and shallow bights on the east coast of Oman, where shelf

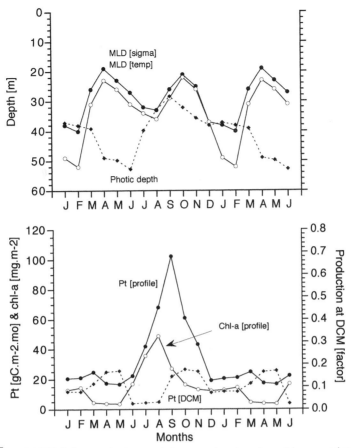

FIGURE 8.5 ARAB: characteristic seasonal cycles of monthly averaged mixed layer and photic depths, chlorophyll at the surface, and rate of primary production, both depth-integrated and at the DCM. Data sources are discussed in Chapter 1.

widths reach about 75 km. The narrow shelf continues along the coast of Iran and western Pakistan to the point where it turns southeast at about 66°E, where it widens significantly into the coastal boundary province of western India (INDW). Across the mouth of the Gulf of Oman lies the Murray Ridge, which is sufficiently shallow (<1000 m) to modify circulation and structure of the upper water column.

Defining Characteristics of Regional Oceanography

Circulation and mixed layer dynamics, which result in upwelling and algal blooms during boreal summer, are a complex response to reversing monsoon wind stress (Fig. 8.4): Entry into the extensive recent studies on this region could well begin with Banse (1984), Burkill *et al.* (1993a), Brock *et al.* (1992), or Schott (1983).

The boreal summer wind field is a mirror-image of the wind field over eastern boundary currents of other oceans. In this case, the coast lies to the west, and the wind maximum is a low-level jet (Findlater, 1969) about 500 km offshore but is poleward rather than equatorward. Wind-stress curl is homologous with that over eastern boundary currents—anticyclonic curl is seawards and cyclonic curl is landwards of the jet (see Charts 46 and 48 in Hastenrath and Lamb, 1979). Ekman dynamics lead to mixed layer deepening seawards (southeast) of the jet and shoaling on the landward (northwest) side. The weaker northeast monsoon jet of boreal winter, conversely, leads to deepening on its landward side and shoaling in the more central parts of the Arabian Sea. In the current geographical scheme, the landward side of the monsoon jet is treated as the coastal boundary province and the seaward side as the oceanic ARAB province.

Because we are interested in the ecological consequences of these changes, it is best to discuss four processes driven by seasonal wind reversal and the topography:

• Winter convective and turbulent mixing: At the onset of the northeast monsoon, southwards flow develops along the western margin of the Arabian Sea from northern Somalia (10 or 11°N, near Ras Hafun and Cape Guardafui) to 2 or 3°S, where the East African Coastal Current (carrying low-salinity water of SEC origin) converges with it, and both turn offshore as the origin of the SECC. North of Cape Guardafui, flow in boreal winter is weak and variable with significant transport into the Gulf of Aden and so into the Red Sea (Halim, 1984). The southward flow of the Somali Current in boreal winter reaches to about 500–600 m (Schott, 1983). The northeast monsoon winds from the uplands of central Asia are cool and dry and induce convective cells in the northern part of the Arabian Sea that entrain nutrients to the surface and hence winter blooms of phytoplankton (Banse and McClain, 1986; Madhupratap et al., 1996; Veldhuis et al., 1997).

• Coastal divergent upwelling: With the onset of the southwest monsoon winds, response and spin-up of the northward, boreal summer Somali Current occurs within a few weeks; the current responds to progressive step increases in wind strength as the monsoon onset evolves in a series of surges, and upwelling is induced in three centers in the northwest Arabian Sea by processes specific to each center (Swallow, 1984). The first indication of the reversal of the monsoon is the onset of onshore winds off southern Somalia in March so that divergence of the current occurs at about 5°N (where the coast changes orientation near Obbia) into northward and southward coastal currents. In April, northward winds develop along the coast south of the equator and within a few days a transequatorial northward coastal current develops, which turns offshore at about 5°N (Schott, 1983). Just to the north of the retroflection, a wedge of cold upwelling water develops along the offshore flow. In May, the main onset of the alongshore southwesterly monsoon occurs quite suddenly: the response of the current is to turn offshore a little further to the south (3 or 4°N). The final development of the southwest monsoon is the extension of alongshore winds to northern Somalia with strong anticyclonic curl offshore of the Findlater jet, usually in June; within about 14 days this situation induces a second, northern gyre to establish the two-gyre system described by Swallow and Fieux (1982). North of the offshore retroflection of the northern gyre, northward flow (though weaker than off Somalia) continues along the coast around the head of the Arabian Sea to the Indian subcontinent (Swallow, 1984; Schott, 1983) so that the Somali Current, which has attracted much attention for its rapid reversal, is actually only part of a much larger boundary current system along which coastal divergence occurs. Finally, in August, the northern eddy intensifies, and the southern eddy moves farther north, so that the two eddies coalesce to become the "Great Whirl," with its northern edge at about 9 or 10°N.

• Offshore vertical Ekman motion: In addition to the coastal upwelling along the Somali coast, Ekman dynamics force a second, offshore region of upwelling in the northwest Arabian Sea (Smith and Bottero, 1977). The low-level atmospheric Findlater jet, after passing along the central line of the Horn of Africa, leaves the coast over Cape Guardafui and proceeds northeasterly toward Pakistan and thus well offshore from the Arabian coast (Brock and McLain, 1992). The area of cyclonic wind curl to the north of the jet creates significant Ekman suction off the Omani continental shelf, whereas the anticyclonic curl field to the south creates Ekman pumping and deepens the eddy known as the Great Whirl (see Chart 48 in Hastenrath and Lamb, 1979). Thus, the total upwelling region off Arabia is very large: 1000 km along the coast and 400 km offshore through which passes 8×106 m^{-3} sec^{-1} at the 50-m level equivalent to a vertical velocity of 1 or 2 m day^{-1} (Swallow, 1984).

• Offshore topography and eddying: The Murray Ridge, lying across the mouth of the Gulf of Oman, induces a persistent field of cyclonic and anticyclonic eddies which intensify during the southwest monsoon. Filaments of cool upwelled water from the Oman coast pass along the flanks of the ridge and the eddy field is more prominent to the north. Wind-induced upwelling occurs along the coast of the Gulf of Oman and of western Pakistan and the surface chlorophyll field suggests that a significant anticyclonic eddy persists in the central Gulf of Oman (Weaks, 1984). The thermocline in the region of the Murray Ridge is shallow but weak in boreal summer and shoals even further above the ridge; in the mouth of the Gulf of Oman it is extremely strong and shallow (23–27°C over about a 5-m depth range below a 10-m surface mixed layer), though farther into the Gulf of Oman the mixed layer deepens progressively, responding to the outflow of highly saline surface water from the Persian Gulf (Owens *et al.*, 1993).

As well as those associated with the Murray Ridge, mesoscale eddies develop along the western coastal boundary zone of the Arabian Sea during the southwest monsoon (Bruce, 1979); as the eddy circulation pattern in the Somali upwelling shifts north during the evolution of the southwest monsoon, much of the upwelled water is retained in these eddies so that offshore thermal discontinuities develop to the east of 55 or 56°E. Eddy formation continues to strengthen through the end of July. By October, the eddy pattern characteristic of the early part of season may be expected to have reformed.

We have already noted the fact that the presence of phytoplankton modifies the heating–cooling cycle of the upper part of the water column. The extremely clear water, and strong DCM, of the Arabian Sea has enabled a calculation to be made of an inverse effect of biology on physics: the relative heating of seawater by differential absorption of short-wavelength solar radiation by pigmented phytoplankton cells in the DCM (Sathyendranath *et al.*, 1991). A maximum rate of 4°C per month (August) was calculated, which is not insignificant compared with upwelling cooling of about 2.5°C per month (July); phytoplankton pigment enhances the rate of heating during the period of warming surface water and reduces the rate of cooling during upwelling periods. This effect should be watched for in other regions, perhaps especially where stratification is initiated in spring-bloom situations.

Biological Response and Regional Ecology

The seasonal reversal of monsoon winds has strong biological consequences over the entire northwest Arabian Sea, and during the southwest monsoon a large area of high surface chlorophyll becomes prominent in CZCS images due to widespread upwelling. These images show that the region occupied by this algal bloom is equivalent to the total area covered by the upwellings in all four eastern boundary currents. Even before CZCS images were available, it was suggested that southwest monsoon bloom of the Arabian Sea might be equivalent in productivity to the four eastern boundary currents together (Smith, 1984). A model of the seasonality in the Arabian Sea is offered by McCreary *et al.* (1996) which diverges principally from previous models by proposing a lesser emphasis on the role of consumption when chlorophyll biomass is decreasing while the rate of primary production remains high.

The reversal of the monsoon winds dominates biological response to mixed layer dynamics in the upwelling province and this coupling is so tight that between-year variability in timing and strength of monsoon winds are directly reflected in the seasonal ecology: The weak monsoon of 1982 generated a coastal phytoplankton bloom having maximum pigment values only about one-fourth of those in a strong monsoon.

During the boreal winter and the northeast monsoon, the northern Indian Ocean, including the Arabian Sea and Bay of Bengal, has low (0.5 μM) or undetectable levels of nitrate. However, in the northern Arabian Sea, winter blooms of phytoplankton occur and are apparently driven both by turbulent wind mixing and by convective overturn after cooling of the sea surface during which convection cells may inject nutrients into the surface layers (Banse and McClain, 1986; Madhupratap *et al.*, 1996). These blooms have high between-year variability, and the highest chlorophyll values are in coastal waters of northwest India and Arabia, reaching at least 5.0 mg chl m^{-3}. The influence of the Murray Ridge can be seen in some CZCS images of this season as a linear patch of relatively low chlorophyll (Owens *et al.*, 1993).

The upwelled water off Somalia may be as cool as 13°C, indicating source depths of at least 200 m. A linear relationship ($r_2 = 0.89$) exists between surface temperature and nitrate content for surface temperatures of 14–27°C (Smith, 1984); average daily warming rates for surface water and the indicated loss rate for nitrate suggest algal primary production of about 98 mg C m^{-3} day^{-1}, which is approximately the same as that observed in experimental data. Rates of 2.5 g C m^{-2} day^{-1} are usual in the upwelling area off the coast of Oman compared with <0.3 g C m^{-2} day^{-1} in the oligotrophic central gyre of the Arabian Sea (Mantoura *et al.*, 1993; Owens *et al.*, 1993).

Surface chlorophyll fields, both from ship-based data (e.g., Halim, 1984) and from many CZCS images (e.g., Banse and McClain, 1986), also match the upwelling pattern computed from wind stress. During the southwest monsoon, many of the anticipated mesoscale circulation features can usually be identified in the surface chlorophyll field: the persistent anticyclonic eddy off northern Somalia, the broad field of upwelling off Arabia, and the lower chlorophyll over the Murray Ridge and in the central Gulf of Oman.

An understanding of algal dynamics in this complex province requires knowledge of the vertical relations of density, nutrients, and light. Chlorophyll profiles respond predictably to deepening of the pycnocline and the level of ambient radiation remaining at the nutricline, which usually lies in the upper pycnocline. Meridional chlorophyll sections through the Arabian Sea in boreal summer show that to the right of the atmospheric jet the pycnocline and nutricline are deep (approx 80–100 m) and are associated with a chlorophyll maximum. To the left of the jet, maximum chlorophyll occurs in the mixed layer, and the nutricline approaches the surface.

The distribution of subsurface light in relation to mixed layer depth can be used to model the seasonality of primary production in the Arabian Sea. During the relatively windless but brief intermonsoon periods, the Arabian Sea becomes oligotrophic with a DCM just above the nutricline at <50 m. The 1% isolume lies deeper than this because water is very transparent with little chlorophyll in the mixed layer (Brock *et al.,* 1993). At the onset of both monsoon seasons, the 1% isolume appears to rise into the mixed layer, as algal growth reduces transparency. In the oligotrophic seasons, production at the DCM exceeds production in the mixed layer.

With the change of the subsurface distribution of chlorophyll, changes in specific composition of constituent algae also occur (Jochem and Zeitzschel, 1993): Picoplankton have greater importance in the oligotrophic period, especially in the mixed layer ammonium-based ecosystem, when *Synechococcus, Prochlorococcus,* and pico-eukaryotes comprise 40–80% of the standing stock of POC (Burkill *et al.,* 1993b; Veldhuis *et al.,* 1997); in the DCM, small diatoms have greater importance, as they do in the near-surface chlorophyll maximum of the upwelling cells close to the coast. In such situations, production rates of around 1 or 2 g C m^{-2} day^{-1} are commonly observed, of which about 50% is generated by the picofraction, about 10% by the nanofraction, and about 40% by larger cells. Offshore, in oligotrophic situations this is reversed: From 60 to 75% of the daily production of 0.4 to 0.5 g C m^{-2} day^{-1} is generated by picoplankton, principally phycoerythrin-rich chroococcoid cyanobacteria and prochlorophytes at concentrations of 107 or 108 cells liter^{-1}. Larger cells (1.2 μm), on the whole, lie deeper than smaller cells (0.7 μm).

As probably occurs elsewhere, the microbial food web here is in very close balance: Cyanobacterial-specific growth rates are on the order of 0.5–1.0 day^{-1}, whereas protistan micrograzers consume 30–70% of the cyanobacteria cells daily. Production and consumption rates are correlated, suggesting very close trophic coupling within this assemblage of very small organisms.

The typical tropical mesozooplankton of the Indian Ocean is modified in the Arabian Sea by the presence of abundant *Calanoides carinatus,* which appear to have a seasonal ecological cycle compatible with observations made on the same species off West Africa (see Chapter 7, Guinea Current Coastal Province). During the southwest monsoon, this is the dominant copepod in coastal upwelling cells and reaches 50–100 m^{-3} (Smith, 1982). Since this is a mean value for 0–200 m, we must assume layers of even greater abundance occur within the mixed layer. A small group of species of copepods, dominated by *C. carinatus* along with *Eucalanus* spp., occurs preferentially in water <25°C, whereas the remainder of the tropical assemblage is equally abundant in and out of upwelling water. Both females and copepodites of *Calanoides* occur together in upwelled water in a spatial sequence from young to older upwelled water, which indicates that reproduction occurs there. Generation time is probably on the order of 25 days, and this suggests that no more than a single generation will occur before the population is forced below the surface by rising temperature as the parcel of upwelled water is advected horizontally. Smith suggests that subsurface circulation is apt to retain resting, lipid-replete stage 5 copepodites (C5's) at >500 m in a trajectory which will position them to return to the surface in a succeeding upwelling cell with reproductive products ready to mature. *Calanoides* consumes small diatoms in the range 25–75 μm (*Rhizoselenia, Nitzschia, Eucampia,* etc.) and Smith (1984) computes the diatom consumption of recently upwelled *Calanoides* as 25–45% of the lower primary production nearshore but only 1–15% of the larger daily rate offshore.

The existence of blooms during both the southwest and the northeast monsoons suggest that *Calanoides* will have a more complex life history here than in simpler upwelling situations. It remains to be seen if this involves a complex pattern of subpopulations specializing in each regional, seasonal bloom or whether (more likely) the deep subthermocline water of the Arabian Sea everywhere carries a sufficient population of resting C5's to seed any upwelling parcel of water. The fact that blooms occur in both seasons perhaps also explains a paradox of the Arabian Sea mesozooplankton: that there is relatively very low variability in overall plankton biomass between seasons (Madhupratap *et al.,* 1996).

Synopsis

ARAB resembles MONS, but seasonality is stronger in ARAB (Fig. 8.5). Southwest monsoon and the spin-up of Arabian Sea gyre and cooling of surface water mass by winter northeast monsoon both deepen the pycnocline. Primary production rate responds rapidly and strongly to coastal upwelling (the local effect of which is not captured in this regional integration) during the southwest monsoon. There is a strong seasonal cycle in productivity at the DCM. Chlorophyll accumulation sufficiently closely matches productivity so as to suggest tight trophic coupling between production and consumption.

Western India Coastal Province (INDW)

Extent of the Province

The INDW province extends from the mouth of the Indus at 25°N to the Gulf of Manar at about 7°N, comprising the continental shelf of western India and southern Pakistan (Colborn: parts of 4 and 11).

Continental Shelf Topography and Tidal and Shelf-Edge Fronts

The continental shelf widens from south to north, reaching about 300 km between about 18 and 25°N, and narrows again only at the westward turn of

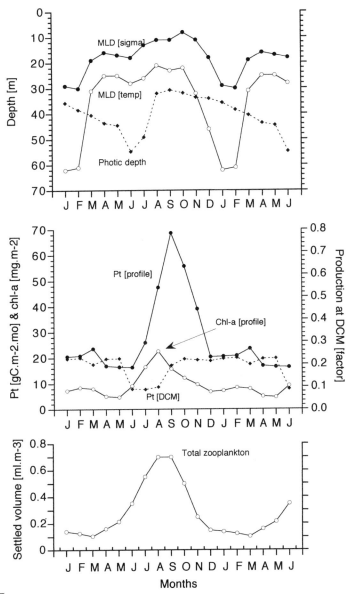

FIGURE 8.6 INDW: characteristic seasonal cycles of monthly averaged mixed layer and photic depths, chlorophyll at the surface, and rate of primary production, both depth-integrated and at the DCM. Zooplankton data are from Menon and George; other data sources are discussed in Chapter 1.

the coast north of Karachi. In the extreme south, the Palk Straits between Cape Comorin and Sri Lanka are uniquely shallow, having a sill depth of only about 10 m. The inner shelf off the alluvial coast of Kerala (8–12°S) is a mobile mud regime, with inshore banks ("chankara") forming during the boreal winter and remobilizing during the summer, when upwelling occurs and alongshore currents and wind mixing are maximal. Shelf-break front upwelling is not indicated by the available CZCS images.

Defining Characteristics of Regional Oceanography

Regional circulation responds both to local and to remotely forced effects of the reversal of the southwest and northeast monsoon winds. Locally, at the coast, the onset of the southwest monsoon occurs first in the south in May and June, then spreads northward, and continues through October or even November. This flow forms the outer edge of the eastern limb of the Arabian Sea gyre and is accompanied by upsloping of the thermocline toward the coast and intrusion of subpycnocline water of low oxygen content onto the shelf, which may occur from April to October (Wooster et al., 1967). From November to April, the province is influenced by lighter, drier northeast winds and some cooling and mixing of shelf water occurs in the northern part of the province off Pakistan. During this season the cyclonic circulation of the Arabian Sea causes downwelling of isopleths near the coast. Coastal currents respond to this local wind forcing with equatorward flow along the west coast of India from February to September or October and poleward during the rest of the year.

However, this apparently simple local process is complicated by distantly forced effects of the spin-up of the whole Arabian Sea gyre, which occurs before the effects of local wind stress are evident. Because a simple model of coastal upwelling forced by Ekman divergence (e.g., Mathew, 1982) did not entirely fit the facts, the relative significance of geostrophic and wind-driven upwelling on this coast was examined by Longhurst and Wooster (1990). Geostrophic upsloping of density contours starts in April, several months before local wind stress could force upwelling, and occurs because offshore isopleths come to slope upwards toward the shore when an anticyclonic gyre is spun up by the southwest monsoon (McCreary et al., 1993). The shoaled pycnocline can then be eroded by several neritic processes. Later in the season, equatorward wind stress advects the now shallower surface layer offshore, leading to Ekman upwelling along the coastline. At such a low latitude the required wind stress is relatively small compared with that for midlatitudes: Off Cochin at 10°N, the same upwelling strength as that off California at 40°N is forced by a much weaker coastwise wind stress. Thus, upwelling is a consequence of both remotely forced baroclinic adjustment (Wyrtki, 1973a) and an equatorward component of wind stress along the coast through the year, intensified during the southwest monsoon.

The Laccadive island chain provides a western barrier to circulation between the southern part of this area and the open Arabian Sea. Colborn (1975) characterizes this province as having a shallow permanent thermocline which breaks the surface during the southwest monsoon—having a mixed layer depth varying from 25 m (southwest monsoon) to 75 m (northeast monsoon).

Biological Response and Regional Ecology

The biological effects of the distantly forced and local wind-driven upwelling should be distinguishable. During the pre-upwelling period, when the pycnocline starts to slope upwards toward the coast, equatorward flow along the coast has a low offshore component, and nutrient levels are low. The mixed layer is thin and subsurface water with low oxygen content has just started to appear on the shelf. When Ekman upwelling begins, surface nutrient levels increase, the mixed layer becomes even thinner, and very low (<0.5 ml liter^{-1}) oxygen concentrations characterize shelf waters below the thermocline.

The upwelling period of the southwest monsoon is associated with algal blooms reaching 8.0 mg chl m^{-3}, related not only to the advection of nutrients in the upwelled water but also to remineralization of nutrients locked in the mobile mud banks after their remobilization. During these near-surface algal blooms, water clarity is very low (chlorophyll maximum = <5 m; water column chlorophyll <200 mg m^{-3}; Secchi disc, 2 or 3 m) with diatom cells at high concentrations of $< 2.9 \times 105$ cells liter^{-1} (Shah, 1973). Because of the pattern of resuspension of liquid mud inshore with the onset of the southwest monsoon, the CZCS images do not present very reliable information on the timing and distribution of algal blooms. Really all that can be said is that they suggest that the most intense algal blooms are neritic and not at the shelf break. Anomalous nitrogen values in boreal spring may accompany *Trichodesmium* blooms, perhaps associated with its nitrogen-fixing capabilities. Though there is no information on seasonal cycles of consumption in this region, Shah suggests that the relatively high ratio of phaeopigments to chlorophyll during upwelling periods indicates strong coupling between herbivores and algal cells.

I now remark on a general characteristic of low-lying coasts in warm seas in which, for various reasons, seasonal diatom blooms are very strong. Such areas may support a pelagic clupeid fish that is unusual in that it depends almost entirely on medium to large diatoms for food, obtained by active filtration on gill-rakers: for example, *Brevoortia* in the western Atlantic, *Ethmalosa* in the eastern Atlantic, and *Sardinella longiceps,* the oil sardine of the Indian Ocean.

The oil sardine is an extremely abundant pelagic fish in this province and supports a fishery sustained since the early 1900s. The arrival of sardines at the coast coincides with the two seasonal blooms of the diatom *Fragillaria* (= *Nitzschia*) *oceanica* and also *Coscinodiscus, Pleurosigma,* and *Biddulphia*. The first bloom, at the start of the monsoon season, coincides with the arrival of prespawning adults, and the second, in September or October, coincides with the main fishery for juvenile sardines, the fish of the year. Dinoflagellates and copepods, together with soft organic-rich material resuspended by tidal streams from the coastal banks of almost liquid mud, appear to sustain the sardines from October to January. The abundance of this rapidly growing fish is therefore closely linked to the productivity of the planktonic ecosystem; having long records of sardine abundance, we can hindcast the relative performance of the plankton over the same period. This was discussed in Chapter 6, in which statistical series of physical time series representing an upwelling index were compared with sardine landings.

Synopsis

Mixed layer depth shoals rapidly in response to spin-up of southwest monsoon and relaxation of winter cooling so that pycnocline becomes illuminated from March to October (Fig. 8.6). Primary production rate responds to coastal upwelling and is tracked by chlorophyll accumulation until buildup of herbivore population causes change of sign, though productivity continues to increase for another 30 days. Production at the DCM is high (20% of total) when productivity is declining and low (5–10%) during population increase.

Eastern India Coastal Province (INDE)

Extent of the Province

The Eastern India Coastal Province (INDE) comprises the coastal regions over and adjacent to the continental shelf of the Bay of Bengal from the east coast of Sri Lanka in the west to the Andaman Islands in the east (comprises Colborn 15 and 19).

Continental Shelf Topography and Tidal and Shelf-Edge Fronts

The shallow water of this province is dominated by the sediment fans off the mouths of the Ganges in the north and off the Irrawaddy delta in the east. In these two areas with low-lying, mangrove coastlines, the shelf has considerable width: Off both Ganges and Irrawaddy the 200-m contour lies as much as 250 km offshore. In both these regions we may expect that tidally forced shelf sea fronts will occur (Sasamal, 1989), with all the consequences discussed in Chapter 7 under Northeast Atlantic Shelves Province and Northwest Atlantic Shelves Province. The continental shelf is much narrower along the Indian coast from 20°N.

Defining Characteristics of Regional Oceanography

Because of the semienclosed nature of the Bay of Bengal, its coastal regions are strongly influenced by the seasonal circulation of the whole bight (see Indian Monsoon Gyre Province). Winter cooling in the north of the Bay of Bengal and the major freshwater signals from Ganges and Irrawaddy that affect the surface salinity of the whole bight are of course strongly implemented in data from this coastal province.

During the northeast monsoon, a western boundary current flows along the east coast of India, observable as a meandering warm-water body in satellite imagery; warm-core eddies are detached at the edge of the current, resembling those of the Gulf Stream (Legeckis, 1987), though temperature gradients are relatively weak at the current edge compared with western boundary currents in the temperate zone. This jet current is about 100 km wide off Madras and flows northwards from February to mid-June; its extent is highly and rapidly variable, and westward intrusions of colder water zonally across the Bay of Bengal may be wind driven or may be compensatory flow for eastward entrainment by the western boundary current, which drives colder water south down the eastern coast of the Bay of Bengal. This western boundary current does not appear to continue around the northern end of the province.

Upwelling-favorable winds force Ekman divergence and upwelling south of Visakhatapam (17.5°N), but farther north, fresh water from the Ganges mouths (especially of the Hoogli) is sufficiently strong to inhibit upwelling (Johns *et al.,*

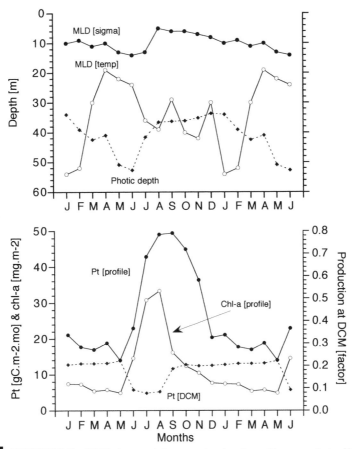

FIGURE 8.7   INDE: characteristic seasonal cycles of monthly averaged mixed layer and photic depths, chlorophyll at the surface, and rate of primary production, both depth-integrated and at the DCM. Data sources are discussed in Chapter 1.

1992). Observations suggest that deep water may be brought to the surface south of Visakhatapam not only by upwelling but also perhaps by vertical mixing in March and April and again in June–August off Waltair (Banse, 1990b). Upwelling on the east coast of India is suppressed during the southwest monsoon by warm, low-salinity surface water. During this season, on the eastern side of the Bay of Bengal the water column may be rendered isothermal over the continental shelf by tidal mixing and thermal convection (Sasamal, 1989).

Biological Response and Regional Ecology

Although the east coast of India is aligned parallel with that of Somalia and Arabia and shares a generally similar seasonal wind regime, the ecological consequences of upwelling during the southwest monsoon off India are in no way comparable with the effect of upwelling in the Arabian Sea. This is evident from a scan of the relevant CZCS seasonal and monthly images which, though they confirm that boreal summer upwelling does result in enhanced near-surface

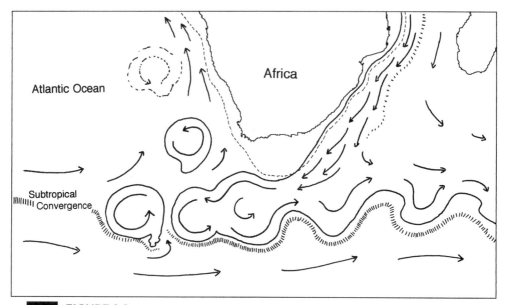

FIGURE 8.8 Retroflection of the Agulhas Current after it leaves the topography of the continental edge (dashed line) and encounters the contrary flow of the South Atlantic Current. The hatched lines represent the Subtropical Convergence front and edge of the longshore jet of the Agulhas Current in the Mozambique Channel. Note the intimate connection between the Subtropical Convergence and the features of the retroflection, often observed in CZCS images. The retroflection represented here is rather extreme, and it frequently occurs somewhat farther to the east.

chlorophyll along the shelf from Madras to about 20°N, indicate that this is a far weaker signal than that in the western Arabian Sea. Even during this season, the water column is sufficiently stabilized that a DCM occurs in all sections of this coast, at about 60–80 m at the edge of the western boundary current and sloping up (and weakening) toward the coast. Maximum chlorophyll values of about 0.75–1.0 mg chl m^{-3} occur in these features (Sarma and Aswanikumar, 1991). There appears to be very other little useful information concerning the pelagic ecology of this long coastline.

Synopsis

INDE has a wide separation of pycnocline and thermocline induced by very shoal brackish surface layer, and lack of water clarity induces shoal photic depth (Fig. 8.7). Though the mechanism remains obscure, it is probably not coincidental that maximum productivity with chlorophyll accumulation occurs during the southwest monsoon, as it does off Arabia.

Eastern Africa Coastal Province (EAFR)

Extent of the Province

EAFR comprises the coastal boundary of the Indian Ocean from northern Kenya to the Cape of Good Hope, including the Mozambique Channel and the east coast of Madagascar. The seawards, eastern limit is the edge of significant flow in the western boundary currents (Colborn: 33 and parts of 27 and 30–32). For convenience, I include the Agulhas Retroflection, south of Africa. Were this province to be subdivided for other purposes into more natural regions, the following would be a suitable primary classification: (i) the equatorial East African coastline—(ii) the subtropical Madagascan region, (iii) the temperate Agulhas region, and (iv) the Agulhas Retroflection. Sufficient differences could be found in their ecologies to support such an arrangement.

Continental Shelf Topography and Tidal and Shelf-Edge Fronts

The continental shelf along most of the eastern coast of Africa is narrow and where there are fringing reefs, as off Kenya and Tanzania, there is no shelf and the coral front stands above deep water. On both east and west coasts of the Mozambique Channel the shelf is wider (40 km wide off western Madagascar and 80 km off the mouth of the Zambezi on the African coast, where it is flat with soft muddy riverine sediments). The shelf is also wider (50 km) in the very obtuse Natal Bight of the Agulhas region. The greatest area of shelf is the flat, triangular Agulhas Bank which lies between the Cape peninsula (19°E) and Port Elizabeth (26°E) and reaches 150 km wide south of Cape Agulhas.

Defining Characteristics of Regional Oceanography

The western boundary current along the east African coast comprises several ecological units (see above) which are discussed together for convenience: (i) the coastal flow at low latitudes of the East African Current, which varies in both direction and strength with monsoon reversal; (ii) the eddying flow in the Mozambique Channel; (iii) the relatively rapid and direct flow along the east coast of Madagascar; (iv) the Agulhas Current flow of subtropical water from south of Madagascar to the Cape peninsula; and (v) the Agulhas Retroflection area to the south of Africa. These individual flows combine to transport water from the SEC up to high southern latitudes and (from the latitude of northern Madagascar) represent the poleward, western limb of the southern subtropical gyre of the Indian Ocean (Wyrtki, 1973b). The seasonal depth of the mixed layer varies rather similarly, but is forced differently, over most of this province. During the southwest monsoon of boreal summer (June–September) there is a general westward tilt of the pycnocline of the Indian Ocean which results in a deepening of the pycnocline along to the coast north of Madagascar, from about 40–60 m to about 80–100 m. Of course, this is also the austral winter so that winter wind mixing and heat loss at the sea surface combine to deepen the mixed layer to about 100 m to the south of Madagascar.

East African Current

This name is often applied to the variable flow along the coast from Somalia to the Madagascar Channel. At about 10°S, to the north of Madagascar, the SEC meets the African continent and diverges north and south throughout the year. During the boreal summer, the northward flow is continuous with, and enters, the deep flow of the Somali Current, which is spun-up by the strong southwest monsoon winds (see Arabian Sea Upwelling Region). In boreal winter, the northward coastal flow of water from the SEC encounters the southward flow of the reversed, slower Somali Current, which during this season is extended along the East African coast: both diverge from the coast at about 2 or 3°S off Malindi and enter the SECC to pass eastwards across the ocean.

Mozambique Channel and Madagascar

The flow in the Mozambique Channel is generally to the south, influenced by a persistent mesoscale anticyclonic gyre field having three major eddies (possibly topographically induced) usually in the northern, central, and southern parts of the channel (Saetre and Jorge da Silva, 1984). Flow from the Mozambique Channel into the Agulhas Current recirculates through these eddies and is therefore intermittent.

Along the remarkably linear east coast of Madagscar there is a minor western boundary current, with strongest flow during the northeast monsoon. The north–south topography of the Madagascar Ridge, to the south of the island, may at times induce bifurcation of this flow directly into the subtropical gyre by retroflection south of Madagascar, but at other times the East Madagascar Current feeds directly into the Agulhas Current at the surface (Lutjeharms, 1988). When it occurs, the retroflection of the East Madagascar Current is accompanied by active meandering and eddy formation.

Agulhas Current

The strength of the Agulhas Current does not vary seasonally for the same reasons as its larger analog, the Gulf Stream of the North Atlantic at similar latitudes (Pearce and Gründling, 1982), where wind stress is transformed into mixing rather than momentum (see Chapter 3). It is, however, the strongest of the western boundary currents and flow frequently reaches 5 knots. This strong southward flow is directly contrary to storm swell generated in the Southern Ocean and propagating northwards and which may be doubled in amplitude by the interaction. These very long wavelength, very high swells are among the most impressive anywhere, and though they may have no ecological consequences we need to worry about here, they do enable albatrosses to put on a very spectacular show.

The origin of Agulhas flow is in the region of variable circulation at the southern end of the Mozambique Channel at about 25°S. Flow is essentially continuous from Delagoa Bay (26°S) to the southern extremity of Africa. As in other boundary currents, three zones can be distinguished: an inshore, coastwise zone of cyclonic shear; a meandering jet of maximum velocities; and a seawards zone of anticyclonic shear. Gradients of velocity and temperature between these zones resemble those of the Florida Current and the Kuroshio. Maximum velocities are over the continental slope, usually about 40–50 km offshore but ranging from 25 to 150 km depending on the form taken by meanders. An inshore countercurrent (or recirculation flow) may be induced (Schumann, 1982), especially where the shelf is relatively wide off Natal. This countercurrent may be associated with pulses of vortex shedding (the "Natal Pulse") from the Natal Bight (Lutjeharms and Connell, 1989).

Flow in the Agulhas Current becomes progressively wider and border perturbations increase in importance downstream so that the dimensions of plumes and eddies associated with them also increase progressively (Lutjeharms *et al.,*

1989). Some meanders are persistent in location, as off Cape St. Francis at 34°S where an episodic meander occurs which may attain a wavelength 100–150 km, moving downstream at 15 km per day (Swart and Gonzalves, 1983). Though mesoscale eddies may be shed by meanders at the seaward boundary of the current at any latitude, they are most frequent and well studied in the terminal Agulhas retroflection area south of South Africa. A persistent anticyclonic eddy (the "Agulhas Eddy") is located 400 km off Natal (Pearce, 1977).

Agulhas Retroflection

This is a unique and persistent terminal feature located where the western boundary current runs out of topography at the southern extremity of Africa and encounters the eastward flow of the circumpolar zonal currents. This feature is difficult to place in our system of provinces, and it is located here simply because its ecological effects (seen as blooms in CZCS images) are continuous with those on the Agulhas Bank. Otherwise, it might be better considered as part of the South Subtropical Convergence (SSTC) province since the flow from the retroflection (the Agulhas Retroflection Current) feeds directly into the easterly jet current flowing around the annular core of the SSTC (Fig. 8.8).

The Agulhas Retroflection differs from that of the Brazil Current, whose flow is forced to turn eastwards by the topography of the northern edge of the shoal water of the Falkland Shelf and plateau. This occurs far to the north of the southern extremity of the continent at Cape Horn. Conceptual models of the Agulhas Retroflection (Lutjeharms and van Ballegooyen, 1984; Peterson and Stramma, 1991) suggest that the scale of the retroflection loop is large (>300 km) and that the flow of the Agulhas Current reverses in this anticyclonic feature by 180° to flow eastwards along 40°S as the South Indian Ocean Current into the zonal circumpolar current system (Stramma, 1992) and the SSTC province. Terminal flow of the Agulhas Current is topographically located along the southern slope of the Agulhas Bank, and the eastward flow from the retroflection loop (as the Agulhas Return Current) is modified by the Agulhas Plateau (21°E,39°S), which lies in its path.

Warm-core anticyclonic eddies are shed from meanders on the western flank of the Agulhas flow after its passage around the Agulhas Bank; some of these, particularly those shed early, pass into the South Atlantic at the origin of the Benguela Current system. Eddies continue to be shed around the retroflection loop, but these normally follow the general flow and pass eastwards to the south of the Agulhas Return Current.

Biological Response and Regional Ecology

Only fragmentary information on the ecology of this province is available, though here the CZCS images should be relatively reliable since over much of the coast the influence of riverborne and suspended sedimentary material should be small; only the outflow from Kipling's "great, green, greasy Limpopo" at the southern end of the Mozambique Channel is well seen in CZCS images.

Along the East African coast (Tanzania and Kenya), downwelling occurs throughout the year but especially during the southwest monsoon when the current is strongest. For this region a few ecological observations exist off

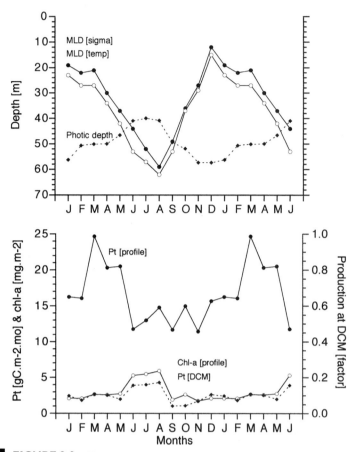

FIGURE 8.9   EAFR: characteristic seasonal cycles of monthly averaged mixed layer and photic depths, chlorophyll at the surface, and rate of primary production, both depth-integrated and at the DCM. Data sources are discussed in Chapter 1.

Dar-es-Salaam and Zanzibar, at 5 and 6°S, which should be typical of the whole of the East Africa Bight (0–10°S). During the southwest monsoon (e.g., May–September) irradiance at the sea surface falls from 520 to about 380 langleys, surface temperature falls from 29.5 to 26°C, and wave height and rainfall are much increased. All this is associated with a deepening of the mixed layer (see above), a reduction in chlorophyll and numbers of larger algal cells, and generally lower abundance of zooplankton. Planktonic primary production is highest and large-celled algae are most diverse off Dar-es-Salaam during the northeast monsoon when stability is strongest in the water column, though the seasonal changes are not large: Mixed layer chlorophyll is 0.3–0.6 mg chl m^{-3} during the southwest monsoon and 0.5–1.0 mg chl m^{-3} during the northeast monsoon. A nitrogen maximum in the premonsoon period (February and March) may be associated with nitrogen fixation by heavy blooms at that season of *O. erythraea*.

Where the converging flows of the East African Current and the reversed Somali Current of boreal winter leave the coast, there is some upwelling in the retroflection area at about 2°S off Malindi (Kabanova, 1968) and an algal bloom associated with this retroflection is seen in some CZCS images, at the base of an eastward filament aligned along the equator (Longhurst, 1993).

Off southern Africa, local topographically driven upwelling occurs in the Agulhas Current, downstream of major coastal prominences at 25 and 34°S. Farther to the north in the same western boundary current system, the East Madagascar Current induces upwelling off the southern tip of Madagascar (Lutjeharms *et al.*, 1981). The same occurs in the Mozambique Current on the African coast during the northeast monsoon when this flow is strongest (Saetre and Jorge Da Silva, 1984). These blooms can be located in CZCS chlorophyll fields, which show enhanced chlorophyll in appropriate locations on the coastal side of the axis of the Agulhas Current. Off the western coast of Madagascar, apparent coastal blooms in level 3 images may simply reflect the effluent from the rivers draining the coastal lowlands on the western side of the island.

In the south, over the Agulhas Bank (35°S) there is good information on the seasonal succession of primary production and how it is partitioned between nanophytoplankton and >15-μm cells of the "net" phytoplankton (McMurray *et al.*, 1993). This is a typical temperate shelf production cycle: In winter, the mixed layer is deeper than the euphotic zone, nitrate and chlorophyll are thoroughly mixed throughout the euphotic zone, and production is light-limited. In summer the euphotic zone is thermally stratified and a nitracline occurs within the shoaler thermocline; a DCM forms at the depth of maximum nitrate gradient, which lies at about the 3–10% isolume. The summer DCM is a typical feature with a shade-adapted flora and accumulation of sinking cells in the density gradient and pertains to a nutrient-limited system; it is approximately coincident with the depth of maximum production rate. Between these two seasons, a spring diatom bloom occurs when irradiance at the sea surface increases, and when turbulence in the mixed layer decreases before the summer, shallow thermocline is established.

During summer on the Agulhas Bank, temporary uplift of the thermocline to depths shoaler than 30 m occurs because of shelf-edge upwelling events and the intrusion of cold bottom water up over the bank. These events result in episodic blooms of diatoms during the normally oligotrophic summer period. During most of the year, 60–95% of all production is contributed by small cells (<15 μm); during the onset of stabilization in spring and the summer bloom events, 60–85% of total production is contributed by larger diatoms (*Chaetoceros, Bacteriastrum,* and *Nitzschia*). The CZCS images suggest that these events produce the most prominent chlorophyll features associated with the Agulhas Current and that they are quite variable between years. In 1981 and 1982, a linear bloom occurred along the outer edge of the bank and was sufficiently persistent to become a prominent feature of the 3-month composite image for austral summer. In 1979, blooms were very small and limited to the neritic zone, whereas in 1980 they were somewhat intermediate.

Another prominent feature is the edge of the eddies shed in the Agulhas Retroflection: Algal growth within these warm-core rings is limited by convective instability, whereas algal growth outside them is light limited. Maximum chlorophyll concentrations around the edges are consistent with ring-induced stability (Dower and Lucas, 1993).

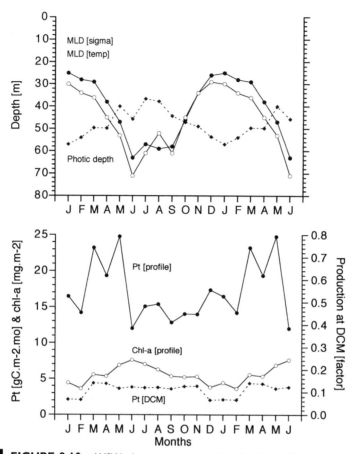

FIGURE 8.10   AUSW: characteristic seasonal cycles of monthly averaged mixed layer and photic depths, chlorophyll at the surface, and rate of primary production, both depth-integrated and at the DCM. Data sources are discussed in Chapter 1.

Synopsis

Weak austral winter deepening of the mixed layer by wind mixing in the south, and westwards thermocline tilt in the tropical region, leads to depression of the rate of primary production and a pycnocline which is deeper than the photic depth from June to September, though chlorophyll seasonality is very slight when integrated over this long province (Fig. 8.9).

Australia–Indonesia Coastal Province (AUSW)

Extent of the Province

Like EAFR, this coastal boundary province is largely a matter of convenience to avoid excessive subdivision in this work. As defined here, it extends from northern Sumatra to southern Australia. Like EAFR, it could readily and logically be subdivided into several primary entities. Were this to be required, a logical first cut would be the following: (i) the tropical southern coasts of

Sumatra, Java, and the Lesser Sunda Islands; (ii) the subtropical western coast of Australia from Cape Leeuwin (35°S) to the Bonaparte Archipelago (15°N) in the southeastern Timor Sea; and (iii) the temperate coast of the Great Australian Bight from Cape Leeuwin eastward to the Bass Strait.

The offshore boundary is identified in the usual manner, where limits of coastal flow can be identified (Colborn: coastal parts of 22, 23, 25, 26, and 34).

Continental Shelf Topography and Tidal and Shelf-Edge Fronts

The southern islands of the Indonesian archipelago have a very narrow or nonexistent continental shelf, except off central to northern Sumatra where the chain of the Mentawai Islands 100–200 km offshore encloses a significant area of shallow water. The shallow shelf of the Timor Sea, not part of this province, is continuous with the wide Sahul shelf off northwestern Australia from the Bonaparte Archipelago to Northwest Cape; the shelf edge runs approximately parallel with the coast, about 150–250 km offshore. Off Western Australia from Northwest Cape to Cape Leeuwin, the shelf is much narrower (50–100 km), especially off Perth (32°S). In the Great Australian Bight, there is a further region of significant shelf, reaching 150 km at about the center of the bight, south of the flatlands of the Nullabor Plains.

Defining Characteristics of Regional Oceanography

It will be convenient to deal with the three subregions previously outlined individually.

South Coast of Sumatra and Java

The Java Coastal Current is reversed by the seasonal changes in orientation of local monsoon winds and the effects of the monsoon rains of boreal summer over the highlands of the southern Indonesian archipelago (Quadfasel and Cresswell, 1992). The ITCZ lies over the southern coasts of the archipelago in its most southerly position in boreal winter so that cyclonicity of monsoon winds approaching the doldrums at the ITCZ requires that wind direction at the coast of Java and southern Sumatra be from the southwest. As the ITCZ moves north to the latitude of the Philippines in boreal summer, monsoon winds become southeasterly during the rainy season from May to September. Precipitation exceeds evaporation (by 15 cm per month) during this period, whereas evaporation dominates (by 5 cm per month) during the dry season of boreal winter.

The low salinity of coastal waters during the wet season induces a cross-shore pressure gradient and so forces flow as a westward jet current at the coast, reinforced by the surface wind stress. This coastal jet forms part of the origin of the South Equatorial Current which is at its most northerly position in boreal summer. The Java Coastal Current is narrowest along central Java and widens toward the west, particularly after it encounters southward drift along the west coast of Sumatra, where both flows turn offshore to enter the SEC and flow together to the southwest along a convergence zone.

The induction of the westward flow of the Java Coastal Current is progressive from east to west, starting at 120–125°E (south of the islands of Sumba and Timor) in February and reaching 105°E (off southern Sumatra) by June. The collapse of this current, however, occurs simultaneously along the whole coastline (Quadfasel and Cresswell, 1992). During boreal winter, when wind stress is toward the east, flow in the same direction is induced along the whole coast from the equator (off central Sumatra) into the western Timor Sea. The current reversal between seasons is influenced not only by local winds but also by distant forcing of strong westerly winds in the central Indian Ocean which force an eastward-propagating equatorial Kelvin wave.

Associated with the season of strongest westwards flow of the Java Coastal Current, a period of significant coastal upwelling occurs from May to September along the coasts of Java and Sumbawa from 105 to 120°E (Wyrtki, 1962). This is a deep upwelling and 200-m temperatures in the center of the upwelling area off eastern Java fall to only 12°C compared with 18–20°C at the same depth in the open ocean between Java and western Australia. At the same season the nutricline rises toward the surface by about 100 m.

Western Australia

The Leeuwin Current is anomalous among all other eastern boundary currents in that it flows polewards, not equatorwards, despite the equatorward wind stress over the eastern Indian Ocean and the general equatorward flow in the eastern limb of the subtropical gyre further offshore. The Leeuwin Current is a surface flow of warm, low-salinity tropical water above an equatorward undercurrent, with the level of no motion between the two flows being approximately 200–300 m. The anomalous poleward flow is forced by the strength of the onshore geopotential anomaly caused by the poleward pressure gradient in the eastern Indian Ocean—this is sufficient to overcome equatorward wind stress (Church et al., 1989). In the coastal current the eastward component of flow arising from the alongshore pressure gradient creates a downwelling situation so that isopycnals slope down toward the coast (Thompson, 1987).

The source waters for the Leeuwin Current are off the northwest Australian coast over the Sahul shelf, where a pool of warm, low-salinity water is formed by the seasonal flow through the Indonesian archipelago. This surface water mass then moves southwards into the Leeuwin flow (Church et al., 1989), which is maximal in late austral autumn and early winter (March–May) when the velocity core of the current is over the shelf edge; in winter it moves farther seawards. The current weakens in austral spring (September–December), when flow returns to the shelf edge, and is weakest in early austral summer (January). These variations are due to changes in local wind stress rather than changes in the longshore pressure gradient (Smith et al., 1991).

The Leeuwin Current is broad and shallow in the north and narrow and deep in the south where it tends to lie beyond the shelf edge, and satellite thermal imagery shows it to be strongly meandering. The eddy field extends at least 250–350 km offshore from the Sahul shelf south to Cape Leeuwin and frequently includes prominent warm filaments exhibiting vorticity, especially south of about 32°S, that extend even farther into the offshore equatorward flow of cool, high-salinity water of the subtropical gyre, sometimes referred to as the West Australian Current. The temperature gradient between Leeuwin flow and the offshore gyral water becomes stronger progressively southwards, despite the progressive entrainment of subtropical water into the southward flow. Finally, as the Leeuwin Current rounds the corner into the Great Australian Bight it narrows and rides up onto the continental shelf.

Great Australian Bight

The coastal boundary of the southern coast of Australia, between Cape Leeuwin in the west and Tasmania in the east, is under the influence of Leeuwin Current water which is topographically locked to the continental edge as it rounds the Cape Leeuwin and proceeds eastwards along the southern Australian shelf.

After rounding the corner, flow follows the shelf break of the Great Australian Bight as far as 130°E (Rochford, 1986) and is maximal in austral winter (May–October). After July, the eddying flow recedes from the eastern part of this coast. However, the warm and saline surface water mass which occupies the central and eastern part of the bight during much of the year is probably water from the Leeuwin Current modified by the arid and evaporative nature of the coastal climatic regime. Cold, high-salinity water of the west wind drift of the Southern Ocean current system lies above the slope year-round, and when the Leeuwin Current flow is relaxed in austral summer this water mass intrudes over the shelf break and may flood onto the continental shelf. Off the eastern shelf, dense high-salinity water may cascade from the shelf and produce temperature inversions (Godfrey et al., 1986).

The shelf edge off this coast is thus a region of very active frontogenesis. The shelf edge current is fast (<1.5 m sec^{-1}) and strongly baroclinic in the west, especially after it rounds Cape Leeuwin, where its dynamics become nonlinear and current speed increases due to a Bernoulli effect. The fronts between warm Leeuwin Current water and the offshore cold water of the west wind drift are sharp and extend to <200 m. Large (200–300 km) cyclonic eddies and sickle-shaped vortex filaments curling back toward the west occur on the seawards side of the shelf edge current and are especially prominent between Cape Leeuwin and Cape Arid, a distance of >800 km. Major eruptions of warm water as filaments (often terminating in eddy pairs) into the west wind drift occur several times a month when flow of the Leeuwin Current is strong (Griffiths and Pearce, 1985).

Biological Response and Regional Ecology

By far the most comprehensive ecological coverage of this part of the ocean was the Australian contribution to the IIOE, which was reviewed in the description of the ISSG province offshore of AUSW. This study (usefully described in CSIRO, 1969) was a meridional section along 110°E, which must have taken it up through that part of the outer eddy field of the Leeuwin Current which lies south of the Sahul shelf (about 25°S) and through the outer part of the Java upwelling from 8 to 10°S. Of the salient findings, those most likely to be of significance in the coastal boundary provinces are the following: (i) The seasonal data showed that surface cooling occurs in winter along the whole section, but that deep winter mixing occurred only farther south than $>30°$S, when 100 m nitrate values were enhanced. This was followed by a winter–spring bloom (indicated by enhanced rates of primary production) at $>30°$S, which evolved into a DCM at 75 m and primary production maximum at about 60 m, weakening as the summer proceeded: (ii) to the south of Java at 8–12°S, there was much cooler water and higher nitrate values at 100 m in all months than along the remainder of the section. In the same region primary production rates were maximal from the surface to 60 m in austral winter (May–November), and during the same period zooplankton biomass was higher in the Java up-welling area than anywhere to the south.

This upwelling along the south coast of Java during the southeast monsoon of austral winter is also reflected in maps of the distribution of nutrients below the photic zone in the eastern Indian Ocean (Levitus, 1982). Though the observations are few, rates of primary production in a coastal cell extending from Java to about 10 or 11°S results in elevated chlorophyll ($<$1.1 mg chl m^{-3}) and rates of primary production (0.7g C m^{-2} day^{--1}) in austral winter (Humphrey and Kerr, 1969; Jitts, 1969). In January and February, values are about 20% of those of the upwelling season May–October.

Along the Sahul shelf of northwestern Australia, tidal mixing is strong and several tropical rivers discharge onto the shelf so that high turbidity and regional algal blooms result. Although there is a seasonal pulse of higher chlorophyll values, especially subsurface and at the shelf break, these are unusual because the source of nutrients on the Sahul shelf (Tranter and Leech, 1987) is neither upwelling (as had been previously thought) nor riverborne nutrients but rather intrusions of cool, nitrate-rich slope water over the shelf edge below the warm, low-salinity surface layer. These intrusions occur most strongly in austral summer when the Leeuwin Current is relaxed, and they result in near-bottom chlorophyll maxima at the break of slope at about 100 m. These chlorophyll maxima ($>$0.5 μg liter^{-1}) are continuous with the offshore DCM, a situation which also occurs below the Florida Current and probably in other boundary currents. Because solar radiation at all seasons on this coast is so strong, the subpycnocline water is sufficiently illuminated for algal growth. In austral winter, stratification is reduced and a "phytoplankton-dispersed" season (April–July) replaces the summer "phytoplankton-stratified" situation (August–March).

Though tidal stream velocities have a fivefold amplitude range between spring and neap tides, they are uniform at depths $>$40 m because solar irradiance is so strong on the Sahul shelf as to impart sufficient stability to the upper water column to resist tidal mixing over much of the shelf. Shelf sea fronts and local eutrophication are therefore not a feature of this otherwise apparently suitable wide and flat shelf (Tranter and Leech, 1987), despite the fact that this is a region of strong potential internal tidal energy (Le Fèvre, 1986).

Further to the south, within the flow of the Leeuwin Current (both before and after its turn to the east around Cape Leeuwin) the thermal signals from the mesoscale offshore filaments and eddies are of similar magnitude to those observed seawards of eastern boundary currents, but here it is not cool upwelled water which it transported away from the coast but rather warm, nutrient-poor water carrying low biomass of pelagic biota. Thus, it is only in the vorticity of the offshore eddy field that nutrients will be advected to the photic zone and where we can expect to observe algal blooms in the CZCS images.

Though there is very little information on ecological seasons south of the Sahul shelf, we have to assume that a temperate cycle develops more typically toward the south, similar to those described in other provinces. This may also be inferred from the section at 110°E which, as noted previously, resembles the seasonal cycle in the Sargasso Sea.

Based on a temperature criterion, the mixed layer depth of the Great Australian Bight deepens from <50 m in summer to 100–150 m in winter. Such a seasonal range must surely be associated with spring bloom and summer oligotrophic conditions perhaps more like those of the North Atlantic model than those observed at the southern end of the 110°E section. However, this remains only speculation.

Synopsis

Shoal, permanent pycnocline is moderately deepened during austral winter so that it lies shoaler than photic depth only in summer (Fig. 8.10). There is some increase of primary production rate to entrainment of nutrients during the deepening phase of mixed layer, with slight response of chlorophyll accumulation during the same period.

9

PACIFIC OCEAN

The Pacific Ocean is clearly a special case, having an areal extent (165×10^3 km^{-2}) which is almost half the total area of all oceans combined. It has a zonal, east–west dimension from 80°W to 130°E across 210° of longitude, about 60% of the circumference of the earth. In the not-so-long-gone days of passenger travel by sea, 21 days nonstop (and always out of site of land except for Pitcairn Island) from Panama to New Zealand gave me a firsthand understanding of Pacific dimensions. The immense size of the Pacific is reflected in strong longitudinal differences in mixed layer depth and other physical circulation features, themselves reflected in the regional phytoplankton ecology.

The Pacific Ocean is also rich in marginal seas, ranging from relatively small, highly evaporative basins (Gulf of California) to large epicontinental seas (Bering Sea). We also have to accommodate the Indo-Pacific archipelago stretching from northern Australia to Luzon and Formosa; flow between two oceans (Pacific and Indian) is exchanged across this complex region of shallow shelves and small, semi-enclosed deep basins.

In the subarctic zone of the North Pacific, unlike its North Atlantic analog, precipitation exceeds evaporation so that a surface layer of low salinity lies above a permanent halocline at 100–150 m. This feature therefore constrains the depth of winter mixing, and so the depth to which winter temperatures penetrate, and also constrains the renewal of mixed layer nutrients during the winter. The North Pacific halocline therefore has important biological consequences which distinguish the subarctic regimes of the two oceans.

The Pacific is not, like the Atlantic, broadly open to the arctic regions in its northeastern quadrant, so the flow of the Kuroshio—the Pacific analog of the

Gulf Stream—does not, like the Gulf Stream, lose part of its flow into boreal and arctic regions. Instead, it flows more directly across the ocean once it has departed from the Japanese mainland. This has important consequences for the comparative oceanography and ecology of the two oceans.

As in the Atlantic, the transition between the biome of westerly winds and the trade wind zone is not sharp. Though we use the zone of the Subtropical Convergence (or, in a more generalized sense, the zone of subtropical convergence fronts between the westerlies and the easterly trades) to locate this transition, we realize that this is imprecise; some winter mixing occurs equatorward of the Subtropical Convergence of the Northern Hemisphere in both oceans. This is, again in a general sense, the zone of winter, rather than spring blooms (McGowan and Williams, 1973; Banse and English, 1993) if this term is relevant to such very weak chlorophyll enhancement.

Perhaps most important, we must accommodate the fact that, more strongly than in the other oceans, Pacific circulation exists in two modes depending on the interannual variability of trade wind stress (see Chapter 6). During El Niño–Southern Oscillation (ENSO) events, when trades weaken and are replaced by westerlies at low latitudes in the western Pacific, the heat balance of the entire ocean at low latitudes is perturbed for periods of a few months to 1 or 2 years. Here, for convenience, we regard this as the anomalous condition and consider the normal condition to be that obtaining when trade winds are well-developed. The between-year, ENSO-scale alternation between two states resembles in many important ways the seasonal response of the tropical Atlantic to between-season changes in trade wind stress. This difference between the two oceans is a consequence of their different zonal dimensions (Philander and Chao 1991; Philander and Pacanowski, 1981; Philander, 1985), a fact which has important consequences for their dynamic biogeography (Longhurst, 1993).

The provinces which we must recognize in the tropical eastern Pacific during normal conditions—the Pacific Equatorial Countercurrent (PEEC), Pacific Equatorial Divergence (PEQD), and North Pacific Tropical Gyre (NPTG) provinces—lose their defining characteristics partially or completely during an ENSO event. During such periods, rather uniform conditions with a deeper, warmer mixed layer obtain across the whole region. We shall refer to this, when convenient, as the Western Pacific Warm Pool (WARM)-ENSO Province while recognizing its ephemeral nature and intermittent manifestation.

Because of the dimensions and hence global importance of the Pacific Ocean for many issues that have evolved over the years, from fisheries potential to ocean–atmosphere interactions, climate control, and global carbon fluxes, we are fortunate that it has attracted at least its fair share of oceanographic effort in recent decades. We are also fortunate that much of this work has been centrally-planned (and peer-reviewed in the planning) so that the issues of the day and—more important from our point of view—the key scientific questions have had a chance of being answered.

Much of this work has been multinational and multiship, an increasingly important aspect in recent years; earlier, there were very extensive explorations by Soviet ships, though these unfortunately have now almost ceased. Quite large regions of the ocean have been studied in a systematic way, along preplanned grids of stations repeated seasonally where required, and it would take too much space to review them all. Because of the planned nature of much of the work that has been undertaken, the most important data sets are already generally available through the national data networks.

One of the minor consequences of all this effort is the fact that it is in the Pacific Ocean that we can best appreciate the gross distribution patterns in the biogeography of pelagic biota. Here, one has the impression that they have room to breathe and are much more readily associated with regional oceanography than in the other oceans where the current systems are less rectilinear. Perhaps the maps of John McGowan (1971) show this best. Here we see superimposed envelopes of distributions of groups of species of pelagic organisms—plankton to tuna—having subarctic, transitional, central, and equatorial distributions that, as we shall find, are in general agreement with the conclusions we reach here on quite different grounds.

PACIFIC POLAR BIOME

North Pacific Epicontinental Sea Province (BERS)

Extent of the Province

The North Pacific Epicontinental Sea Province (BERS) comprises the Bering Sea and the Sea of Okhotsk, enclosed by the arcs of the offshore Kuril and Aleutian Islands, respectively. There is considerable commonality between these two seas, but in a more dispersed classification of coastal provinces it would be logical to separate them.

Continental Shelf Topography and Tidal and Shelf-Edge Fronts

The shelf of the Bering Sea is a flat, featureless plain (as was noted in 1778 by Captain James Cook) and occupies most of the northeastern quadrant of the sea (Coachman, 1986); there is a strong break of slope at the shelf edge, with depths of 1000 m occurring only tens of kilometers beyond the 200-m contour. The shelf of the Sea of Okhotsk is less wide, the shelf break much less prominent, and the continental slope less steep. These facts have significance for the occurrence of a relatively strong slope current in the Bering Sea and an unusually complex system of shelf break and shelf fronts there, though tidal fronts are known to also occur on the continental shelf of the Okhotsk Sea (Zhabin *et al.*, 1990).

Three convergent fronts, defining three shelf domains, are normally identifiable in the open-water season over the continental shelf (Coachman *et al.*, 1980; Coachman, 1986) of the Bering Sea. An inner, coastal front lies above the 50-m isobath from Bristol Bay to about 60°N, where it passes east of St. Lawrence Island and so north to the Bering Straits. This coastal front encloses unstratified low-salinity coastal water, where the depths of wind and tidal mixing overlap. A middle front lies along the 100-m isobath and separates a mid-shelf region, which stratifies in summer, from the outer shelf region where a third, middle layer occurs in the water column. Stratification breaks down in the middle shelf region after severe storms or during extreme winter cooling

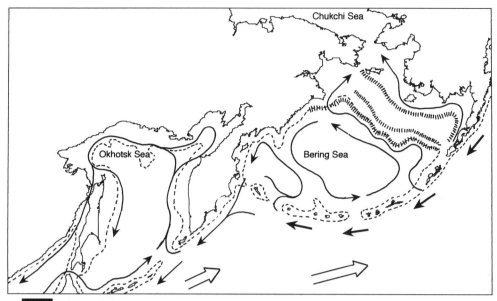

FIGURE 9.1 Idealized circulation of the Bering Sea and Okhotsk Sea, the 200-m isobath being the dashed line. The inner, middle, and shelf-break shelf fronts associated with the wide shelf of the Bering Sea are shown, and flow is shown to illustrate the significance of the Anadyr Current feeding Bering Sea productivity north into the Chukchi Sea and of the warm Soya Current passing from the Sea of Japan into the rim current of the Okhotsk Sea. The CZCS images (Fig.1.4 4) show how the Kuril and Aleutian island arcs generate almost permanent coastal blooms, as do the shelf edges of both seas.

(Schumacher and Reed, 1983). Still further seawards, a rather diffuse (50-km wide) shelf-break front occurs along the 200-m isobath and can be traced for 1000 km into the western Bering Sea; this front marks a transition between the isosaline water mass of the deep basin and shelf waters having important horizontal salinity gradients. Seawards of the shelf-break front, flow is consistently to the northwest around the central gyre, as part of the Bering Slope Current, whereas inshore of this front the flow is dominated by tidal forcing and is primarily cross-shelf. Figure 9.1 shows some of the features of this complex region.

Defining Characteristics of Regional Oceanography

Both of these marginal seas have seasonal ice coverage, especially in their northeastern quadrants, so each has a seasonally migrating marginal ice zone (MIZ) with all that that implies for their ecology. In the Bering Sea, locally-formed sea-ice covers the shelf from November to May, and at the end of winter it extends over 75% of the entire sea. Coverage of the Okhotsk Sea varies annually from complete (to the line of the Churl Islands) to <25% coverage.

The surface circulation of each sea is dominated by a central cyclonic gyre situated over the deeper parts of the basin, the gyres of the two seas being connected by southward flow of the East Kamchatka Current along the east coast of the peninsula (Dodimead *et al.*, 1967) which subsequently passes around the central gyre of the Okhotsk Sea to augment the southward flow of the cold Oyashio.

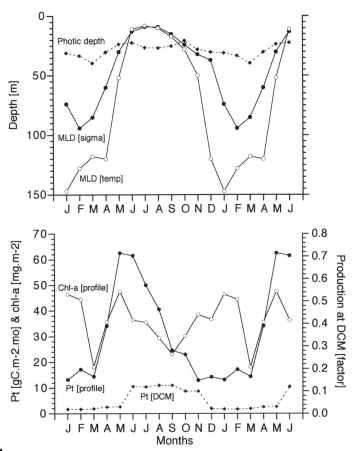

FIGURE 9.2 BERS: characteristic seasonal cycles of monthly averaged mixed layer and photic depths, chlorophyll at the surface, and rate of primary production, both depth-integrated and at the DCM. Data sources are discussed in Chapter 1.

Surface water flow into the Bering Sea is largely via the Alaska Coastal Stream, water from which passes across the shelf into the coastal current through the eastern passages between the Aleutian Islands, especially at Unimak Pass. Flow from the Alaska Stream into the central gyre of the Bering Sea occurs through the western passages; through Near Strait, the flow is almost directly into the East Kamchatka Current, whereas the Bering Slope Current, the deep eastern limb of the cyclonic gyre, is fed through Amchitka Pass (Favorite, 1974; Stabeno and Reed, 1994). Flow from the Bering Sea northwards into the Arctic Ocean originates in continental shelf water passing around the northern limb of the central gyre and into the Gulf of Anadyr, then northwards along the Asian coastline. In 1990, the axis of the Alaska Stream shifted southward, flow through the Near Strait into the western Bering Sea ceased almost entirely, and the East Kamchatka Current was seriously weakened. This entirely unexpected change in the oceanographic regime must have modified significantly the ecology of the whole Bering Sea basin because subsurface water masses must have cooled significantly.

Although the Alaska Stream loses its characteristic highly stratified structure passing between the Aleutian Islands, dispersion of low-salinity shelf water from the eastern quadrant results in strong stratification of the central gyre. Thus, the surface water mass of the Bering Sea, as in the Sea of Okhotsk, has low salinity and lies above a stable halocline at 100–300 m (Dodimead *et al.*, 1967; Takenouti and Ohtani, 1974). In both seas, there is a significant input of fresh water over the inner continental shelf both from ice-melt and from coastal runoff and rainfall.

Biological Response and Regional Ecology

In these two marginal seas, algal growth is both light and nutrient limited so that blooms are induced in spring and summer by a variety of mechanisms which bring nutrients into the seasonal euphotic zone. The Coastal Zone Colour Scanner (CZCS) images show apparently higher chlorophyll values over the shelf than the deep basin and evidence of a shelf-break bloom. Unfortunately, because the Sea of Okhotsk is almost entirely enclosed by the coasts of Russia, very little has been published concerning its ecology that is accessible. The best I can do is to predict that it will resemble to some degree what is known for the shelf domain of the Bering Sea. High chlorophyll values occur along the northern shelf during the summer (Ivanenkov and Zemlyanov, 1985), though the CZCS seasonal images suggest that a spring bloom occurs, apparently along the break of slope, on the eastern coast (i.e. , to the west of Kamchatka), whereas the northern and western parts of the sea are relatively oligotrophic. In summer, the situation is reversed with a bloom in the west, along the coast of the long island of Sakhalin; the north coast at Magadan appears relatively oligotrophic during the whole ice-free season.

The fisheries resources of the Bering Sea are very productive and it has long been assumed that they are supported by a productive shelf ecosystem. For this reason, the seasonal plankton succession in the Bering Sea has been well worked out by investigations in two fisheries-oriented programs: the PROBES and ISHTAR investigations (McRoy *et al.*, 1985; Hansell *et al.*, 1989). The original supposition has been confirmed and several mechanisms for intense growth of phytoplankton have been identified, as has the fact that the levels of production exceed the demand of the primary planktonic herbivores and so support a very rich benthic ecosystem—the food base for the fisheries resources. The identification of several characteristic oceanographic domains within the Bering Sea has proved to be a key feature in understanding the planktonic ecosystem.

Shelf Region

Epontic algal growth occurs below sea ice as early as reduction in snow cover and increase in sun angle will permit, and ice-edge blooms occur as soon as meltwater from the receding ice and increasing irradiance induces stability in the water column. These blooms may be terminated by wind-mixing events. Such processes vary in location and timing from year to year. In coastal water, enclosed by the inner front, a single spring bloom occurs early in the season, probably fueled by nutrients remineralized during the preceding winter. This bloom is initiated during May and continues for about 1 month while its depth of maximum chlorophyll values deepens by about 1 m day^{-1}; it moves progressively seawards as stability is induced in deeper water across the middle and outer shelf domains. Ephemeral blooms occur throughout the summer, especially in the middle shelf, whenever wind events disrupt the two-layered density structure and allow nutrient-rich bottom water to be mixed into the euphotic zone; even during the bloom the rate of primary production is much less (<20 g C m^{-2} day^{-1}) than that over the slope. Nutrient pumping in the bottom water onto the shelf is induced by the passage of low-pressure systems, which also enhance the cross-shelf advection rates; despite such events, new, nitrate-fueled production is minimal over the shelf and maximal along the slope. The same wind events may churn bottom sediments at depths <90 m and transport finer particles seawards in suspension: Such events will confuse interpretation of CZCS images. On the southeastern shelf, a high rate of primary production persists during the summer in the lee of the Aleutian Islands rather than in the highly turbulent flow through, for instance, Unimak Pass.

In the inner and middle shelf domain the zooplankton fauna is characterized by the copepods *Calanus pacificus* (= *C. glacialis* of Heinrich, 1962a; Motoda and Minoda, 1974), together with several species of *Pseudocalanus* and *Acartia longiremis* and other biota. The euphausiid *Thysanoessa raschii* dominates the early spring zooplankton, followed by *C. pacificus* during early summer. In this area, *C. pacificus* passes the winter as stage 5 copepodites (C5's), presumably in deep water beyond the shelf edge. Springer *et al.* (1989) computed that the demand of these herbivorous copepods, and their associated biota, was such that they were capable of modifying standing stocks of algae and at times required the total daily production to satisfy their needs.

Slope and Northern Bering Sea

The shelf-break front is associated with a major, linear area of high chlorophyll that can be traced north to the Bering Straits, with maximum rates of about 70 g C m^{-2} day^{-1}. Production in the shelf-break front begins about the same time as the spring bloom on the inner shelf and continues until wind stress in the autumn breaks down the pycnocline and deep mixing occurs. The chlorophyll profile follows a typical seasonal evolution: After the near-surface spring bloom, algal growth during the summer (a 120-day growing period) continues in a deep chlorophyll maximum (DCM) near the pycnocline, fueled by nutrients in the deeper water of the Bering Slope Current.

The DCM of the Slope Current bifurcates around St. Lawrence Island. The east part subsequently spreads over the northeastern continental slope, inducing rich benthic fauna, and exists as a near-bottom chlorophyll maximum. The west part is significantly enriched in the western jet current (Anadyr Current) which flows through the Bering Straits north into the Chukchi Sea. The shoaling of this flow as it enters the shallow water of the Anadyr Strait injects Slope Current nutrients into the photic zone as a cross-shelf flow (Springer and McRoy, 1993), inducing a plume of intense growth of chain-forming diatoms (reaching 16 g C m^{-2} day^{-1} and <600 mg chl m^{-2} in the center of the straits) which are transported into the Chukchi Sea through the Bering Straits. Springer and McRoy describe the Anadyr Current as a "north-flowing river of oceanic water . . .

maintains a portion of these shelf waters in a eutrophic bloom summer-long," and Sambrotto *et al.* (1984) stated, "this phytoplankton production system from June through September is analogous to a laboratory continuous culture." In short, nitrate is continually supplied as slope water passes up onto the shelf and cells are continually lost by sinking, to fuel high benthic production.

This domain of the Bering Sea supports an expatriate subarctic zooplankton assemblage, advected through the Aleutian chain and dominated by North Pacific biota: *Neocalanus plumchrus, Neocalanus cristatus, Eucalanus bungii, Metridia pacifica* (which comprise 70–90% of the copepod biomass), together with small copepods such as *Pseudocalanus* spp., the euphausiids *Thysanoessa* spp., the chaetognath *Sagitta elegans,* and others. These zooplankters support the immense flocks of diving planktivorous seabirds (mostly auklets, *Aethia* spp.) most of which nest in the northern part of the Bering Sea and whose numbers are evocative of a rich food supply. The larger copepods pass through one (*N. cristatus, E. bungii,* and *M. pacifica*) or two (*N. plumchrus*) generations a year (Heinrich, 1962b) and their abundance peaks during the summer.

Unlike the *C. pacificus* association of the eastern shelves, these oceanic zooplankton, which pass through the Slope Current and into the Anadyr Strait—that is, they are advected through the highly productive shelf-edge front and Anadyr Straits algal blooms—have a food demand which is far lower than the algal production rate, so the blooms are uncontrolled by herbivore grazing pressure. In fact, the advection of expatriate North Pacific planktonic biota into the Anadyr Straits is itself a significant carbon flux, headed for the deposits of the Chukchi Sea: Springer *et al.* (1989) compute this flux as 1.8×10^{12} g C during a single summer.

Deep Basin of Bering Sea

Because of the fisheries missions of PROBES and ISHTAR, little modern research has addressed the central cyclonic gyre since the review of Motoda and Minoda (1974), who gathered together scattered, earlier observations. It is clear from this study that (as could be expected) the populations of large oceanic copepods in the open gyral region resemble those advected into the Slope Current (see above) with biomass (20–40 g wet weight m^{-2}) not different from values over the shelf; only in the Slope Current itself are values significantly higher. A zooplankton profile in the eastern part of the gyre, but seawards of the shelf-edge front, shows a biomass maximum at the pycnocline at about 45 m which suggests that the DCM described previously extends clear across the gyre in summer. The standing stocks of diatoms near the surface in the central gyre in early summer are lower by about one order of magnitude than at the shelf break (e.g., 10^6 compared with 10^7 cells m^{-3}).

Synopsis

Winter mixing occurs to only 100 m, with a rapid near-surface stabilization in April and May, and is not captured by archived data (Fig. 9.2). Pycnocline shoaler than photic depth only in summer during declining algal biomass phase. Primary production rate conforms initially to irradiance cycle, but nutrient-limitation is imposed in May. Chlorophyll accumulation tracks productivity until August, whereas apparent accumulation in boreal winter probably represents sediments suspended during a period of deepening of mixed layer depth.

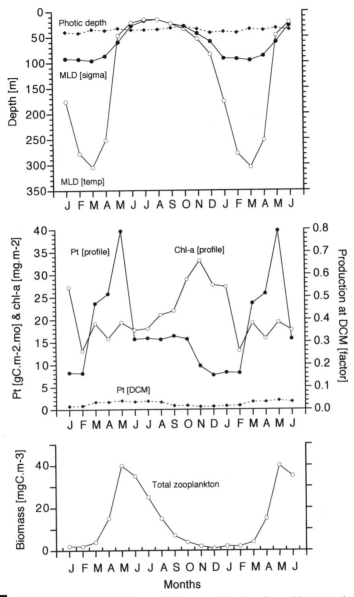

FIGURE 9.3 PSAGE: characteristic seasonal cycles of monthly averaged mixed layer and photic depths, chlorophyll at the surface, and rate of primary production, both depth-integrated and at the DCM. In PSAGW, the spring maximum for the Pt DCM factor is higher and stabilization of the water column more rapid. Zooplankton data from Frost (1987); other data sources are discussed in Chapter 1.

PACIFIC WESTERLY WINDS BIOME

Pacific Subarctic Gyres Province (East and West) (PSAG)

Extent of the Province

The Pacific Subarctic Gyres Province (east and west) (PSAG) is a rather large province, but I have chosen to discuss its two parts together so as to emphasize similarities, which are more important than their differences. It might be equally logical, though not as convenient, simply to treat them as two entities.

To the east and north, this province is enclosed by the offshore boundaries of the Alaska Downwelling Coastal Province (ALSK) coastal province and to the north and west by the boundary of BERS along the line of the Aleutian and Commander Islands, then seawards of the edge of the East Kamchatka Current, and finally along the Kuril Islands to the eastern cape of Hokkaido. To the south, I have taken the divergence of surface flow along about 45°N (Uda, 1963; Ware and McFarlane, 1989), where North Pacific Current and West Wind Drift waters diverge northwesterly and southeasterly. The boundary between the larger eastern and the smaller western subarctic gyres occurs at about 170–180°E, south of the westernmost group of the Aleutian Islands.

Continental Shelf Topography and Tidal and Shelf-Edge Fronts

There are none in this province. The continental shelves are entirely within the coastal boundary provinces whose shelf-break fronts bound this province. It is along the Kuril Islands at the extreme western end of the province that the effect of the proximity of coastal processes may be most significant.

Defining Characteristics of Regional Oceanography

We are fortunate in having three definitive reviews of this region (Dodimead et al., 1967; Uda, 1963; Favorite et al., 1976). I use them heavily in what follows. The subarctic (or subpolar) region comprises two partially-isolated cyclonic gyres; the whole province is encircled by flow of warm Kuroshio water along the West Wind Drift (or North Pacific Current) that passes into the Alaska Coastal Current and subsequently into the Alaska Stream along the Aleutian chain. The basin-scale cyclonic circulation is completed by cold water passing south out of the Bering Sea to the Oyashio/Kuroshio conjunction at about 50°N. This essentially closed circulation loses water only to the Bering Sea (see North Pacific Epicontinental Sea Province) and to the California Current: a round-trip takes about 4 or 5 years. Balance between flow from the North Pacific Current and West Wind Drift into the gyre and flow from all sources into the California Current system cannot be balanced without the addition of another term, the upward entrainment of subhalocline water into the surface layer. This process has significance for the biological dynamics of the province through the continual supply of deep nutrients to the photic zone.

Within this general cyclonic circulation are embedded two essentially independent, smaller gyres. The larger of the two, the Alaska gyre, is centered at about 55°N 155°W and its northern limb comprises flow of the Alaska Coastal Stream. The smaller, western gyre is centered at about 50°N 175°E. The southern limit of the province occurs at the northern limit of the Polar Frontal Zone (or the West Wind Drift), which shall be discussed in detail later); briefly, it is the zonal flow of water formed by mixing at the confluence of the Oyashio and Kuroshio Currents. To the north of this frontal zone lies subarctic water and to the south, subtropical water. The best marker of the edge of the subarctic water is the 33.0% isohaline in the salinity front at the northern border of the Polar Frontal Zone, where the halocline of the subarctic comes to the surface.

The eastern, Alaskan gyre has been the site of intensive biological studies at Ocean Weather Station (OWS) P (50°N, 145°W), whereas the western gyre is less well-known. The Alaska gyre has an elongated (southwest–northeast) cyclonic dome in the halocline, shoaling to 75 m in the center of the gyre. To the southeast of this dome, heavy precipitation accumulates low-salinity water by dilution. The dome, which is axial to circulation within the eastern gyre, can be identified by a surface salinity maximum in the range 32.8–33.0%. Seasonal changes in the depth of the surface layer of low-salinity water are slight, and it is effectively isolated from subhalocline water by the permanent halocline at 100–150 m. However, within the surface layer a thermocline is established at 30–60 m in early summer and remains at about the same depth until progressively deepening during fall and early winter. Between this thermocline and the permanent halocline, the salinity increases slowly but progressively during summer, suggesting some exchange across the top of the halocline. Recent calculations (reviewed by Miller *et al.*, 1991) suggest that Ekman suction of 1.5–3.0 m month^{-1} is the mechanism for a continual supply of new nutrients into the surface mixed layer. Above the seasonal thermocline, transient thermoclines occur with increasing frequency during the summer, mixing down within a few days to join the main seasonal thermocline.

The western gyre is smaller, triangular in outline (examine the triangular space between the Commander and Kurile Islands), and also has a central halocline dome. However, the halocline is deeper and extends to 200–300 m at 180°W. The western, equatorward limb of this gyre is a coastal jet, the Kuril Current, heading southwest and topographically locked to the seawards edge of the Kuril chain of islands; farther south, this is the Oyashio Current of the Japanese fisheries and oceanographers, which passes into a region of complex interaction with various branches of the warm subtropical Kuroshio flow east of Hokkaido (see Kuroshio Current Province) and is retroflected eastwards back across the ocean in the North Pacific Transition Zone Province (NPPF) province. This conjunction is productive of both warm and cold eddies with appropriate vertical density structure.

Though we shall discuss nutrients in greater detail later, it will be useful here to note that the southern boundary of the province is marked by a strong surface gradient in mixed layer nitrate during winter: A contour of 0.4 μM passes across the whole North Pacific basin, just beyond the southern boundary of the PSAG province, which is almost entirely occupied by values in the range 5–20 μM. Winter mixing is progressively deeper across the province to the west and reaches 150 m from 145°E to 180°W in the central part of the western gyre. Late spring mixed layer depths are approximately similar in both eastern and western gyres. Maximum mixed layer winter nitrates (15–20 mmol m^{-3}) occur south of the western Aleutians and also to the south of Alaska (Anderson *et al.*, 1969; Glover *et al.*, 1994).

Biological Response and Regional Ecology

Perhaps because it is in the Pacific that biogeographic patterns of plankton are most clearly expressed that the distribution of some individual species confirm the reality of the boundaries of PSAG. Between the northern edge of the Polar Frontal Zone and the northern coasts, McGowan's (1971) analysis lists an assemblage of subarctic biota, whose distributions very closely match the boundaries of PSAG with the single and important exception that they also occupy (given the transport patterns, they could hardly do otherwise) the adjacent coastal provinces ALSK and California Current Province (CALC) and also the polar province BERS. The most important species included in this assemblage are some copepods that will be discussed later (*N. plumchrus, N. cristatus,* and *E. bungii*) together with the chaetognath *Sagitta elegans,* the euphausiids *Thysanoessa longipes* and *Euphausia pacifica,* and the molluscs *Lima helicina, Clio polita,* and *Clione limacina.* This does not mean that deep tows will not encounter expatriates of some of these species in subsurface water masses far to the south; rather, the Pacific subarctic is their center of distribution. I have already mentioned *E. bungii* deep in the equatorial Pacific, and *N. cristatus* is transported equatorwards at 600–800 m with the submergence of Oyashio water below the Kuroshio off Japan: who knows how far expatriate individuals of these species may go?

More important, this province is the site of two of the great enigmas in biological oceanography today: (i) Why should there not be a spring bloom (in the sense of accumulation of chlorophyll) here as there is in the North Atlantic? and (ii) Why does active primary production during summer not utilize all the available mixed layer nitrate?

It is perfectly clear from the CZCS seasonal images that high-chlorophyll areas are restricted in the North Pacific to the coastal boundary at seasons when the central, open North Atlantic supports high levels of chlorophyll. Though the relatively shallow depth of winter mixing in the North Pacific, constrained by the halocline, is fundamental to this question, it is not alone sufficient to account for what is observed since nitrate remains unutilized under conditions when generally it would be expected to be wholly taken up by algal cells. For at least the past 10 years a debate has, if not exactly raged, then been actively engaged between two schools of thought: That grazing by herbivores is so closely coupled with algal growth that accumulation of cells does not occur in spring or, alternatively, that a Liebigian limitation to growth is imposed by a lack of Fe in the surface water mass. The latter was the hypothesis of the late John Martin in 1988, who later came to extend it to other "high-nitrate, low-chlorophyll" areas of the ocean (see Chapter 4). Unfortunately, as Banse (1990a) pointed out, the results of Martin's original flask experiments were equivocal.

In the open-ocean environment of the subarctic North Pacific the seasonal production cycle is reasonably well-known: Almost all accounts emphasize the "lack of a spring bloom," or rather the lack of seasonal accumulation of chlorophyll. This seems, however, a misuse of the term: more correct is Parsons' (1966) account of "the advent of the spring bloom," by which he meant the increase in the rate of primary production that had already been observed. This rate increase is initiated in the spring exactly as predicted by the Sverdrup model, when the mixed layer depth passes up through the critical depth in March and April. Recent time-series observations have shown this to be essentially correct, though there are some doubts about the absolute values of the rates achieved in summer. Production rates increase from a winter low of <100 mg C m^{-2} day^{-1} in March progressively to a summer maximum that extends from May until October having values in the range 200–400 or 400–1500 mg C m^{-2} day^{-1} in two sets of data, the latter being ultraclean observations made during the SUPER program.

Should the higher values be supported, they are as high as any observed in an open oceanic environment (Welschmeyer *et al.*, 1993). Nevertheless, the seasonal range of 0–50 m integrated chlorophyll is a relatively meager increase from about 12 to about 15 mg chl m^{-2} (McClain *et al.*, 1996). The integrated annual productivity was modeled to change by no more than 5% between years over a 30-year period, and it appears that variability in Ekman suction, mixed layer depth, and surface nitrate concentration had little effect on this result. A few anomalous high values during summer in the time-series noted previously may be accounted for by the observations that, sporadically during summer, ephemeral population pulses of large diatoms do occur at OWS P.

As in the North Atlantic, small nanoplankton (2–5 μm) dominate both autotrophic and heterotrophic biota (Booth *et al.*, 1993), though—also as in the Atlantic—autotrophic cells of 5–10 μm often contribute about half of all primary production. Though the partitioning of total biomass among autotrophic biota is the same (15 μg C liter^{-1}, 0–60 m) , the total biomass of all biota except *Synechococcus* is about half of the North Atlantic average. Curiously, noncolonial motile cells of the colony-forming *Phaeocystis pouchettii* are unusually consistent in North Pacific samples. Heterotrophic biota, mostly protists, have a total biomass closer to the North Atlantic average (30 μg C liter^{-1}, 0–60 m) and their computed grazing rate often exceeds the computed primary production rate.

Despite this close coupling between auto and heterotrophic microbiota, the subarctic ecosystem also supports an abundant population of several species of large herbivorous copepods, though their life histories do not resemble those of *Calanus finmarchicus* of the North Atlantic. *Neocalanus plumchrus, N. cristatus,* and *E. bungii* are the most important of these forms and comprise 80–95% of total mesozooplankton biomass. The last has a 2-year life cycle, but the others undergo maturation from C5's to adults below the surface layer without the long overwintering interval of *C. finmarchicus*. Consequently, these large copepods are already abundant in surface waters in spring when the phytoplankton bloom is initiated.

It will be useful to review the life cycle strategies of the dominant large copepod grazers of the North Pacific to compare them with those of the North Atlantic open ocean ; what follows is intended to enable a comparison between life histories at OWS Papa (North Pacific) with those of *C. finmarchicus* at OWS India (North Atlantic; see Chapter 7, Atlantic Subarctic Province) and is taken from Miller *et al.* (1984). *Neocalanus plumchrus* adults are present throughout the year at 400 m and some reproduction occurs in all months, peaking in winter. Larvae rise to the surface and pass through copepodite stages above 100 m, with the C5's descending into diapause at 500–800 m. Maturation of these into adults may be rapid or delayed, depending on the season. Reproduction is sufficiently pulsed to ensure highest numbers of copepodites near the surface in May and June, with lowest numbers in late summer. To emphasize the frailty of our contemporary nomenclature of important biota, I note (Miller *et al.*, 1994) that there are two forms of this "species," perhaps seasonal morphs: Both are lumped together as *Calanus tonsus* Brady (*f. typica* and *f. plumchrus*) in the Russian zooplanktological literature which forms an important part of the whole.

Female *N. cristatus* lie deeper (800–900 m) and reproduce in all months, though peak egg production occurs in November, and larvae pass to the surface where development through the copepodite stages occurs before copepodite 5's descend into diapause at 1000–1500 m. Once again, egg laying is sufficiently pulsed to ensure highest numbers of copepodites in surface waters in the first half of the year, peaking in May–July. *Eucalanus bungii* has a 2- or even 3-year life cycle, the complex details of which place the whole population in the upper 100 m in summer and from 250 to 500 m in winter. Development from eggs to C4's occurs in the first year, with the 3's and 4's overwintering at depth to go as far as C5 during their second summer, to overwinter as C5's during their second winter; maturing at depth, the new adults produce a fresh generation of eggs near the surface. Some of the females may enter diapause for a third winter and produce a second batch of eggs near the surface the following spring.

The important result of all these complexities is that some individuals of each species are present in the surface layers at all seasons. It is this fact which prompted the suggestion that sufficient grazing pressure was already available to suppress the accumulation of phytoplankton biomass during the spring increase in production rate. This hypothesis became the textbook example of the control of phytoplankton by mesozooplankton grazing. However, it has not stood up to careful examination. Careful grazing rate experiments made during the SUPER investigations showed conclusively that the total copepod community present in spring 1983–1988 at OWS P had a consumption of only 6–15% of the daily rate of primary production. Though capable of higher ingestion rates, the small size of autotrophic cells and their low-standing stocks keeps copepod grazing rates relatively low compared to rates (same species and same ocean) in the ALSK coastal province.

In light of this information, other solutions must be found to explain the residual nitrate. Zooplankton grazing releases ammonium, preferentially removes large cells, and must force more of the production/consumption cycle to pass around the microbial loop. Therefore, possibly NH_4 is used preferentially relative to its availability as a N substrate by at least the smaller autotrophic cells and this fact may be invoked to explain the residual NO_3: As Wheeler and Kokkinakis (1990) questioned, Does ammonium recycling inhibit nitrate use? By careful budgeting, and investigating temporal variability in NH_4 and NO_3 concentrations, it was concluded that this might be a significant factor by imposing an upper limit on NO_3 uptake in nutrient-rich, grazing-balanced ecosystems.

To complete the enigma, there remains the possibility that Fe-limitation establishes a phytoplankton community deficient in large cells, as suggested by Martin *et al.* (1989), who demonstrated a correlation between regional eolian Fe inputs, nitrate depletion, and community cell size in the Gulf of Alaska. In vitro cultures also showed complete nitrate uptake, compared with controls, with the addition of relevant amounts of Fe. It may be, as suggested by Banse (1990), that the Fe limitation applies only to large cells (small surface area/volume ratio in the presence of low concentrations of Fe), which normally can outcompete nanophytoplankton for nitrate because of their very rapid doubling rates. These large cells are not consumed by the smaller, very numerous micrograzers, and in the absence of large herbivores (as in the North Atlantic in early spring) they may escape consumption and instead accumulate or sink.

These are enigmas which will not be resolved quickly, probably not until an iron-enrichment experiment is successfully performed *in situ* at sea at OWS P (see PEQD for discussion of the successful IronEx experiments performed elsewhere). It must be pointed out, as an additional caution, that though we have come to take OWS P as a model for the subarctic domain of the North Pacific, it does in fact lie in one of the two areas of maximum winter mixed layer nitrate; we would be discussing levels of winter nitrate about one-third lower were OWS P to have been placed only a few degrees to the west and south at, for example, 45°N 160°W.

If eolian Fe-limitation is determined to be a critical factor in determining the composition of the dominant autotroph community in this province, it will no doubt be found to have a more significant effect in the Alaska rather than in the western gyre: Eolian dust over the North Pacific comes from the exposed loess deposits of eastern Asia and is attenuated eastwards (Duce and Tindale, 1991; Duce *et al.*, 1991).

We have less information on the western gyre as a unit, though transpacific sections contain information concerning the southern limb of the gyre. End-of-winter nutrient levels in the surface mixed layer are somewhat higher in the western (>25.0 μM) than in the eastern gyre (about 15.0 μM), and this difference is maintained through the summer, when the nitrate contours between 100 and 200 m slope down consistently eastwards. Surface observations across the subarctic biome show similar chlorophyll values between the two gyres in the first half of the year, though significantly higher values of both chlorophyll and productivity occur in summer in the west, especially close to the Kuroshio–Oyashio convergence (Venrick, 1991). A strong accumulation of chlorophyll in the open oceanic western gyre is clearly seen in the CZCS images which I attribute to the relaxation of grazing pressure, whereas in the Oyashio cold current along the western edge of the province winter mixing, a spring bloom of diatoms and a summer oligotrophic profile with dinoflagellates and a DCM have been described: This sequence is also supported by the CZCS images. Nitrate is reduced to 0.3–2.6 μM, which is much lower than normal summer levels in the eastern gyre.

Synopsis

There is moderate boreal winter mixing and pycnocline within photic zone from May to October (Fig. 9.3). The pulse of primary production occurs in March–May, during which chlorophyll accumulation is depressed by herbivores. Chlorophyll tracks productivity rate decline until September, when herbivores descend, so that strong accumulation occurs until November, when productivity takes winter values and chlorophyll is progressively lost, perhaps mainly by sinking. Cycles are similar in eastern and western parts of province.

Kuroshio Current Province (KURO)

Extent of the Province

The Kuroshio Current Province (KURO) comprises the flow of the Kuroshio from its origin east of the Philippines at about 15°N to the Shaskiy Rise at

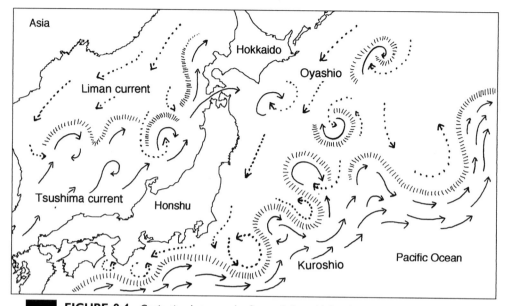

FIGURE 9.4 Conjuction between the flows of the cold Oyashio and warm Kuroshio Currents. Hatched lines show the ideal location of fronts between the cold (dotted) and warm (solid arrows) streams both in the Sea of Japan and to the east of the Japanese islands. The complexity of the system illustrated here is the source of the rich pelagic fisheries, with each arm of both cold- and warm-core eddies having its characteristic pelagic nekton (e.g., sauries, squid, skipjack, and sardine), well-known to regional fishermen.

about 160°E, which is a convenient marker for the termination of the Kuroshio Extension. Also included is the deep basin of the Japan Sea, excluding the shelf areas along the Chinese coastline which I place in the China Sea Coastal Province (CHIN) coastal province. The landward, warm-core eddy field along the shelf edge of the East China Sea and the flows from the Kuroshio over the shelf are considered part of the coastal boundary province of the East China Sea, whereas the eddy-field of cold-core rings seawards of the axis of the Kuroshio is considered to the part of this province. Also included in this province is the field of both warm-core and cold-core eddies from 35 to 40°N, east of Honshu and southern Hokkaido, where the flow of both Oyashio and Kuroshio turns definitively eastwards. Other arrangements could be considered: Perhaps because this eddy-field is the root of the eastwards flow in the NPPF province it might have been just as logical to assign it to that province. Figure 9.4 shows some of the complexities of this region.

Other arrangements were also considered for the Sea of Japan; the obvious alternative, which some may prefer, is to include it in its entirety as part of the CHIN coastal boundary province. The logic for retaining the eastern part in KURO is simply that this region is occupied by flow of unmodified Kuroshio water passing west around the island of Honshu. Because this province spans a wide range of latitudes (10–45°N) it might, for some purposes, be desirable to subdivide it: In this case, I propose that a latitudinal division be made at the Bashi Channel (about 25°N) north of Taiwan.

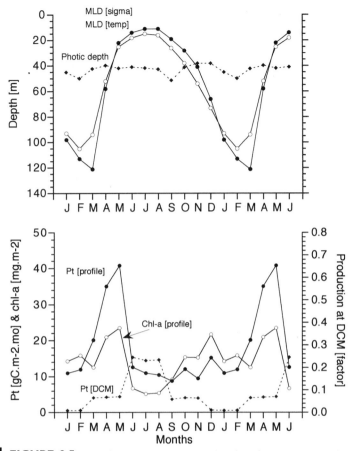

FIGURE 9.5   KURO: characteristic seasonal cycles of monthly averaged mixed layer and photic depths, chlorophyll at the surface, and rate of primary production, both depth-integrated and at the DCM. Data sources are discussed in Chapter 1.

Continental Shelf Topography and Tidal and Shelf-Edge Fronts

The continental shelf of the Japanese islands and the Sea of Japan is very narrow, and it is not useful to designate a separate coastal boundary province since the whole of the Kuroshio–Oyashio region is dominated by interaction with the topography of the islands and adjacent continent. Though the flow of the Kuroshio is for much its length locked to topography, it is only in the sector of the East China Sea that this topography is the edge of a shelf having significant width. The Sea of Japan has a small tidal range (0.2 m) and even on the Pacific coast of Japan the range is not large.

Defining Characteristics of Regional Oceanography

Much of what is reviewed here is derived from the work of Japanese oceanographers and fishery scientists, for whom the "black stream" of the Kuroshio has special significance. I have relied heavily on the many reports of the Cooperative Studies of the Kuroshio (CSK).

The Kuroshio is the western boundary current of the North Pacific subtropical gyre and invites comparison with its Atlantic homolog, the Gulf Stream. The beginning of the Kuroshio can be traced to the divergence of the North Equatorial Current (NEC) as it encounters the western margin of the ocean east of Mindanao at about 10°N (Nitani, 1972). The flow to the south passes around the persistent cyclonic Mindanao Eddy, and the flow to the north passes into an intensified western boundary current that flows northwest along the continental edge east of the Philippines, leaving a persistent anticyclonic eddy to the east at about the latitude (17 or 18°N) of Luzon, from whose shelf edge the velocity core separates before passing north to Taiwan. This separation is the site of active upwelling from October to January that is apparently unconnected to local wind forcing. As it goes along, the Kuroshio is progressively strengthened by lateral entrainment of water from the subtropical gyre and the pycnocline slopes upwards consistently toward the west. The main thermocline lies at 75–200 m near the shelf edge at 25°N and at 200–350 m out over the deep water.

The boundary current is intensified along the steep-to continental edge east of Taiwan and then follows the topography of the continental edge of the East China Sea, thus passing through the Bashi Channel and landwards of the long Ryuku chain of islands. The main flow then reaches the open ocean again through the Tokara Strait south of Japan (Taft, 1972; Nitani, 1972) but only after a major bifurcation of flow has taken place. The Tshushima Current carries warm Kuroshio and cooler shelf water into the Sea of Japan through the Korea Straits; this flow is at a minimum in boreal winter and is reinforced by summer polewards winds (Isobe *et al.*, 1994). The surface water mass of the Sea of Japan, like that of the shallow Yellow Sea, cools more strongly in winter than the main core of the Kuroshio that passes to the east of the Japanese islands. Warm water from the Tsushima Current is carried back to the open ocean through the strait between Honshu and Hokkaido as the Tsugaru Warm Current: "warm" because on reentering the Pacific Ocean it encounters the cold southwards flow of the Oyashio and turns south landward of this stream to enter the "perturbed region." Within the Sea of Japan, cold water enters from the north between Sakhalin Island and Siberia so the whole basin, which has no major capes or islands, has a gyre-like cyclonic flow: polewards flow of warm water up the Japanese coast and equatorward flow of cold water down the Sibero-Korean coast. The result is that the body of the sea is occupied by many wandering cold- and warm-core mesoscale eddies.

Velocity of the Kuroshio flow is intensified when a sharp angle is encountered after passing north of Taiwan. The upwelling in shallow water that is forced by this effect will be discussed in the account of the adjacent coastal boundary province (see China Sea Coastal Province). There is a minor gyral circulation over the Okinawa trough between the Ryuku Islands and the shelf edge so that counterflow runs south along the western margin of the Ryukus. Passage through the Tokara Strait generates a persistent field of mesoscale eddies on both sides of the velocity core, and these pass along with the flow close to the coast of Honshu.

Important and well-studied variation occurs in the main stream which passes south of the Japanese islands. South of southern Honshu (at about 135–

140°E) a variety of paths are taken by the core, principally in two modes: a simple alongshore path (i) flowing along the shelf edge or (ii) passing around a large cyclonic meander extending as much as 500 km offshore. In the latter case, which occurs during periods of high transport, the return limb of the meander lies along the western side of the topography of the Izu–Ogasawara Ridge, and the meander itself encloses a cold-core cyclonic eddy adjacent to the coast (Taft, 1972). Alternation between the two circulation modes occurs on the decadal scale and has important biological consequences (see, for example, Sherman and Alexander, 1989).

Retroflection occurs at a coastal prominence (Cape Inubo-Saki) at about 35°N (Su *et al.,* 1990). At this point, the current passes northeastwards in strongly meandering flow adjacent to the southerly flow of cold Oyashio water, and it has a cold wall—like the Gulf Stream—on its landward side. This is the Kuroshio Front, which frequently has a double-thermal structure and is continuous with the complex of fronts in the conjunction with the Oyashio northeast of Japan (Nagata *et al.,* 1986). The meanders and flow of Kuroshio water can then be traced eastwards into the North Pacific Current which is the frontal jet of the Transition Zone, or the NPPF province; the boundary between the Kuroshio and the NPPF may be set at about 155°E, above the topography of the Shatskiy Rise.

The retroflection eastwards of the Kuroshio, and its encounter with the cool Oyashio flow, is a region of great complexity, though now rather well described (Kawai, 1972). The warm core of the Kuroshio (about 20°C) turns offshore near Tokyo and enters a region of meandering interaction where we can identify (i) two intrusions or very large meanders of Oyashio water, (ii) the flow of the Tsugara Warm Current from the Sea of Japan through the straits between Honshu and Hokkaido, and (iii) the "perturbed region" described by Kawai where modified water masses occupy the region between the offshore trending Kuroshio and the coast. Fronts between the cool Oyashio, the perturbed area surface water, and the warm Kuroshio surface water are well-marked and form the origin of the two fronts associated with the oceanic Polar Frontal Zone as it passes east across the North Pacific. As the combined Kuroshio/Oyashio flows move eastwards they pass around a pair of persistent eddies (one anticyclonic and one cyclonic) that span almost 5° of latitude and longitude. As noted previously, I include this whole interaction region for convenience within this province.

The Kuroshio, like the Gulf Stream, is a region of anomalously high values for the kinetic energy contained in the mesoscale eddy field (Dickson, 1983), and this is undoubtedly the reason for the high variability noted in shipboard measurements of the chlorophyll field (Taniguchi and Kawamara, 1972). The eddy field of the Kuroshio resembles that of the Gulf Stream: Cyclonic cold-core eddies are shed to seaward (Kawai, 1979) and warm-core anticyclonic eddies to landward (Kitano, 1979) of the meandering velocity maximum of the current. Both kinds of eddies then move westwards, contrary to the direction of the current, for periods of up to several months. Although rings are perhaps not as numerous in the Kuroshio region as in the Gulf Stream, as many as 90 cold-core rings have been identified in historical data, many originating in the region of the major Kuroshio meander south of Honshu.

Warm rings occur mainly in the transition area between the mean positions of the Oyashio and Kuroshio Fronts and also about 120 km to the east of Hokkaido. The perturbed area between the cold and warm currents is rich in mesoscale features, including not only eddies but also warm streamers and filaments carried around cold eddies and secondary fronts (Kawai and Saitoh, 1986) and cold filaments carried around warm-core eddies (Sugimoto and Tameishi, 1992). Ephemeral episodes of warm water and strong currents in inshore waters and even bays along the south coast of Japan must reflect the incidence of warm-core eddies or meanders encountering the coastline. These Kyucho events are therefore the equivalent of the irruptions of warm water at the Nova Scotia coast (see Chapter 7, Northwest Atlantic Shelves Province) when a Gulf Stream eddy founders over the shelf.

Biological Response and Regional Ecology

Because of the great latitudinal extent of the province we should really discuss two regimes: (i) one that is tropical and probably very similar to the WARM province from 10 to 25°N, and (ii) the one having a temperate, winter-mixing regime from Taiwan northwards. Compared with the oceanographic information and with what is known of pelagic fisheries ecology (tuna, sardines, saury, etc.) there are few modern studies of the primary production and consumption regimes. Most of what is available to discuss refers to the region north of Taiwan; therefore, for the southern region, the "beginning of the Kuroshio" of Japanese oceanographers, I shall simply refer the reader to the discussion of production seawards of the current, in the adjacent province (NPTG).

For the northern region, I have been able to locate no time-series observations that would enable synthesis of a satisfactory seasonal cycle of production and consumption. There is a clear seasonal cycle in the mixed layer depth (summer, 15–25 m; winter, 80–100 m) and because there is extensive information on the presence and differentiation of a DCM during summer, we can reasonably assume that a spring bloom–summer oligotrophy situation is normal.

The CZCS images support this suggestion and show that the northern Kuroshio/Oyashio interaction region, east of Japan, has strong seasonal algal blooms, though these are more restricted spatially than those of the North Atlantic, with the regions of higher production values lying adjacent to the coastlines (Saijo *et al.*, 1970). Here, CZCS chlorophyll values are highest in spring, lower in summer, and (except along the shelf edge of the East China Sea) lowest in winter.

The strength of the summer DCM varies regionally, with the strongest features between Taiwan and the south coast of Honshu, where they are deep (100–150 m), near the bottom of the euphotic zone, and contain chlorophyll maxima in the range 0.5–0.6 mg chl m^{-3}. In this region, daily primary production reaches 0.4 g C m^{-2} day^{-1}. Close to the Tokara Straits themselves, DCMs are generally shallower and may extend to the bottom over the continental shelf of the East China Sea. In both spring and summer *in situ* observations show that the Kuroshio main flow corresponds with a linear chlorophyll feature, and this is supported by CZCS imagery. Chlorophyll sections across the stream southeast of Honshu in June show that the DCM has higher values within the axis of flow, and that it lies at a remarkably uniform depth (about 75 m and at the 1% isolume) from the coast to 400–500 km offshore. Predictably, the DCM lies the whole length of these sections in water with undetectable nitrate but immediately above a strong nitracline (Takahashi *et al.,* 1985).

Initiation of blooms and chlorophyll enhancement are induced by eddy vorticity, upwelling along the shelf edge topography, mixing with nutrient-rich coastal water, and the encounter with Oyashio water in the "perturbed area" between the cold and warm currents east of Hokkaido. The margins of warm-core rings may be associated with enhanced chlorophyll zones and higher than background abundance of other pelagic biota (Yamamoto and Nishizawa, 1986). This appears to result from the entrainment of slope-water biota in cold filaments drawn around the individual warm-core eddies.

The annual and decadal shifts in the core flow of the Kuroshio off the south coast of Honshu have important biological consequences that we can trace through their effects on the recruitment of the small clupeid fish that collectively comprise the Iwashi fishery: principally *Sardinops melanosticta, Engraulis japonicus,* and *Etrumeus micropus.* These three planktivorous fish are not always present in the same relative or absolute abundance, and major fluctuations have been observed since the historical fishery was started in the 1500s. These are major shifts in resources, the latest downswing being from about 1.1×10^6 tons in the 1930s to a very small catch in the 1960s. *Sardinops* is typical in its food requirements, is able to take large diatoms but cannot subsist on them entirely, and utilizes progressively larger zooplankton during its growth. *Engraulis* can utilize much smaller plankton organisms by active gill-raker filtration. It has long been known that these two biota tend to be mutually exclusive: many sardine and few anchovies or vice versa. It has also long been known that a major population of sardines would only occur when warm oceanographic conditions obtained on the coast of Honshu. It has been realized recently that the relative abundance of one or the other species is controlled by the path taken by the Kuroshio loop off Honshu. During the 1964–1971 period of rapidly declining *Sardinops* abundance, the Kuroshio meander to the southeast of Honshu formed a strong arc enclosing a cold cyclonic eddy at the coast where conditions for the survival of sardine larvae were poor for almost a decade.

Related to these observations is the fact that the conjunction between warm Kuroshio and cold subarctic water is also a boundary zone for the distributions of many pelagic species; the same must occur at the north wall of the Gulf Stream off eastern Canada. However, off Japan, the regional cultural interest in pelagic fisheries has led to a much greater knowledge of the distribution of pelagic invertebrates than in the Atlantic. For instance, maps of the distribution of the eggs and larvae of the Pacific saury *Cololabris saura* show this to be restricted strictly to the warm Kuroshio water. Similarly, two species of chaetognath show specialization to cold and warm water: *Pterosagitta draco* in Kuroshio water and *Sagitta nagae* in coastal water. For these two species the axis of the Kuroshio is a rather watertight fence, with only isolated patches occurring on the inappropriate side of the axis.

The major species of mesozooplankton resemble those of the open North Pacific, though the biomass is about one order of magnitude higher than that in the open ocean of the PSAG province. The copepod fauna is dominated by *N. cristatus, N. plumchrus, E. bungii, M. pacifica,* and species of *Pseudocalanus,*

Oithona, and *Euchaeta/Pareuchaeta.* These perform the seasonal and ontogenetic migrations described for the open-ocean provinces of the North Pacific and during summer they are aggregated from 10 to 50 m below the surface.

Synopsis

Moderate winter mixing occurs to 100–125 m so that pycnocline lies within photic zone for 6 months from April to October (Fig. 9.5). There is a strong, brief increase in production rate in spring (April and May) which is tracked by simultaneous chlorophyll accumulation and loss and a weaker autumn October–December increase. Both blooms are probably nutrient-induced and nutrient-limited. This model is strongly weighted by events in the Kuroshio/Oyashio retroflection area.

North Pacific Transition Zone Province (NPPF)

Extent of the Province

This is a linear zonal province lying across the North Pacific, from the end of the Kuroshio Extension over the Shatskiy Ridge at 145°E to about 145°W, where divergence of the surface flow commences into the coastal currents of North America. It is bounded by two oceanographic fronts at approximately 30–32°N (Subtropical Front) and 42–45°N (Polar or Subarctic Front) in the central Pacific. Because the term transition zone has been so loosely used, it is important to clarify what is not intended here: This is neither the transition zone of Roden (1970) or of McGowan (1971), who used this term for the broad flow of cool water from the North Pacific Current as it turns south into the California Current system, nor is it the 50 to 150-m deep halocline layer in the Alaska Gyre of Uda (1963).

Continental Shelf Topography and Tidal and Shelf-Edge Fronts

These are not relevant to this entirely oceanic province.

Defining Characteristics of Regional Oceanography

This province forms a transitional zone between the cyclonic subarctic and anticyclonic subtropical gyres of the North Pacific, and therefore includes the convergence between the westerly winds of the Temperate Zone and the subtropical easterlies which—farther equatorwards—are the trade winds. Roden (1970) reminds us that though climatic maps of wind stress show the westerlies to be a persistent flux toward the east, instantaneous maps will show them to be a series of depressions marching toward the east at the crests of long planetary waves, of which two or three usually occupy the air mass over the North Pacific. Now that satellite images are commonplace, it is well to remember what the single image is likely to show. We shall encounter the same phenomenon in boundaries between zonal flows of equatorial currents. Converging Ekman transport between the zones of westerlies and easterly trade winds forms an area of subtropical convergent fronts (Roden, 1975) at about 28–35°N across the ocean. This is our Subtropical Front or convergence, its position varying seasonally with the meridional migration of the wind systems; it marks the

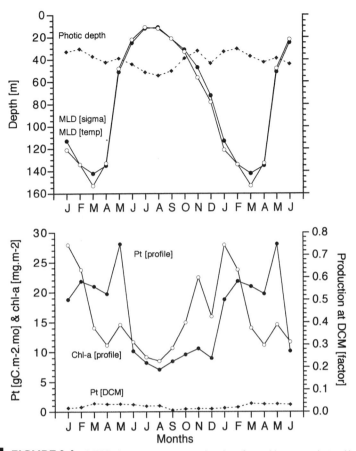

FIGURE 9.6 NPPF: characteristic seasonal cycles of monthly averaged mixed layer and photic depths, chlorophyll at the surface, and rate of primary production, both depth-integrated and at the DCM. Data sources are discussed in Chapter 1.

equatorward limit of deep mixing in winter and hence an alternation between mixed and stratified conditions.

The Transition Zone comprises the area of meandering flow across the ocean which originates at the Kuroshio/Oyashio confluence. As these currents turn eastwards the Kuroshio flow enters the northern limb of the subtropical gyre, whereas the Oyashio water enters the subarctic gyre to pass eastwards as the North Pacific Current. These two streams mingle and are bounded by the two fronts, separated by about 10–15° of latitude. Because of the isothermal character of subarctic surface water (see Pacific Subarctic Gyres Province) the northern front is marked only by a salinity gradient and is the front noted in discussing the southern boundary of PSAG, for which the 33.0 isohaline is a good indicator. Because of the seasonal thermal response of the ocean, the southern front is a temperature and salinity front in winter but only a salinity front in summer, and the surfacing of the 35 isohaline is a useful indication of its location during all seasons.

Particularly to the west of 170°W, the Transition Zone is energetic, and although density gradients across the feature are small, baroclinic shear is large, suggesting strong vorticity. Baroclinic eddies on the scale of 100–1000 km are frequent in the western half of the ocean. The Transition Zone also differs from the subarctic and subtropical water bodies between which it passes in having relatively low stability of the water column. To the north, the subarctic domain has a permanent shallow halocline (see Pacific Subarctic Gyres Province) which is extremely resistant to winter mixing, whereas to the south in the subtropical domain the combined effects of the thermocline and halocline confer high stability. In the Transition Zone, on the other hand, until a summer thermocline is established, convective motion can be significantly deeper than the 50 to 100-m winter mixing which is imposed by stability to the north and south of the Polar Frontal Zone (Roden, 1970, 1975). We can expect that these factors will have biological consequences distinguishing this province from those adjacent to it.

Biological Response and Regional Ecology

If indeed the Transition Zone is not only transitional between relatively high chlorophyll concentration to the north and low concentration to the south but also (as its analog, the Southern Subtropical Convergence Zone clearly does) has ecological dynamics which set it apart, the first place to look for evidence is in the CZCS images. The images for boreal spring 1979–1986 reported by Glover *et al.* (1994) show a chlorophyll front at 32–35°N, at the southern front of this province, but no enhancement within the province. However, the autumn and winter images from a reduced 1° grid set used to compute global primary production (Longhurst *et al.*, 1995) for the same set of years do suggest that an oceanwide band of patches of higher chlorophyll (in the range 0.7–1.5 mg chl m^{-3}) occurs from 35 to 40°N. The meridional late-summer URSA MAJOR chlorophyll section near 150°W shows two features of interest to this discussion: a rapid change to lower values in the DCM at 34 or 35°N (near the southern boundary as anticipated) and a discontinuity at 42 or 43°N in both slope of the DCM and the values within it.

The zonal sections for temperature, chlorophyll, and nitrite (NO$_2$) from INDOPAC (July 1977) along 45°N show a clear difference between the east and west parts of the ocean (divided at about 160°W), even though they cut along the northern front of this province: Thus, (i) while the western half has a few large (diameter, 5° of latitude) mixed layer bowls (<250 m), the eastern part has a more or less uniform mixed layer depth at 75–100 m; (ii) while the western section has several deep excursions of high values of mixed layer chlorophyll to 200 m, the east has a single bowl-shaped DCM at 75–150 m, which is deepest at 140°W where the mixed layer is deepest; and (iii) the nitrite section has high values associated with the high chlorophyll values in the west, indicating rapid remineralization of DOC, whereas in the oligotrophic east values are very low. Passage over deep-ocean ridges (e.g., the Emperor Seamount Chain) modifies the characteristics of the western ocean.

Since this province has not been investigated comprehensively, the previous data are perhaps less than convincing. However, it is here that we may finally take some comfort from biogeographical data because the province does support a small but very characteristic assemblage of species for which it is the unique distribution area and this must surely support the concept of a unique ecosystem.

One of the clearest demonstrations of this are the ZETES winter 1966 and the URSA MAJOR summer 1964 sections along 155°W (Venrick, 1974). Group analysis of 66 diatom taxa identified recurrent groups whose distributions matched subarctic and subtropical gyres and also a group of three species endemic to the "transitional domain" (our Transition Zone at about 35–42°N) in winter and two groups of species comprising 99% of the diatom stock in this zone in summer. Because the two meandering boundary fronts must be leaky, it is no surprise to find that the species groups endemic to the adjacent domains to the north and south overlap into the Polar Frontal Zone province. However, it is the relative fidelity of the "transition species" to their zone that is significant.

We have even better evidence from zooplankton. An early review of the biogeography of the Pacific Ocean (Reid, 1962) revealed a remarkable zonal strip of high biomass of mesozooplankton across the ocean at about 38–45°N that was very clearly separated by low biomass occupying most of the subarctic gyre from another zone of high values in the Bering Sea. Even more striking are the envelopes for the distribution of 12 "transition zone" species assembled by McGowan (1971); these lie in a tight group across the ocean with average meridional extents from 35 to 45°N. They represent a wide range of taxa (euphausiids, copepods, molluscs, and foraminifera) and some, such as *Nematoscelis difficilis,* have a most remarkable fidelity to this province. These species "go with the flow" into the coastal currents off western North America and they will by discussed further under the CALC province. How these species maintain populations in this zonal river in the ocean is an unanswered question but not pertinent to our current problems.

Synopsis

The mixed layer deepens moderately in winter and the pycnocline lies within the photic zone only during summer (June–October) (Fig. 9.6). The rate of primary production is lowest in summer but increases progressively as the mixed layer deepens from October until February and remains high thereafter until May, when the summer mixed layer is established. Productivity in the DCM is always low (Pn/Pt = <0.1). Accumulation of chlorophyll occurs during the early part of productive period; thereafter, the loss rate matches the production rate.

Tasman Sea Province (TASM)

Extent of the Province

The Tasman Sea Province (TASM) lies between the Subtropical Convergence Zone across about 45°S and the Tasman Front to the south of the Coral Sea at about 32°S and is enclosed by the coasts of Australia and New Zealand. This is a relatively small province but because of its unique characteristics it is useful to recognize the Tasman Sea as an entity.

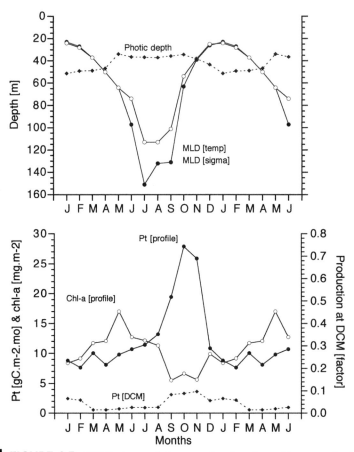

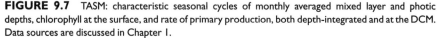

FIGURE 9.7   TASM: characteristic seasonal cycles of monthly averaged mixed layer and photic depths, chlorophyll at the surface, and rate of primary production, both depth-integrated and at the DCM. Data sources are discussed in Chapter 1.

Continental Shelf Topography and Tidal and Shelf-Edge Fronts

There are none in this province: the eastern Tasman Sea coast of Australia is in AUSE and the western coastline of New Zealand is allocated to the New Zealand Coastal Province (NEWZ). Shelf-edge fronts may, however, represent the landward boundary along these two coasts and may be especially prominent along the Westland upwelling strip of South Island.

Defining Characteristics of Regional Oceanography

This is a complex region of interaction between the general anticyclonic gyral flow and the instabilities induced therein by its topographic boundaries.

The southern limit of the province is the edge of the biological enhancement associated with the South Subtropical Convergence. This front passes from the east coast of Tasmania to the Snares Bank and south of New Zealand; many maps show it passing northeasterly across the southern Tasman Sea to Cape Egmont, but CZCS images (Comiso *et al.*, 1993) and modern survey data (Garner, 1967; Heath, 1981) show that it passes to the south of New Zealand and the Snares plateau, though its position is variable (see New Zealand Coastal Province). As the convergence passes across the southern part of the Tasman Sea it lies in a northward arc reaching, in the center of the Tasman Sea, as much as 40–42°S, or about halfway up the South Island of New Zealand. The eddy-field associated with the northern edge of the convergence adds to the mesoscale activity of the central Tasman Sea and the location of the convergence forces the general anticyclonic flow of the Tasman Sea to encounter the Westland coast of South Island where flow diverges north and south (see New Zealand Coastal Province).

The oceanography of the Tasman Sea is dominated by the consequences of the retroflection of the East Australia Current, which is the poleward western boundary current of the east coast of Australia. The boundary current continues polewards down the east coast of the North Island of New Zealand, which requires that it leave the continent near Sydney (32–34°S) and flow eastwards across the Tasman Sea. This flow forms a frontal zone, within which is embedded a frontal jet and across which the topography of the thermocline changes abruptly: Warm Coral Sea water lies to the north and cool Tasman Sea water to the south (Andrews *et al.*, 1980). This Tasman Front coincides with a zone of seasonal Ekman suction where upwelling reaches 2–4×10^{-5} m sec^{-1} during winter, though vertical motion in summer is weaker (Comiso *et al.*, 1993).

Meanders within the Tasman Front shed equatorward cyclonic meanders and poleward anticyclonic meanders, and these in turn shed warm and cold-core rings. The climatological origins of the eddies are topographically determined by the shoalest points of Lord Howe Rise and the West Norfolk Ridge (Hamilton, 1992), and hence preferred locations can be mapped for cold and warm features along the front. The actual location of the front as it passes across the Tasman Sea varies from 30°S to as much as 38°S (Mulhearn, 1987). Also, its meanders are large (commonly extending across as much as 5° latitude), but the axis of the front is most commonly near 34°S, before it rounds Cape North (35°S) to enter the poleward flow of the east New Zealand coastal currents.

The retroflection of the East Australian Current has major consequences for the Tasman Sea apart from bounding it to the north by the Tasman Front. The meandering flow along the front generates a field of Rossby waves which pass westwards toward the Australian coast and induce the shedding of very large warm eddies: two or three are shed per year, and a standing population of six to eight named eddies in the Tasman Sea is normal. Within these eddies the seasonal cycles of mixing and stabilization is imposed on their initial structure. The surface thermal signature and shallow thermocline of Coral Sea water is lost during the first winter of the existence of an eddy, and in subsequent summers the surface mixed layer established over the whole Tasman Sea passes across the subsurface warm core of the eddy.

Biological Response and Regional Ecology

The Tasman is a windy sea. Winter wind stress and thermal convection drives the mixed layer down to 300 m in the southern half, though the regional mean depth of winter mixing is only about 125 m. This mixing delivers end-of-winter mixed layer nitrate in the range 2.0–4.0 μM. Incursions of subantarctic water have slightly higher (4.0–5.0 μM) values and those of subtropical water are nitrate-deficient.

Though the southwest Tasman Sea exhibits a typical spring bloom (Harris *et al.*, 1987) regular seasonality does not extend across the whole region. Monthly CZCS images indicate that higher surface chlorophyll occurs patchily across the whole southern geographical Tasman Sea (though not the TASM province) through austral summer and autumn associated with frontogensis and other bloom-inducing processes in the Subtropical Convergence [see Chapter 10, South Subtropical Convergence Province (SSTC)]. Because of the current difficulty of precisely defining boundaries for computations based on the CZCS data, the seasonal estimates of chlorophyll biomass for this province used here are probably too strongly weighted by events in the SSTC zone.

The spring bloom has been best described east of Tasmania and is a typical but rather irregular Sverdrupian process. However, though strong westerlies mix water regionally to 300 m, the actual depths achieved are also dependent on the nature of the subsurface water mass, having high interannual variability as the location of the Subtropical Convergence shifts and as mesoscale eddies come and go. Blooms occur intermittently as the intermittent westerlies (there is a 40-day periodicity in wind stress in the westerly wind field) slacken and permit stabilization of the upper water column. The bloom and onset of nutrient depletion varies between years by as much as 4 months at a fixed station and is usually about 1 month earlier inshore than in the open Tasman Sea. Austral winter chlorophyll values of <0.5 mg chl m^{-3} rise to 2.0–2.5 mg chl m^{-3} at the peak of the early summer bloom, usually between August and October, and then decline in late austral summer and early autumn from January to March. Nutrients become wholly depleted by about February. Primary production is enhanced and surface chlorophyll values are higher than background at major sources of vorticity (as at the Tasman Front) and these effects are discernible in CZCS images. In fact, one of the "classical" CZCS images shows the seas around the island of Tasmania populated by a crowd of fronts, dipoles, squirts, and eddies, all clearly evident in the chlorophyll field, and the result of the dynamic interaction between the Subtropical Front and the complex geography of the southern Tasman Sea.

That this should be a very dynamic and changeable part of the ocean is obvious from its boundary conditions, lying between the Southern Ocean and the Coral Sea and adjacent to the great landmass of Australia. This supposition is confirmed in the observations of variability not only within seasons but also between seasons (Harris *et al.*, 1988, 1991). Though most relevant observations have been made on the east Tasmanian coast, I accept them as generally applicable to the Tasman Sea. The observations compare the effects of different wind-stress during two 2-year periods. In 1986 and 1987, both summers were cool and windy, and in 1988 and 1989 they were warmer and quieter. The cool years more closely resembled the subantarctic (deep winter mixing and high nitrate) and the warm years subtropical conditions (shallow winter mixing and low nitrate). Particle size of all planktonic components indicated a trend toward subtropical oligotrophic conditions in 1988 and 1989. Primary production was reduced, small copepods dominated, and all large zooplankters (especially salps and *Nyctiphanes australis*) were eliminated: These appear to be dependent on large-cell new production rather than on small-cell regenerated production. In cool years, large populations of euphausiids occur, whose swarms support a rich *Trachurus* fishery, which collapsed during the two warm years. As Harris *et al.* suggest, this entire chain of events appears to be a regional manifestation of the Pacificwide ENSO warm event of 1988.

Synopsis

There is moderate winter mixing so that the pycnocline is within the photic zone only during the austral summer (November–April) (Fig. 9.7). The rate of primary production increases progressively during the winter mixing period until late spring without significant accumulation of chlorophyll, which is presumably inhibited by consumption. The data for Fig. 9.8 are unduly weighted by the effect of processes associated with SSTC frontogenesis.

PACIFIC TRADE WIND BIOME

North Pacific Tropical Gyre Province (NPTG)

Extent of the Province

This very large province is the equatorward part of the North Pacific central or subtropical gyre lying between the Subtropical Convergence at about 30–32°N and the northern Doldrum Front (Roden, 1975) at about 10 or 11°N and is often referred to as the central region. The western boundary of NPTG is taken as the offshore edge of the Kuroshio flow after the bifurcation of westward flow when it encounters the Philippines. In the east, the boundary is the edge of the offshore California Current system (for definition, see California Current Province). Large-scale ecological gradients are lacking, as indicated by the depth and kind of the DCM, compared with the adjacent province north of the Subtropical Convergence; the 24°N zonal trans-Pacific sections of chlorophyll and primary production (Venrick, 1989, 1991) have much greater uniformity than zonal sections along 47°N. In meridional sections (e.g., Venrick *et al.*, 1973), the DCM is seen to conform to the general northward upsloping of the thermocline northwards across the province. I find no reason, therefore, to subdivide this province, though it is one of the largest presented in this book.

Continental Shelf Topography and Tidal and Shelf-Edge Fronts

These are not relevant in this oceanic province.

Defining Characteristics of Regional Oceanography

In the normal (non-ENSO) condition, the westwards flow of the NEC lies between the Subtropical Convergence and the northern Doldrum Front of Roden (1975) along 10°N, where salinity decreases abruptly to the south into equatorial water. West of 180° latitude, the southern part of this limb of the gyral circulation passes progressively under the influence of the heavy precipitation which characterizes the western Pacific "warm pool"; where these conditions are fully developed (south of 10°N and west of 160°E) I have suggested

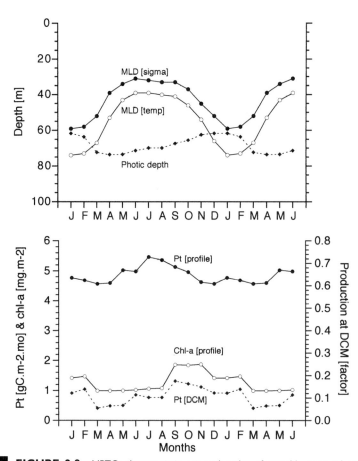

FIGURE 9.8 NPTG: characteristic seasonal cycles of monthly averaged mixed layer and photic depths, chlorophyll at the surface, and rate of primary production, both depth-integrated and at the DCM. Data sources are discussed in Chapter 1.

we should define a special province (see Western Pacific Warm Pool Province). Some of these characteristics will nevertheless be relevant to the western part of NPTG. To the east, the boundary of NPTG against the offshore-flowing extension of the California Current system (CALC) is definable only with difficulty, and I consider the weak terminal region of the Subtropical Convergence, which Roden (1975) traces southeastwards toward the tip of Baja California, to be the eastward limit of NPTG.

Flow perturbation past the Hawaiian island chain causes the formation of a complex version of the von Karman vortex street, generating alternate cyclonic and anticylonic eddies similar to those downstream of the Canaries (see Chapter 7, North Atlantic Subtropical Gyral Province). However, in the case of the Hawaiian Islands it is likely that eddies are also formed by the effect of curl of the wind stress which induces upwelling and downwelling of surface water at the wind-shear lines by strong Ekman pumping. Off Hawaii it is suggested that this process induces sufficient vertical nutrient flux to induce features in the chlorophyll field (see the review of island eddy wakes by Aristegui *et al.*, 1997).

The Subtropical Convergence (at about 30–32°N in midocean and 20–25°N in the west), which forms the northern boundary of this province, lies below the seasonally migrating conjunction of the westerly winds and the trades. This province, then, lies below the subtropical easterlies but nevertheless winter mixing, forced by the frequent extratropical cyclones passing eastward (see North Pacific Transition Zone Province), extends as far south as the Hawaian Islands at 20–25°N. At 22°N, the mixed layer deepens seasonally, from 35–40 m in summer to 80–90 m in winter. The permanent pycnocline lies still deeper and a permanent nitracline occurs across its density gradient. These features are continuous and at a nearly uniform depth (incidentally, they are close to the 1% isolume) from the eastern edge of the province (from where they slope up toward the coast) to at least the longitude of Hawaii. They are deepest in the northern part of the province and shoalest in the south.

Biological Response and Regional Ecology

We are fortunate that for this province we have two sets of excellent time-series studies and some useful meridional and zonal sections. Many expeditions, mostly from Scripps Institute of Oceanography in California, have investigated their CLIMAX area (28–30°N) from 1968 to 1987 (see Hayward, 1987, for entry to the publications). At the ALOHA station, at 22°N, the recent HOT (Hawaii Ocean Time Series) investigations took place, which were managed by the University of Hawaii (see the 1996 special issue of *Deep-Sea Research II 43*, Parts 2 and 3). What follows relies very heavily on this work.

The ridge–trough topography of the zonal equatorial current system dominates the stability and vertical eddy diffusivity of the regional water column and, hence, the long-term supply of nutrients to the photic zone. The deep, nitrate-depleted euphotic zone of this province in summer, often exceeding 100 m, lies above a permanent nutricline (Venrick, 1991). The vertical ecological structure is modified by the development of the seasonal shallow thermocline (Hayward *et al.*, 1983); this process follows the general model for the subtropical ocean with moderate winter mixing that was shown in the Sargasso Sea (see Chapter 7, North Atlantic Tropical Gyral Province).

Year-round, and each year, a DCM occurs at about 100–120 m, just above the nitracline and the permanent thermocline and close to the 1% isolume; the DCM and nitracline remain at the same depth when the summer thermocline is established at 35–45 m. The summer DCM contains maximum values of around 0.2–0.4 mg chl m^{-3}, whereas the winter DCM is a weaker feature, usually about 0.1 mg chl m^{-3}.

Primary production rate is maximal, on the other hand, in the upper mixed layer at 10–25 m. Primary production rates increase somewhat in winter, but the HOT time series show that interannual differences are at least as great as seasonal differences: The seasonal/interannual range is between 200 and 700 mg C m^{-2} day^{-1}. The apparent vertical diffusive nutrient flux can only be partially satisfied by observed nitrate uptake, and the mechanism by which calculated new production is sustained above the summer thermocline remains unexplained in the presence of a stable vertical ecological structure since we would expect any upward physically-driven nitrate flux across the nitracline to be taken up in the DCM (Hayward, 1987). The curious observations of Young *et al.* (1991)

of enhanced production rates after ephemeral falls of dust particles from Chinese loess fields, during which the depth of enhancement progressively deepens, require explanation. Interpretation by the authors simply as release of the autotrophs from Fe-limitation seems inappropriate in this nitrate-depleted system: Perhaps it is more probable that the observed modification of near-surface density structure by the wind events that accompanied the dust-fall induced an ephemeral modification to the nitrogen budget. Nevertheless, such events must be examined closely, especially in the adjacent nitrate-replete regions to the north.

In this context, we should take note that in this province the presence of autotrophic cells (diatoms *Rhizoselenia cyclindricus* and *R. hebetata*) having nitrogen-fixing symbionts (cyanophyte *Richelia intracellularis*) have been commonly reported. Venrick (1974) surveyed the occurrence and potential productivity of this symbiosis at the CLIMAX site and found it to be most abundant in summer and computed a potential carbon fixation rate (37–77 mg C m^{-2} day$^-$) that was about 30–60% of the difference between total winter nonbloom and total summer bloom production at this station. Villareal *et al.* (1993) have investigated diatom "mats" composed of several species of *Rhizoselenia* which are very abundant at CLIMAX (containing >90% of all biogenic silica); the mats appear to have the ability to modify their sinking rate from negative to positive buoyancy, and shuttle between the upper euphotic zone and the nitrate-replete deeper zone. It is suggested that these mats may be an important vector of new nitrogen inputs to the euphotic zone, representing as much as 50% of the required flux.

Profiles of microplankton biomass (mostly protists, monads, flagellates, and naked dinoflagellates) do not have a subsurface maximum corresponding to the DCM but rather a broad depth range of relatively high abundance (0–100 m, 5–10 μg C liter^{-1}) above a deeper zone of lower abundance (100–200 m, 1–5 μg C liter^{-1}). Separation of vertical habitat occurs among mesozooplankton (Ambler and Miller, 1987). Some species specialize in the DCM (presumably specialists in the consumption of larger cells), whereas others occupy preferentially the upper mixed layer where primary production rates are maximal. In the western part of the province (at 28°N 136°E), Tsuda *et al.* (1989) have observed that grazing by microplankton in the DCM (70–120 m) contributes 60–100% of total consumption and that this is balanced by production within a time-scale of several days. Mesozooplankton were found to contribute only <5% of all consumption; at shallower locations than the DCM, these organisms are largely secondary predators, consuming microplankton, though at the DCM they contributed to the consumption of algal cells.

The envelopes of the distribution of 10 "central water mass" species plotted by McGowan (1971) provide an interesting point regarding the definition of this province: In the east of the ocean, and as far west as about 170°E, they match remarkably well the limits of this province as defined here. To the west of this longitude, the envelopes widen and eventually encompass the whole western Pacific from New Guinea to southern Japan. It is certainly possible that further analysis of the biological oceanography of the western Pacific (than can be done with) existing studies will suggest that the western province WARM should be much more extensive.

Synopsis

Mixed layer deepens modestly in winter but is always well within the photic zone (Fig. 9.8). A weak maximum for primary production rate occurs in midsummer with some accumulation of chlorophyll in early summer during the period of deepening mixed layer.

North Pacific Equatorial Countercurrent Province (PNEC)

Extent of the Province

The North Pacific Equatorial Countercurrent Province (PNEC) lies between the northern and southern Doldrum salinity fronts at about 5 and 10°N in the central ocean, widening toward the east. PNEC includes the triangular region of weak currents between the influence of the equatorial divergence and the California Current extension into the NEC. To the west, a useful limit can be set at about 180°W.

Continental Shelf Topography and Tidal and Shelf-Edge Fronts

These are not relevant for this province.

Defining Characteristics of Regional Oceanography

This province does not correspond with a single feature of the surface circulation nor does it encompass the whole of the most prominent feature that dominates it. This is, nevertheless, a region of the ocean that can usefully be defined as an ecological unit.

The North Equatorial Countercurrent (NECC) flows above the north slope of the equatorial thermocline ridge, from 120°E across the ocean to Central America where it turns northward. Flow of the equatorial countercurrent is weaker and more intermittent progressively eastwards; 25 Sv flows at 135°E but only 10 Sv near the termination in the eastern Pacific. Accordingly, mixed layer depth shallows also toward the east. Despite the trans-Pacific nature of this flow, west of 180° longitude it lies below the western Pacific atmospheric convection cells where excess precipitation over evaporation induces a characteristic and different ecological regime which we recognize as the WARM province. We therefore set the western boundary of the current province at the dateline. The southern boundary is the Equatorial or Doldrum Front that crosses the equator near the Galapagos and is marked at the surface by a discontinuity in temperature, salinity, and mixed layer nitrate originating in the equatorial divergence.

Generic NECCs are the product more of wind stress curl than of wind stress (Sverdrup, 1947; Philander, 1985), and when the Intertropical Convergence Zone (ITCZ) is in its northerly, boreal summer position, the NECC lies below a zone of exceptionally strong positive wind-stress curl that lies across the whole ocean, from the Indo-Pacific conjunction all the way east to Central America above the thermal ridge associated with flow of the NECC (Lagler and O'Brien, 1980). From July to January, positive values of Ekman vertical velocity are higher in the eastern part of the ocean (from 140°W to the American continent), reaching local maxima of 60×10^{-6} m sec^{-1}.

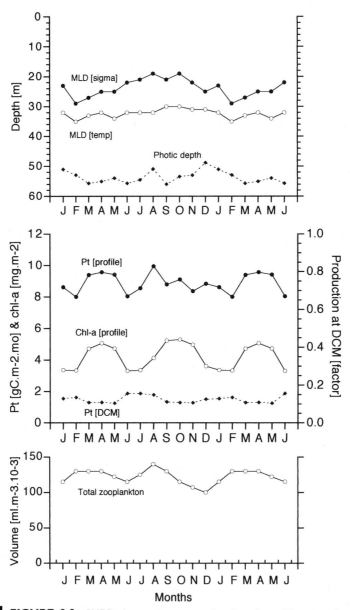

FIGURE 9.9 PNEC: characteristic seasonal cycles of monthly averaged mixed layer and photic depths, chlorophyll at the surface, and rate of primary production, both depth-integrated and at the DCM. Zooplankton data are from the EASTROPAC monitoring cruises; other data sources are discussed in Chapter 1.

For these reasons, flow of the NECC is strongly seasonal in the eastern part of this province and at least at the surface. During the boreal winter when the atmospheric ITCZ is at its most southerly, the countercurrent can still be detected in thermocline topography, but it is sufficiently weak that flow at the surface does not occur east of about 110–120°W (Philander *et al.,* 1987; Tomczak and Godfrey, 1994). In the eastern tropical Pacific, the ridge–trough system of the thermocline occupies a wider, triangular region between the coast and the convergent fronts of the North and South Equatorial Currents where they turn westwards away from the continent. Within this region of weak flow and shallow mixed layer, the eastward countercurrent bifurcates at the American continent and the influence of the continent on regional wind stress and curvature results in several cyclonic domes of ecological significance: the Costa Rica Dome offshore in the PNEC province and the Tehuantapec and Panama Bight domes of the coastal boundary province (Blackburn, 1983). The quasi-permanent doming of the thermocline off Costa Rica, with seasonal intensification, has regional ecological importance.

Biological Response and Regional Ecology

The curvature of wind stress not only forces the flow of the NECC but also induces enhanced algal growth in that flow (Longhurst, 1993), and the existence of Ekman suction in the NECC is perhaps a sufficient explanation for the zone of chlorophyll enhancement associated with the oceanic NECC in CZCS images. The effect of maximum curl values in winter is a vertical velocity at the pycnocline of 0.75 m day^{-1}, which is a significant velocity where the mixed layer depth is 25–50 m. The surface chlorophyll enhancement seen in CZCS images in the eastern part of the NECC does not extend west of about 110°W under normal conditions, and this is consistent with the progressive westwards deepening of the mixed layer. We may expect enhancement of subsurface blooms along the NECC toward the west but have no evidence for this.

Fiedler (1994) has analyzed monthly and between-year variability for the countercurrent from CZCS and wind-field data; he confirmed that wind stress, Ekman suction, and chlorophyll are maximal in winter, in the period December to April or May. He also showed that there were significant between-year differences in 1979–1985. The 1983 ENSO warm event was marked by weak winds and low pigment from 90 to 105°W, whereas the period 1984 and 1985 was marked by the return of winds and a very strong chlorophyll enhancement from the coast out to 98°W. Investigations performed at sea in single seasons must be interpreted in light of the existing condition of the Southern Oscillation index. CZCS seasonal images also show that to the west of the region investigated by Fiedler, the NECC bloom exists in summer when it is absent farther to the west. In boreal summer the NECC bloom is truncated at 120°W but extends westwards to about 145°W.

These CZCS observations are consistent with the enhancement of chlorophyll both at the surface and in the DCM at 10–12°N and 90–120°W seen in some of the EASTROPAC chlorophyll fields in boreal winter rather than summer. The EASTROPAC seasonal monitoring cruises in the NECC also confirm a boreal winter primary production maximum (300–400 mg C m^{-2} day^{-1}) and a rather long period of high column-integrated chlorophyll (February or March to June or July) reaching values of 25 mg chl m^{-2} (Blackburn, *et al.*, 1970).

Within this general region of seasonal enhancement lies the Costa Rica Dome (CRD), which is a quasi-permanent cyclonic feature in the eastern part of the province, usually centered near 9°N,88°E though it is constantly shifting its location and strength in response to changes in atmospheric forcing and

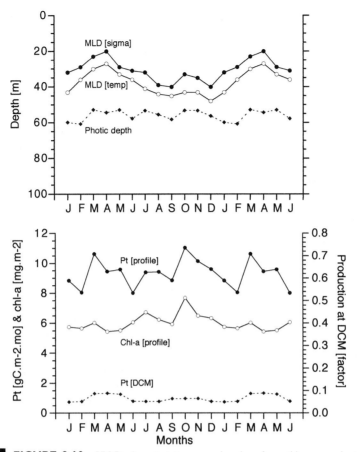

FIGURE 9.10   PEQD: characteristic seasonal cycles of monthly averaged mixed layer and photic depths, chlorophyll at the surface, and rate of primary production, both depth-integrated and at the DCM. Data sources are discussed in Chapter 1.

current variability. The divergent upwelling effect may be so strong as to remove completely the mixed layer laterally and thus exposing the strong tropical thermocline (and its associated nutricline) at the surface in the center of the dome with significant biological consequences. The CRD evolves rapidly in boreal spring and early summer as the ITCZ and the associated wind-stress curl moves to the north and as the flow of the NECC is strengthened. The CRD reaches its greatest development in late summer and erodes rapidly with the retreat southwards of the ITCZ in boreal autumn (Umatani and Yamagata, 1991). The mechanism of its erosion is associated with the passage through the region of anticyclonic vorticity and eddies forced in the coastal region by transmontane northerly winds.

A typical expression of the ecology of the CRD was investigated in March and April 1981 (Herman, 1989; Longhurst, 1985; Sameoto *et al.*, 1986), before the importance of picoautotrophic plankton was understood. The main gradient of the thermocline mixed layer was at 20–50 m, the DCM was at 15–20 m with a maximum concentration of 3.0 mg m^{-3}, and primary production was maximal at 5–15 m with a maximal rate of about 8.0 mg C m^{-2} hr^{-1}. Copepods were the principal mesozooplankton grazers and these had their depth centroids closer to the depth of maximal primary production than to the DCM. There was, however, a broad mesozooplankton zone of high abundance from the surface down to about 50 m, below which a planktostad was interrupted only by layers of diel migrants by day at 200–250 m, which in this case is well above the top of the deep oxygen minimum zone. Repetitive profiles were greatly more variable within than beyond the CRD.

To the west of the CRD, the same arrangement was found at the BIOSTAT station chosen to be representative of the unperturbed shallow-mixed layer of the eastern tropical ocean. Here the standing stocks and rates of all biological variables were lower, and features were deeper in the water column. The mixed layer was about 30 m deep, the DCM lay at about 40 m (<1.0 mg m^{-3}), the primary production maximum was just slightly shoaler at 35 m (<2.0 mg C m^{-2} hr^{-1}), and copepods occupied most of the mixed layer above, though their depth centroid lay close to the depth of maximum production. Mesozooplankton could be partitioned among small herbivorous copepods (grouped just shoaler than the DCM), *Eucalanus* spp. (grouped mostly below the DCM and within the thermocline), and omnivorous copepods [grouped below the thermocline and well into the upper oxygen-depleted (<1.0 μl liter^{-1}) zone]. Predators (e.g., *Euchaeta* spp.) specialized in each of these depth zones. The overall number of species was greatest in the pycnocline, where greatest stability (Brunt–Väisälä frequency <30 cph) favored niche specialization among species-groups both within genera and between related genera.

Synopsis

There is almost invariant depth of mixed layer so that photic depth is always deep in pycnocline (Fig. 9.9). There is trivial seasonality in any of the archived data for primary production rate or chlorophyll accumulation, except for an indication of inverse relationship between chlorophyll and herbivore abundance.

Pacific Equatorial Divergence Province (PEQD)

Extent of the Province

This province corresponds with the zone of nitrate-replete water of the eastern tropical Pacific and is aligned along the equator south of about 5°N from 180°W to the Galapagos Islands, where it joins the coastal boundary province (see Humboldt Current Coastal Province). It is not symmetrical about the equator, except from 180 to 115°W where it is at its narrowest (<5°N to <5°S), but east of 115°W the nitrate-replete water is biased toward the south, reaching about 20°S at its eastern boundary with the regime of the coastal boundary current. Its westward extent is time-dependent, and 180°W is set simply as a climatological average condition.

Continental Shelf Topography and Tidal and Shelf-Edge Fronts

These are not relevant in this province, though within the wider eastern part of this province, the presence of the Galapagos Islands results in important anomalies in the surface chlorophyll field. Under normal conditions, the island wakes to the west and northwest are marked by plumes of high chlorophyll induced by the turbulent wake processes. Under ENSO conditions, when flow at the surface may be reversed, plumes occur to the east and northeast of the islands.

Defining Characteristics of Regional Oceanography

The surface circulation is dominated by the flow of the South Equatorial Current (SEC) westwards across the ocean. This stream extends from the Southern Hemisphere across the equator to about 5°N and thus requires a ridge in the topography of the thermocline below the equator; it flows along the poleward slope of this ridge in each hemisphere. Within this ridge, and embedded in the upper thermocline, is the narrow (<200 km), long (14,000 km) jet of the Equatorial Undercurrent (EUC) described by Townsend Cromwell prior to his untimely death in an air crash on his way to join a Scripps ship on the Mexican coast. In this province the EUC may flow within 40–50 m of the surface, whereas further west it lies at 100–300 m.

Within this general circulation occurs a strong linear divergence and upwelling along the equator, reaching no more than 2° of latitude from it. This divergence is forced by the change of sign of Coriolis force at the equator polewards in both hemispheres for a westward flow. In many transequatorial sections from 135 to 160°W (Wyrtki and Kilonsky, 1984; Colin *et al.*, 1987; Carr *et al.*, 1992) the divergence—as indicated by the surfacing of isotherms and the nitrate distribution—is aligned almost precisely along the equator. East of 120°W, however, the divergence at all seasons lies somewhat to the south of the equator; the EASTROPAC sections show that 25°C water is exposed at the surface from 2°N to 5°S in August and from 1°N to 3°S in February. Satellite thermal imagery shows that the boundary between upwelled and surface water carries 1000-km wavelength instability waves which propagate westwards. These fronts may be spectacular (the word is carefully chosen) at the surface and mark a transition across a few tens of meters at the conjunction between cold, clear upwelled water in the SEC and warm, green water of the NECC. These fronts are well-known to navigators and visible to Space Shuttle crews. They may also be marked by a field of whitecaps and an aggregation of floating *Thalassiosira* mats. Very high concentrations of chlorophyll (background × 3) and extremely high rates of carbon uptake (1.4–1.8 g C m^{-2} day^{-1}) have been observed along this convergent front at 140°W (Barber, 1992) apparently associated with current shear and eddying.

The mixed layer normally deepens progressively westwards during relaxation of ENSO conditions, whereas zonal wind stress decreases progressively westwards. Poleward from the divergence are basin-scale zonal fronts, particularly for salinity. The strongly convergent southern Doldrum Front of Roden (1975) usually lies at about 2–4°N and contains ishalines for approximately 34.4–34.7, whereas further south (but not always at the edge of the divergence zone) there is usually a second salinity front containing the isohalines for 35.0–35.4. These indicate transitions between low-salinity water of the Equatorial Countercurrent and the denser water of the SEC.

During well-set-up ENSO events, the defining characteristics of the province do not exist as equatorial divergence ceases, and warm, deepened thermocline conditions obtain over the whole eastern tropical Pacific.

Biological Response and Regional Ecology

Unlike many other provinces, for PEQD the available studies are, quite literally, voluminous: A good starting point for the literature would be the *Deep-Sea Research* special issue (**42**, Parts 2 and 3, 1995) on the EqPac studies of 1992. The reason for the abundance is simple: This is one of those enigmatic high-nitrate, low-chlorophyll (HNLC) regions where for several decades biological oceanographers have been trying to figure out why phytoplankton, given adequate stability in the water column and sufficient irradiance, do not utilize all the available mixed layer nitrate but persist at low population levels in nitrate-replete water. In short, "Why isn't the equator greener?" was the question posed formally by the WEC88 group at 150°W and by many others before and since. Fortunately for both writer and reader, we do not have to review all the competing theories because as of 1996 the enigma was satisfactorily resolved.

One of the EqPac studies provides direct support for the ecological reality of the boundaries suggested for this province. Optical studies of microorganisms along a time-series section (12°N–12°S at 140°W) revealed three "biohydrographic regimes" (Chung *et al.*, 1996). Characteristics of beam-attenuation due to particles (C_p), dominated by heterotrophic bacteria, prochlorophytes, cyanophytes, and eucaryotes (all <3.0 μm), showed that the area from 7°N to 7°S could be clearly distinguished from the more poleward parts of the section: this, of course, corresponds quite well with the region here proposed as the PEQD province. In addition, it was only in the central regime that the particle population responded significantly to the 1992 El Niño event; a 30% increase in beam C_p and depth-integrated C_p were noted only in the central regime, corresponding to our PEQD. We should not expect, of course, that these boundaries, even if real, should be neatly aligned parallel with the equator in the real ocean: EqPac investigators also noted the cusp-like form of the boundary of cooler equatorial water which is frequently also observable in the CZCS chlorophyll fields.

The biota are arranged in relation to the stratification of the water column in a reasonably simple and well-known arrangement. As the mixed layer deepens toward the west across the ocean and polewards from the equator, it carries with it the permanent DCM and associated vertical layering of zooplankton, which has its depth of greatest abundance near the depth of maximal primary production rate in the upper mixed layer and significantly shoaler than the DCM. At 155°W the core of the DCM (0.2–0.3 mg chl m^{-3}) deepens from 50 m under the equator to 100 m at 15°S, whereas the rate of normalized primary production is maximal in the upper 10 m at about 3 or 4 g C (g chl)$^{-1}$ hr^{-1}. Diel migrants (euphausiids and metridiid copepods) are at 400–500 m by day and join the epiplankton in the upper 50 m at night.

There is much evidence that the major fluxes in the pelagic ecosystem are—as we have come to expect in recent years—dominated by flux through the small cells of the microbial loop (Coale *et al.*, 1996). The basic composition of the autotrophs is approximately as follows (expressed as μg C liter^{-1}): *Synecho-*

coccus, 2.0; prochlorophytes, 4.5; prymnesiophytes, 3.0; autotrophic dinoflagellates, 5.0; pennate diatoms, 2.0; and *Phaeocystis,* 0.5. With these occur the protistan grazers: heterotrophic and dinoflagellates, 7.0; and mixotrophic ciliates, 0.25. This is a recipe for a tightly coupled production/consumption system in which the population size of the grazers can respond equally as fast as that of the autotrophic organisms. We may presume that the flux of this system, together with the associated bacterial flora and based on internally regenerated ammonium, is little mediated by the grazing of mesozooplankton. The most likely mediation is through removal of the sparse diatoms and application of mortality to the protistan grazers, as well as some role in the recycling of nitrogen.

To fully understand the nature of the nitrate enigma, it is necessary to consider the three-dimensional distribution of nitrate (Thomas, 1972, 1978). We must examine both the eastern and western parts of the province: In the east, at 110–120°W, the EASTROPAC surveys found a zone of nitrate-replete water between about 4°N and 10–15°S, or far to the south of the divergence zone. North of this region, only in the CRD is 10-m nitrate higher than 0.1 μM. A nitrate section across the province shows that the mixed layer nitrate from about 8°S is separated from the deep nitracline (values of >10.0 μM at 150 m at 10°S sloping down to 350 m at 29°S) by nitrate-depleted water (<0.1 μM). Therefore, the nitrate of the upper layer (6.0 μM near the equator and 1.0–2.0 μM near 12°S) is transported poleward in the diverging surface layer from the equatorial upwelling, which passes above the more saline nitrate-depleted water, rather than originating in vertical eddying across the nitracline *in situ.*

This distinction in the east of the province is important in understanding why the near-surface water should remain replete in nitrate. To the west, at 140°W, the nitrate-replete near-surface water is not separated by a nitrate minimum layer below, and the meridional extent of the nitrate-replete zone is narrower; here, the 2.0 μM contour intersects the surface at only about 5 or 6°S.

The following explanations have been advanced over the years for the existence of the nitrate-replete surface water: (i) The multiplication of autotrophic cells may simply be constrained by the rapidity of response of the grazing organisms; (ii) the availability of a preferred nitrogen source, ammonium, may constrain the uptake of nitrate; (iii) nitrate is simply upwelled faster than it can be used; (iv) sufficient nitrate is deposited in eolian dust to account for nitrate repletion; and (v) some other unknown molecule may be limiting in the sense of Liebig's law of the minimum. All these suggestions describe processes that are real and important, but none except the final one offer a sufficient explanation (Chavez *et al.,* 1991; Chisholm and Morel, 1992; Cullen *et al.,* 1992). As early as the EASTROPAC cruises of the 1960s, Bill Thomas of Scripps was seeking a missing micronutrient (I remember one squashed copepod per carboy as his starting point), but it was not until the now-famous suggestion of the late John Martin and of John Fitzwater in 1989 that Fe might be the missing element that things really started to come together. We too easily forget that this recalled earlier suggestions along the same line: For instance, in 1934, T. J. Hart, of the Discovery Expeditions, suggested that this element might limit algal growth in the Southern Ocean.

This is not the place for a history of how this idea was tested, but things moved very rapidly thereafter, first with ultraclean incubation experiments both in the Southern Ocean and in PEQD. These suggested that the addition of suitable quantities of Fe did indeed result in enhanced algal growth, but because bottle incubations have so many ways of going wrong, and because the enigma was so important, a proposal to perform *in situ* iron-enrichment experiments at sea was funded. The first experiment (IronEx I in 1993), led by John Martin in the PEQD, was only a partial success: Fe was released into the ships wake around a 8×8-km patch to raise mixed layer Fe concentrations from ambient 0.05 nM to experimental 4.0 nM. Response of the autotrophic biota was instantaneous, with photosynthetic efficiency peaking after 2 or 3 days, when chlorophyll concentration and primary production rate had doubled. Unfortunately, after 5 days the patch was subducted below fast-moving low-salinity surface water and lost.

IronEx II in 1995 (also in the PEQD), after John Martin's death, was a resounding success, reported on by Coale and 14 others in 1996. Repeated injections of Fe in the ships wake (tracks 400 m apart) while the 70-km^2 patch drifted 1500 km, always at the surface, maintained for about 1 week a concentration of 2.0 μM within the patch. The results were dramatic: Nitrate was drawn down from an initial >10 μM to <5.0 μM, chlorophyll increased from <0.2 mg m^{-3} to >3.0 mg m^{-3} in the patch center, and pCO$_2$ decreased from >530 to <450 μatm as carbon was required for photosynthesis. When injection ceased, the patch began to weaken both through diffusion and by the loss of Fe, which had been expected to be conserved longer. In control patches, there was no response to injection of biologically-inert molecules.

The autotrophic biota showed differential responses that make perfect sense. The cells which responded most rapidly and completely were diatoms which increased in abundance by a factor of 85, whereas the picoautotrophs responded only by doubling their number. The micrograzers responded in step with the picoautotrophs, but mesozooplankton responded very little—as indeed their long generation time would ensure. The increase that did occur was probably a result of modification in diel migration depths mediated by decreasing water clarity. No matter the reason for the increase, the resulting imbalance between copepods and diatoms allowed a bloom to occur that was analogous to the imbalance in a high-latitude spring bloom. As has also been suggested for other HNLC areas, Price *et al.* (1994) characterize this as an "Fe-limited, grazer-controlled ecosystem." This concept is also advanced by Banse (1996) to explain the HNLC areas of the subantarctic water ring [see Chapter 10, Subantarctic Water Ring Province (SANT)].

The only remaining enigma, then, is why is there a lack of Fe in the upwelled water? In other words, if equatorial divergence is approximately symmetrical about the equator, why is Fe lacking to the south while to the north (in PNEC) there is apparently a sufficient supply to permit nitrate depletion? The answer may be sought in what we dimly know of the pattern of deposition of eolian dust at the sea surface in the eastern tropical Pacific. The ITCZ, separating the dusty air mass of the Northern Hemisphere and the clear air of the Southern Hemisphere, lies diagonally across the eastern part of PEQD, having a climatological position not very different from the boundary of our province. There is plenty of evidence that African dust from the desertification of the Sahel is carried into the Caribbean and across the isthmus: 25–37 million tons of dust are computed to pass westwards at 3-km altitude over Barbados an-

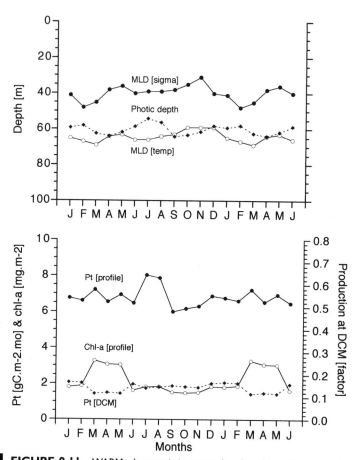

FIGURE 9.11 WARM: characteristic seasonal cycles of monthly averaged mixed layer and photic depths, chlorophyll at the surface, and rate of primary production, both depth-integrated and at the DCM. Data sources are discussed in Chapter 1.

nually (Szekielda, 1978), which is much more than halfway to the Caribbean. In fact, the stratified yellow dust veil is easily observable from commercial aircraft passing across the Caribbean area—sufficiently striking to cause me to take a couple of photographs of it on one occasion. Duce and Tindale (1991) suggest that this puts 1–10 mg Fe m^{-2} $year^{-1}$ onto the surface of PEQD and that the <1.0 contour runs zonally along the equator, which is where the hypothesis predicts it to be. There is direct evidence (Maenhaut *et al.*, 1983) that in PEQD at about 115°W the trades of the Northern Hemisphere carry a significantly higher concentration of crustal components than the trades of the Southern Hemisphere.

Synopsis

A weak but consistent seasonal deepening of shallow mixed layer occurs in response to geostrophic adjustment during boreal summer, but photic depth remains well below mixed layer at all seasons so that upper pycnocline is well illuminated (Fig. 9.10). There is very low variability, and insignificant seasonality, in primary production or chlorophyll biomass.

Western Pacific Warm Pool Province (WARM)

Extent of the Province

This province lies under the low-pressure region of the atmospheric Walker circulation—that is, westwards from the date line to the Indo-Pacific archipelago—and has a maximum meridional extent (at about 160°E) between 10°N and about 20°S. The surface 29°C isotherm is a convenient practical boundary.

Continental Shelf Topography and Tidal and Shelf-Edge Fronts

There are scatterings of atolls and numerous high islands and island groups throughout this province, around which consequences for phytoplankton growth may be looked for, and perhaps observed, in satellite chlorophyll fields, though the effect is multiple and complex (Dandonneau and Charpy, 1985).

Defining Characteristics of Regional Oceanography

The diverse circulation features of the western tropical Pacific are included within a single province because of the dominant role of regional meteorology in regulating stratification in the ocean below the heavy convective cloud cover of the low-pressure cell of the Walker circulation of the tropical atmosphere. The zonal, near-equatorial cloud band of the ITCZ (between the northeast and southeast trades) and the cloud band of the South Pacific Convergence Zone (between the southwest and southeast trades) meet the west winds coming out of the Indo-Pacific archipelago in this region and this conjunction forms the largest region of persistent clouds in the tropics. The resulting heavy rainfall leads to a positive precipitation/evaporation ratio of $+50$ to $+150$ cm year^{-1}, and a lens warm, brackish surface water, which may be further diluted by the eastward advection of low-salinity surface water from the Indonesian archipelago. The Western Pacific Warm Pool is the largest such feature in the oceans, and similar conditions occur elsewhere under normal non-ENSO conditions only in two much smaller warm pools, to the west of Central America and in the northern Caribbean. As defined by the 29°C surface isotherm, the Western Pacific Warm Pool varies in size on the decadal scale (Yan *et al.*, 1992).

The brackish surface isohaline layer overlies a halocline at an average depth of 30 m within a deeper thermostad that on the average extends from the surface down to about 75 m (Lukas and Lindstrom, 1991; Sprintall and Tomczak, 1992). The Ekman layer, therefore, corresponds with the brackish surface layer and the deeper part of the thermostad forms a "barrier layer" between halocline and thermocline. Since there is no vertical temperature gradient across the halocline, vertical heat flux does not occur across it (Lindstrom *et al.*, 1987), and because of the relatively calm winds over the western Pacific the diurnal temperature cycle may be quite pronounced and may exceed 3°C in the near-surface layer (Soloviev and Lukas, 1997). The recent WEPOCS surveys have changed our classical view that the Pacific mixed layer deepens progressively from east to west. In fact, the effective mixed layer deepens only to about the date line, then remains at about 100 m into the western Pacific. At 15–20% of the WEPOCS stations, however, recent wind-bursts had mixed the brackish water down to the thermocline, which is also the primary pycnocline: In these circumstances the barrier layer is eroded, some flux of nitrate passes surfacewards, and the profile resembles the classic view of western Pacific stratification.

Such westerly wind-bursts, generated north of New Guinea, occur mostly in boreal winter and by impulsive forcing they generate equatorial Kelvin waves. If these waves straddle the equator, convergence and equatorial down-welling are induced in just a few days, with upwelling occurring along the pycnocline ridges at 2 or 3° from the equator, especially to the north. When the Southern Oscillation is relaxed and westward trade wind stress is persistent, divergence occurs at the equator as in the eastern Pacific. Because the nitracline is so deep, equatorial divergence only enhances nutrients significantly in the euphotic zone during periods of strong westward wind stress; obviously, during El Niño events when trade winds are reversed or absent in this region, equato-rial upwelling is suppressed (Blanchot et al., 1992). It is also in this province that the eastward wind anomalies that mark the onset of an ENSO event first develop, and during such strong negative anomalies of the Southern Oscillation the conditions characteristic of the WARM province may come to lie as far east as 180° latitude and we may find it convenient to evoke the concept of a trans-Pacific WARM-ENSO province (see Chapter 6). During at least some ENSO events, as occurred during 1986 and 1987, the effect of the Rossby upwelling wave in the WARM province is to shoal the thermocline until it is coincident with the bottom of the surface mixed layer; when this occurs, a "barrier layer" no longer exists and the nutricline comes to lie at the bottom of the wind mixed layer so that vertical flux of nutrients into the euphotic zone must be enhanced (Radenac and Rodier, 1996).

It is the pattern of zonal currents that induces the ridge–trough system of the regional thermocline upon which is imposed the special stratification which defines the province. A weakly flowing South Equatorial Countercurrent (SECC) at about 7–10°S (at 165°E) is scarcely observable as a slope in the thermocline topography (Godfrey et al., 1993; Gouriou, 1993), which is other-wise dominated by the westwards flow of the SEC and the consequences of this flow occur across the equator in both hemispheres. This flow requires the exis-tence of a pycnocline ridge aligned along the equator.

Biological Response and Regional Ecology

Because mixed layer chlorophyll concentrations are so low (0.05 mg m^{-3} is a typical value), the euphotic zone is deep and generally comprises two ecolog-ically distinct depth strata: an upper nitrate-limited, cyanobacteria-dominated zone and a deeper light-limited zone dominated by eucaryotic microalgae (Le Boutiller et al., 1992). There is some evidence for midday photo inhibition of primary production in the extremely clear water typical of this province. The nitracline commences at about 75 m below values typically in the range 0.006–0.008 μM, and by 100 m nitrate is typically 9.0 μM. The weak chlorophyll maximum (<0.5 mg chl m^{-3}) lies on the upper part of the nutricline rather shoaler at 50–100 m. The seasonal CZCS images show that near-surface chlorophyll values are somewhat higher a few degrees on either side of the equator, and this effect is most clear at the eastern and western sides of the province but is weaker at 150–160°E. At the equator, when divergence is sufficiently strong the thermocline and halocline may occur together at only about 30 m, with the DCM still remaining deeper at 40–100 m. During those ENSO events which bring the nutricline to lie at the bottom of the wind mixed layer, we may reasonably assume that this is associated with the 25–50% increase in average primary production that occurs in the WARM province in these years (Barber and Chavez, 1990).

The cyanobacteria-dominated assemblage has its maximum cell abundance in the upper DCM (80 m), whereas the microalgal assemblage is most abundant on the lower slope of the DCM at about 100 m. Biomass of mesozooplankton and micronekton is concentrated in a broad zone from the DCM to the surface, biased toward the DCM depth especially at night; biomass is relatively low. In both ENSO and relaxed conditions, zooplankton biomass from 20°S to 6°N takes low values in the range 0.5–1.0 g dry weight m^{-2}; at the equator, when westward trade wind stress is strong and upwelling occurs, zooplankton biomass rises to 2.5 g dry weight m^{-2}. At the same time, integrated primary production at the equator shifts up from 0.7 to 1.7 g C m^{-2} day^{-1}, and the depth of the maximum rate rises from 60 to only 20 m.

We have little information on ecology of zooplankton from this region, save for euphausiids (Hirota, 1987). These have the anticipated distribution with strong diel vertical migration of several genera between great depths by day (as would be anticipated in such clear water) of around 400–500 m and at the DCM at around 100 m at night. This is the pattern for several species of *Thysanopoda, Euphausia,* and *Nematoscelis,* but several species of *Stylocheiron* remain within the DCM by day and scatter both up and down at night. These results, species for species, follow the same pattern as those of Sameoto *et al.* (1986) and Brinton (1962) in the eastern tropical Pacific.

An enigma associated with this province is that it should be so productive of tunas of several species. The distribution of these open-ocean predatory fish is probably better known globally than that for any other pelagic organisms: Modern Japanese, Korean, and American long liners, bait boats, and purse seiners have efficiently explored all tropical and subtropical oceans and several international fishery commissions have meticulously recorded their catch rates. Much better than any comparable ocean basin-scale maps of the distribution of plankton organisms, the maps of the catch rates of skipjack, yellowfin, albacore, and bluefin tuna give confidence that the represented distributions are real. Even these must, of course, be read with caution. Is the edge of a species distribution natural or the effect of fishery regulations, as occurs in the eastern tropical Pacific, or is it the effect of a deepening thermocline on the efficiency of purse seines as may occur toward the west of the ocean?

Nevertheless, quite consistently in all Pacific maps of tuna distribution the boundaries of the WARM province enclose the region of greatest abundance in the entire Pacific of skipjack (*Katsuwonus pelamis*) and yellowfin (*Thunnus albacares*), which are the characteristic species of the trade wind zone (see, for example, Bayliff, 1980; Sund *et al.,* 1980). For both these species, and also for bigeye (*Thunnus obesus*), the greatest concentrations of larvae occur in WARM, though they are also widely distributed across the other trade wind biome provinces. Though adult yellowfin and skipjack are also widely distributed in PNEC, PEQD, and the warmer parts of South Pacific Subtropical Gyre Province (SPSG) and NPSG, it is in WARM province that the most persistent high concentrations occur; for skipjack, this population center extends seasonally into KURO as far as the Japanese Islands.

There seems to be no simple explanation for the paradox that such anomalously high concentrations of tuna should occur in such an oligotrophic province. It has been noted that skipjack occur preferentially in very warm water, but it seems most unlikely that a difference of <2°C in surface water temperature in this region compared with other Pacific tropical regions could be physiologically of such advantage as to select for this otherwise apparently unsuitable area, at least in terms of food supply. I suggest that the explanation may perhaps lie in the unique character of the province and its extraordinarily strongly stratified water column with a boundary layer. Such a feature may have two consequences for tuna: The multiple pycnoclines may serve to aggregate layers of food organisms, thus simplifying their location by foraging tuna (which are able to tolerate cold temperatures during deep hunting forays), or perhaps the very stable water column provides invariant and predictable conditions for first-feeding tuna larvae. The latter suggestion is made in the light of well-known studies of the differential survival of fish larvae when their prey abundance matches their hatching date and when concentrated layers of prey organisms are disrupted by wind-mixing episodes.

Synopsis

This is an anomalous case in which the photic depth coincides more closely with the thermocline, but the halocline lies shoaler; all are essentially seasonally invariant (Fig. 9.11). A minor rate increase in primary production occurs for 2 months in the middle of the rainy season (July and August) accompanied by episodic accumulation of chlorophyll; this may not be generally observable and is perhaps a singularity of the archived data.

Archipelagic Deep Basins Province (ARCH)

Extent of the Province

The Archipelagic Deep Basins Province (ARCH) is, admittedly, something of a grab-bag, though I provide some justification for this in the sections which follow. I have assembled all the minor deep seas enclosed or partially enclosed by the island chains and archipelagoes that lie between the Indian and Pacific Oceans, separating these from the intercalated continental shelf areas which have been assembled in the Sunda–Arafura Shelves Province (SUND), named for the Sunda Islands. Therefore, ARCH comprises several deep seas within the Indo-Pacific Archipelago, or adjacent to it but enclosed by island arcs: the Andaman, South China, Sulu, Celebes, Molucca, Flores, and Banda Seas and, further to the east within the arcs of the Bismarck, Solomon, and Vanuatu Islands, the Bismarck, Solomon, and Coral Seas. The last of these is by far

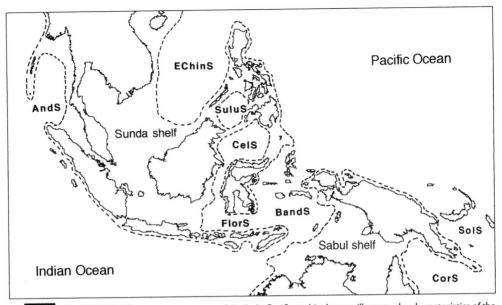

FIGURE 9.12 Bathymetric map of the Indo-Pacific archipelago to illustrate the characteristics of the ARCH and SUND provinces. The dashed line represents the 200-m isobath, enclosing the SUND province, whereas the deep basins of the ARCH province are indicated by abbreviations representing individually the Andaman, East China, Sulu, Celebes, Flores, Banda, Solomon, and Coral Seas. Note that the East China Sea is inhabited by a central region of coral atolls that may be visible in the CZCS chlorophyll field.

the largest, and perhaps it would be better to consider it as a province in its own right.

Continental Shelf Topography and Tidal and Shelf-Edge Fronts

In this complex region, it is not useful to describe in detail the relations between shelf areas and oceanic depths; these can be better understood simply by examining Fig. 9.12. The differences between deep and shallow channels should be noted, as should the relative sill depths of deep basins, because these determine circulation patterns.

The Sulu Sea is the deep basin most constrained by shallow sills (<450 m), whereas the Celebes Sea, though open to the Pacific, has only narrow deep channels to the Indian Ocean by way of the Macassar Straits. The Banda and Molucca Seas, of the strictly archipelagic deep basins, together have the most open connections to both oceans. The Coral, Bismarck, and Solomon Seas to the east and the Andaman and South China Seas to the west are embayments of the open ocean or lie partly isolated behind island arcs.

Defining Characteristics of Regional Oceanography

The circulation of the shelves and deep basins comprising the Indo-Pacific Archipelago are components of a single hydrographic and atmospheric system, the Australo-Asiatic Mediterranean Sea of Dietrich *et al.* (1970) or the South-

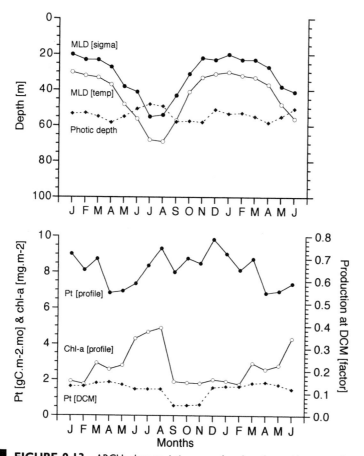

FIGURE 9.13 ARCH: characteristic seasonal cycles of monthly averaged mixed layer and photic depths, chlorophyll at the surface, and rate of primary production, both depth-integrated and at the DCM. Data sources are discussed in Chapter 1.

east Asian Waters of Wyrtki's *Naga Report* of 1961. However, the differences between the shallow shelf and the ocean-depth basins are so great that it is misleading to unify the ecology of these disparate seas into a single ecological province. Because of its complex geography, the archipelago could properly be considered as a coastal boundary region, and the dimensions of the deep ocean basin are such that lateral boundary effects are pervasive. The necessary discussion of circulation through the archipelago, and its consequences, is intended to apply to both the coastal (SUND) and oceanic (ARCH) provinces, together comprising the Indo-Pacific archipelagic region.

The regional oceanography of the archipelago (of which the most complete description remains that of Wyrtki) is unique and dominated by the effects of the reversing monsoon winds over the archipelago and the currents consequently driven between the islands and from ocean to ocean. Additionally, the major zonal current systems of the two oceans influence the flow through the archipelago: In general, the channel geography of the region is more open on the Pacific side and more closed on the Indian Ocean side.

In boreal winter, the ITCZ, the equatorial front between the north and south trade wind systems, lies south of the equator along the southern coasts of the Java and the Lesser Sunda Islands (at about 10°S) so that the main archipelago is entirely under the influence of the northeast trade winds (which also cross the Malayan isthmus to the Andaman Sea); south of the equator, wind vectors become progressively zonal so that the Arafura shelf and the eastern seas lie under increasingly eastward winds. In May, the northeast monsoon collapses as the ITCZ moves northwards, south winds prevail to the equator, and the southeast trades (Beaufort force 4 over the open sea) develop over the whole archipelago, and by August the southeast monsoon is fully developed. Northeast winds are initiated in boreal autumn in the East China Sea, strengthening as the ITCZ returns south of the equator in November.

Seasonal alternation of wind stress, as in the Arabian Sea, forces reversing current systems through the channels of the archipelago. The strongest flows in both monsoons are in the South China Sea, due to western intensification along the coast of Vietnam, and through the Molucca Sea, because it affords the most open connection between the Pacific and Indian Oceans.

In February, at the peak of the northeast monsoon, flow down the west coast of the South China Sea continues to the southeast, through the Java Sea, to join the flow from the Molucca Sea. This combined stream then passes east and northeast through the Banda Sea to rejoin the Pacific water masses around New Guinea. Flow into the Indian Ocean occurs through all the available passes, particularly persistently through the Malacca and Sunda Straits. In August, the southerly monsoon reverses the western coastal jet of the South China Sea, and flow is driven to the northwest through the Java Sea and to the west through the Banda Sea by the westward component of the southerly monsoon in the southeastern part of the archipelago. Part of the water entrained from the Pacific into the Celebes Sea around Mindanao returns to that ocean along the north coast of Celebes, while some passes south toward the Indian Ocean through the Straits of Macassar (see Plates 1–6 in Wyrtki, 1961). Where the westward flow of the Pacific SEC meets the landmasses of the archipelago, seasonal gyres occur: the Mindanao Dome (2–4°N) and cyclonic gyral motion around the West Fiji Basin (10–12°S).

This complexity of flows is associated with a great diversity of hydrographic processes likely to have biological consequences which are summarized for each significant deep basin, largely from the review of Wytrki (1961).

• *Andaman Sea:* Flow is persistently from the north as part of the Bay of Bengal gyres and is the root of the southward current that flows along the western coast of Sumatra, to diverge into the SEC. During the northeast monsoon, upwelling occurs on the east coast of the Andaman Sea along the coasts of Thailand and Burma from December to March and is accompanied by an increase in surface salinities (Wyrtki, 1961; LaFond and LaFond, 1968).

• *South China Sea:* In the South China Sea, the western jet current is associated across a convergent boundary with a countercurrent offshore in both monsoons, and a basin-scale gyral circulation develops especially during the southerly monsoon.

• *Straits of Macassar:* Flow is persistently to the south with maximum velocities in February and March and July–September. Flow is into the Java Sea during the southerly monsoon and into the Flores Sea during the northerly monsoon.

• *Flores Sea:* Flow is predominantly to the east, though a cyclonic gyre forms during the southerly monsoon with westwards flow to the south of Celebes. The flow passes through the Lesser Sunda Islands and to the east of Timor, leaking surface water to the Indian Ocean at all seasons.

• *Banda Sea:* The deep basin of the Banda Sea, especially toward the east, experiences regional upwelling during the southerly monsoon, and downwelling during the northerly monsoon, to satisfy balance between the archipelagic flow-through and the SEC of the Indian Ocean (Wyrtki, 1961). The Banda Sea is a major pass for flow from the Pacific Ocean to the Indian Ocean past Halmahera and Ceram in the north, and by the Flores Sea and Timor to the south. Currents are weak and variable in intermonsoon periods.

• *Sulu Sea:* Apart from local wind drift, little flow occurs through the Sulu Sea and there is little change in its mixed layer depths between the two monsoons; this is a relatively quiet and self-contained hydrographic basin.

• *Celebes Sea:* Anticyclonic gyral flow persists during all seasons, though it extends farther west during the northeast monsoon; strong flow passes south of Mindenao, returning to the Pacific (to enter the SECC) together with flow leaving the Molucca Sea around the Halmahera gyre. During boreal summer, when the gyre is displaced to the east, surface drift is received from the Sulu Sea; in winter, this drift through the Sulu archipelago is reversed.

• *Coral Sea:* The SEC of the Pacific Ocean flows westward into and through the Coral Sea, diverging as it meets the Australian continent at about 18 or 19°S during the dry season, though during the monsoon season the divergence migrates equatorwards to at least 14°S. At this divergence, to the east of the 200 m topography of the Queensland Plateau, there is a feed into the East Australian Current to the south and–to the north–a cyclonic circuit around the Gulf of Papua and the Solomon Sea (Andrews and Clegg, 1989) which transports $10–15 \times 106$ m^{-3} sec^{-1} into the Indonesian archipelago. In the area of the divergence itself, currents are weak and include a persistent cyclonic eddy in which water is transported onto and over the Great Barrier Reef. The Coral Sea is a uniformly oligotrophic region in which water column stability is very strong and the surface layers are strongly nitrate deficient. Only a pair of persistent eddies at 12 and 17°S seem potentially likely to enrich the euphotic zone with nitrate-replete water, though their presence is not detectable at the surface.

• *Solomon and Bismarck Seas:* From the southeastern tip of New Guinea, a coastal jet current flows along the north coast and passes into the southern Bismarck Sea inside New Britain. The Bismarck Sea is occupied by a weak cyclonic gyral system, associated with westward flow along the mainland coast.

In some parts of the province, seasonal variability of mixed layer depth is slight (Wyrtki, 1961) as in the Sulu, Celebes, and Flores Seas. Elsewhere, it responds to changes in monsoon wind stress: In the central and northern parts of the South China Sea, mixed layer depths are shallow (30–40 m) in the southerly monsoon, deepening (to 70–90 m) in the northerly, boreal winter monsoon. By the end of the northerly monsoon, the mixed layer has deepened to 100 m. In the Banda Sea (and also in the shelf areas of the Arafura Sea), the changes are even greater, upwelling under the influence of the southerly monsoon forcing a 2°C temperature drop at the surface and thinning the mixed layer to about 20 m or less. This upwelling is principally limited to May–August. Other monsoon-driven, persistent upwelling regions within the province are predicted by Wyrtki (1961) to be (i) off the Macassar peninsula; (ii) along the coast of Vietnam, where temperature drops of >1°C occur during the southerly monsoon; (iii) on the coast of Sarawak; and (iv) south of Hong Kong. In the Sulu and Celebes Seas, upwelling is unlikely because of continuously high sea levels. It is now possible to examine all these suggestions by means of satellite thermal imagery.

The consequences of the unique meteorology (heavy rainfall and intense cloud cover) over Indo-Pacific Archipelago are significant for biological oceanographic processes: The rivers of the archipelago itself discharge 3.0×10^9 tons of sediment annually into coastal water, or about twice the sediment discharge of the Amazon. The coastal rivers of Southeast Asia discharge another 4.1×10^9 tons. Together, this is more than twice the sediment discharged from all other rivers (Milliman and Meade, 1983). Though apparently not described, it would not be surprising to find examples of barrier layers (shallow halocline above a deeper thermocline) here, as is typical of the WARM province.

Biological Response and Regional Ecology

Through much of this province, the complex topography both of the land and of the seabed is likely to produce many small and nonpersistent hydrogeographic instabilities, themselves likely to be associated with nutrient transport to the photic zone and biological enhancement. Such features are difficult to predict or map comprehensively, even with satellite imagery, because of the extensive and pervasive cloud cover of the region, but evidence of their existence is available for some areas.

In the Bismarck Sea, meridional island wakes up to 300-km long have been observed as high-contrast features in CZCS images, and these represent the effects of turbulent plumes downwind of mountains of New Britain and high islands such as Sakar Island to the east of the Vitiaz Straits (Wolanski *et al.*, 1986). These plumes appear to represent cooler, more transparent water surfacing in a region where the surface water is highly turbid, either with phytoplankton or suspended sediments. In either case, phytoplankton growth will be enhanced in the plumes.

Fields of large internal waves in the southern Bismarck Sea appear in the same images and probably represent visualization of deeper, clearer water being brought to the surface (Wolanski *et al.*, 1986). We can anticipate that such features will be revealed throughout the province as examination of further images becomes possible; we can also expect to find that the orientation and location of the features will depend on seasonal and shorter changes in wind direction and strength.

In the Indonesian seas, recent observations have confirmed, in general, earlier predictions; chlorophyll enhancement during the southeast monsoon is from 0.25 to 2.5 mg m^{-3} above levels occurring during the northwest monsoon. This seasonal buildup of biomass, associated with an increase in the rate of primary production, is greatest in the eastern seas (Banda, Flores, and Seram) and decreases westwards through the archipelago, becoming a very slight effect in the Sulawesi Sea and the Makassar Strait (Kinkade *et al.*, 1997).

Though there are interannual differences, the predicted downwelling–upwelling seasonal cycle in the Banda Sea has been observed (Zijlstra *et al.*, 1991), with predictable consequences for photic zone nitrate (Wetsteyn *et al.*, 1990) and seasonal primary production (Zevenboom and Wetsteyn, 1990; Gieskes *et al.*, 1990). In general, the degree of seasonal chlorophyll enhancement throughout the archipelago seems to be correlated with sea-surface temperature and hence the relative strength of the seasonal upwelling effect. In August, during the (upwelling and light-limited) southerly monsoon the surface layer of the Banda Sea is homogenous down to 60 m and primary production rate is highest at the surface, whereas during February, under the influence of the (downwelling and nutrient-limiting) northerly monsoon, stratification is established and a DCM forms at 40–80 m. Primary production rate is higher during the southerly than the northerly monsoon (1.85 and 0.9 g C m^{-2} day^{-1}, respectively). Patches having much higher production rates (<7 g C m^{-2} day^{-1}) occur to the south of New Guinea, where upwelling occurs over the outer slope of the Papuan Barrier Reef (Furnas and Mitchell, 1996).

The southern part of the Solomon Sea, and the internal basin of the Louisiade Archipelago, which forms the southern margin of this body of water, also has relatively high productivity. As in other oligotrophic regions, there is evidence in the stratified region that the cells of the DCM are more dependent on nitrate than those closer to the surface, and that the picofraction generally dominates the assemblage of autotrophic cells. Most of the primary production occurs shoaler than the midday 20% isolume, and very little is lost to sinking so we can assume that production and consumption of small cells are very closely in balance. The mesozooplankton biomass of the Banda Sea is augmented strongly at the onset of the upwelling process by the rise from depth of large populations of *Calanoides philippinensis* together with *Rhincalanus nasutus,* which reach a stock of 4000 m^{-2} grazing on phytoplankton (and microzooplankton), when chlorophyll accumulation reaches 3 or 4 mg m^{-3}. Again as expected, vertical excursions during diel migration are reduced in the less clear water during the southerly monsoon. A consequence of this change in behavior is a stronger diel difference in gut chlorophyll in copepods between the two seasons; though it should also be noted that, once again, tropical mesozooplankton herbivores account for only a small part of primary production (5–25% daily) or of standing stock (2–6% daily).

There are few modern biological oceanographic observations of this region, except for the Solomon Sea, for which there are observations of a shade-adapted DCM (0.6 mg chl m^{-3} and 0.15 mg C mg chl hr^{-1}) at about 75 m and at the 1.2% isolume, containing almost half of the total chlorophyll in the 0 to 200-m water column. In the Coral Sea, as noted previously, temporary zonal ridges in the variable circulation have been observed at 12 and 17°S, with consequences for mixed layer nutrient levels (Rougerie and Henin, 1977). During the southerly monsoon the mixed layer of the Solomon Sea also deepens and chlorophyll profiles have a DCM with a shade-adapted flora at around 75 m, or at about the 1% light level (Satoh *et al.*, 1992). There is a significant coastal upwelling cell during the southerly monsoon in the coastal jet which carries flow from the SEC along the north coast of New Guinea. The Coral Sea and the other eastern basins share some biological characteristics with the oceanic South Pacific, including the persistent occurrence of blooms of nitrogen-fixing tufts of cycanobacteria. Very large ($<9 \times 10^4$ km^2) blooms of these organisms, identified as *Trichodesmium* by its chromatic signature, have been observed at 10°S 165°E in the ocean around New Caledonia in the surface chlorophyll field observed by the CZCS satellite (Dupouy *et al.*, 1988). Such occurrences have major significance for basin-scale budgets of inorganic nitrogen.

Synopsis

Very weak deepening of mixed layer occurs during the boreal summer monsoon, carrying pycnocline below the photic zone for 3 months (Fig. 9.13). Weakly enhanced primary production rate is correlated with seasonal mixed layer deepening. Chlorophyll biomass tracks primary production rate in early part of period of higher productivity and then declines rapidly to very low values.

South Pacific Subtropical Gyre Province (North and South) (SPSG)

Extent of the Province

The SPSG province comprises the central and southern part of the subtropical gyre of the South Pacific Ocean, omitting the extreme northwestern part which lies equatorward of the Solomon Islands and New Guinea: This is recognized separately as the WARM province. SPSG is bounded to the south by the far-field effects on chlorophyll enhancement of the Subtropical Convergence Zone (see SSTC) and to the north by the southern edge of the chlorophyll enhancement caused by equatorial divergence (see Pacific Equatorial Divergence Province). To the east, SPSG is bounded by the offshore eddy field of the Humboldt Current (see Humboldt Current Coasta Province) and to the west by the 29°C surface isotherm at the edge of the western Pacific warm pool and the line of the New Hebrides (now Vanuatu) which enclose the Coral Sea. For some purposes, and when the region is better explored, the province could probably be subdivided meridionally at about 30°S.

Continental Shelf Topography and Tidal and Shelf-Edge Fronts

These are not relevant to this province. However, the presence of the Polynesian Islands from Easter Island at 120°W to the Samoas at 170–180°W must be noted: the Marquesas, the Tuamotos of French Polynesia, the Cook Islands, Pitcairn, Ducie, and many others populate this otherwise empty ocean.

Defining Characteristics of Regional Oceanography

This is the least well-described region of the ocean, as one can see by glancing at any oceanographic atlas having global maps showing where oceanographic observations have been made of any variable. Were this very large

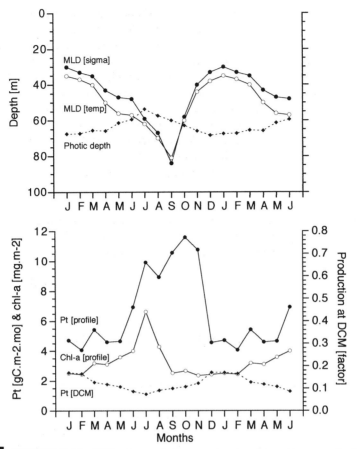

FIGURE 9.14   SPSG: characteristic seasonal cycles of monthly averaged mixed layer and photic depths, chlorophyll at the surface, and rate of primary production, both depth-integrated and at the DCM. Data sources are discussed in Chapter 1.

province better known it might be desirable to subdivide it, probably meridionally to recognize the fact that its poleward regions lie under seasonal westerly winds and must undergo some winter mixing. Although current models of the anticyclonic surface circulation occupying this province treat it as a single gyre centered at about 25°S, this is probably a simplification of the real surface circulation.

Be that as it may, this is the most uniform and seasonally stable region of the open oceans and (as noted by Tomczak and Godfrey, 1994) the origin of the name of the Pacific Ocean. Surface winds are as stable as the permanent atmospheric high pressure centered approximately above Easter Island (10°S,110°W), and the trades are weak but remarkably constant year-round so that seasonal maps of wave height are almost invariant. The likely line of subdivision of the province would be at 30°S, under the divergence between trades and subtropical westerlies; these are almost invariant between seasons as far south as the subtropical convergence, where the wind systems of the Southern Ocean are encountered. The dry descending air mass that comprises the eastern termination of the Walker circulation cell (the wet, ascending air mass over the WARM province comprises the other pole of this atmospheric cell) maintains an evaporation–precipitation index of 40–80 cm year^{-1} and thus warm, salty surface water.

An atmospheric convergence (the South Pacific Convergence Zone) between the two major wind systems lies northwest–southeast from New Guinea to about 30°S, 120°W, south of Easter Island (Barry and Chorley, 1982). This atmospheric convergence is associated with an oceanwide line of relatively heavy cloud cover and is a feature in the distribution of wind divergence over the ocean (Lagler and O'Brien, 1980), but its consequences for the surface oceanography of the South Pacific appear not to have been investigated. It does align with a thermocline ridge in winter, indicating some divergence of surface water, and in the climatic global CZCS chlorophyll field, higher chlorophyll values also occur along this feature.

The permanent pycnocline is bowl shaped with its maximum gradient lying at about 300 m at 124°W, at the center of the bowl, and rising to 150 m from 150°W to the western edge of the province near the dateline; consequently, as in the South Atlantic Province, we should note the general baroclinic upslope of nitrate isopleths toward the edges of the gyre. A thermocline (0.5°C criterion) occurs at 25–50 m over the whole province during austral summer, but to the south of 30°S it deepens to 75–150 m during austral winter.

The canonical single gyre model has the SEC passing along its equatorward side (and flowing more strongly in boreal summer) and a South Pacific Current along its poleward side, associated with the oceanic Subtropical Convergence Zone. The western part of the gyre includes circulation into the Coral Sea and the Western Pacific Warm Pool so that in the present regional classification of the ocean, these parts of the anticyclonic gyre are not included in the SPSG province. An SECC, apparently weaker and more variable than its counterpart of the Northern Hemisphere, flows across the northern part of the province, embedded within the westward flow of the SEC from 7 to 14°S at 155°W, being farther south and stronger in austral winter (Wyrtki and Kilonsky, 1984; Eldin, 1983).

Biological Response and Regional Ecology

An interesting meridional phytoplankton section along 115°W has been reported (Hardy *et al.*, 1996), which tends to confirm the first assumption made here that this province should not be subdivided zonally, as might otherwise seem more reasonable. Phytoplankton species composition was remarkably invariant from about 11 or 12°S, a line coinciding with the southern edge of the PEQD province, right down to 36°S at the edge of the Subantarctic Convergence. Beyond these two boundaries, species composition, carbon biomass, chlorophyll, and primary productivity all take higher values. In fact, principal component analysis of the whole section from 10°N to 60°S showed five groups for relative abundance of phytoplankton taxa: The meridional separation of these groups corresponds very well with the five provinces PNEC, PEQD, SPSG, SSTC, and SANT. There were, of course, smaller cells in the group corresponding to SPSG (or, at any rate, lying within its boundaries as defined here).

Nitrate-replete water lies to the north (in PEQD) and to the south (in SSTC) of this province, whereas nitrate-depleted surface water occupies the whole of the center of the gyre, having an annual mean value of <0.5 μM from the surface to 150 m. The nitracline follows the pycnocline very closely so that the 0.3 μM isobar lies just deeper than the depth of sharpest gradients in the pycnocline: that is, about 100 m in the east (at about 85°W), about 175 m in the center of the gyre, and about 50 m in the western, poleward limb of the gyre at 155–165°W. Care must be taken in interpreting the nitrate field because of the potential effect of another limiting element constraining the biological uptake of nitrate (see Pacific Equatorial Divergence Province). However, the generalized chlorophyll field clearly shows the consequences of the baroclinicity of the nitrate isopleths: The lowest values consistently occur above the deepest mixed layers and where the nitracline lies furthest removed from wind-induced mixing at the surface.

There is some evidence of seasonality in the surface chlorophyll field from ship observations in the western part of the gyre and into the adjacent Coral Sea. There is an austral winter (May–July) enrichment (in the range 0.15–0.4 mg chl m^{-3}) polewards of 20°S to at least 35°S and presumably further south, though these were the limits of the data set of Dandonneau and Gohin (1984). It is likely that this process occurs across the province, although the evidence from CZCS images is very weak. The most that can be said is that the "big blue hole" of the South Pacific in these images appears to be slightly reduced in intensity and area in austral winter. During summer, the *Trichodesmium* blooms observed in the same area by Dandonneau and Gohin are also probably a feature of much of the central, island-studded part of the province.

Apart from the EASTROPAC voyages of the 1960s in the northeast corner of the province, there are no comprehensive ecological studies of the open-ocean plankton ecosystem and the best that can be done is a brief extrapolation from other regions. It is probably safe to say that the observations discussed below for the CLIMAX and HOT site in the North Pacific subtropical gyre will obtain here—only in a more extreme sense. It is also safe to assume that a subsurface maximum chlorophyll layer lies across the province just shallower than the nitracline and on the upper part of the density gradient and that it acts to trap and utilize any nitrate mixed up across the upper pycnocline. Finally, I also assume that maximum primary production rate will occur much shallower; that nitrogen-fixing organisms will be prominent; that the microbial loop, or regeneration of ammonium as a substrate for algal growth, will be very active and mediated by bacterioplankton, picofraction cycanobacteria, prochlorophytes, and microflagellates; and that most consumption will be by a complex protist community. Such mesozooplankton as occur will be small, except for diel migrant metridiids, which will rise at night, some to the DCM and some to graze on protists in the mixed layer.

For the EASTROPAC area, which extended only to 20°S at 90–125°W, some of these suggestions are generally confirmed. A DCM slopes downwards in the upper thermocline, weakening to the south so that by 20°S it lies at about 120 m in February and March at the same time that the depth of maximum primary production rate is at about 20 m (Longhurst, 1976). The same pattern is repeated in August data, though the DCM may be a little deeper. Whereas mesozooplankton profiles in PNEC and PEQD to the north are strongly structured with near-surface and near-DCM layers of maximum abundance, in SPSG their layering is relatively weak with a very broad zone of relative abundance progressively weakening downwards to the midpycnocline.

Although coral reefs are not part of our discussion, a passing reference to their place in this ocean microcosm is in order because the myriad islands inhabiting the province owe their existence to the activities of reef-forming biota. If the growth of many of these reef-forming corals and molluscs is supported by photosynthesis of symbiotic algae within their tissues, a source of nitrogen still has to be found. This nitrogen comes from three sources, the balance between which is complex to compute. In part, this is provided by the activity of nitrogen-fixing blue-green algae, in part by capture of the sparse oceanic plankton drifting over the reef, and in part—perhaps in some cases most important—by endo-upwelling within the fractured carbonate rock of the reef platform. Very slight geothermal heat flux from the earth's crust is sufficient to maintain a slow upward movement of water within fractured rock, drawing nitrate-replete water in at subpycnocline depths and releasing it at the surface. For this system to function another requisite is perfect clarity of the surface water to obtain maximum solar energy within the symbiotic tissues. These facts answer Darwin's paradox: Why do we find the most exuberant growth of corals in the clearest water?

Therefore, where reef-forming organisms occur, we can be sure that seasonal blooms of planktonic algae and upwelling processes are not significant. This suggestion is confirmed by the observations of Rougerie and Rancher (1994), who found that in the immediate vicinity of numerous atolls the vertical structure of the oligotrophic profile is preserved intact. Right up to the coral front the nitracline (150–200 m), the pycnocline (100–175 m), and the DCM (125 m) remain undisturbed by the proximity of the reef platform. Weak, irregular flows past the atolls produce little turbulence or chlorophyll-enhanced wakes downstream of the atolls. Even in strong wind, outbursts (tropical cyclones) do not appear to modify the strongly stratified oligotrophic profiles.

It is interesting to note that once again, McGowan's biogeography and Brinton's euphausiid maps support at least some of our conclusions. His map of the distribution envelopes of 10 central water biota (*Stylocheiron suhmii, Euphausia brevis, Euphausia mutica, Sagitta serratodentata*, etc.) is a mirror image of the Northern Hemisphere pattern (see NPSG) but with the importance difference that their pattern is more diffuse in the eastern part of the ocean. This is exactly what we would expect from the hemispheric differences between the circulation patterns. Furthermore, the distribution of the subtropical albacore tuna (*T. alalunga*) from catch-rate maps for the South Pacific appears to correspond very well with the limits of this province.

Synopsis

There is a very weak deepening of mixed layer in austral winter, carrying pycnocline below photic zone for only 3 months (Fig. 9.14). Significantly enhanced primary production rate in austral winter and spring is correlated with mixed layer deepening. Data indicate a short pulse of chlorophyll accumulation at the start of the productive period which is rapidly removed as consumption comes to balance production.

PACIFIC COASTAL BIOME

Alaska Downwelling Coastal Province (ALSK)

Extent of the Province

The ALSK province comprises the coastal boundary region from 53°N (Queen Charlotte Sound) to the end of the Aleutian chain of islands at about 180°W, and from the velocity maximum of the Alaska Current to the coastline.

Continental Shelf Topography and Tidal and Shelf-Edge Fronts

Along the north–south fjord coastline the continental shelf is very narrow or absent. A greater width occurs where the coastline turns west toward the Aleutians, and the island of Kodiak stands on a shelf about 150 km wide, though this narrows progressively and rapidly toward the southwest.

Defining Characteristics of Regional Oceanography

This coastal boundary province comprises the rim current (the Alaska Current) of the Alaska gyre from its origin at the divergence north and south where the North Pacific Current encounters the American continent approximately at Vancouver Island, Canada; this province is similar to the Alaska Current System of Favorite *et al.* (1976) and the Alaska Downwelling Province of Ware and McFarlane (1989). The velocity maximum of the Alaska Current lies above the steep-to continental slope, and velocities are greater in winter than in summer because of the winter intensification of the Aleutian atmospheric low-pressure cell (Schumacher and Reed, 1983; Thompson, 1981).

Flow of the Alaska Current, and a general downwelling tendency at the coast, is maintained by the density distribution within the coastal flow, resulting from dilution of the surface water mass, from wind-curl stress, and from a longshore wind-induced slope in sea level. Flow is therefore strongest during early winter when river effluent is greatest. This situation obtains along most of this coast: Only off the Alaska peninsula in summer (near Kodiak Island) do local winds force any significant upwelling of isopleths and weak upwelling lasts only from June to August. Along the Alaska peninsula and the Aleutian chain to the southwest, the strong boundary current, here called the Aleutian Steam, continues the northern limb of the Alaska gyre and is a typical meandering coastal jet, topographically locked and associated with meanders, eddies, and filaments. Though the Aleutian Stream normally flows close along the topography to the end of the Aleutian Islands and is the dominant source water for the Bering Sea gyre, anomalous separation of the current from the topography at about 170°W can occur, and this condition may persist for many months (Stabeno and Reed, 1992).

The permanent halocline of the Gulf of Alaska extends through this coastal province, though it is shoaler than out in the open gulf. The pycnocline places a cap on winter mixing at about 50–60 m, though convective cooling extends much deeper. The stability induced by the halocline constrains the vertical entrainment of nitrate during winter and it is suggested that the spring bloom would be as brief as it is offshore (see Pacific Subarctic Gyres Province) if the interaction between the Alaska Current and coastal topography, where the shelf

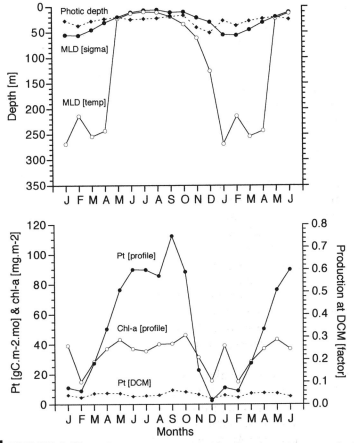

FIGURE 9.15 ALSK: characteristic seasonal cycles of monthly averaged mixed layer and photic depths, chlorophyll at the surface, and rate of primary production, both depth-integrated and at the DCM. Data sources are discussed in Chapter 1.

is narrow and steep-to, did not create conditions for some entrainment of nitrate throughout the summer.

Biological Response and Regional Ecology

The permanent halocline of the Alaska gyre extends into the coastal boundary province and dominates vertical mixing. There is little information on the biological consequences of circulation features within this province (Anderson *et al.,* 1977), save that production in shelf waters is thought to occur at about twice the open-ocean rate (200–300 compared to 70–100 g C m^{-2} year^{-1}). Where the flow becomes approximately zonal, along the Alaska peninsula, instability develops and the stream may separate from the continental edge or may regain stability by anticyclonic meandering and eddy shedding (Okkonen, 1992).

The ecology of the Strait of Georgia, at 48°N between Vancouver Island and the Canadian mainland, will serve as a useful model appropriate to the sounds inside the islands that were given regal names by their navigating discoverers (Queen Charlotte, Prince of Wales, Alexander, and Prince William Islands) up to 60°N in Alaska (Harrison *et al.*, 1983). These passages are dominated by semidiurnal tidal pulses which may be very fast (up to 6 m sec^{-1} in the Straits of Georgia) and are associated with tidal fronts at mouths of bays and rivers. This is a region of heavy clouds and rainfall, with two consequences for the biological cycle: (i) Circulation generally resembles a positive estuary with seawards flow of brackish water at the surface, and (ii) irradiance has a very strong seasonal cycle from 70 W m^{-2} on a rainy winter day to 1400 W m^{-2} under a sunny sky in summer—moreover, water clarity is reduced in winter by silt effluent from flooding rivers (Secchi depths in river plumes of 0.1 m in winter to 3.0 m in summer are not unusual). Primary production is light limited in winter and nutrient limited in summer.

A strong spring bloom is initiated in March (about 15 mg chl m^{-3}) and phytoplankton biomass, measured by chlorophyll, declines throughout the whole summer to reach very low values (<1 mg chl m^{-3}) in November. A succession of diatom species ????? the seasons: *Thalassiosira* and *Skeletonema* dominate the spring bloom, followed by *Chaetoceros, Ditylum, Nitzschia,* and *Leptocylindricus* in summer and *Coscinodiscus* in autumn. Nanoflagellates dominate during winter but form only 10% of the >2- to 4-μm phytoplankton in summer. During the summer, some shallow stratification is imposed and a very shallow DCM develops at 10–20 m, with a primary production peak at about 5 m. At other times primary production takes highest values just under the surface.

The winter mesozooplankton is dominated by small copepods (*Pseudocalanus, Oithona,* and the rest), but their development in spring is swamped by a massive influx of larvae and early copepodite stages of the offshore *Neocalanus plumchrus* together with immigration, also from the ocean, of *Calanus pacificus, C. marshallae,* and *M. pacifica*. A summer community of predators also joins the throng: *Sagitta* and amphipods. *Neocalanus plumchrus* dominates biomass in spring and *C. marshallae* dominates somewhat later in summer. Most copepods remain in the upper tens of meters of the water column, with smaller numbers at about 50 m, whereas *Sagitta*, amphipods, coelenterates, and Euphausiids form deeper layers fueled largely by sinking organic material.

We can be encouraged in a belief that the ecology of the Strait of Georgia is appropriate to this province generally, modified more or less in timing of events by latitude and water clarity, and in the knowledge that in Prince William Sound at 60°N, *N. plumchrus* together with *N. cristatus* and *E. bungii* are the dominant (their biomass is >25–35% of all >0.333-mm net plankton) spring–summer copepods inshore during those months when they are not overwintering in deep water beyond the shelf (Cooney, 1986). Apparent abundance tends to be reciprocal between the shelf and deep water; when each of these species is abundant on the shelf it is sparse offshore and vice versa. Appearing in spring as nauplii and copepodites over the shelf, these copepods descend to overwintering depths beyond the shelf edge in October where maturation and reproduction takes place to produce the early larvae of the new generation. It has been speculated that the convergent front between the Alaska Coastal Current and the shelf water inshore is a zone of enhanced zooplankton abundance (individuals would be expected to resist sinking at the front and so be aggregated there) in summer along the route of juvenile salmon, spawned in rivers to the south but bound for offshore feeding grounds to the north. Their trip takes 1 or 2 months, during which time considerable growth as well as locomotion is accomplished by the young fish.

Synopsis

The permanent halocline caps winter mixing and entrainment of nitrate from below, so the spring–summer bloom is thought to be supported by nitrate entrained during this period by topographic coastal processes (Fig. 9.15). Chlorophyll consistently tracks primary production rate and nitrate-based production is assumed to dominate during the whole productive period.

California Current Province (CALC)

Extent of the Province

The CALC province comprises the California Current from the bifurcation of the eastwards flow of the North Pacific Current south to the convergent front which lies southwest off the tip of Baja California at the root of the NEC. The California Front is the seaward limit, though this front is progressively less well defined toward the south. For this reason, the canonical boundary of the California Current is often set at 1000 km offshore, a definition I use where necessary.

Continental Shelf Topography and Tidal and Shelf-Edge Fronts

The continental shelf is narrow, with the 200-m isobath occurring as little as 10 km from the shore in southern California (or even closer, where Scripps Canyon runs into the beach) to a distance of about 75 km off Oregon and Washington. A continental borderland with deep (approx 200 m) basins, shallow banks, and islands occupies the bight (including the Santa Barbara Basin) south of Point Conception (34.5°N). Tidal fronts have not been described, but the entire coastal boundary province is populated by upwelling fronts and fronts associated with meanders of the coastal jet and their associated cyclonic eddies.

Defining Characteristics of Regional Oceanography

The California Current is the most intensively investigated of the eastern boundary currents, especially because of the results from the long time series (1950–present) of the California Cooperative Fisheries Investigations (Cal-COFI) managed jointly by federal, state, and university research groups. Much of what follows is drawn from reports, atlases, and research papers arising from CalCOFI.

The California Current takes its source in the divergence of the west wind drift water of the North Pacific Current as this approaches the American continent within what Favorite *et al.* (1976) term the dilution zone, where precipitation greatly exceed evaporation. The eastward flow diverges northwards into the eastern limb of the Alaska gyre and southwards into the California Current as a cool, low-salinity, high-nutrient water mass entraining biota with the flow.

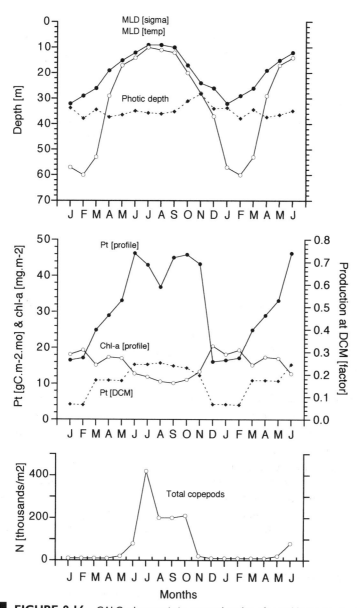

FIGURE 9.16 CALC: characteristic seasonal cycles of monthly averaged mixed layer and photic depths, chlorophyll at the surface, and rate of primary production, both depth-integrated and at the DCM. Data sources are discussed in Chapter 1.

The following account of the California Current describes the normal condition, when the trade wind regime is fully developed; during ENSO events, when trade wind stress is relaxed in the western Pacific, El Niño conditions prevail in the eastern boundary currents of the Pacific (see Chapter 6). The characteristics of the province derive from the conjunction between the southward geostrophic current, the coastal boundary, and an alternating wind regime—equatorward upwelling winds in summer and poleward downwelling winds in winter.

The offshore California Current is a shallow, complex southerly flow of cool, low-salinity water extending about 1500 km seawards off central California but progressively closer to the coast toward the south: 850 km off Cape Mendocino and 500 km off Cape San Lazaro. The main core of the current occurs 200–400 km offshore, where southward wind stress is greatest, especially at 30–45°N and during summer. The zone of maximum wind stress coincides with zero wind-stress curl, which is anticyclonic (convergent) to the west of this line and cyclonic (divergence) to the east (Bakun and Nelson, 1991). A separate and narrower maximum flow occurs within 100–150 km of the coast as the inshore California Current, which partially reverses with the seasons.

The southward flow of the offshore current includes mesoscale eddies and meanders 120–150 km in dimension; these are anticyclonic on the seaward side and cyclonic on the landward side of the flow. Among the mesoscale features of the offshore current are intense jets and plumes originating at the coast, which entrain upwelled water and advect it far offshore (Burkov and Pavlova, 1980; Mooers and Robinson, 1984). At about 1500 km from the coast, mesoscale eddying gives way to flow to the south, which is less complex and slower. This transition is often marked by a salinity front which is continuous with the Subarctic Front and is often termed the California Front.

Upwelling at the coast is forced by coastal winds, modified by their response to blocking of the zonal westerlies by the coastal mountain chain. Thus, the coastal wind regime is not uniform: A local maximum of cyclonic curl is associated with the Southern California Bight, and a lobe of anticyclonic curl frequently reaches the coast at Punta Baja, where longshore equatorward wind stress is maximal. Off Oregon and Washington wind-stress curl is variable, with frequent brief episodes of anticyclonic curl related to storm tracks (Bakun and Nelson, 1991) which are reflected in event-scale variability of circulation within the coastal boundary province.

It remains critical to understanding circulation in this province, however, to distinguish the flow of the coastal current from the low-salinity velocity core of the offshore California Current. Equatorward transport in the separate coastal velocity core may be interrupted by discontinuous northward flow beginning seasonally in August–October. In early winter (November–January) continuous poleward flow occurs from the Mexican border almost to Cape Mendocino, with this continuity breaking down in February and March with the development of frequent cyclonic eddies between Point Conception and Cape Mendocino. In March or April the summer pattern of eddying southward flow becomes fully established. (Wyllie, 1966). When reversal of coastal flow is most strongly developed during early spring, the poleward current lies closest to the coast.

When transport is integrated over 200 or 500 m, the northward counterflow below the inshore California Current is continuous. Therefore, especially during spring and early summer, there is a shear zone between coastal poleward and offshore equatorward flow about 100–125 km offshore. Geostrophy requires that a pycnocline ridge (marked at the surface by a divergent front) should occur along this line of zero flow to compensate a depression in the sea surface, and this has consequences for the supply of nutrients to the photic layer. There is also, especially off Oregon, an upwelling front that occurs 5–15 km offshore during the summer coastal upwelling season which is forced by cross-shelf circulation over the very narrow continental shelf (Peterson *et al.*, 1979).

The southward flow of the coastal current in summer is a meandering jet with persistent eddies and cool filaments that may extend 200 km offshore and originate at prominent capes. Some of these topographically locked eddies, especially those at the Straits of Juan de Fuca, at Hecata Bank, and at Capes Blanco and Mendocino, often form cold filaments that may extend several hundred kilometers offshore, having a dipole termination, representing the development of a pair of counterrotating eddies. The actual development of these features is associated with variability of coastal winds. Upwelling winds normally occur for periods of 1–3 weeks, with relaxation periods of 2 or 3 days, and individual cold filaments may transport water from more than a single upwelling event (Lagerloef, 1992; Traganza *et al.*, 1981).

Coastal upwelling cells in the Southern California Bight are small and localized east of Point Conception, at Point Dume and in the bight lying between La Jolla and San Clemente; they are effective only 25–30 km from the coast. These cells, in the Southern California Bight, exhibit simultaneous upwelling (marked by dense sea fog and increased sales of wet suits in the beach towns of surfers' paradise) several times each summer, forced by events in the local wind field (Dorman and Palmer, 1981). Along the coast of Baja California, coastal upwelling cells occur during summer at Punta Eugenia, Punta Abreojos, Cape San Lazaro, and Punta Tosca.

Biological Response and Regional Ecology

The effects of wind-stress curl on nutrient dynamics are similar in each of the four principal eastern boundary currents (Bakun and Nelson, 1991). Though the paradigm of eastern boundary current nutrient dynamics is usually described simply as "coastal upwelling," the reality is far more complex and a variety of enrichment processes occur: (i) coastal upwelling cells, due to Ekman divergence at the coast, usually of small dimension and often topographically locked; (ii) continued upwelling in offshore-trending filaments of cool coastally upwelled water; (iii) upwelling in cyclonic eddies shed from meanders of the coastal current; (iv) and upwelling in the offshore divergent front at the shear zone between inshore poleward and offshore equatorward flows.

Of these, the offshore divergent front may be the most difficult feature to identify and is likely to be represented by a weak coastwise field of high chlorophyll but connected in various ways with the offshore-trending chlorophyll fields arising from eddies and filaments. It is often associated with the location of an offshore zooplankton biomass maximum.

Seasonal variation in algal growth within cyclonic eddies can be associated with seasonal variation in Ekman suction (Robinson *et al.*, 1993). However, in the offshore California Current, nutrient input appears to be principally advective in the cool, low-salinity, high-nutrient core of the southward flow. Both off California (especially at 35–42°N) and off northern Oregon–British Columbia (46–48°N and beyond), a winter bloom occurs which is unrelated to the effects of coastal upwelling plumes and filaments. Here, from October to March, chlorophyll values >1.5 mg chl m^{-3} occur over wide areas as far as 300 km from the coast. Between these two regions, chlorophyll values offshore are generally lower even at this season, and a "blue hole" region (covering several degrees of latitude) is persistent offshore, centered at about the Columbia River.

Because the processes leading to upwelling are not uniform throughout the coastal area, it is useful to review the processes that are characteristic of four compartments: Oregon–British Columbia, Point Conception to Cape Mendocino, the Southern California Bight, and the Baja California coast.

• *Oregon–British Columbia (42–48°N):* Winter storms are strong and frequent and seasonal current reversal occurs regularly; primary production is strongly seasonal. Upwelling occurs in summer at a coastwise front about 10 km offshore as well as at the coast itself, and response to wind events typically results after 4 or 5 days in the development of an upwelling cell, whose seawards front moves progressively offshore during development and returns shorewards during subsequent relaxation of upwelling. During this process, mean offshore flow is restricted to the surface 20–30 m (Brink, 1983; Smith, 1981). Upwelling occurs in persistent, topographically locked gyres during summer (e.g., at Juan de Fuca). During both summer and winter, frequent cold tongues of upwelled water on the scale of hundreds of kilometers extend westwards from the continent across and beyond the relatively broad continental shelf. Relatively high levels of nutrients occur throughout this region offshore, entrained toward the south from the subarctic zone.

• *Point Conception to Cape Mendocino (33–41°N):* Upwelling is strongest here but primary production is markedly seasonal. CZCS images show persistent offshore meanders, shed eddies (anticyclonic to the north and cyclonic to the south) and offshore cool filaments entraining coastally upwelled water. These most frequently occur in summer and south of Cape Mendocino, Point Arena, and Point Sur. Upwelling also occurs at the shallow banks off Point Reyes.

• *Southern California Bight (32 to 33°N):* The upwelling plume from Point Conception is frequently observed in summer, curving to the south around the outer limb of the eddy which occupies this bight. This plume has consistently lower surface chlorophyll values than those north of Point Conception or off Baja California. Off Southern California, the winter wind regime is established only relatively briefly, though (because the bight is wholly occupied by a quasi-permanent cyclonic eddy) poleward flow along the coast is more continuous here than elsewhere. Upwelling occurs as small coastal cells in summer (Point Dume and Del Mar) and at offshore islands and banks during all seasons. The inshore flow around the southern limb of the eddy may be marked as a chlorophyll feature.

• *Baja California (22–31°N):* The coastal wind regime is weaker but apparently favorable for upwelling year-round, though some seasonality in upwelling is evident in the chlorophyll field. Upwelling cells south of prominent capes are persistent: Cape Colonet, Punta Baja, Cape San Quintin, Punta Eugenia, Punta Abreojos, and Cape Falso all may generate such cells. The seasonal CZCS pigment fields (e.g., Thomas *et al.*, 1994) suggest that jets and filaments of upwelled water pass farther offshore from coastal upwelling cells from late summer (August) through early winter (November) and that south of Cape San

Lazaro even coastal upwelling ceases in autumn and winter (September–January).

Because of the large number of published studies of the California Current, in reviewing this province it is easier to concentrate on the individual processes rather than to see the whole. For instance, it is easy to lose sight of the fact that there is a seasonal cycle in the depth of the pycnocline that obtains over the whole area of the province (summer, 20–25 m; winter, ~75 m). At all seasons, if one ignores the effects of mesoscale features, the thermocline slopes downwards to the west, offshore, as it must. In upwelling cells, the density profile may be relatively featureless but wherever a significant mixed layer exists, a DCM occurs on the density gradient, usually with the depth of primary production a few meters shoaler. The offshore region has a seasonal cycle typical of subtropical oceans: Primary production rate and chlorophyll accumulation begin as soon as the mixed layer begins to deepen in the autumn. As noted previously, in the offshore areas chlorophyll values peak in midwinter, and primary production rate slows again with the shoaling of the thermocline in spring. Analysis of chlorophyll fields, integrated for the whole province, shows that this process—not summer upwelling at the coast—dominates the seasonal cycle. This observation recalls earlier suggestions that between-year variability in biological properties was forced primarily by changes in the advection of nutrients from the source of the California Current rather than by variation in the nutrients brought to the surface by coastal upwelling (Chelton, 1982).

Coastal upwelling is, nevertheless, a special case which merits special attention. The upwelling cells generate diatom–copepod assemblages of remarkably low diversity: In repeated net tows in one assemblage such as a few kilometers off Baja California (25°N), I could find no more than 29 species of zooplankton (and no more than 20 in any one haul), or about one-quarter the number taken 25 km farther offshore and examined with equivalent attention. Large, filter-feeding copepods, especially *C. pacificus* (which you will find described as *C. helgolandicus* in much of the Californian literature), at 115 ind m^{-3} comprised 77% of zooplankton dry weight at the coast. The seasonal ontogenetic migrations of this species cause the deep basins on the continental shelf to trap large concentrations of overwintering stage 5 copepodites; early in the winter these aggregate near the bottom, but the layers progressively shoal as oxygen concentration in the bottom water progressively declines, eventually forcing them over the sill depth of the basin (Osgood and Checkly, 1997). Compare this situation with that of *C. finmarchicus* in the deep basins of the Scotia Shelf in the Northwest Atlantic Shelves Province.

A few large species of diatoms (*Coscinodiscus*, *Nitzschia*, and *Tripodonesis*) formed 81% of algal cell volume. However, these rich diatom crops are also utilized by a very unusual organism—a bright red, swimming galatheid crab, *Pleuroncodes planipes*. These (in their pelagic phase) crowd into the surface layer off Baja California where they tail-flip up to the surface and then parachute down again with outstretched legs, filtering actively with their maxillipeds; this is a remarkable sight against the rich olive-green upwelled water. These crabs (at one per 3 m^3, each capable of clearing diatoms from 3 or 4 liters^{-1} hr^{-1}) may comprise 90% of the total zooplankton/nekton biomass in upwelling cells and contribute 85% of all zooplankton/nekton grazing pressure. *Pleuroncodes* is directly preyed on, and a preferred food of, yellowfin tuna in the same region, so this is a remarkably direct link from diatoms to your table.

How copepods exploit upwelling cells off Oregon has been most elegantly described by Peterson *et al.* (1979), who were able to use the three-dimensional differential distribution of the five abundant species to clarify the details of water movement during upwelling. Over a very narrow continental shelf (100 m is 10 km from the coast) the pycnocline lies at 20–50 m, sloping up toward the beach: in an upwelling episode the pycnocline intersects the surface 5–10 km seawards and nitrate-replete deep water is brought to the surface. Within this system each copepod has a narrowly defined and specialized distribution, in which it is maintained by details of the circulation pattern and its reproductive behavior: *Acartia clausii* is restricted to the upper 5–10 m and within 5 km of the shore; *Acartia longiremis* occurs 10 km offshore and similarly near the surface; *Pseudocalanus* occurs out to 15 km from shore but also only within the pycnocline at 10–20 m; *Oithona similis* occurs at similar depths but not in the first 10 km offshore; and *C. pacificus* has wider ranges for both depth and distance offshore. In fact, off California the endemic *Calanus* has a life history like that of *Calanoides* elsewhere; during winter and other periods when upwelling is not active, it descends to 400–600 m as a population of copepodite 5's and remains dormant in the oxygen minimum layer which underlies the coastal California Current.

I have already discussed some of the consequences for the ecology of this province of an ENSO event in Chapter 6. Because the El Niño phenomenon was first described in relation to the Peruvian coastal region I shall reserve for the HUMB province a general account of the processes involved. Much of what appears in that section is relevant in general terms to the California Current.

Synopsis

Archived data does not capture details of upwelling cells, so the winter deepening of the mixed layer below the photic zone represents the seasonality of the California Current generally (Fig. 9.16). Offshore, an autumn–spring progressive increase in primary production rate is accompanied only initially by accumulation of chlorophyll. Herbivore abundance appears to be a negative correlate of chlorophyll biomass.

Central American Coastal Province (CAMR)

Extent of the Province

The Central American Coastal Province (CAMR) extends from the tip of Baja California at Cape San Lucas in Mexico to the Gulf of Guayaquil in Ecuador. For convenience, it also includes the long, narrow epicontinental sea of the Gulf of California which lies behind the peninsula of Baja California.

Continental Shelf Topography and Tidal and Shelf-Edge Fronts

The west coast of America is an active continental margin in the geological sense, so the continental shelf is narrow, usually a few tens of kilometers wide.

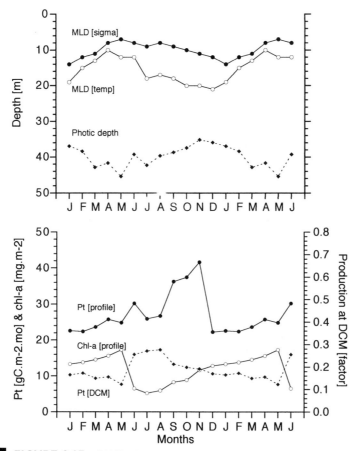

FIGURE 9.17 CAMR: characteristic seasonal cycles of monthly averaged mixed layer and photic depths, chlorophyll at the surface, and rate of primary production, both depth-integrated and at the DCM. Data sources are discussed in Chapter 1.

Only in the Gulfs of Panama and Guayaquil and at the head of the Gulf of California are there somewhat wider shelf areas. Hermatypic corals are minor reef-forming agents in this province, which biologically resembles the coast of intertropical West Africa (Longhurst and Pauly, 1987).

Defining Characteristics of Regional Oceanography

This region lies between the two equatorial fronts which separate the tropical surface water mass from the flow of the Peru and California Currents as they pass offshore toward the west. Because drainage from the sierra mountain chains is mainly to the east, few significant rivers discharge onto the continental shelf, the Gulfs of Panama and Guayaquil (and especially the latter) being the sole regions to receive much fresh water and silt. Longshore currents are variable, though generally dominated by the reflux of the NECC after encountering the west coast of the continent. Mixed layer depth is generally shallow (30–40 m) and the permanent thermocline is sharp. A very shallow brackish surface layer may occur during the wet season of boreal summer when southerly monsoon conditions occur. The northern and southern equatorial fronts are sharp features, linear near the coast and becoming increasingly meandering toward the open ocean (Griffiths, 1968).

The Gulf of California (the "Vermilion Sea" of John Steinbeck) is a long, narrow epicontinental sea which includes deep basins, has steep-to coasts and continental slopes, and numerous islands. It is convenient to include the bight lying between Cape San Lucas and Cape Corrientes within the gulf. The northern end of the gulf is desertic, whereas the southern end lies below the boreal summer southerly monsoon rainfall conditions; total rainfall depends on the incidence of tropical storms, or Chubascos. Circulation is complex and dynamic due to the interaction of the coastal wind regime with mountainous terrain between the gulf and the open Pacific Ocean. Exchange with the open Pacific is minor and is largely driven by the fact that the upper gulf is an evaporative basin (Bray, 1988); the general circulation thus resembles that of a small Red Sea, except there is no shallow sill across the mouth. Residual, tidally forced circulation is strongly modified by local, temporary wind stress. Land–sea breeze effects are important and regional northwesterly winds dominate in October–May, and weaker southwesterlies dominate during boreal summer.

In the northern gulf, the water column is fully mixed by the boreal northeasterlies, assisted by convective overturning to a depth of 100 m (Roden, 1964). Over the shallow northern shelf, seasonal heating and cooling of the highly saline estuarine water of the Colorado River forces a complex surface pattern of turbidity gyres (Lepley et al., 1975). This dense, cool water is the origin of the deep water outflow of the gulf.

Biological Response and Regional Ecology

This region has a generally stable oceanographic regime with rather few locations where enhanced vertical flux of nutrients is induced. The exceptions are the following:

• *Gulf of California:* Tidal and drift currents can be strong through passes between islands, especially in the northern gulf around Angel de la Guardia Island, where the Ballenas Channel (to the west of the island) is isolated by relatively shallow (450 m) sills (Alvarez-Borrego, 1983). This is a persistent pool of cool water and is associated with tidal fronts whose orientation depends on wind direction (Badon-Dongon et al., 1985). The alternation of southwesterly winds in boreal autumn and northwesterly in spring force coastal upwelling plumes from opposite sides of the gulf: on the west coast in autumn and the east coast in spring. In the Gulf of California, wind-driven circulation changes are reflected in seasonal blooms, especially in autumn and winter when sea-surface temperature cools (from 30 to 18°C) and CZCS chlorophyll increases (from 0.25 to 2.5 mg chl m^{-3}), with diatom blooms initiating the sequence, followed by foraminifera and coccolithophores and the bloom ending in winter with a population of silicoflagellates and some diatoms (Thunell et al., 1996).

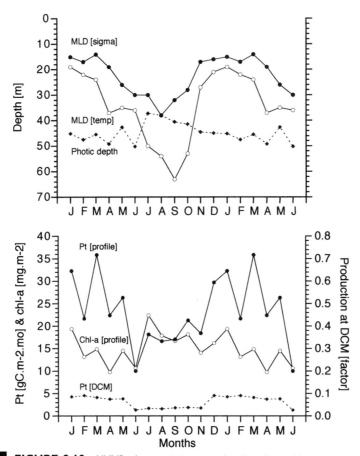

FIGURE 9.18 HUMB: characteristic seasonal cycles of monthly averaged mixed layer and photic depths, chlorophyll at the surface, and rate of primary production, both depth-integrated and at the DCM. Data sources are discussed in Chapter 1.

These data were obtained from sequential sampling from sediment traps and must be presumed to integrate the individual local processes induced by topography referred to previously.

• *Gulf of Tehuantapec:* This gulf is important oceanographically because it lies west of a major gap in the Central American sierra chain, the Chivela Pass, through which the transmontane winds from the Gulf of Mexico (only 150 km to the east) have easy passage (Blackburn, 1962). These winds, prevailing northerly in direction, force surface flow in the gulf to the south. Interaction between this and the northward flow along the Mexican coast generates a meridional thermocline (pycnocline and nutricline) ridge. In boreal winter, the mixed layer above this feature may be completely eroded by wind stress, and surface temperature, salinity, and nutrient levels resemble values in the pycnocline. After extreme transmontane wind events, a plume having such characteristics may stretch several hundred kilometers to the southwest from the gulf. Surface productivity is high (<150 mg C m^{-3} day^{-1}) and the plume, then lying above the Tehuantapec Ridge, becomes a feature of the surface chlorophyll field.

• *Gulf of Panama:* Upwelling occurs for the same reason as in the Gulf of Tehuantapec and at approximately the same season, though the occurrence of upwelling in one region does not necessarily denote upwelling in the other. Upwelling can in some years produce a large feature in the sea-surface thermal field, reaching seawards to the Galapagos region. In such events, the northern thermal front off Panama carries 500-km waves which pass westwards along the front at about 30 km day^{-1} (Legeckis, 1988). Similar upwelling, due to transmontane winds, occurs in the Gulf of Papagaya but apparently not in the Gulf of Chiriqui.

Beyond these exceptional events and locations, the ecology of the coastal regions of Central America can best be regarded as comprising two systems: those off arid coasts with little freshwater input (such as much of the Mexican coast and particularly the Gulf of California) where water clarity is high, and those within the area of tropical rainfall where coasts are invested with mangrove, there is much river effluent, and water clarity is seasonally very low. The inshore Gulf of Panama will serve as a model for the latter type of pelagic ecosystem: Here the ecological balance and biota appear to closely resemble those of the mangrove coasts of the Gulf of Guinea (see Chapter 7, Giunea Current Coastal Province). Here, there is an algal bloom, apparently dominated as in the Gulf of Guinea by large diatoms and at the same season (January–March), though wind-forced upwelling at the mouth of the Gulf of Panama also appears to be implicated and perhaps the relaxation of river flow and consequent reduction in water clarity is less important than in that the Gulf of Guinea. Less than 10% of daily production by phytoplankton can apparently be consumed by the standing stock of mesozooplankton; since these are nitrate- and diatom-dominated blooms, it seems likely that modern investigations of the role of picoplankton and their protist grazers will modify the models based on older techniques less than they will in the open ocean. Another case of apparent surplus production, requiring an export process or a sink for organic matter, is the Gulf of Guayaquil, to the south of the Gulf of Panama.

Synopsis

Almost invariant mixed layer depth occurs with an early winter increase in primary production rate reflecting winter wind events in both the Gulf of California and the Gulf of Tehuantapec (Fig. 9.17). Chlorophyll does not accumulate significantly.

Humboldt Current Coastal Province (HUMB)

Extent of the Province

This province is defined as extending from the coastline to (and including) the offshore anticyclonic eddy field. To the south, it is defined by the divergence zone at about 40–45°S and to the north at its separation from the coast at about 5°S. It has by far the greatest north–south extent of any of the coastal boundary current systems and for that reason alone—if not for better reasons discussed below—it would be desirable in some circumstances to subdivide this province meridionally into Chilean and Peruvian provinces, separated at about the bend in the coast near the international border near 18°S. Because I do not make this separation, I revert to the time-honored name of the Humboldt Current for this province.

Continental Shelf Topography and Tidal and Shelf-Edge Fronts

The continental shelf is generally very narrow along the whole west coast of South America. The break of shelf occurs at 200 m, and the slope itself is relatively steep-to. Only off northern Peru, from 6 to 10°S (off Trujillo and Chimbote) is the shelf relatively wide, reaching a maximum of 125 km. From Callao (12°S) in Peru to Valparaiso (33°S) in Chile the 200-m isobath lies within 10–20 km of the shoreline. From Valparaiso southwards, the shelf again widens and reaches 50 km off Concepcion and Chiloe Islands, but over this region the continental edge at 200 m is as topographically complex as the coastline itself. Upwelling fronts are numerous, complex, and highly variable: We may distinguish coastal fronts no more than 10–20 km offshore, outlining clear upwelled water; plume fronts outlining tongues of green upwelled water passing offshore and encountered up to 30–50 km offshore; and an oceanic front separating the coastal regime from warmer, blue oceanic water, often marked by the 15°C isotherm, which may meander 100–300 km offshore.

Defining Characteristics of Regional Oceanography

This is the eastern boundary current of the southeastern Pacific and is the Pacific homolog of the Benguela Current on the southwest African coastal boundary, with which it shares many characteristics. Forming the eastern, equatorward limb of the South Pacific subtropical gyre, it takes its source in the divergence zone where the West Wind Drift of the Subantarctic Current reaches the South American continent; this occurs at about the latitude of Chiloe Island (42°S), where transport (with very sharp curvature of flow) is partially directed to the north into the Humboldt Current system and partially to the southwest along the fjord coast of southern Chile as the coastal continuation (the Cape Horn Current) of the West Wind Drift. The alignment of the flow into the divergence zone is from the northwest rather than from the southwest as is classically supposed (Silva Sandoval and Neshyba, 1977).

It has been understood since the earliest scientific observations (e.g., Gunther, 1936) that northwards flow in the Humboldt Current system (which is most consistent during austral fall) was complex and has been described as a system of interleaving "currents" from single-ship geostrophic sections. Now that regional synoptic fields can be captured from satellite images, these complex currents appear to comprise temporary features of mesoscale eddies and meanders, though modern data do confirm that the two principal streams in the Humboldt Current system suggested by early observations are indeed real. These have usually been referred to as the oceanic and coastal branches of the Humboldt Current (e.g., Sievers and Silva, 1975).

Off Peru, equatorward flow is shallow (about 25 m), and below this depth flow is poleward, with maximum velocities at about 100 m. Countercurrent flow thus occurs in the upper layers of the oxygen-depleted zone but shallower than the depth of maximum nitrate depletion in the anoxic layer. Because of the shallow depth of the equatorward "Peru Current" flow, it is water of the countercurrent which is upwelled at the coast during normal periods (Brockman *et al.*, 1980).

At the latitude of northern Chile, the oceanic and coastal flows diverge: The oceanic component enters the eastern limb of the subtropical gyre and transports cool subantarctic water into the equatorial system of zonal currents. Anomalous anticyclonic wind-stress curl in the Arica-Iquique Bight may cause a breakdown in the coastal flow which is otherwise continuous from Chiloe Island (42°S) to northern Peru at about 5°S. Between the coastal and oceanic equatorward flows, there may be some surface transport of subtropical water toward the south (Guillen, 1980; Robles *et al.*, 1980) and from about 10 to 20°S, poleward flow frequently also occurs at the coast (Bernal *et al.*, 1983). North of Arica, and off southern Peru, equatorward flow is again separated into coast and oceanic components by weak, intermittent poleward flow at the surface, except in boreal spring when poleward flow is mostly subsurface (Guillen, 1980).

The pattern of wind stress and its curl over the province resembles the general pattern established for eastern boundary currents (Bakun and Nelson, 1991) but with some special characteristics due to the alignment of the coast and the coastal mountains and also because of the greater latitudinal range of this province compared with other eastern boundary currents. Cyclonic curl adjacent to the coast extends to 37°S near Concepcion and takes maximal values in two areas: off Peru from 9 to 15°S and off Chile at about 25°S in winter. The Peru maximum is seasonally stable in winter, whereas off Chile it moves poleward in summer to about 32°S and simultaneously strengthens. In the bight at 18–20°S (Arica to Iquique) there is anomalous and persistent anticyclonic curl probably because the bight is open to wind from the south. As a consequence of the specific characteristics of the wind field, the regional partitioning of positive and negative wind-stress curl is significantly less organized into inshore cyclonic and offshore anticyclonic areas, especially during austral winter (April–September).

Coastal divergence and cyclonic wind-stress curl near the coast give rise to extensive upwelling features resembling those of the California Current and equally prominent in satellite imagery of chlorophyll fields. These images show that upwelling occurs seasonally in two primary regions: along the Peruvian coast at 5–15°S and along the Chilean coast from 25 to 40°S. South of 40°S, in the poleward Cape Horn Current, summer blooms which extend 200–300 km offshore are perhaps unrelated to upwelling processes but are probably simple spring–summer midlatitude blooms.

The Chilean and Peruvian regions have different upwelling characteristics. Upwelling-favorable winds are relatively lighter along the Chilean coast, though still persistently favorable, and their maximum potential for forcing upwelling at the coast occurs in the vicinity of Valparaiso. Despite this relative lack of seasonality of upwelling-favorable winds, the CZCS images for the coast of southern Chile show higher chlorophyll values, extending father offshore, in winter than in summer. Though the cloud cover is greater, and images therefore fewer, the available monthly CZCS images for the Peruvian sector suggest that upwelling chlorophyll plumes do not extend as far offshore as those on the Chilean coast. The most persistent chlorophyll feature is the Cape Nazca upwelling cell at 15°S.

Though there are apparently regional differences in the depth of the source water in upwelling cells, the canonical view is that upwelling is restricted to relatively shallow depths and that there is an offshore flow of bottom water over the continental shelf (e.g., Codispoti *et al.*, 1982). Upwelling occurs as numerous small coastal upwelling cells, from which relatively small plumes (<100 km) of cool water may be entrained offshore. These cluster in the following preferred locations on the Peruvian coast:

• *Point Negro (5°S):* Here, the trend of the coast changes abruptly at a major prominence, north of which lies the tropical surface water of the Gulf of Guayaquil. This prominence, which lies south of Paita, appears to be a major and persistent upwelling center, and filaments offshore may be aligned with the convergent front that marks the separation of the Peru Current from the continent. This front is a major hydrographic discontinuity.

• *Chimbote (9°S):* Though there is no major topographic feature here, this seems to be a consistent upwelling center. The upwelling plume proceeds seawards over the widest region of shelf on the Peruvian coast.

• *Callao (12°S):* The coastal prominence at Callao and Lima is a sufficient feature in the coastal topography to significantly focus the upwelling processes.

• *Pisco (14°S):* In the bight north of Pisco, surface water is consistently cool and a major upwelling center appears to be topographically linked to the Paracas peninsula.

• *Cape Nazca–Point Sta. Ana (15°S):* The upwelling center off San Juan is the most consistent and strongest upwelling center on the Peruvian coast and was known to early investigators (Gunther, 1936). Much of what is known about upwelling dynamics on the Peruvian coast, and about the ecological physiology of phytoplankton during upwelling, is owed to work at this site by international programs (e.g., JOINT I and II of CUEA and the Coastal Upwelling Ecosystem Analysis program; see MacIsaac *et al.*, 1985).

Despite the existence of these persistent upwelling centers off Peru, as many as 15–20 small (<25 km) cool water cells may occur simultaneously along the upwelling coast from San Juan to Paita (Longhurst and Pauly, 1987) early in a period of upwelling-favorable winds. Because appropriate wind stress may be imposed for long periods off Peru, the upwelling front may move far offshore and become indistinct (Brink, 1983). At such times the whole coast from 5 to 15°S may present a single inshore cool zone from Arica to Paita, with a weak temperature front parallel to the coast and about 100–125 km offshore though progressively farther offshore to the north (Bohle-Carbonell, 1989). This zone may even extend southerly into the corner of the Arica Bight at about 18°S.

In the Chilean upwelling region the situation is different. Here, especially in winter and spring (as indicated by the level 3 CZCS images), a wide zone is populated by meanders, eddies, and very prominent cool filaments within which vorticity occurs and which extend 200–300 km offshore (Fonseca and Farias, 1987); these mesoscale features resemble the better known eddies and cool filaments of the northern part of the California Current off Oregon and Washington. Five major, semipermanent upwelling centers have been identified on the Chilean coast from analysis of CZCS images (Fonseca and Farias, 1987). To these upwelling centers can be traced the offshore mesoscale tongues of cool water that are a more prominent feature of upwelling in the Chilean than the Peruvian sector of the Humboldt Current system. All the persistent centers are apparently topographically locked to the vicinity of capes and submarine topographic features, though their individual characteristics are at the present time little known:

• *Point Patache (20°S):* South of Iquique, major upwelling events frequently occur and appear to be distinct from the southernmost upwelling on the Peruvian coast, separated by the circulation features noted previously.

• *Mejillones–Antofagasta (23°S):* The coastal prominence lying between Punta Tetas and Punta Angamos forms a major topographic feature and is a site of persistent upwelling.

• *Point Lengua de Vaca (30°S):* South of Coquimbo, this cape and the bight to the north of it appear to locate a persistent upwelling feature.

• *Point de Curaimilla (33°S):* This, and the Gulf of Arauco south of Valparaiso, is the best known of the persistent upwelling centers and the maximum potential for upwelling-favorable winds on the Chilean coast occurs here. This is the only location where the dynamics of upwelling on the Chilean coast have been studied in detail (Johnson *et al.,* 1980), and the data from these studies are consistent with the existence of a classical double-celled cross-shelf circulation as upwelling proceeds. Upwelled water originates in the high-salinity, low-oxygen coastal undercurrent and the upwelling effect reaches to a maximum depth of about 250 m. Short-term variations in upwelling wind intensity result in variations in the location of the undercurrent on the shelf and in pulsations in upwelling intensity.

• *Point Lavapie (33°S):* This peninsula south of Concepcion is the site of persistent upwelling events and the most southerly location where they occur with regularity (Arcos *et al.,* 1987). Upwelling response to favorable winds may occur in response to short-period (1 or 2 days) wind events and also to those of longer duration (6 or 7 days).

Upwelling water, and hence nutrient advection, differs in the Peruvian and Chilean sectors of the Humboldt Current system: Off Peru, equatorial subsurface water flowing south in the subsurface countercurrent is upwelled in normal years, whereas off Chile it is subantarctic water of the equatorward coastal current which upwells, though countercurrent upwelling has been observed at Valparaiso (Johnson *et al.,* 1980). The boundary between these two source-water regimes lies at about 15°S according to Wyrtki (1963).

It is believed that there is a relatively weak relationship between wind stress and surface nutrients in the Peru Current because the generally persistent, generally equatorward winds drive only a shallow surface layer northwards along the coast and this flow is rather frequently interrupted, as noted, by surfacing of the polewards subsurface current. Nevertheless, because nutrient levels below the photic zone are relatively high in the Pacific, the nutrient content of upwelled water is high, at least in normal years (Codispoti *et al.,* 1982). Especially in the Peruvian sector, or wherever countercurrent water is upwelled, the oxygen content of upwelled water is relatively low.

The existence of offshore flow both at the surface and on the bottom, with onshore flow toward the site of upwelling, has consequences for nutrient regeneration over the shelf. Sinking organic material is transported across the shelf and sequestered in the bottom layer of oxygen-deficient water. In these circumstances, this organic material is no longer available for regeneration (Codispoti *et al.*, 1982).

Biological Response and Regional Ecology

The regional chlorophyll field as seen in CZCS images shows the anticipated general relationship between upwelled water and high chlorophyll values. Though fewer images are available than for other upwelling coasts, mesoscale features apparently resemble those better known for other eastern boundary currents. Rapidly changing cool filaments and plumes, exhibiting vorticity, are normal features of the chlorophyll field of the southern, Chilean sector of this province; development and decay of such features occurs on the scale of a few days to a few tens of days.

The biological evolution of a parcel of upwelled water, which requires about 10 days to complete, has been well worked out (see MacIsaac *et al.*, 1985; Harrison *et al.*, 1981) in the Humboldt Current, in a location (Cape Nazca–Cape Santa Ana) having consistent upwelling for several weeks. In the actual upwelling zone <7 km from the coast, nitrate remains relatively constant in the upwelling water at about 20 μM, whereas silicate supply varies according to its source. The evolution of phytoplankton growth passes progressively through several phases: A small inoculum of phytoplankton cells is upwelled from 50 to 60 m depth, the origin of the seed stock being cells sinking from farther seawards; these cells are physiologically conditioned to low light levels and therefore have low rates of nutrient uptake and growth. In addition, the upwelled water itself may require conditioning by exposure to light to permit active growth of algae. If this occurs, it consists of progressive modification of trace metal chemistry, exchanging available copper and manganese ions.

Progressively, the cells entrained in the upwelled water shift up to increased physiological rates so that as stratification is induced by solar heating in the upwelled parcel, the entrained cells are adapted to high irradiance at high nutrient levels. The now fast-growing cells, held in a very shallow (<10–15 m) Ekman layer, reduce ambient nitrate levels quite rapidly and there is a massive increase in standing stock of cells. Rates of primary production are maximal at 0–10 m, whereas two depths of high concentration of chlorophyll may occur— the first near the depth of maximum production rate and the other, a DCM, near the bottom of the thermocline. This may represent unconsumed, sinking cells. Subsequently, in response to nutrient depletion in the Ekman layer, the cells respond by limiting some of their cellular processes, sequentially slowing nutrient uptake, photosynthesis, and storage of carbohydrates and lipids. Because most of the growth can be attributed to diatoms, it can happen that the limiting nutrient is silicate rather than nitrate. In the maximum growth phase, a typical upwelling plume has a phytoplankton species assemblage dominated by no more than 10 species of diatoms (*Detonula, Chaetoceros, Hemiaulis,*

Rhizosolenia, and *Thalassiosira*) together with dinoflagellates and <10-μm flagellates. The total cell number of diatoms is usually about the same as the total of the other two groups.

Herbivorous mesozooplankton consume only a small fraction (10% is a typical finding) of the daily primary production. The herbivores are dominated by three copepods (*Calanus chilensis, Centropages brachiatus,* and *Eucalanus inermis*) of which only the last continues to perform diel migrations while it is entrained in an upwelling feature. These species aggregate close to the surface in an upwelling cell and their layer depth coincides with the lower part of the upper chlorophyll layer—perhaps where cells are fresh, near the depth of maximum production rate (Herman, 1984).

It was the variable and sometimes disastrous changes in weather patterns on the coast of Peru that first attracted the attention of oceanographers and fisheries scientists to the ENSO phenomenon. Obviously, we cannot end the discussion of the Humboldt province without some consideration of how an ENSO event is manifested. There is a sufficient number of easily available accounts of the phenomenon not to require another full-blown description here: If you are not familiar with the issue, I recommend starting with Tomczak and Godfrey (1994) for the physics and with Pauly and Tsukuyama (1987) or Pauly *et al.* (1989) for the ecology and, particularly, for the disastrous fisheries consequences which may ensue.

Briefly to recapitulate the effects of the ENSO cycle in the HUMB province, the onset of an ENSO event is heralded by the slackening of upwelling intensity and the reduction of the number and extent of upwelling centers as indicated by surface water temperatures. This most often occurs in the autumn or winter (see Tomczak and Godfrey, 1994, p. 364, for an interesting account of how this process was misnamed by oceanographers for El Niño, the child Christ) and accompanied by progressive warming of the surface water, in which Chile lags Peru. For the years 1948–1985, three major and three minor ENSO events were correlated by Chavez *et al.* (1989) with deepening of the 14°C isotherm (from 80 m to 100–175 m) by reduction in 60 m nitrate (from 25 to 14–22 μM) and by a decrease in new, nitrate-based primary production rate (by about 1.0 g C m^{-2} day^{-1}). Curiously, the rate of upwelling at the coast may even show an increased rate during ENSO periods because the coastal wind system is enhanced by lessened cloud cover even as the offshore trades, which regulate the depth of the thermocline, are at their weakest: What is upwelled to the surface is, of course, warm tropical or subtropical surface water, already fully nutrient depleted, rather than nutrient-replete subthermocline water.

The consequences of the 20-fold decrease in productivity and phytoplankton biomass that may occur during ENSO events are catastrophic and cascade all the way from the phytoplankton through herbivores to the disappearance of pelagic fish stocks and great perturbations in the fishmeal trade and soybean futures. The details of this ecosystem collapse have been sufficiently described that I do not need to provide a litany of loss of reproduction in seabirds, starvation of marine mammals, and the disruption of the hake, jack mackerel, shrimp, and sardine fisheries because the changed distribution of the stocks put them beyond the traditional ambit of the fishery, whereas the Peruvian anchoveta (*Engraulis ringens*) stock, already stressed by heavy exploitation, may collapse. From 1961 to 1970 this species—dependent directly largely on diatoms for food—maintained a stock off Peru of 10–21 million tons. In 1971 the population crashed to about 2 million tons and has made only faltering recoveries since, each time crashing again after an ENSO event apparently because of recruitment failure. It is too simplistic, as has been done, to ascribe this collapse simply to the 1970–1971 ENSO event. It is beyond the scope of this book to analyze the possibilities, but you are urged to explore the potential scenarios leading to recruitment failure discussed in the references noted previously.

Synopsis

The biological seasonality in this archive is dominated by processes off Peru, where austral spring–summer upwelling is the principal seasonal feature in the province (Fig. 9.18). Mixed layer seasonality is probably dominated by offshore and southerly winter wind mixing.

China Sea Coastal Province (CHIN)

Extent of the Province

The CHIN province includes the continental shelves of the Bohai Sea, Yellow Sea, East China Sea, and the eastern part of the Sea of Japan through the Formosa Strait to about 22°N, the latitude of Hong Kong. Because the Kuroshio passes along the edge of the continental shelf along the east coast of Taiwan, only the west coast of this island facing the continent should generally be included in this province. The offshore limit is taken to be the edge of the continental shelf or the edge of the landward eddy field of the Kuroshio Current if that is farther seawards; thus, flow of the Kuroshio across the shelf is considered part of this province, whereas the core and offshore eddy field of this current comprise a separate province (see Kuroshio Current Province).

Continental Shelf Topography and Tidal and Shelf-Edge Fronts

This is one of the largest areas of shallow shelf anywhere; the Yellow Sea/Bohai Gulf basin is 38,000 km² with an average depth of only 44 m, with much of the western part of the basin being occupied by the Great Sandbank. This basin is partially enclosed by a series of transverse shallow banks on either side of the mouth of the Yellow Sea. A critical characteristic of the inner shelves of this province is the effect of the discharge of fresh water and sediments from the great rivers draining the coastal plains of China, principally the Yellow River opening into the Yellow Sea and the Yangtse into the northern East China Sea. The resultant coastal and shallow water sediment regime is a paradigm for a coast dominated by mobile, thixotropic mud banks. The similar cases along western India and the Guiana coast of South America are minor in comparison. The sedimentary load (suspended sediment may be on the order of 20 kg m⁻³ near the mouths of the Yellow River at flood tide but less in the Yangtse) leads to the development of extensive intertidal and offshore mudflats, continually shifting by tidal resuspension (Wang, 1983). These lunate features lie normal to the coast and extend up to 75 km offshore off Jiangsu province (about 32°N).

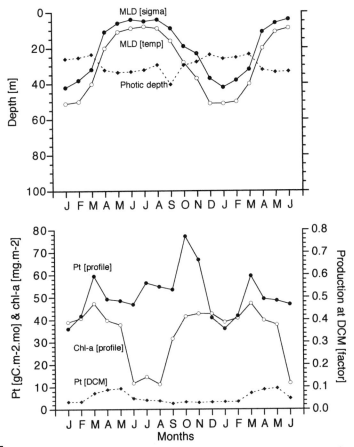

FIGURE 9.19 CHIN: characteristic seasonal cycles of monthly averaged mixed layer and photic depths, chlorophyll at the surface, and rate of primary production, both depth-integrated and at the DCM. Data sources are discussed in Chapter 1.

Shelf sea fronts appear not to have been described even over the wide flat shelf areas that occur in the southern part of the East China Sea, which is not surprising because the tidal range in the Sea of Japan is very small (>0.2 m), so tidal frontal development will be very weak.

Defining Characteristics of Regional Oceanography

The East China and Yellow (Huanghai) Seas are large epicontinental shelf-depth seas. The East China Sea is open to the Pacific Ocean except for the presence of Taiwan and the chain of Ryuku Islands, whereas the Yellow Sea is enclosed on three sides by the Asian continent. The Bohai Sea is a smaller, very shallow (<20 m) basin at the head of the Yellow Sea.

Nontidal circulation in these epicontinental seas is forced by the strongly seasonal regional wind systems, the presence of the seasonally invariant presence at the shelf edge of warm Kuroshio flow, and the southward flow of cold water along the Manchurian coast from the Sea of Okhotsk passing down the

Gulf of Tartary inside Sakhalin and through the Soya Straits. The cold Oyashio lies farther offshore, seawards of the Kurils and Hokkaido, and does not directly affect circulation in this province.

Like New England and Atlantic Canada, this region has an extreme contrast between winter and summer conditions. Winter cooling of surface water occurs throughout the province, forced by the cold, dry northerly winds of the Siberian anticyclone (Barry and Chorley, 1982). During boreal summer, the advance of the southerly monsoon winds begins in early May and by mid-July these reach the northern Yellow Sea so that the whole province lies under the influence of tropical air.

Winter cooling at the surface is by about 10°C, even at Hong Kong at the southern end of the province, though mixing extends to the bottom only inshore. The summer thermocline deepens in winter from 30–50 m to about 100 m at the shelf edge but intersects the bottom only at 20 km of the coast (Watts, 1972; Chan, 1970). Farther north, in the Yellow Sea, a very cold surface water mass develops during winter, reaching 0°C in the inner Bohai Sea, and stratification is broken down; summer warming reaches 28°C in the central Yellow Sea but a cold bottom water mass of 4 or 5°C lies below a remarkably steep thermocline below the seasonal warm layer (Wang and Zhu, 1991). The summer mean temperature of the cold subthermocline water mass is related to regional air temperatures of the previous winter.

The eastern boundary of this province is the warm flow of the Kuroshio, the velocity maximum of which is topographically locked to the continental slope (which is steep-to around Taiwan) and then continues northwards through the Bashi Strait between Taiwan and the Ryuku chain and along the western slope of the Okinawa trough that here forms the edge of the continental shelf of the East China Sea. During its passage northwards outside the shelf break, some Kuroshio surface water leaves the main axis and passes onto the continental shelf of this province as two principal flows: (i) around Taiwan at the root of the Taiwan Warm Current that then passes north through the East China Sea and (ii) south of the islands of Cheju and Shikoku into the Yellow Sea and along its north coast, as the Yellow Sea Warm Current (Su *et al.*, 1990). Part of this flow also passes northwards through the Tsushima Straits into the Sea of Japan as the warm Tsushima Current along the west side of the Sea of Japan and within the KURO province as defined here.

The flow from the Kuroshio around the south of Taiwan onto the shelf of the East China Sea occurs during winter and joins a northward drift along the coast from the South China Sea through the Taiwan Strait (Fan, 1982). Return flow of modified coastal water from the western side of the Yellow Sea occurs as the southward China Coastal Current which, under the influence of the Siberian northerlies, enters the landward side of the Taiwan Strait in boreal autumn and winter (Shaw, 1992) at the same season as some Kuroshio flow occurs on the eastern side of the straits. This southward flow is reinforced by discharge of fresh water from the Yellow and Yangtse Rivers and is associated with a persistent cyclonic mesoscale (100 to 200-km diameter) eddy located about 150 km south of Cheju Island; this eddy is well predicted by geostrophic calculations and has cool, high-salinity water at its center (Mao *et al.*, 1986). The flow of cold subarctic coastal water from the Sea of Okhotsk as the Liman and Korean coastal currents forms major cyclonic eddies on the western side of the Sea of Japan and encounters Kuroshio water on the eastern side of the Korean peninsula, forming a prominent thermal front across the shelf.

Along the continental margin, the Kuroshio flow meanders and induces mesoscale eddies and cool filaments that represent shelf water being advected seawards; concurrently, eddies of warm oceanic water flood intermittently across the shelf just as they do from the Gulf Stream over the Scotia Shelf. The meanders propagate to the north along the edge of the shelf (Shibata, 1983) with a wavelength of about 300 km. A distinct shelf-edge front develops on the landward side of such meanders and the cyclonic eddies which they shed (Chen et al., 1992). Shelf-edge eddies normally appear in satellite images as warm, cyclonic filaments of Kuroshio water outlining cold cores of shelf or slope water. Upwelling from 400 to 500 m occurs within the cold cores (Zheng, 1992).

Apart from the vorticity in shelf-edge meanders and cyclonic eddies, upwelling occurs along the Zhenjiang coast at about 26–30°N (Guan, 1984), on the northwest coast of Taiwan, and in the quasi-permanent cyclonic eddy south of Cheju Island. As the Kuroshio rounds Taiwan, it encounters the sharply curved and steep topography of the continental shelf of the East China Sea; a westward loop current resembles the loop current of the Gulf of Mexico, but unlike that feature, it does not detach from the meandering jet (Hsueh et al., 1992). The consequence of these processes is a regional upwelling in the vicinity of the Penghu Islands where a cold thermal anomaly is frequently observed at the shelf break; upwelling here is episodically enhanced by the onset of the northerly monsoon winds in boreal autumn. Nitrate-replete water reaches the surface from thermocline depths during these episodes (Fan, 1982). This is a very dynamic place, and Liu et al. (1992) have categorized several modes of eddy-filament formation from AVHRR thermal imagery.

Biological Response and Regional Ecology

Because of the enormous silt loads in inshore water, we cannot infer much about the nature of the seasonal blooms from CZCS images. We have to expect that three principal processes shall regulate primary production: (i) The provincewide winter cooling and mixing of the water column alternating with summer stratification will generate a spring bloom whose timing will depend both on the seasonal irradiance cycle and on seasonal differences in water clarity; (ii) the coastal zone, especially in front of the major river mouths, will have a turbidity front, beyond which there will be a zone of enhanced primary production at the appropriate season; and (iii) there will be enhanced production in response to the various local eddying and upwelling processes described previously. Examination of monthly and seasonal CZCS images supports these statements: The Yellow and Bohai Seas are dominated by sediment load almost year-round, whereas the Sea of Japan—having no major river effluents except the far-field effects of the Amur late in the year—shows evidence of a spring bloom from February to May, after which oligotrophy appears to become established.

From surface observations, a general boreal spring increase in chlorophyll of surface waters is noted in the East China Sea (Li and Fei, 1992), and enhancement of phytoplankton growth has also been observed in association with persistent eddies and upwelling. The river mouths, especially that of the Yangtse, and the cyclonic eddy south of Sheju Island were noted by Guo (1992) as supporting enhanced algal growth, and neritic diatoms (*Coscinodiscus, Skeletonema, Melosira,* and *Chaetoceros*) are prominent in these situations. The intrusions of Kuroshio water onto the shelf as shelf-edge eddies transport very significant amounts of nitrate onto the shelf; one such eddy was computed to carry more nutrients than the combined annual river discharges (Chen *et al.*, 1992). Where the intrusion of these eddies over the shelf break results in locally strong vertical stratification, blooms and the development of strong subsurface chlorophyll maxima may occur in all seasons.

Mesozooplankton biomass increases by a factor of about two seasonally (in summer), and the large herbivores (*C. sinicus*) and omnivores (*E. pacifica*) which dominate were the food of the large stock of Pacific herring that inhabited the Yellow Sea until it was fished out during the 1980s (Sherman and Alexander, 1989).

Synopsis

Shallow winter mixing, and the winter chlorophyll signal, may be biased by the increased silt load in the east China Sea (Fig. 9.19).

Sunda–Arafura Shelves Province (SUND)

Extent of the Province

The SUND province comprises the continental shelves from Burma and South China down to the north coast of Australia, west of Torres Strait. There are several major embayments, including the Gulf of Tonkin (behind Hainan Island) and the Gulf of Thailand, as well the smaller shelf areas throughout archipelagic Malaysia, Indonesia, and the Philippines. The deep water of the Flores Sea separates the northern shelf areas from the southern shelves of the Timor and Arafura Seas and the Gulf of Carpentaria. The many islands, large and small, and island arcs that stand on the shelf regions or arise from deep water all have a coastal fringe of shallow water, about which very little is known that is relevant to this study.

As will become very clear, this province could probably be subdivided into four or five uniform and logical subunits; the problem is that we currently lack the information on which to base such a logical partitioning of this complex and beautiful region, which remains so poorly explored scientifically.

Continental Shelf Topography and Tidal and Shelf-Edge Fronts

This is the largest region of continental shelf anywhere in the ocean, the scale of which is not often appreciated (see Fig. 9.13). The Arafura and Sunda shelves are together equivalent in area to about 10 Grand Banks of Newfoundland and the total extent of shelf area in this province (about 4.5×10^6 km^2) is about 18–20% of the total shelf area of all oceans. There are three major blocks of shelf topography, each representing recently drowned coastlines (Ben-Avraham and Emery, 1973): (i) the shallow shelves fringing continental Southeast Asia (Gulf of Tonkin, Vietnam, Gulf of Thailand, eastern South China Sea,

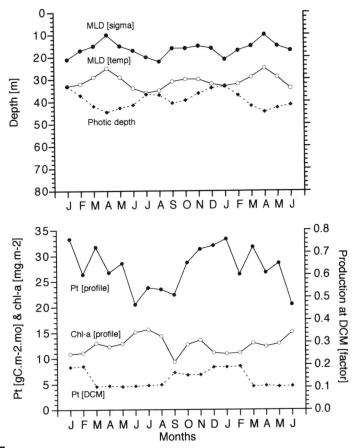

FIGURE 9.20 SUND: characteristic seasonal cycles of monthly averaged mixed layer and photic depths, chlorophyll at the surface, and rate of primary production, both depth-integrated and at the DCM. Data sources are discussed in Chapter 1.

Malacca Straits, and Andaman Sea), (ii) the shelves of the western part of the archipelago (western South China Sea, Sunda Shelf, and the Java Sea), which are separated by deep water from (iii) the shelves of the eastern archipelago and Australia–New Guinea (Arafura Sea, Timor Sea, and the Gulf of Carpentaria). Obviously, this is a region ripe for further logical subdivision, and these three entities are a useful start.

This shelf province is unique in other ways than mere dimension: The unusually shallow areas of shelf which nevertheless have great width, the complex tidal dynamics due to the interaction of the tides of two adjacent oceans, and the sediment input from rivers (7.2×10^9) tons annually all combine to make this a most unusual area. For instance, the central parts of the Sunda Shelf are less than 100 m, and the southern parts, between Borneo and Sumatra, are only 10–40 m deep and the submerged valleys of the rivers that previously drained the shelf area still exist and some have been deepened by tidal flows. Some of the passes between the islands opened only during the historical era and are still very shallow indeed.

The eastern complex of shelves is dominated by soft muddy deposits more frequently than the western regions, and the strong tidal streams (>5 knots through the Torres Straits) generate sufficient bottom friction in many places to overcome water column stability. In such places, we may expect tidal fronts and associated biological enhancement. The Malacca Strait and the Arafura Sea–Gulf of Carpentaria region have been identified as potentially productive of such features; the area between Australia and the Lesser Sunda Islands ranks second among 50 shelf areas (for all seas) in dissipation of tidal energy per unit length of coastline (Hunter and Sharp, 1983; Forbes, 1984). Tidal mixing is especially strong around headlands in the Gulf of Carpentaria and also along the whole south coast of New Guinea. Fronts between tidally mixed and thermally stratified areas may develop only in austral summer, during the southerly monsoon, when wind stress is lowest in the eastern archipelago and when thermal stratification can occur over the shallow shelves.

Defining Characteristics of Regional Oceanography

This region is wholly under the influence of the seasonally reversing monsoon winds and of the circulation thereby generated, which has been described already (see Chapter 8, Indian Monsoon Gyre Province; Western Pacific Warm Pool Province) and will not be reviewed again here in detail. It suffices to say that flow between the Indian and Pacific Oceans is generally reversed with the monsoon winds. It must also be emphasized that in thinking about the oceanography of this province, you should remember that it spans the equator and should expect the consequences of seasonality in the two hemispheres to be observable.

The whole region must maintain a permanent tropical shallow mixed layer, except where tidal mixing breaks this down, and this is confirmed by the splendid regional synthesis of Wytrki (1961), from which I have drawn much of the following account. Over much of the province an Ekman layer of 30–50 m exists during all seasons, and only in the northern part of the South China Sea (north of Hainan) and in the seas north of Australia does this deepen in boreal winter to 70–90 m; in the former case, this is due to wind mixing (see China Sea Coastal Province), and in the latter it is caused by the seasonal alternation between upwelling and relaxation in the Banda and Flores Seas (see Archipelagic Deep Basins Province). A simple comparison between potential mixed layer depths and the depth of water over the shelves shows that much of the southern Sunda Shelf between Borneo and Sumatra must be mixed to the bottom, and here we can expect active development of shelf sea fronts. The temperature of the mixed layer thus varies more at the northern and southern ends of the province, whereas salinity is more variable seasonally in the central regions, though actual seasonality is not uniform, following rainfall and river effluent patterns.

As might be predicted, the mixed layer is nutrient depleted and lies above a nutricline which is coincident with the thermocline; Wyrtki comments that the input of nutrients from rivers is (was, in the 1950s before serious clear-cut logging?) remarkably low and that flux of nutrients into the mixed layer seems to occur mostly in tidally mixed areas or where seasonal upwelling of subthermocline water occurs. Major coastal fisheries have recently collapsed, mainly because of overfishing but perhaps, it is thought, it is also related to environmental change if there is also evidence of a concurrently changing phytoplankton ecology. In such places, concern is currently expressed at the amount of nitrogen and phosphorus (and also, in some places, the biological oxygen demand of the discharge) entering some of the embayments. The Gulf of Thailand is the best example of this, or at least it is the most investigated. Here, there are several episodic blooms during the year without a very clear seasonality, and the frequency of these is said to have significantly increased in recent decades and to be associated with the discharge of nitrate from the major rivers entering the gulf.

Biological Response and Regional Ecology

Though heavy cloud cover renders this one of the least observed regions by the CZCS sensors, the regional, seasonal images do show what could be predicted from the previously discussed information: There is an almost continuous coastal belt of highly turbid water throughout the province around even quite small islands, especially prominent in the Arafura Sea and Gulf of Carpentaria in the east and over the Gulf of Tonkin, the western South China Sea and the Straits of Malacca in the west. To what extent this signal can be referred to biological activity has not been investigated except in the Bismarck Sea (see Archipelagic Deep Basins Province). However, one aspect of the ecology of this province can be observed, though with some technical reservations at these low latitudes, with CZCS images: This is the occurrence over the tidally mixed shelf areas outlined previously of coccolithophore blooms, presumably of *Emiliania huxleyi* (Brown, 1995).

The locations of coastal upwelling features, forced by interaction between seabed and terrestrial topography and the local wind strength and direction, have been generalized in the discussion of the deep basins within the archipelago (see Archipelagic Deep Basins Province); this is a matter of convenience since these features may be sufficiently large as to occur over both a shelf and an adjacent deep basin. A prominent feature of this kind is the front aligned with the long shelf edge lying zonally across the South China Sea between the 1000 to 1500-m deep "Dangerous Grounds" and the Sunda Shelf; flow approximately normal to this shelf edge generates a meandering front with associated uplift of the nutricline by as much as 90 m (Lim, 1975).

Over the continental shelf of the Arafura Sea, upwelling occurs extensively during the southerly monsoon (Wetsteyn *et al.*, 1990). The mechanism appears to be that water from 100–150 m over the eastern slopes of the Aru Basin spreads up over the Arafura shelf reaching several hundred kilometers from the shelf edge; this phenomenon carries cool, saline, high-nutrient slope water (>10 mmol NO_3) far up over the shelf both north and south of the Aru Islands. This transport is supported by a slow offshore Ekman drift of coastal, low-salinity water. These processes are insufficient to upwell the slope water to the surface, and increased levels of primary production during boreal summer (Ilahude *et al.*, 1990) and higher cell counts of large phytoplankters (Adnan, 1990) suggest that upwelled nutrients are utilized. The observations in the regions also suggest that riverborne nutrient inputs, though utilized biologically, are less effective in mediating algal dynamics over the Arafura Shelf.

Although I know of no comprehensive surveys of regional primary productivity, it is evident that it is relatively high in the coastal gulfs compared with the open shelf, and this was already evident from the very first voyages deploying the ^{14}C production method in the 1950s. Platt and Rao (1975) list average figures for the Gulf of Thailand and the Vietnam coast as about 1.2–1.3 g C m^{-2} day^{-1} compared to 0.3–0.6 g C m^{-2} day^{-1} for the remainder of the province, though it is not easy to separate data from the deep basins and the continental shelves.

Since I have been unable to locate any comprehensive study of the integrated planktonic ecosystem, I shall assume that it must be similar to that described for the coastal regions of the Gulf of Guinea (see Chapter 7, Guinea Current Coastal Province).

Synopsis

The SUND province is essentially invariant seasonally in all features, with a very shoal pycnocline because of the extent of the surface brackish layer of river water encountered everywhere in this province (Fig. 9.20).

East Australian Coastal Province (AUSE)

Extent of the Province

Two ecologically different coasts are aggregated in the East Australian Coastal Province (AUSE) province for convenience, and they could easily be separated into two entities for other purposes: (i) the Coral Sea coasts of Australia from Cape York (11°S) to Sandy Cape (25°S) and of New Caledonia, and (ii) the Tasman Sea coast of Australia from Sandy Cape to, and including, eastern Tasmania to Southeast Cape at about 45°S.

Continental Shelf Topography and Tidal and Shelf-Edge Fronts

The continental shelf off northeast Australia is almost wholly occupied by the Great Barrier Reef which is both remarkably wide (250 km at 20–22°S) and long (1750 km from Torres Strait to Sandy Cape). The southern coast of New Caledonia carries a similar barrier reef, though this is much narrower (10–25 km) but extends northwest of the island to the Entrecasteaux reefs, so the total length of the barrier is about 600 km. South of Sandy Cape, off the New South Wales coast, the shelf is narrow and is rarely more than 10–20 km wide.

Defining Characteristics of Regional Oceanography

North of 18°S, where westwards flow across the Coral Sea diverges north and south on encountering the continent, the flow at the shelf edge is highly variable and unpredictable, though it is generally equatorward (Sculley-Power, 1987). South of this latitude, the East Australian Current forms a western boundary current with predictable polewards flow from the central Great Barrier Reef south to about 30°S, where retroflection occurs into the eastwards frontal jet within the Tasman Front (see Tasman Sea Province). The poleward flow of the East Australia Current is alongshore, above the continental slope, and a convergent front occurs offshore with equatorward counterflow (Church,

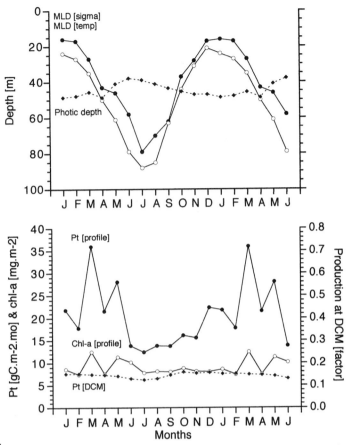

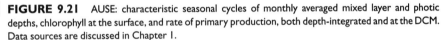

FIGURE 9.21   AUSE: characteristic seasonal cycles of monthly averaged mixed layer and photic depths, chlorophyll at the surface, and rate of primary production, both depth-integrated and at the DCM. Data sources are discussed in Chapter 1.

1987). The East Australia Current system is seen as a of series southward meandering anticyclonic eddies, with consequences for shelf-edge effects and advection of nutrient-rich water onto the shelf and through the reef systems.

Biological Response and Regional Ecology

The two sections of this province have such different nutrient regimes and dynamics that they are dealt with separately.

Coral Sea Coasts

The CZCS images show two features along the Australian coast of the Coral Sea: a coastal zone of generally high chlorophyll, whose outer boundary appears to approximate that of the Great Barrier Reef, and, offshore and separated from that feature, mesoscale patches which may represent vorticity in the meanders of the poleward East Australia Current. Two components are visible in the chlorophyll field in the region of the Barrier Reef and these exhibit seasonality. During the dry season the very clear water of the lagoon inside the Barrier Reef separates a near-shore zone of high chlorophyll from the lower chlorophyll zone over the reef itself. The latter signal probably emanates at least in part from the benthic algal mats and other components of the reef flora: Such a suggestion is reinforced by the clear image given by the carbonate shoals of the Bahama Banks, identifiable as a persistent feature surrounded by water of high clarity in many published Atlantic CZCS images. During the wet monsoon season the effects of the higher nutrient loading and wind mixing in the lagoon generate a bloom that obscures the distinction between the near-shore and Barrier Reef chlorophyll signals (Gabric *et al.*, 1990).

The regions where reef biota are most strongly developed on the northwest Australian continental shelf are those where the shelf edge is steepest; this early led to interesting suggestions that the richness of the Great Barrier Reef might be a response to upwelling in the tropical western boundary current (Orr, 1933). In fact, the meandering East Australian Current has been found to have an oscillatory intensification with a period of about 3 months, pumping nutrient-rich deep water to the shelf break. Periodic reversals in the longshore wind component as pressure systems drift across the western Coral Sea, then force Ekman suction so that nutrients are advected inshore, and algal blooms occur rapidly inshore of the shelf break. These intrusions of upwelled water penetrate the whole width of the reef facies and extremely high inputs of nitrate ($20 \ \mu M$ year^{-1}) are introduced into the reef system by this mechanism.

Tasman Sea Coast

Apart from the general existence of a winter–spring bloom, stronger toward the south, the western coast of the Tasman Sea is dominated by two processes causing biological enhancement: coastal upwelling cells and vorticity in the offshore eddy field (Rahmstorf, 1992). The narrow shelf of eastern Australia has several sites where coastal upwelling cells are persistent, notably at Evans Head (27°S), Laurieton (31°S), and Sydney (34°S). In part, the location of these cells is due to topographic influence, such as the narrowing of the shelf just north of Laurieton (Rochford, 1975) or where the flow of the East Australia Current separates from the coast (Tranter *et al.*, 1987). Nutrient-rich water may uplifted from 125 to 275 m (at 15–17°C) either to a bottom Ekman layer over the shelf, as is usually the case off Sydney, or to the surface, as occurs farther north. The major wind events on this coast ("Southerly Busters") generate equatorward stress and downwelling and are of short duration. On the other hand, strong current events are of longer duration. Therefore, bottom Ekman layer forcing produces upwelling events of long duration, whereas wind stress produces downwelling of shorter duration (Griffiths and Middleton, 1992).

A special case is the biological enhancement along the shelf-break front east of Bass Straits and the Victoria coast: Here, the shelf-break front is the so-called Bass Strait Cascade at which cold, dense straits water slips below lighter Tasman Sea water and there is a linear zone over the upper slope of enhanced nutrients, chlorophyll, and zooplankton. This ecological feature survives the cessation of algal growth over the shelf region at the end of austral winter in September (Gibbs, 1991).

The offshore eddy field of the East Australian Current is relatively weak during austral winter (May–July) but at other seasons its intensive meandering induces enrichment of the coastal water mass. When a meander or eddy approaches the shelf break, slope water is driven onto the shelf by the peripheral anticyclonic current associated with the feature. Such onshore flow would create an Ekman boundary layer that would move equatorwards, displacing the surface coastal waters as these move offshore at locations where the East Australia Current diverges from the coast. Local blooms would result and are observed in CZCS images (Tranter *et al.*, 1987). In addition, some coastal upwelling appears to be wind induced, especially along those parts of the coast having a southwest/northeast orientation. Finally, a few cyclonic frontal eddies may spin off from the landward side of the velocity maximum of the coastal current and become entrained in the coastal water mass (Tranter *et al.*, 1987). Similar elongated cyclonic features are termed "shingles" on the landward side of the Florida Current by Atkinson *et al.* (1984) and are an important source of nutrients over the shelf.

Warm-core eddies containing water originating in the Coral Sea may have complex horizontal distribution of surface chlorophyll, often near their western edge. One such eddy, tracked over a 19-month period, had persistent integrated chlorophyll of 60–80 mg chl m^{-2}, of which 60–70% was contributed by very small cells of the pico- and nanoplankton; this contribution reached 90% in some of the more oligotrophic stations worked within the eddy. Edge enrichment where warm-core eddies entrain crescents of cooler, nitrate-replete water (about 50 μM) may occur in the coastal regions, especially where the edge of the eddy is comparatively straight. Here, populations of the upwelling copepod *Calanoides carinatus* may occur and the general biological enhancement (NO_3, 50–100 μM; chlorophyll, 1.5 μg liter^{-1}; copepods, 50–70 m^{-3}) may attract schools of the southern bluefin tuna (Tranter *et al.*, 1983).

Synopsis

In AUSE there is moderate winter mixing of the very shoal summer mixed layer, with a late austral summer maximum for primary production rate and no significant chlorophyll accumulation (Fig. 9.21).

New Zealand Coastal Province (NEWZ)

Extent of the Province

The NEWZ province comprises the continental shelf around New Zealand, together with the area within the 1000-m isobath on the Chatham Rise. To the south and east, the offshore boundary is the edge of the Subtropical Convergence Zone, whereas to the west and north it is the shelf-edge front.

Continental Shelf Topography and Tidal and Shelf-Edge Fronts

The 200-m isobath lies 75–100 km offshore on the east side of South Island, widening northwards to meet the shallow topography of the Chatham Rise. The west coast of the South Island is especially steep-to, and to the south of Abut Head, along the Fjordland coast, the 1000-m contour is within a few kilometers of the shore. The most extensive areas of shallow shelf enclosed by the 200-m isobath are along both coasts of the North Island and around the long, southward-pointing Snares Shelf south of Stewart Island, which itself lies

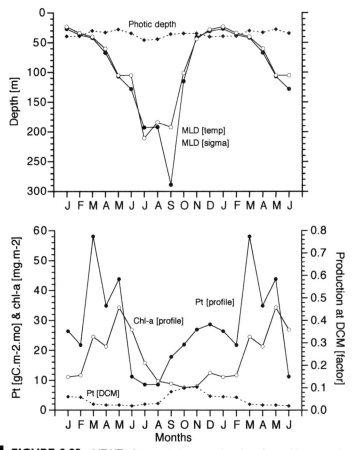

FIGURE 9.22 NEWZ: characteristic seasonal cycles of monthly averaged mixed layer and photic depths, chlorophyll at the surface, and rate of primary production, both depth-integrated and at the DCM. Data sources are discussed in Chapter 1.

off the south coast of South Island. Cook and Foveaux Straits afford east–west passages between North and South Islands and Stewart Island.

Defining Characteristics of Regional Oceanography

The geographical location of New Zealand and the relatively shallow topography of the Campbell Plateau and the Chatham Rise strongly modify the zonal arrangement of the circumpolar frontal systems, especially the Subtropical Convergence and the Subantarctic Front of the Polar Frontal Zone (see Chapter 10), and determine circulation around the islands. A confluence region between the Subtropical Convergence and the western boundary current occurs to the east of New Zealand and recalls in several ways the confluence between the SSTC and the Brazil Current at the Falklands Plateau (see Chapter 7, Southwest Atlantic Shelves and Brazil Current Coastal Provinces). Much of the following account of both the physical and the biological oceanography of New

Zealand waters is drawn from Heath (1981, 1985), Heath and Bradford (1980), Vincent *et al.* (1991), and Butler *et al.* (1992).

The frontal jet of the Tasman Front rounds Cape North and, topographically locked, passes around North Island carrying subtropical water originating in the Coral Sea. This flow forms a relatively high-velocity stream in rounding East Cape before passing down the east coast of North Island to the region (at about 42–44°S) between Cook Strait and the Banks Peninsula on South Island and north of the Chatham Rise. Here it meets the Subtropical Convergence Zone (the frontal jet of which is here represented by the cool Southland Current) and this front, after passing across the southern Tasman Sea, encounters the shallow water south of Stewart Island and then, topographically locked, rounds South Island to flow north along the shelf edge of the east coast as the Southland Current. Offshore, beyond the Southland Current is the main frontal zone of the SSTC, which passes eastwards again toward the open South Pacific along the edge of the topography of the Chatham Rise. When released from the topographic constraint of the rise, a persistent meander to the south is formed in the Subtropical Convergence as it passes east across the ocean at about 40°S.

East of the Cook Straits and over the western part of Chatham Rise the confluence zone is replete with mesoscale eddies, plumes, and fronts, and in this complex region there is frequent cross-frontal exchange of subtropical and subantarctic surface water bodies. At this conjunction, the distinction between subtropical and subantarctic water masses is especially clear and the interleaving water parcels are typified by their origins: mixed layer depth, chlorophyll profiles, and nitrate levels in the surface water are all clearly different.

Along the Westland of South Island the coastal current diverges at about 44°S, where the westward limb of the gyral circulation of the Tasman Sea encounters the coastline of New Zealand. From here, flow bifurcates northwards along the Westland coast of South Island to the west coast of North Island and southwards along the Fjordland coast down to Cape Providence, which it rounds and provides an inshore component for the Southland Current. Within these flows occur several sites of coastal upwelling: (i) at Cape Farewell (about 40°S) at the northern tip of the South Island if the intermittent coastal current is intensified by an onshore wind (Shirtcliffe *et al.*, 1990); (ii) the same intensification induces a sea-surface slope which accelerates water over the Kaharungi Shoals, thus creating a strong upwelling source near Kaharungi Point; and (iii) south of Cape Providence (46°S) there is some uplift of water from the deep salinity maximum over the continental slope of the Puysegur Bank resulting in high nitrate concentrations inshore (Bradford, 1983).

However, even more important is the persistent upwelling in the open bight between Cape Cascade (44°S) and Cape Foulpoint (42°S). This simple wind-induced coastal divergence is limited by the shelf-edge front of the upwelled water which bears a series of wave-like, regularly spaced plumes of cooler water (about 13°C) in the warmer subtropical water mass of the Tasman Sea. Seawards of Abut Head the plume may reach 70 km offshore and carry much of the discharge from the Buller River. Inshore, both here and on the east coast, freshwater runoff from these rainy islands creates a sandwich of multiple, overlying surface water masses identifiable in the density profiles of the upper 25 m.

In the seas west of New Zealand, wherever frontal processes are weak, the mixed layer depth can be modeled simply by local irradiance and wind stress. The seasonal range in mixed layer depths is large—about 20 m in summer and 140–200 m in winter. This cycle is largely convective, being generated more by variance in irradiance (simply, sun angle and cloudiness) than by variance in wind stress.

Biological Response and Regional Ecology

This is a very difficult region in which to generalize the ecological processes, though a winter–spring bloom, forced by the seasonal cycle of irradiance and changes in mixed layer depth, is the dominant signal in the general chlorophyll field around the islands.

Nitrate is often in excess of limiting concentration except in the more open ocean areas adjoining the TASM province, where a simple Sverdrup model is probably appropriate and where nitrate reaches limiting levels in surface water in summer. To the south and east of New Zealand, the surface water masses are distinguishable by their NO^3 levels: 4–7 μM in subtropical and 12 μM in subantarctic water. Elsewhere in the coastal regions, values of 0.5 to about 3.0 μM are normally encountered. Water brought to the surface in the Westland upwelling strip is rapidly reduced to within this range of nitrate values by biological uptake.

The CZCS images show surface chlorophyll enhancement in this province principally where the Subtropical Convergence Zone encounters topography; that is, to the south of New Zealand from the Foveaux Straits to the nose of the shallow bank, and again over the deeper shoals to the east of New Zealand. The CZCS images suggest that chlorophyll accumulation occurs principally during the same season (March–November) as that to the south of the Subtropical Convergence (Comiso et al., 1993). Also observable in the seasonal images is the linear upwelling along the Westland coast, especially in austral spring. However, as Vincent et al. (1991) note, a simple statistical correlation cannot be found between physical and nutrient variables at groups of individual stations and biological variables at the same stations; moreover, Chang et al. (1992) suggest that strong winds and unusually constant deep wind mixing around the New Zealand coasts may constrain both the accumulation of phytoplankton cells and the complete utilization of nitrate. This is too dynamic a region for simple relationships between nitrate and chlorophyll to be evident in survey data, and the relationships that might exist with phytoplankton growth rate are, as Vincent et al. point out, probably masked by the fact that phytoplankton biomass is a secondary variable representing (as I have noted several times) merely the resultant between accumulation and loss. Nevertheless, high F ratios (new/regenerated or nitrate/ammonium utilization ratios) reaching 0.9 are often encountered inshore in the Westland upwelling region.

Here, as in so many other sea areas in the past 10 years, we have confirmation of the paramount importance of the picophytoplankton, both prokaryotes and eukaryotes. In the upwelling region of the Westland coast, Hall and Vincent (1994) have demonstrated the relatively important role these organisms play in the whole phytoplankton community. In summer the <2-μm fraction accounted for 73% of phytoplankton particulate nitrogen, whereas in winter 40–80% of particulate nitrogen, 55% of chlorophyll, and 45–70% of primary productivity are attributable to the picofraction. At single stations (e.g., at the edge of the upwelling plumes) prokaryotes contributed 99% of all cells. These studies also reinforce the general observation of the high variability in abundance of the small photosynthetic cells: Prokaryotes are normally within the range $5–75 \times 10^7$ cells liter^{-1}, which is at least twice the relative variability normally observed for the small eukaryotes. In general, and as occurs elsewhere, maximum prokaryote cell numbers occur deep in the mixed layer and in association with a nitracline. In the inshore phytoplankton, biomass is more evenly distributed among the size classes. Chang *et al.* (1992) suggest that the 20- to 200-μm fraction contributes here as much as 55% of total phytoplankton nitrogen because high nitrate levels favor the growth of diatoms. Cell division rates among large cells are actually higher here than for the smaller fractions: At surface irradiance, the 20- to 200-μm fraction = 0.5 doublings day^{-1}, the 2- to 20-μm fraction = 0.22 doublings day $^{-1}$, and the <2-μm fraction = 0.18 doublings day^{-1}.

Both the Westland and the Cape Kahurangi (Cook Strait) upwelling blooms support populations of herbivorous macrozooplankton, as might be anticipated. Here, crustacean herbivores are dominated by small copepods (species of *Acartia, Paracalanus, Clausocalanus, Oithona,* and *Centropages*) and a euphausiid (*Nyctiphanes australis*), which is an important component in the pelagic food web both here and in the AUSE province. At Cape Kahurangi, *Acartia* may form up to 60% of net zooplankton biomass, and on the Westland coast it dominates the inshore crustacean plankton; offshore in the upwelling water a mixture of oceanic and neritic species is normally encountered. Off Westland, zooplankton grazing (calculated for organic nitrogen) is 0.3–20% of algal production, with consumption being relatively higher inshore. *Nyctiphanes* manages its diel migration in the inshore–offshore transport associated with upwelling cells so as to place its eggs and larvae inshore, whereas the main feeding population of older larvae and adults remains offshore. In the Taranaki Bight, south of the upwelling at the Cape, *Nyctiphanes* may comprise more than half of all zooplankton biomass (James and Wilkinson, 1988; Bradford and Chapman, 1988).

On the east coast of North Island, where conditions are dominated by the effect of the subtropical water of the western boundary current, especially during ENSO events, there may be extensive coastal microalgal blooms of "nuisance" species: *Mesodinium* (a red tide organism), *Gonioceros,* and *Aulacodiscus* are the main microalgae in these blooms. In Tasman Bay, similar blooms occur after especially heavy coastal rainfall and runoff from the land during summer.

Synopsis

Deep excursion of mixed layer occurs in winter, and pycnocline is illuminated only briefly in summer (Fig. 9.22). Spring bloom, as an increase in productivity, is initiated early (September) and is sustained throughout austral summer and autumn (May); productivity is generally tracked by chlorophyll biomass and nitrate-based production is assumed to dominate.

10

THE SOUTHERN OCEAN

The annular Southern Ocean is unique in having no continental barrier to continuous wind-driven flow eastwards around the globe. Only the bottleneck of the Drake Passage between South America and the Antarctic Peninsula constrains this flow, pinching together the components of the circumglobal current structure but not eliminating them. The lack of land barriers permits the strong westerlies of the Southern Ocean to build up a fully developed field of wind waves having higher mean height than anywhere else; this is the windiest and roughest part of the oceans. It goes without saying that the poleward parts of the Southern Ocean are seasonally ice covered, and the Antarctic coastline bears large permanent ice shelves.

The Southern Ocean is also unusual in the richness of the ecological studies performed there; although it is distant from most oceanographic centers, the lure of the rich whaling grounds early attracted the attention of biologists (really government departments would be more accurate). The list of research ships that worked the Southern Ocean is a roll call of early oceanography, and I refer you to a fascinating chapter about this era by Deacon (1984), from which I especially relish the fact that the German leader (Captain Thaddeus von Bellingshausen) of Tsar Alexander I's Russian expedition of 1819–1821 aboard *Vostok* (East) and *Myrni* (Peaceful) made the first net tows which demonstrated the diel vertical migration of planktonic animals and the existence of the Southern Ocean krill, *Euphausia superba*. From then on, we have a cascade of biological studies up to the *Discovery II* expeditions organized by the British Colonial Office a century after Bellingshausen (read Alistair Hardy's 1967 account for the full flavor of these expeditions). The recently celebrated voyages of the German research ship *Polarstern* shows that the tradition is alive because this fine ship carried the Joint Global Flux Study (JGOFS) 1992 oceanographers who successfully demonstrated why algal growth should be constrained in the NO_3-replete circumpolar water masses. It has been one of the enigmas of biological oceanography that over much of the Antarctic Polar Current, in the traditionally biologically rich Southern Ocean, surface water should be blue and clear and nitrate should remain unutilized.

Associated with the vertical motion and banded structure of the Antarctic Circumpolar Current (ACC) which flows perpetually eastwards in the Southern Ocean, a series of oceanic fronts within the current form prominent, permanent, and almost continuous circumpolar features, which partition the surface water masses, their flow, and their ecological characteristics into zonal bands. This complexity is usually resolved partially by reference to two "branches" of the ACC, north and south of the fast frontal jet at the Polar Front along about 50°S. Any good oceanographic text will have a three-dimensional analysis of the frontal systems and the vertical water flux which maintains them. Figure 10.1 shows the relationship between the oceanographic fronts and the ecological provinces of the Southern Ocean which shall be discussed in this chapter.

Only south of the most polewards of the fronts is there contrary, westwards transport in the East Wind Drift or ACC. The Southern Ocean has, perhaps more than other oceans, frequently been partitioned into "zones" or "biomes" because of the existence of these relatively easily identified circumpolar features. Unfortunately, this complexity has led to much duplication of terminology; therefore, to make things subsequently simpler, it will be convenient here to summarize these fronts and the zones between them using the set of definitions of Tomczak and Godfrey (1994, p. 76), which I shall use subsequently throughout this chapter. The following definitions will be extended in the discussions of each province:

- *Subtropical Convergence* (at about 40°S): Lies between the subtropical and subantarctic surface water masses. On the polewards side of the Subtropical Convergence, the cold, low-salinity Antarctic Intermediate Water mass converges with, and slides beneath, the warm, high-salinity subtropical surface water at the Subtropical Front. Here, the whole convergence zone is proposed as the South Subtropical Convergence Province (SSTC).
- *Subantarctic Water Ring* (at about 45–50°S): This is the most complex entity to grasp. It comprises the subantarctic surface water mass between the Subtropical Convergence Front and the Polar Front. This is where the confusion of names is likely to intervene in the readers mind because the Polar Front is one feature in what was previously described as the Antarctic Convergence (e.g., Deacon, 1984) or the Polar Frontal Zone (Anonymous, 1989); this entity, no matter what you choose to call it, is embedded in the Subantarctic Water Ring together with the two fronts which define it—the Subantarctic Front (45°S) and the Polar Front (50°S), though the latter may exist as a multiple front (Read *et al.*, 1995) with active meandering and eddy shedding (Veth *et al.*, 1997). Here, because the Subtropical Convergence Zone (STC) is such a prom-

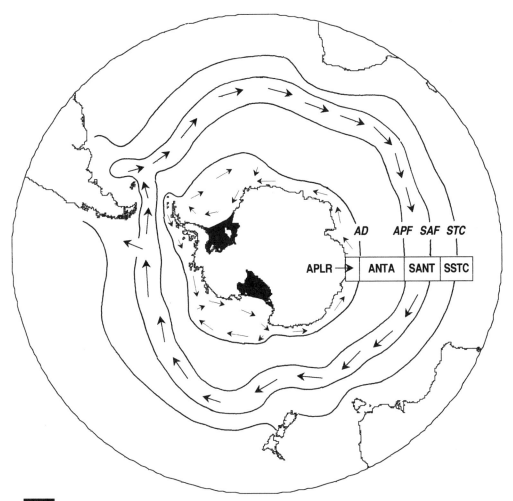

FIGURE 10.1   Diagram of the oceanography of the Southern Ocean illustrating the relationship between oceanic frontal zones, landmasses, and the boundaries of the secondary ecological provinces of the austral regions. AD, Antarctic Divergence; APF, Antarctic Polar Front; SAF, Subantarctic Front; STC, Subtropical Convergence. APLR, ANTA, SANT, and SSTC are abbreviations for province names (see text). The flows of the ACC and of the Antarctic Coastal Current are indicated as are the permanent ice shelves of the Ross and Weddel Seas.

inent feature in the chlorophyll field, only that part of the Subantarctic Water Ring which lies poleward of the biological enhancement forced by the dynamics of the STC is defined as the Subantarctic Water Ring Province (SANT).

• *Antarctic Divergence or Continental Water Boundary* (about 65°S): This front is the boundary between eastward flow in the ACC and westward flow close to the Antarctic coast. Divergence represents the surfacing of North Atlantic Deep Water to complete the global thermohaline circulation. From this front to the Polar Front the Antarctic Surface Water mass is defined as the Antarctic Province (ANTA), and between this divergence and the coast of Antarctica the westward flow of the ACC (or East Wind Drift) of continental water is the Austral Polar Province (APLR).

• *Coastal Convergence:* The edge of the coastal zone of permanent ice is aligned with a coastal convergence where the Antarctic Bottom Water is formed by near-boundary convection and begins its descent into the bottom circulation of all oceans.

These suggestions for an ecological classification of the Southern Ocean accommodate the fact that neither the Subtropical nor the Antarctic Convergences are simple fronts: Rather, they are complex frontal zones, some degrees of latitude wide, with characteristic ecology resulting in relatively high chlorophyll biomass compared with the clearer water to the north and south of each. To be consistent with a global ecological partitioning of the ocean, it differs only slightly from recent suggestions, such as the partitioning of the open ocean south of the Subantarctic Front by Tréguer and Jacques (1992) into a Polar Frontal Zone, a Permanently Open Ocean Zone, and a Seasonal Ice Zone. The coastal region is then divided into an ice-free and a permanently ice-covered zone. The differences between the two schemes are not substantive.

The division of the Southern Ocean into four provinces is consistent with new information on algal ecology that has been presented since the scheme was first drafted. It seems to be the distribution of the essential micronutrient Fe that regulates algal growth in each of the annular zones of the Southern Ocean (de Baar *et al.,* 1995): Fe-deficient water in part of the Atlantic Subarctic Province (SARC; oligotrophic ecosystem), Fe-replete water in the Polar Frontal Zone (also part of SARC; rich diatom growth), and Fe-limiting water in the southern component of the ACC or the ANTA province (moderate diatom growth). The Fe-replete water of the coastwise, East Wind Drift APLR province (rich diatom blooms) and the horizontal exchange of water types within the Subtropical Convergence (episodic blooms of calcium-secreting plankton) of the SSTC province complete the logic. The same pattern, at a larger scale, corresponds with the sediment-type below the annular zones. South of the Polar Frontal Zone, siliceous ooze originates in diatom blooms in Antarctic surface water, whereas north of this zone and under the Subantarctic Water Ring, the sediments are calcareous and originate in an oligotrophic phytoplankton assemblage.

At the Drake Passage, between Cape Horn and the South Shetland Islands, the West Wind Drift is restricted to a narrow passage of only about 300 km compared with the 1200 km separating South Africa from the Antarctic continent. This requires that the zonal bands south of the Subtropical Convergence Zone be squeezed together as they pass through the straits and there is some flux from the ACC into the southern part of the Humboldt or Chile–Peru Current up the west coast of the continent. Nevertheless, the zonal bands retain their relative positions, and only the Subtropical Convergence itself is not identifiable here. East of the Drake Passage, the Antarctic and Subantarctic Water Ring Provinces separate, and in the Atlantic sector of the Southern Ocean the zonal boundaries take their most northerly position. Further east, the narrowing of the passage south of Africa again somewhat constrains the northern limits of the banded flow of the ACC.

ANTARCTIC WESTERLY WINDS BIOME

South Subtropical Convergence Province (SSTC)

Extent of the Province

Though this is by definition a boundary, the meridional extent and strength of biological enrichment within the feature as seen in all seasons in the Coastal Zone Colour Scanner (CZCS) chlorophyll field requires that the convergence zone be recognized as a province in the present context. It can perhaps best be defined operationally by a chlorophyll contour, probably 0.2–0.3 mg chl m^{-3}.

Continental Shelf Topography and Tidal and Shelf-Edge Fronts

There is no coastline in this province.

Defining Characteristics of Regional Oceanography

The Subtropical Convergence divides the anticyclonic circulations of the southern Atlantic, Indian and Pacific Oceans from the cyclonic circulation of the ACC and can be traced almost continuously around the globe at about 35–45°S, and in each sector its location is predictable within narrow limits. It turns equatorward when it encounters shallower water and poleward over deep water.

The Subtropical Convergence is marked by a sharp decrease in the westerly winds of the Southern Ocean, lies below a permanent line of zero wind-stress curl, and is therefore the equatorward limit of northward Ekman drift in the subantarctic water and of southward drift in the subtropical gyre: Strong downwelling consequently occurs here. The convergence is marked by sharp surface thermal gradients (14–18°C summer, 11–15°C winter, and about 0.5, over a distance of <10 miles) separating warm, salty subtropical water from the cooler, fresher water to the south. The frontal zone is sufficiently dynamic to have an eddy field associated with it, and it may include more than one surface discontinuity front in the same sector (Deacon, 1984; Colborn, 1975; Lutjeharms, 1985).

The location of the Subtropical Convergence south of Africa is determined by the position of the retroflection zone of the Agulhas Current as the warm Agulhas water thrusts south of the Cape of Good Hope, pushing a retroflection loop westwards (Lutjeharms *et al.,* 1985; Lutjeharms and van Ballegooyan, 1984) in one of the most dynamic regions in the oceans, associated with an active eddy field (Deacon, 1984; Colborn, 1975). Large meanders form between the warm Agulhas water and the cooler water of the Subantarctic Water Ring Province; on the coastal side of the Agulhas Current, eddies are to some extent topographically controlled in relation to the Agulhas Plateau, south of the Cape (Lutjeharms and van Ballegooyan, 1984; Lutjeharms and Valentine, 1988).

The southern tip of the American continent, reaching to 57°S, is the sole obstacle to circumpolar flow and forces the ACC to the south and through the narrow Drake Passage. Only here is the location of a Subtropical Convergence undefined. To the west of southern Chile the convergence weakens and trends increasingly northwards in the equatorwards flow of the Humboldt Current. To the east of Cape Horn, flow in the ACC turns northwards (as the Falkland

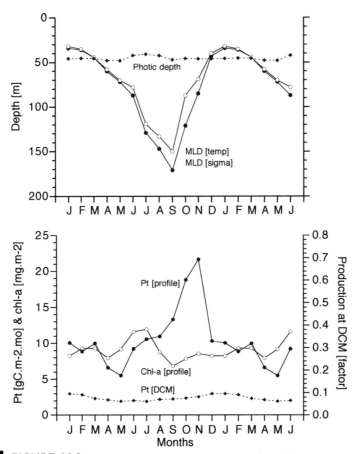

FIGURE 10.2 SSTC: characteristic seasonal cycles of monthly averaged mixed layer and photic depths, chlorophyll at the surface, and rate of primary production, both depth-integrated and at the DCM. Data sources are discussed in Chapter 1.

Current) across the Argentine Basin, and the Subtropical Convergence is re-established along the southerly loop of the front which forms between the eastwards-turning subtropical water of the Brazil Current and the colder sub-antarctic water (Peterson and Whitworth, 1989; Peterson and Stramma, 1991). This occurs at about 40–45°S. The Subtropical Convergence then proceeds across the South Atlantic, reaching its most northerly position close to 35°S at the Greenwich meridian.

In the Tasmania–New Zealand sector, the Subtropical Convergence conforms to topography, looping northwards along the continental edge east of Tasmania (Harris *et al.*, 1987) and south and east of New Zealand. It is only across the Tasman Sea and then to the eastwards from the Chatham Rise that the convergence again becomes an oceanic feature.

Biological Response and Regional Ecology

The Subtropical Convergence forms a mixed layer nutrient discontinuity, where sharp gradients in nitrate concentration occur, from <0.5 μM in the subtropical water to the north of the SSTC to around 8–10 μM in the subantarctic water to the south, though silicate values remain low and do not increase until the Antarctic Polar Front is crossed at about 50°S.

The Subtropical Convergence is routinely observable in sea-surface color images as an enhancement of chlorophyll biomass, and several mechanisms may explain enhanced chlorophyll at or just north of the SSTC (e.g., Lutjeharms *et al.*, 1985; Furuya *et al.*, 1986; Yamamoto, 1986). Physical aggregation of biota may occur, as in all convergent fronts, but frontal processes are likely to establish near-surface stability in mixed water columns and initiate accumulation of plant biomass: Where the front slopes downwards, this may provide sufficient thermal stability to initiate and maintain a bloom as at some shelf-slope fronts (Marra, 1982). Otherwise, mixing of warm, nutrient-poor subtropical water south across the SSTC may produce sufficient thermal stability, and a sufficient nutrient supply, to establish a bloom. Finally, the transport of nutrient-rich water north across the front from a well-mixed to a more stratified zone may again produce conditions leading to a bloom. The latter may be the dominant mechanism and lead to the observed chlorophyll enhancement zone, usually strongest just north of the SSTC. Such processes are best developed to the south of Africa, where the Aghulas Return Current joins the convergence zone; here, CZCS images show very large regularly spaced meanders as high-chlorophyll features in the flow proceeding eastwards across the Indian Ocean sector.

A frontal bloom, apparently typical of such events, was observed at 150°E in austral summer (Furuya *et al.*, 1986) during which the warm, salty subtropical water was locally overlain by colder, less salty subantactic water, producing a temperature inversion and a strong near-surface pycnocline, thus confining the colder water assemblage in a well-lit shallow mixed layer rich in nutrients. The biota contributing to such episodic blooms originate from two assemblages of phytoplankton which converge here and from each of which the characteristic diatom species contribute most to the new (nitrate-fueled) production: from the south, a community which comes to be diatom dominated (90–92% diatoms, as carbon biomass determined by fluorescence microscopy, and including *Chaetoceros, Nitzschia, Thalassiothrix, Rhizosolenia*, etc.) in the convergence, whereas from the north a more mixed community (only 25–50% diatoms) incorporating a deep chlorophyll maximum (DCM) at 75 m. Once incorporated in the front, even the warm-water assemblage of unicellular cyanophytes and small flagellates (again as counted by fluorescence microscopy) were never more than 15% of autotrophic carbon. Silicate depletion is likely to occur in these blooms and will result in an overall reduction in new production rate and progressive evolution of the algal assemblage to one dominated again by dinoflagellates and microalgae as in the parent stocks.

The same process has been observed in the Indian Ocean sector, where Soviet investigations concluded that the strongest enhancement of productivity occurred at intrusions of warm into cool water. In the Atlantic sector, near the Greenwich meridian, though there is a strong chlorophyll peak at the convergence in early summer, this is not necessarily the case at the end of summer and this probably occurs in all sectors of the Indian Ocean at 40–60°S. Though nutrient gradients were strong across the convergence here, they were patchy and negatively correlated with chlorophyll.

In addition to chlorophyll enhancement, there is also enhancement of other biological properties because the SSTC supports concentrations of large pelagic fish such as the mackerel *Trachurus picturatus murphyi,* which is endemic to the SSTC and maintains a large population from Chile to 160°W (Parrish, 1989; Evseenko 1987). This province is also the home range of the warm-blooded southern bluefin tuna (*Thunnus maccoyi*), which exploits the relatively high biomass of small fish and squid that must occur here; this species leaves the SSTC only to enter warmer water, principally to the northwest of Australia, to spawn in the austral winter, at which time fish appear in Tasman Sea Province and New Zealand Coastal Province as far north as the tropical convergence that passes across the northern Tasman Sea. The region of the SSTC adjacent to the Agulhas Retroflection appears to carry an especially large population of southern bluefin.

It should also be noted that this province coincides not only with an often-observed increase in the abundance of zooplankton but also with the southern "Transition Zone" of McGowan's Pacific biogeography. The same group of 12 species of zooplankton that we encountered in the North Pacific Transition Zone Province (NPPF) also occur here in the South Pacific equivalent of the North Pacific transition zone between subtropical and subpolar gyres. Representative species are the euphausiid *Thysanoessa gregaria,* copepods *Eucalanus bungii, E. elongatus, E. longiceps,* the chaetognath *Sagitta scrippsae,* and the mollusc *Limacina helicina.* However, here is a good example of the dubious support given to ecological geography by taxonomic geography—or classical biogeography: Consider the euphausiid species pair *Nematoscelis difficilis* and *N. megalops.* In his monumental work on Pacific euphausiids, Brinton (1962) considered the former characteristic of the North Pacific transitional zone (the NPPF province of this work), and *N. megalops* (morphologically differing only in the details of the male copulatory organ) to be characteristic of the southern transitional zone (the SSTC province). He goes on to tell us that *N. megalops* occurs widely in the North Atlantic (an identification supported by later writers), in the Mediterranean, and in the Benguela Current, whereas *N. difficilis* is isolated in the North Pacific. One may be able to infer from such a distribution something about the evolution of the taxon represented by this species pair, and that in one ocean there is some ecological commonality between the NPPF and the SSTC provinces. One is constantly reminded, when one examines the factual base of biogeography, of its limited relevance for ecological geography. Nevertheless, I take comfort from Brinton's global distribution of *Thysanoessa gregaria,* which has three components: (i) a circumglobal distribution in the SSTC, following its pattern closely in each ocean though with some drift into the eastern boundary currents; (ii) distribution in the transitional zone of the North Pacific, where it closely matches the distribution of the NPPF and northern California Current Province; and (iii) distribution in the North Atlantic, where it inhabits the terminal part of the Gulf Stream Province and the whole of North Atlantic Drift Province across the ocean.

Synopsis

There is a moderate mixed layer depth increase in winter, with rapid recovery in austral spring (October) (Fig. 10.2). Primary production rate begins to

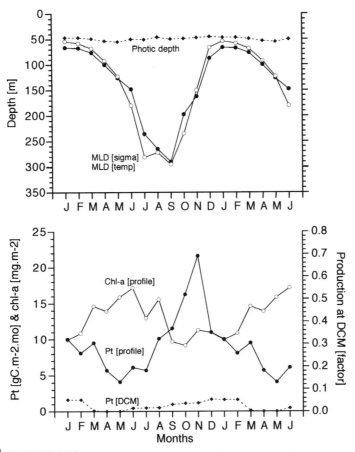

FIGURE 10.3   SANT: characteristic seasonal cycles of monthly averaged mixed layer and photic depths, chlorophyll at the surface, and rate of primary production, both depth-integrated and at the DCM. Data sources are discussed in Chapter 1.

increase progressively in response to dynamic frontal processes when pycnocline starts to descend, continuing until summer pycnocline is reestablished. Chlorophyll accumulates only briefly at the start of the period of high productivity and production in the DCM is always small.

Subantarctic Water Ring Province (SANT)

Extent of the Province

For a note on the problems concerning how best to define this province, see the beginning of this chapter. Simply, it occupies the zone between the southern edge of the biological enhancement in the Subtropical Convergence (about 35°S) and the Antarctic Polar Front at about 55°S, this being the southern limit of the Polar Frontal Zone. I emphasize that this province comprises two distinct ecological zones which will certainly be better separated into individual entities for some purposes: the Polar Frontal Zone itself, in which algal growth is enhanced by a variety of processes, and the remainder of the Subantarctic Water Ring which lies to the north and exhibits persistent oligotrophy. Banse (1996) refers to these as the northern and southern hydrographic regimes, but I prefer the more descriptive "oligotrophic regime" and "frontal regime" and shall use those terms.

Continental Shelf Topography and Tidal and Shelf-Edge Fronts

There is no coastline in this province, though the northern branch of the Subantarctic Circumpolar Current does encounter shallow water as it passes through the Drake Passage.

Defining Characteristics of Regional Oceanography

This province is occupied by the northern branch of the ACC, whose general characteristics were noted briefly at the beginning of this chapter, and its velocity maxima lie between Subantarctic and Polar Fronts. Each of these is associated with a strong frontal jet (Veth *et al.*, 1997) which together comprise the main stream of the ACC.

Frontal Regime

Almost half of the latitudinal extent of this province is occupied by the Polar Frontal Zone, the oceanography of which has biological consequences similar to those of the Subtropical Convergence front. The frontal zone is a complex feature, bounded by two fronts (Subantarctic and Polar Fronts) and covers about 4° latitude, at about 50°S. The mean position of the Polar Front is remarkably constant in the Atlantic sector: Of 200 observations, 50% were within 100 km of the mean latitude, whereas only 10% differed by more than 200 km (Deacon, 1984). At this front, Antarctic Surface Water slips below and mixes with warmer water to the north, forming the Atlantic Intermediate Water which then proceeds northwards and down to occupy much of the interior of the ocean. Thus, the front is usually marked by a surface temperature discontinuity, but in cases in which this is not evident, the difference between stratification of the Antarctic province to the south and well-mixed water in the frontal zone to the north is usually clear (Anonymous, 1989). The strongest subsurface expression of the southernmost part of the Polar Front is usually 50 km north of the surface feature (Lutjeharms *et al.*, 1985).

The flow of the velocity maximum of the ACC is embedded in the frontal zone at an average latitude of 53°S (Carmack, 1990), where a zone of intense baroclinicity occurs at the transition between Subantarctic and Antarctic surface water characteristics. The jet current is especially well developed above the southwest Pacific midocean ridge, in the Drake passage, and south of the Agulhas Retroflection loop current. The ACC exists as a banded structure, highly variable over short time intervals and associated with strong lateral density gradients at the two major fronts (Nowlin and Klinck, 1986; Veth *et al.*, 1997).

Compared with the other circumpolar fronts, the Polar Frontal Zone is relatively undistorted by topographic influences, lying approximately halfway between Antarctica and the southern tips of the other continents (Tréguer and Jacques, 1992). To the east of South America the two fronts defining the Polar Frontal Zone separate more widely than elsewhere. Here, the Subantarctic

Front passes northwards close around the Falkland Islands, looping around the Argentine Basin to meet the southerly flow of the Brazil Current at about 40°N, whereas the Polar Front passes directly north between the Ewing Bank and the Falkland Plateau to approach the Subantarctic Front again, after which both pass eastwards across the ocean defining the frontal zone lying between them.

Oligotrophic Regime

The Subantarctic Water Ring Province equatorwards of the Polar Frontal Zone is characterized by deep (175–225 m in the north and 300 m in the south) winter mixing (April–October), a thermocline with an extremely weak thermal gradient, and a shallow (35–50 m in the north and 50 m in the south) summer thermocline (November–April). Deepest winter mixing (>500 m) occurs at about 50°S and prevents the formation of a permanent thermocline in the southern part of the zone. To the south of the Subtropical Convergence there is a meridional gradient in surface nitrate values across the Subantarctic Water Ring from 1–3 μM just south of the convergence to 10–15 μM in the Polar Frontal Zone; these values show little seasonality. No discontinuity in the surface nitrate gradient is associated with the Polar Frontal Zone at the southern boundary of the province.

Biological Response and Regional Ecology

In the oligotrophic regime of both this province and that of the ANTA province to the south of the Polar Front, the phytoplankton is dominated by very small cells despite the relative absence in these cold seas of the procaryotic picoplankters. There are still few assessments of the relative abundance of autotrophic flagellates in the Southern Ocean and there is a potential for misleading counts because it is possible that extracellular chloroplasts from damaged cells may continue to fluoresce and be mistaken for monads for some days at these very cold ambient temperatures. Nevertheless, it seems inescapable that the old "diatoms–krill–whales" model for the ecology of the Southern Ocean really carries only a small fraction of the total flux of biological material and energy. Estimates for the contribution of nanoflagellates and very small diatoms (3–20 μm) to total water column chlorophyll and primary production are on the order of 50–75 and 50–65%, respectively. Since small cells are able to utilize available irradiance more efficiently than large cells, it should be no surprise that, generally, nanoplankton (fueled by the microbial loop) contribute more in a well-mixed water column, whereas micro- or net plankton contribute more as soon as stability is introduced at periods longer than the potential time scale of bloom development (Weber and El-Sayed, 1987; Jochem et al., 1995) or if trace nutrients should cease to limit nitrogen uptake.

The set of factors that constrain primary production which we must consider for the Southern Ocean are more complex than those in many other regions. These factors must include not only irradiance and nitrate flux but also silicate flux and potential eolian or benthic sources of Fe. This is because parts of the Southern Ocean, and in particular the Subantarctic Water Ring, represent the third high-nitrate, low-chlorophyll region of the oceans, probably because of an insufficient Fe supply. Moreover, because parts of the Southern Ocean do support unusually dense diatom blooms, we must also keep in mind the relative availability of silicate and nitrate. Finally, because at least part of the Southern Ocean lies poleward of the Antarctic Circle, and is therefore dark for part of the year, we must also pay more attention compared to lower latitudes to the possibility that light limitation may occur even at the surface.

Is there a simple explanation for the oligotrophic state of the northern regime between Subantarctic Front and the Subtropical Convergence? It would be convenient if we could ascribe oligotrophy and partial utilization of nitrate in these zones simply to micronutrient limitation by insufficient Fe, which certainly takes lower ambient levels there than in the frontal zone. However, as Banse (1996) explains with conviction, it is really more complex. It is necessary not only to account for the lack of a significant spring–summer bloom but also for the continued and unusually high (for the season) winter levels of chlorophyll, and, unfortunately, a simple Fe-limitation hypothesis cannot address both issues. Because the cell division rate of the predominantly small cells is relatively high, perhaps the effect of Fe limitation is to prevent the development of diatom blooms and the population of small cells is regulated by the rapidly responding protistan grazing community. The *in vitro* Fe-enhancement experiments appear to support such an interpretation, and this seems to be another "grazer-controlled population in an iron-limited ecosystem." All this is consistent with the conclusions from incubation experiments (see de Baar, 1994, for an overview) in which Fe addition always enhanced algal growth; but since the controls consistently outgrew chlorophyll levels in the sea it was inferred that bottle incubation excluded some factor that limits algal growth in the sea.

The Polar Frontal Zone lies above a circumpolar band of silica-rich deposits and it is, like the Subtropical Convergence, the site of enhanced biological production. Lateral density and nutrient gradients are strong (the strongest gradient in surface silicate in the Southern Ocean occurs here), mesoscale eddies are numerous and active, and subpycnocline ammonium levels are high. Sediment traps and estimates of primary production suggest rates of $<80 \text{ g C m}^{-2} \text{ year}^{-1}$, or about twice as high as the Antarctic average and an order of magnitude higher than that in the oligotrophic zone of the Subantarctic Water Ring (Wefer and Fischer, 1991; Tréguer and Jacques, 1992). Here, microphytoplankton ($>20 \ \mu m$) are dominant.

The key to this puzzle was provided by the JGOFS section at 6°W which showed that the core of the flow in the Polar Frontal Zone at this longitude carried Fe from shelf sediments acquired by the zonal current to the west: The core of Fe-replete water at 60–100 m had values of 2.0–4.0 nM Fe prior to the spring bloom compared with subnanomolar concentrations (>0.17 nM Fe) in the Antarctic water to the south (de Baar *et al.*, 1995). However, these high values in the front in the Atlantic sector cannot be extrapolated to the whole of the annular current because comparable inputs are unlikely to occur until a nearly complete circuit of the Southern Ocean has been made. Confirming this, a downstream Fe decline of almost 1.0 nM has been observed between 23 and 15°W in the Polar Frontal Zone. Be this as it may, in the Atlantic sector the high values of Fe in the core of the frontal current are spatially associated with a subsurface spring bloom that increased from >0.8 to $<2.4 \text{ mg chl m}^{-3}$ over about 30 days, with rates of primary production in the range of 1 or 2 g C m^{-2} day^{-1}. Despite these indications of an important spring bloom in the Polar

Frontal Zone, this is not a prominent feature in the CZCS chlorophyll field. Although some frontal chlorophyll enhancement can be traced in the seasonal images, especially in the Atlantic sector, it is a very weak feature (<0.3 mg chl m^{-3} g chl m^{-3} at the surface) compared with the continental ice-edge blooms (Lutjeharms *et al.*, 1985; Comiso *et al.*, 1993) and those that develop in spring and summer along the coasts of the Antarctic Peninsula and around the island arcs north to South Georgia and which are the dominant chlorophyll features in the Southern Ocean.

These findings have been expanded by the 1992 JGOFS austral spring investigations aboard *Polarstern* (Smetacek *et al.*, 1997) which encountered persistent diatom blooms—identifiably dominated by *Fragilariopsis kerguelensis*, *Corethron inerme*, and *C. criophilum* with standings stocks of 177, 223, and 277 mg chl m^{-2}, respectively, which is an order of magnitude higher than those in the oligotrophic regions. These blooms required ambient Fe levels of around 1 nM that were found to be maintained by basinwide upwelling in Polar Front water, but (at primary production rates < 3 gC m^{-2} day^{-1}) were unable to take macronutrients to depletion.

The Polar Frontal Zone (corresponding with the less precisely defined Subantarctic Convergence of earlier writers) is thus a major discontinuity in the Southern Ocean, well expressed in an evocative description (Deacon, 1984) of what it feels like to cross the convergence on returning from the polar regions, because the air smells again as air should smell but does not in the high Antarctic. It is also a significant border in species distributions and a major feeding zone of seabirds. For example, *Euphausia vallentini* and *E. longirostris* occur strictly to the north and *E. frigida* occurs strictly to the south of this zone, and for recent investigations of copepod distributions in relation to Antarctic fronts, see Errhif *et al.* (1997).

It is, of course, in this province that we first meet the dominant Antarctic herbivores, comprising copepods (*Calanoides acutus*, *Calanus propinquus*, *Metridia gerlachi*, and *Rhincalaus gigas*), the chaetognaths *Sagitta gazellae* and *Eukrohnia hamata*, and swarming tunicates (*Salpa thompsoni*). Though implicated in regulating diatom blooms (and feeding whales and seabirds), the crustacea of this assemblage are also facultative predators and (in the case of some species) able to feed on detritus and algae from the undersurface of ice floes. Given our modern knowledge of the relative abundance of nanoplankton in the Southern Ocean, we should not be surprised that salps which use extremely fine-mesh filters to obtain algal cells should be a prominent feature of the zooplankton here, reaching 7 or 8% of total mesozooplankton numbers in the Indian Ocean sector (Longhurst, 1985a), though recent investigation (see Antarctic Province) have suggested their greater importance in the oligotrophic regions to the south of the Polar Front.

The 1992 JGOFS investigations suggested that despite their abundance, herbivorous copepods (mostly *C. acutus* and *C. propinquus*) consumed only 1% of the daily primary production of the Polar Front diatom blooms (Dubischar and Bathman, 1997). A high export rate of sinking organic material and biogenic silica is indicated.

The seasonal vertical migrations and the ontogenetic migrations of copepods, which carry them alternately between near-surface feeding and deep over-

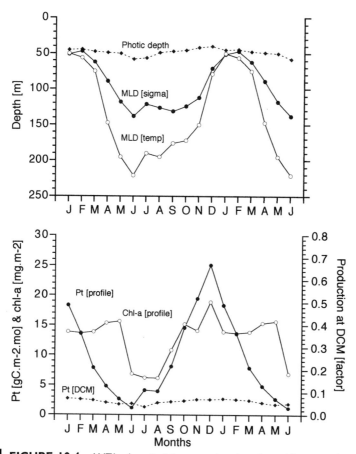

FIGURE 10.4 ANTA: characteristic seasonal cycles of monthly averaged mixed layer and photic depths, chlorophyll at the surface, and rate of primary production, both depth-integrated and at the DCM. Data sources are discussed in Chapter 1.

wintering, have been studied in detail. Two related species of calanoids have fundamentally different strategies: I deal with them together for comparison, though it must be noted that these species have somewhat different zonal distributions. *Calanus propinquus* is the large copepod most typical of the SANT province, whereas *C. acutus* is the more polar species, at home also in both ANTA and APLR provinces farther to the south (Mackintosh, 1937).

In line with this observation, and perhaps related to the distribution of the genus which otherwise occurs in low-latitude upwellings, note that *C. propinquus* departs from the normal high-latitude norm and stores its lipid reserves as triglycerides rather than as wax esters as do *C. acutus* and most other high-latitude calanoids. Both species migrate deep in autumn, but with differences appropriate to their zone. *Calanoides acutus* descends to great depths (500–1000 m) to overwinter in diapause to conserve lipid reserves required for reproduction at the surface in spring. *Calanus propinquus* goes less deep (100–500 m) and does not enter physiological diapause; some feeding, often by predation on microzooplankton, may continue and reproduction occurs over a longer period than that for *C. acutus*. Both species are concentrated near the surface in summer (usually at 0–50 m), and *C. acutus* both ascends and descends earlier than *C. propinquus*. The ascent of *C. acutus* females is timed to coincide with the onset of sea-ice melting or the very earliest austral spring bloom (October) in the permanently open water, and eggs are matured and shed quite soon. In the austral autumn (March), *C. acutus* descends (as copepodite 4's or 5's) while there is still sufficient phytoplankton in the surface waters for *C. propinquus* to continue feeding (Smith and Schnack-Schiel, 1990; Schnack-Schiel *et al.*, 1991).

In contrast to the two calanoids, *R. gigas* is equally at home in both SANT and ANTA/APLR provinces and may produce two generations per year (in spring and autumn). It also lies somewhat deeper in summer (50–100 m) than *C. acutus* and *C. propinquus,* and in winter it descends to 500–1000 m. Several species perform regular diel migrations in this water body between the near-surface chlorophyll layer at night and the warmer subsurface layer by day at >200 m. These are familiar genera: *Pleuromamma robusta* and *Euphausia* (*E. triacantha, E. vallentini,* and *E. frigida*).

Though I have introduced some of the principal pelagic biota of the Southern Ocean, I emphasize that the generality is as Hempel (1985) describes the plankton ecosystem of the ACC: a rather typical open-ocean oligotrophic plankton, limited by low levels of primary production at all seasons. The eutrophic situation at the Subantarctic and Polar Fronts is a special case.

Synopsis

Primary production rate progressively increases during winter mixing through early summer (November), though chlorophyll accumulation does not match this pattern, with greatest accumulation during early austral fall, perhaps when ontogenetic migrants descend to overwintering depths and production at the DCM is relatively low (Fig. 10.3). This province provides probably a grazer-controlled population in an Fe-limited system.

ANTARCTIC POLAR BIOME

Antarctic Province (ANTA)

Extent of the Province

This is a simple annular province lying between the Polar Front at about 50°S and the Antarctic Divergence at about 60–65°S. It is synonymous with the southern branch of the ACC.

Continental Shelf Topography and Tidal and Shelf-Edge Fronts

There is no coastline in this province, but the island arcs between the Antarctic Peninsula and South America (South Orkneys, South Sandwich, and South Georgia) intruding into this province should be noted. I propose that they should be included in the APLR province because although they do not lie within the East Wind Drift of that province, their ecology is (like the rest of APLR) dominated by coastal and ice-edge effects, and thus differentiated from the open-ocean remainder of ANTA, which is inhabited only by smaller oceanic islands lying farther to the east (Bouvet, Kerguelen, and Heard of the Atlantic and Indian Ocean sectors).

Defining Characteristics of Regional Oceanography

The Antarctic Province has two components which could perhaps be treated as distinct provinces but are retained together here: a zone of permanently open water and a zone carrying seasonal pack ice. I prefer to place boundaries in relation to fronts rather than at the edge of the variable sea ice as was done by Tréguer and Jacques (1992), who partitioned both this and APLR East Wind Drift together into a seasonal ice zone and a permanently open-ocean zone. These two entities would perfectly match the two polar biome provinces if ice cover was perfectly constrained poleward of the Antarctic Divergence, which is only approximately the case. Seasonal ice reaches its greatest extent in August–October extending to about 58°S and farthest from the continent in the central Atlantic and Pacific sectors. I assume that the influence of lenses of meltwater extends throughout the ice-free zone in summer so that this province matches my general definition of the polar ecology biome (see Chapter 4).

Above the Antarctic Divergence, the oceanic westerlies and the coastal easterlies meet and there is a line of zero wind-stress curl (Deacon, 1984). Ekman transport diverges on either side of the front and density surfaces slope downwards both to the north and south; a convenient way of locating the axis of upwelling is a 200-m salinity maximum. The upwelled water has a component of North Atlantic Deep Water, which is Fe-replete, and this fact has significance for the ecology of the province. The winds over the divergence are variable, as is upwelling, but progressively to the north across the province the westerly winds become more persistent. Recall that since wind stress introduces kinetic energy to the Ekman layer at low latitudes, but potential energy at high latitudes, it will be no surprise to find that the field of mesoscale eddies in the ACC is unusually complex and active.

Biological Response and Regional Ecology

A characteristic of this province that I must note at the outset is that there are only slight gradients in surface nitrate. During the open-water season, surface values of about 10–15 μM occur wherever data are available (Foster, 1984), though nitrates tend to be lower in the Indian than in the Atlantic and Pacific Ocean sectors. South of a discontinuity at the Polar Frontal Zone, silicate values rise progressively southwards across this province to the Antarctic Divergence, beyond which there are uniformly high mixed layer values of about 50 g liter^{-1} down to the Antarctic continent. Everywhere except at the Antarctic Divergence, along the poleward margin of the province, the near-surface concentrations of Fe are very low, usually on the order of 0.1 nM. The linear upwelling along the divergence brings North Atlantic Deep Water to the surface which, prior to a spring bloom, contains Fe on the order of 0.75–1.0 nM, which (you will recall) is lower than that in the Polar Frontal Zone, for which a benthic source of Fe is suggested, but somewhat higher than that in both the northern, oligotrophic regime of the Subantarctic Water Ring

Province and the oligotrophic open-water regime of the current province. Zonal variability in available Fe may result in the zonal differences in primary productivity that we observe.

The pelagic ecology of the permanently open-water zone of the Antarctic Province is, in its essential respects, similar to the oligotrophic communities of summer conditions in warm, stratified oceans and characteristic of a retention community in the sense of Peinert *et al.* (1989). It is dominated by small cells with an appropriate population of heterotrophic protists and bacteria. Low chlorophyll values during all seasons result in water of exceptionally high clarity and Secchi disc depths of >60 m have been recorded. Only 5% of chlorophyll values from ship data in this zone exceed 1.0 mg m^{-3}, and most are on the order of 0.25 mg m^{-3} (Trégeur and Jacques, 1992). There is some evidence that the higher values occur during summer, peaking in March, but there is none for the existence of a spring bloom comparable to that of the North Atlantic or of the zone of seasonal ice cover to the south. The sedimentation rate from the summer algal growth is at least two orders of magnitude lower than that which occurs in the marginal ice zone (Smetacek *et al.*, 1990). The JGOFS 1992 investigations showed that up to 90% of total chlorophyll in this province resides in pico- and nanophytoplankton, with very small contributions from diatoms, dinoflagellates, and prymnesiophytes of chlorophytes (Peeken, 1997). Nanoprotozoan biomass (especially heterotrophic dinoflagellates) was equivalent to about 70% of their total potential food—that is, of bacterial and nanophytoplankton cells (Becquevort, 1997). Nitrogen metabolism is based largely on ammonium, and there is heavy grazing pressure by small copepods (Smetacek *et al.*, 1990), though this is likely to remove no more than about 30% of the daily primary production; *Salpa thompsoni,* on the other hand, is present in sufficient numbers to remove up to 100% of the daily production (Dubischar and Bathman, 1997). Under the conditions for investigations at sea in the Southern Ocean, one should not expect to be able to balance global production/consumption budgets with great accuracy; undoubtedly, however, we may assume very rapid removal of autotrophic cells in the oligotrophic regime.

These observations from ship data are confirmed by the CZCS image analysis of Comiso *et al.* (1993). It is only in the Scotia Sea—eastwards from the Drake Passage and within the island arc—that a strong spring bloom occurs in this province, and this can be attributed to the existence of marginal ice-zone conditions rather than to open-water processes (Trégeur and Jacques, 1992; Comiso *et al.*, 1993).

The zone of seasonal pack ice (Tréguer and Jacques, 1992) occupies approximately the southern half of this province, except in the western Pacific/Drake Passage sector where the permanently open Antarctic surface water is very narrow and seasonal pack ice approaches the Polar Frontal Zone. The seasonally migrating edge of the ice cover, especially when it is retreating during the summer, forms a local linear zone of near-surface stability by the release of low-salinity meltwater. However, we should note the recent and interesting observations made by *Polarstern* scientists in 1992: The spring breakup and retreat of the oceanic marginal ice zone (MIZ) in this province—essentially in the southern branch of the ACC—does not result in strong diatom blooms presumably because of constraints by low-ambient Fe levels.

Associated with the MIZ at the Antarctic Divergence there is local chlorophyll enhancement with values reaching about 4 mg chl m^{-3} at the subsurface chlorophyll maximum (Smith and Nelson, 1985). Sediment trap data indicate that the very sharp spring bloom is dominated by diatoms in December and January and is followed by a late-summer season of regenerative algal growth of small cells from which sedimentation rates are two orders of magnitude lower (Smetacek et al., 1990).

Though krill (*Euphausia superba*) occur everywhere in this province in small numbers, their highest concentrations are clearly rooted in the coastal enrichment at the tip of the Antarctic Peninsula and the islands to the north. Because they are the defining organism of the APLR province to the south, I shall discuss their ecology when reviewing that province. They do not extend equatorwards of the Polar Front and are more abundant (by an order of magnitude) eastwards of a line joining the tip of the peninsula to South Georgia, passing by way of the island arcs. This zone of high abundance passes eastwards over deep water to about 30°W (thus covering about 20% of this zonal province). Spotty patches of lower abundance may be encountered around the remainder of the southern branch of the ACC (Mackintosh, 1937). It is the presence of krill in the Southern Ocean ecosystem that most distinguishes it from that of the Arctic where, at high latitudes, euphausiids or their equivalent are insignificant or absent. The other major herbivore of Antarctic seas is *S. thompsoni;* during years of salp abundance, this species may be capable of consuming the entire daily primary production and so preventing the accumulation of phytoplankton biomass. This supposition is consistent with the rather trivial role played by herbivorous copepods, though capable of consuming only about 1% of the daily production rate (Smetacek et al., 1997).

Synopsis

There is moderate winter mixing to about 200 m, with a rapid near-surface stabilization in September and October not captured by archived data (Fig. 10.4). Photic depth does not normally illuminate pycnocline. Primary production rate conforms to irradiance cycle and is tracked by chlorophyll except late in summer when secondary accumulation occurs, although herbivores, especially krill, do not consistently descend at this season as in the boreal polar provinces.

Austral Polar Province (APLR)

Extent of the Province

This is the "real Antarctic" of oceanographers, whalers, nineteenth-century navigators, and twentieth-century tourists. It comprises the seasonally ice-covered sea from the coasts of Antarctica (including the peninsula) north to the Antarctic Divergence at about 65°S, and consequently much of this area lies polewards of the Antarctic Circle and has a period of winter darkness. Because of their ecological similarity to the rest of APLR, though their conditions are less extreme, I include in this province the arcs of islands between the Antarctic Peninsula and South America (South Orkneys, South Sandwich, and South Georgia).

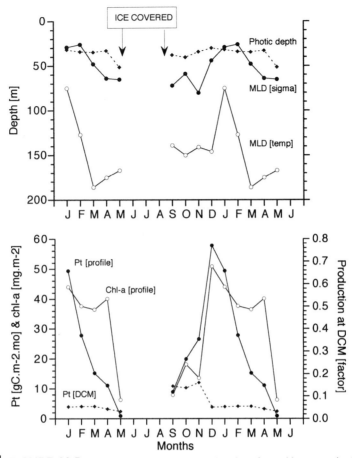

FIGURE 10.5 APLR: characteristic seasonal cycles of monthly averaged mixed layer and photic depths, chlorophyll at the surface, and rate of primary production, both depth-integrated and at the DCM. Data sources are discussed in Chapter 1.

Continental Shelf Topography and Tidal and Shelf-Edge Fronts

The Antarctic shallow shelf is represented by only a few small patches where water depth is <200 m at about 30°E: Elsewhere, water depth of 200–500 m is found within a few miles of the coastline. Like Labrador, this deep continental platform represents a submerged continental shelf, its submergence caused by the pressure of the 4000-m ice cap resting on the Antarctic continent. A subsurface thermal front of about 2°C at 200 m (the Continental Water Boundary) occurs close to the continent: This is not simply a typical shelf-edge front but rather a convergence of significance in the global thermohaline circulation that I shall discuss briefly below.

The unique characteristic of the coast of Antarctica must be, of course, that each of the major shallow embayments on the coast except the rather open Bellingshausen Sea is partially covered by a permanent ice shelf or ice barrier that is anchored at the coast and extends several hundreds of kilometers seawards across shallow water. The Ross Sea ice shelf extends about 800 km seawards from the termination of the Axel Heidberg glacier, whereas in the Weddell Sea the edge of the Ronne ice shelf lies 300 km offshore. This shelf is currently reported to have developed major cracks and is perhaps in the initial stages of breaking up.

Defining Characteristics of Regional Oceanography

This is a region of westward-moving Antarctic surface water: The East Wind Drift, which is up to 300 km wide, is ice covered and dark in winter but has some open-water areas in summer between the Antarctic Divergence and the Antarctic continent. The area covered by ice is a direct function of sun angle, and a simple sine wave represents the sea-ice coverage from a minimum (5×10^6 km^{-2}) in February to a maximum (19 or 20×10^6 km^{-2}) in August. In austral summer, significant sea ice is restricted to the Weddell Sea and the sector between the Ross Sea and the Antarctic Peninsula, though even in these sectors there is much open water. Elsewhere, sea ice is restricted to a narrow strip along the shores of Antarctica.

The Antarctic Divergence (see Antarctic Province) lies at about 65°S and marks the transition between the oceanic west winds and the east winds nearer the Antarctic continent. The mean position of the Antarctic Divergence is rather variable, responding to weather patterns (Deacon, 1984). South of the divergence the water above 500 m is cold (<2°C in summer and colder in winter) and thermally unstratified except in ice-free areas where slight surface warming may occur, resulting in a mixed layer depth of about 25 m. Winter mixing may extend to 200–400 m, though winter data are very sparse. Flow in the East Wind Drift is meandering, with eddies being generated by the topography of the coast and effects of high latitudes on the strength of the Coriolis force and hence on the wind-generated Ekman spiral.

The Coastal Convergence Front marks the boundary between very cold and very saline ice-shelf water and the warmer circumpolar deep water upwelled at the Antarctic Divergence which otherwise occupies the surface of this province. At this convergence, the very dense (cold and salty) ice-shelf water is transported downwards into the interior of the ocean. Recall that it is only in regions where surface evaporation from cold dry winds occurs simultaneously with ice formation and brine rejection that sufficiently dense water can be evolved to sink and form the bottom water mass of the oceans. The Antarctic bottom water so formed along the coast of Antarctica is one of the few roots of the global thermohaline circulation of the oceans (Lutjeharms, 1985). The only other sources of true oceanic bottom water are in the Labrador and East Greenland Seas.

There are two major gyres in the westwards flow around Antarctica, occupying the Weddell and Ross Seas, and each is confluent with the flow of the southern branch of the ACC in the ANTA province to the north. The Weddell Sea gyre comprises very cold coastal water (−1.8°C in the center of the gyre in summer), occupies the area enclosed by the Antarctic Peninsula, and extends eastwards to about 10–15°W before rejoining the westwards flow along the coast into the southern Weddell Sea. The flow in this gyre eastwards from the tip of the peninsula is often called the Weddell Current and forms a confluent front (at the Antarctic Divergence) with the ACC. The smaller Ross Sea gyre is generated by the coastal prominences on either side of this embayment.

Because so much of the research performed in Antarctic seas has been centered on a few areas, principally the major embayments, it is useful to discuss them in more detail.

The Larsen ice shelf on the east coast of the Antarctic Peninsula and the Filchner–Rønne ice shelf at the head of the bay occupy a significant part of the surface of the Weddell Sea. The coastal current of the East Wind Drift enters the Weddell Sea at 20°W and contributes to its cyclonic circulation (clockwise in the Southern Hemisphere), whereas near Halley Bay (25°E on the southeast coastline) the incoming coastal current passes through a divergence zone which somewhat separates the ecology of the sea into a northeastern and southern component. The cold shelf water (-1.8°C) is separated from warmer (0°C) Weddell gyre water along the regional manifestation of the coastal water boundary both in the southeast and along the coast of the peninsula. The gyral circulation passes toward the east at the tip of the Antarctic Peninsula and can be traced to about 15°W where it returns south toward the coast. Weddell Sea gyre water has anomalously high silicate (70–80 μg liter^{-1} near the peninsula) and nitrate (25–30 μM) compared with the water beyond the gyre.

Although the Weddell Sea is in the zone usually mapped as "permanently ice covered," much open water does occur seasonally within the confines of its gyral circulation. Along the southeast coast a persistent polynya is maintained by offshore winds; this polynya may be 800 km long even in winter, and in summer it expands into a mosaic of shoreleads and patches of open water that persist for several weeks. Persistent open water also occurs in the much-studied Bransfield Sound and in the passages between the islands of the South Orkneys.

The Ross Sea gyre is small and narrower than the Weddell gyre but has many of the same properties; a wide area of sea ice continues to invest the eastern coast and partly encloses the Ross Sea so that even in austral midsummer the sea is open to the ocean only toward the northwest, off Cape Adare in Victoria Land. The Bellingshausen Sea is hard to define, lying in the very open embayment west of the Antarctic Peninsula; most of the embayment is ice covered even in summer and the southern edge of eastwards flow in the ACC is unusually far south and lies quite close to the ice edge in summer (Read *et al.*, 1995).

Biological Response and Regional ecology

Analysis of CZCS pigment fields in this province is relatively satisfactory despite extensive cloud and ice cover because air clarity is a feature of the Antarctic (Comiso *et al.*, 1993). Seasonal distribution of chlorophyll obtained in this way shows a very strong pigment enhancement between the coast of Antarctica and the Antarctic Divergence, principally along receding ice fronts, and also around the Antarctic Peninsula and the islands to the north. Blooms are initiated close to the coast in the Ross Sea polynyas; these bloom subsequently propagate seawards along with a massive effluent of meltwater in the surface layer. However, all such data are incomplete in one important respect: Because seasonal ice is not very thick in these seas, we must not only measure phytoplankton and contend with ice algae growing on the underside of the pack ice but also quantify the algal "superblooms" that occur below continuous ice cover, at least in the coastal regions of the Weddell Sea. The fact that the MIZ of this province supports much higher concentrations and productivity of diatoms than does the MIZ of the open ocean recalls the very early suggestions of Hart (1934, 1942) that it is the continental source of Fe, lacking beyond the Antarctic Divergence, that is the key to the productivity of this province. This is a suggestion that was all but lost in the recent debate about the role of Fe in oceanic productivity, and I am indebted to Smetacek *et al.* (1997) for the reminder.

Although most of the investigations of the algal production in this province emphasize the critical role of "net" diatoms, from 20 to 200 μm in size or even larger, these do not reveal the whole picture. Recent studies (see, for example, those listed by Weber and El Sayed, 1987; El Sayed and Taguchi, 1981) suggest that small phytoflagellates and 5- to 10-μm diatoms (e.g., *Nitzschia curta* and *Chaetoceros neglectus*) are here, as in the ANTA and SANT provinces to the north, dominant for much of the year and over much of the open-water area of the province. Typical values for <20-μm cells are in the range of 50–90% of total water column chlorophyll.

In this province, a grid of stations around Elephant Island (61°S) show that <20-μm cells consistently comprise 55–100% of total chlorophyll. In spring, prokaryote contribution is negligible, but in summer these cells represent about 5–10%, though at anomalous stations with very low chlorophyll values they may contribute <50% of total chlorophyll. In many areas, these small cells are probably comprised largely by motile cells of *Phaeocystis* (<3 × 10⁶ cells liter⁻¹) released from macroscopic colonies of this prymnesiophyte. Several other investigations close by in the Weddell Sea have shown the existence of a rich protist fauna of heterotrophic flagellates, choanoflagellates, dinoflagellates, and ciliates whose abundance is correlated with autotrophic cells and is direct evidence of a well-developed microbial loop. Everywhere, we can expect this association to have greater biomass and flux near the ice edge and progressively less biomass out into the open water.

Superblooms were investigated by Smetacek *et al.* (1992) in very early spring (October and November) along the eastern coast of the Weddell Sea in a feature which left a brown wake behind the icebreaker *Polarstern* as it passed through the continuous pack ice. This continuous ice cover was only 0.2 m thick but was underlain by a low-salinity layer of nitrate-depleted water from one to several meters deep. This layer was rich in unconsolidated ice platelets (1–3 mm × 2⁻¹⁵ cm) and irradiance was reduced by as little as 75% by the 20 cm of superjacent ice. Within the stratification afforded by the platelets there was a miniature chlorophyll profile having maxima in the range 7–36 mg chl m⁻³ and with POC maxima of 250–1250 mg chl m⁻³. Chlorophyll concentrations were highest below freshly frozen leads. The bloom was composed largely of centric diatoms, and the only herbivore of consequence was a large heterotrophic dinoflagellate. As nitrate became exhausted, the nutrient-depleted diatoms sank, despite the platelets, to form a miniature DCM at the bottom of the platelet layer and hence at a nitracline. Such bloom conditions, underneath continuous ice cover, extended over an area of 20,000 km² and contrasted with the oligotrophic conditions then obtaining throughout all the open waters to the north. Other investigations of the below-the-ice autrotrophs have revealed abundant pico- and nanophytoplankton along with the overwintering very early larval stages of copepods.

These under-ice cell concentrations are the resource for a variety of organisms, some having life cycles closely adapted to these singular conditions (the small copepod *Stephos longipes* being one such organism). This species manages the alternation between ice-covered and open-water conditions as follows: There are two concurrent generations of which the younger overwinters within and below the mass of unconsolidated platelets below the fast ice as nauplii and early copepodites while the older generation lies in the water column just below the ice as C4's which become adults in very early spring as the ice starts to break up. The summer population, in the upper 50 m of ice-free water, comprises mostly C1–C3's. In autumn, the adults from this generation enter the freezing ice to generate eggs and nauplii, whereas the slightly younger copepodites remain in the water column to overwinter as C4's, closing the cycle (Schnack-Schiel *et al.*, 1995). Using such an apparently unlikely stratagem, the equally improbable ice-platelet bloom is utilized.

Open-water blooms occur most often near receding ice fronts because the fresh water released by melting sea ice induces stability in the water column which may extend many tens of kilometers beyond the ice edge; such MIZ conditions are a common feature in the passages and sounds between the Antarctic islands. However, not all surface-stabilized layers induce a bloom, and not all blooms occur in places with density stratification at an appropriate depth in relation to the photic zone. A section in open water beyond the receding ice of the Weddell Sea from Graham Land (67°S) to the Filchner ice shelf (77°S) in February and March reported by El-Sayed and Taguchi (1981) illustrates the typical development of a spring bloom in open water in this province. A pycnocline at 50 m was progressively sharper toward the south, where wind mixing was weaker among scattered ice; south of 71°, at the transition between deep and shelf water (recall that the shelf is 600 m deep around Antarctica), sufficient stability had been induced by irradiance as to cause a bloom having a DCM at about 15–25 m and its maximum production rate a little shallower. Here, mixed layer chlorophyll reached 3.0 mg chl m^{-2} and primary production about 0.4 g C m^{-2} day^{-1}, with some nitrate depletion. Similar MIZ sections elsewhere, as in the Scotia Sea and western Ross Sea, confirm that this is a characteristic evolution and that chlorophyll concentrations in the range 2–8 mg chl m^{-3} are typical, with a shallow DCM at about 25 m. Size-fractionated studies during ice melt in the Lazarev Sea showed a seasonal shift from nano- and picoplankton toward microphytoplankton as the bloom is induced and integrated primary production is doubled; concurrently, partitioning of grazing pressure between protozoans and mesozooplankton shifted accordingly (Froneman *et al.*, 1997).

CZCS imagery shows that such blooms are very extensive and confirms the supposition that it is the Ross Sea that is the site of the greatest and most consistent phytoplankton biomass in the Southern Ocean, marked by deep deposits of siliceous sediments. Coastal blooms along Victoria Land are consistently diatom dominated, whereas offshore blooms are more likely to be dominated by *Phaeocystis antarctica* (DiTullio and Smith, 1996); an event in the coastal polynya of the western Ross Sea contained intense blooms (up to 10–40 mg chl m^{-3} over an area of 106,000 km^2) and occurred before the polynya opened up. Later, about 10 mg chl m^{-3} was seen over an area of 126,000 km^2 (Arrigo and McClain, 1994). Diatom blooms of this kind may be terminated rapidly by gales which destroy the stability on which they depend. These images obtained in December and January, later in the season, show the very great extent of the *Phaeocystis* blooms that occur here and are advected into McMurdo Sound after the spring bloom of centric diatoms is over. This appears to be a characteristic sequence in many regions in this province.

An apparently paradoxical situation was found in the Bellingshausen Sea during the JGOFS expedition of 1992 (see the special 1995 issue of *Deep-Sea Research II* **42**, 4–5). A strong zonal bloom was encountered along a section in early austral summer (November and December) from open water into the ice which did not appear to require density stratification of the water column; further, where low-salinity surface did induce stratification close to the ice, there was no bloom, even though conditions otherwise appeared to be suitable. A chlorophyll bloom (3.0–3.5 mg chl m^{-3} at 10–70 m) occurred north of 68°S; from about 68 to 69.5°S there was density stratification with a 75-m pycnocline (but very low chlorophyll), and beyond 69.5°S there was continuous ice cover and a mixed water column. This was interpreted as a frontal bloom associated with a southern branch of the Polar Front which here approaches the ice edge very closely prior to all the frontal systems bunching as they pass through the Drake Passage.

The dynamics of the consumption of Antarctic blooms has become clearer from research in recent years. Burkill *et al.* (1992) showed from the UK-JGOFS Southern Ocean studies that even here in the APLR province we cannot ignore the protist microplankton, as has been done in the past. Along the Bellingshausen Sea section mentioned previously, protistan microplankton were sparse below continuous ice cover but increased in abundance seawards in the open water, with abundance being positively correlated with chlorophyll biomass. Where maximum numbers occurred (54 µg C or 17,000 organisms liter^{-1}) the microplankton was dominated by naked oligotrich ciliates (*Strombidium*, *Tontonia*, and *Lohmaniella*) and heterotrophic dinoflagellates (*Gyrodinium*, *Cochlodinium*, *Torodinium*, and *Pronoctiluca*) and was equivalent to about 25% of the phytoplankton biomass. The feeding activity of these protists required from 3 to 40% of the phytoplankton biomass daily, which was between 21 and 270% of the concurrent daily production. Other studies suggest that this investigation of the protist population revealed a situation typical in the Southern Ocean as elsewhere (see Garrison, 1991).

The role of mesozooplankton in the consumption of phytoplankton in this province is not clear. Some studies suggest that only an insignificant part of the daily production during diatom blooms is consumed by meso- and macrozooplankton (e.g., 0.6% daily in the Drake Passage and 8% in the Bellingshausen Sea) but in other situations, as in the Ross Sea, it is suggested that grazing is a major factor in constraining the accumulation of cells. In the oligotrophic phase they take relatively more. For example, in the Drake Passage, where phytoplankton biomass is low (1 or 2 g C m^{-2}) and copepods are numerous (about

0.8 g C m^{-2}), the latter consume 50% of daily primary production (Schnack *et al.,* 1985). Whatever their relative importance, the mesozooplankters involved in the reduction of diatoms blooms are dominated by *C. propinquus,* *C. acutus,* and *R. gigas,* and *M. gerlachi* is the principal diel migrant. These copepods rise to the surface from overwintering depths to encounter the spring bloom, with *C. acutus* appearing a little later in the season than the others. This species has a life cycle resembling that of the arctic calanoids previously discussed (see Chapter 7, Borea Polar Province) though the life cycle appears to be completed within 1 year in most of this province; however, in those areas where the summer season of open water is shortest, as in the eastern Weddell Sea, a small proportion of the population takes 2 years to reach maturity (Atkinson *et al.,* 1997). Though these are the dominant species, it must be remembered that even here, in such high latitudes, lists of dominants conceal the real complexity of the mesoplankton that consume both phytoplankton and protists: In the western Weddell Sea in March, at the end of summer, a total of 113 mesoplankton species, of which 31 species were encountered in the upper 50 m, were listed by Hopkins and Torres (1988).

It has been recently suggested that the massive *Phaeocystis* blooms that are typical of some parts of this province are susceptible to grazing by copepods, despite previous opinions that neither protists nor macrozooplankton were capable of consuming this organism except for the consumption of the motile cell stage by some protists. However, sedimentary material suggests that most losses from the colonial phase comprise unconsumed sinking aggregates (see DiTullio and Smith, 1996).

This province is, of course, also the kingdom of the krill, though this organism is not uniformly abundant; the Ross Sea, for example, despite the very high concentrations of diatoms noted previously, has rather few krill. Despite exaggerated claims concerning their fishery potential, krill do represent an enormous resource and aggregate in sufficient concentrations as to inflict "grazing events" on diatom blooms, leaving behind water almost completely void of cells. *Euphausia superba* has a life strategy more like a fish than a planktonic crustacean, being long lived (5–8 years), during which period it reproduces two or three times, usually late in the summer. Krill occur in small to very large swarms that resemble those of clupeid fish, though fish shoals are more uniform in terms of size of individuals, swimming speeds, and maturity stage: Krill swarms are not sorted according to such criteria. Such swarms may themselves aggregate into "superswarms" comprising more than a million tons of biomass spread over several hundred kilometers^{-2}, though these swarms occurs mainly in the APLR province in the vicinity of coastal chlorophyll enrichment.

Eggs of krill sink rapidly (out of harms way from the grazing activities of their myriad parental stock) to as much as 1000 m prior to hatching. The nauplii and young larvae rise progressively toward the surface so that *E. superba* is typically much more abundant over deep than shallow water. Close to the coast, and especially in shallow water as in the inner Weddell Sea, *E. superba* is replaced largely by *Euphausia cristallophoria,* whose life cycle is adapted to this environment.

During normal swimming, krill are oriented with their dorsal side uppermost, but when food is encountered and filter-feeding begins, individuals may be oriented in any direction. Feeding is most active in darkness and swarms undergo diel vertical migration between the surface and about 200 m. One reason for the success of these organisms is their ability to feed on a very wide range of food particles from nanoflagellates to crustacean nauplii. Since krill have such a patchy distribution, attempts to balance their consumption against production in the blooms, or their browsing on under-ice algae, have been quite unsatisfactory on a regional basis. Nevertheless, it is calculated that up to 60–80% of daily production may be consumed by krill in regions in which they are generally abundant. The consumption of diatom blooms by krill can be very spectacular: A summer diatom bloom in the Weddell Sea was grazed down in 20 hr from 4.0 to 0.5°1.0 mg chl m^{-3}, this remnant apparently representing the unconsumed nanophytoplankton component of the original bloom.

The Antarctic krill is the keystone organism for the biomass builtup in this province by baleen whales, crab-eater seals, and many seabirds which directly forage for it by a variety of feeding techniques. One has the impression that if *E. superba* did not exist, it would have to be invented. It is hard to imagine a more efficient organism to harvest the spring and underice blooms of this province and to directly transfer the harvest to mammals and seabirds, enabling them to establish the truly astonishing populations that existed before whaling and other forms of exploitation were begun.

Krill dominate the diets of large vertebrates in the Antarctic polar seas but compete with salps for the available phytoplankton cells on which each (in their different manners) feed. These two species, *E. superba* and *S. thompsoni*, exist in a dynamic balance, with the relative abundance of one being associated with relative scarcity of the other (Loeb *et al.*, 1997). At Elephant Island, near the tip of the Antarctic Peninsula, this balance is determined each spring by the extent of sea-ice cover during the previous winter. In standard surveys of the U.S. Antarctic Marine Living Resources Program from 1975 to 1996, krill density in summer varied from 5 to 510 individuals 1000 m^{-3} and salp density from 4 to 3489 individuals 1000 m^{-3}. Large salp populations occur after winters with small ice coverage followed by an early open-sea spring bloom which then permits an early and rapid increase of the overwintering salp population. Krill, as previously noted, are long-lived and have population dynamics resembling fish so that their stock size is determined by reproductive success the previous year. Heavy recruitment of first-year krill to the population occurs 1 year after a heavy ice winter which promotes early spawning by the adult population and, incidentally, inhibits a spring outburst of salps. Two consecutive heavy-ice years enhance the effect. In the seas around Antarctica, the consumption of phytoplankton production by both salps and krill is reported to be significantly lower than that in the ANTA province to the north; salps in APLR are reported to consume only 19% of daily primary production in years of high abundance (e.g., 1994) and <1% in poor salp years (e.g., 1995), but there is a correlation between computed salp/krill consumption and between-year variability in satellite-observed sea-surface chlorophyll in summer. The balance between krill and salps is not just a result of competition for food because it has recently been shown that salps may be a preferred food of adult krill, which may consume them at a rate of more than 0.5 individuals per day and in laboratory tests showed a preference for salp extract over material extracted from phytoplankton, krill, or polychaetes (Kawaguchi and Takahashi, 1996).

We would not expect the dynamic balance between krill and salp populations to be stable, and such is not the case. During the period of routine observations a trend to more frequent years with light ice cover and lower krill densities has been detected; this trend can be extrapolated securely back to 1947 when routine observations of relative winter ice cover were established. The inferred progressive decrease in the abundance of krill during a period of 50 years is cause for concern not only for the direct fishery for *E. superba* of about 1.5×10^6 tons year^{-1}, managed by the Commission for the Conservation of Antarctic Marine Living Resources, but also for the marine mammals and birds dependent on krill for their livelihood. In fact, very recent evidence suggests that this trend can be extrapolated back to the early 1930s in the records of whaling ships (de la Mare, 1997). This is a problem we shall hear more about in the coming years.

Synopsis

In this province there is weak, variable wind mixing with brief shoaling in summer (Fig. 10.5). Pycnocline is always deeper than photic zone. Primary production rate conforms closely to irradiance cycle with relatively strong secondary chlorophyll accumulation in austral autumn (March and April), when productivity is declining perhaps due to release of razing pressure.

BIBLIOGRAPHY

Ackefors, H. (1966). Plankton and hydrography of the Landsort Deep. *Veroff. Inst. Meeresforsch. Bremerhaven.* **2**, 381–386.

Adnan, Q. (1990). Monsoonal differences in netphytoplankton in the Arafura Sea. *Neth. J. Sea Res.* **25**, 523–526.

Aiken, J. (1981). The UOR.Mk.2. *J. Plankton Res.* **3**, 551–560.

Aiken, J., Moore, G. F., and Holligan, P. M. (1992). Remote sensing of oceanic biology in relation to global climate change. *J. Phycol.* **28**, 579–590.

Allison, S. K., and Wishner, K. F. (1986). Spatial and temporal patterns of zooplankton biomass across the Gulf Stream. *Mar. Ecol. Prog. Ser.* **31**, 233–244.

Alvarez-Borrego, A. (1983). Gulf of California. *In* "Ecosystems of the World. 26. Estuaries and Enclosed Seas," pp. 427–450. Elsevier, Amsterdam.

Ambler, J. W., and Miller, C. B. (1987). Vertical habitat partitioning by copepodites and adults of subtropical oceanic copepods. *Mar. Biol.* **94**, 561–577.

Anderson, D. L. T., Leetma, A., and Molinari, R. (1981). The Somali and Brazil coastal currents. *In* "Recent Progress in Equatorial Oceanography," pp. 453–466. Nova University, Dania, FL.

Anderson, G. C., and Munson, R. E. (1972). Primary productivity studies using merchant vessels in the North Pacific Ocean. *In* "Biological Oceanography of the North Pacific Ocean" (A. Y. Takenouti, Ed.), pp. 245–251. Idemetsu Shoten, Tokyo.

Anderson, G. C., Parsons, T. R., and Stephens, K. (1969). Nitrate distribution in the northeast Pacific Ocean. *Deep-Sea Res.* **16**, 329–334.

Anderson, G. C., Lam, R. K., Booth, B. C., and Glass, J. M. (1977). Description and numerical analysis of factors affecting the processes of production in the Gulf of Alaska. *Spec. Rep. Univ. Washington* **76** (ref. M-77-40).

Andrews, J. C., and Clegg, S. (1989). Coral Sea circulation and transport deduced from modal information models. *Deep-Sea Res.* **36**, 957–974.

Andrews, J. C., Lawrence, M. W., and Nilson, C. S. (1980). Observations of the Tasman front. *J. Phys. Oceanogr.* **10**, 1854–1869.

Anonymous (1973). Continuous plankton records: A plankton atlas of the North Atlantic and the North Sea. *Bull. Mar. Ecol.* **8**, 1–174.

Anonymous (1989). "Ocean Circulation." Pergamon, Elmsford, NY.

Antoine, D., André, J.-M., and Morel, A. (1996). Oceanic primary production 2. Estimation at global scale from satellite (CZCS) chlorophyll. *Global Biogeochem. Cycles* **10**, 57–69.

Arcos, E., Nuez, S. P., Castro, L., and Navarro, N. (1987). Variabilidad vertical de chlorofila—A en un rea de surgencia frente a Chile. *Central Invest. Pesqu. Chile* **34**, 47–55.

Aristegui, J., *et al.* (1997). The influence of island-generated eddies on chlorophyll distribution: A study of mesoscale variation around Gran Canaria. *Deep-Sea Res.* **44**, 71–96.

Arntz, W. E. (1986). The two faces of El Niño, 1982–83. *Meeresforschung* **31**, 1–46.

Arrigo, K. R., and McClain, C. R. (1994). Spring phytoplankton production in the western Ross Sea. *Science* **266**, 261–263.

Atjay, G. L., Ketner, P., and Duvigneaud, P. (1979). Terrestrial primary production and phytomass. *In* "The Global Carbon Cycle" (B. Bolin, E. T. Degens, S. Kempe, and P. Ketner, Eds.), pp. 129–181. Wiley, Chichester, UK.

Atkinson, A., Schnack-Schiel, S. B., Ward, P., and Marin, V. (1997). Regional differences in the life cycle of *Calanoides acutus* within the Atlantic sector of the Southern Ocean. *Mar. Ecol. Prog. Ser.* **150**, 99–111.

Atkinson, L. P., O'Malley, P. G., Yoder, J. A., and Paffenhofer, G. A. (1984). The effect of summertime shelf break upwelling on nutrient flux in southeast USA continental shelf waters. *J. Mar. Res.* **42**, 9679–993.

Badon-Dongon, A., Koblinsky, A., and Baumgartner, T. (1985). Spring and summer in the Gulf of California: Observations of surface thermal patterns. *Océanol. Acta* **8**, 13–22.

Bainbridge, V. (1960a). Occurrence of *Calanoides carinatus* in the plankton of the Gulf of Guinea. *Nature (London)* **188**, 932–933.

Bainbridge, V. (1960b). The plankton of inshore waters off Sierra Leone. *Col. Off. Fish. Pubs.* **13**, 1–48.

Bakun, A., and Nelson, C. S. (1991). The seasonal cycle of wind-stress curl in subtropical eastern boundary current regions. *J. Phys. Oceanogr.* **21**, 1815–1834.

Banse, K. (1984). Overview of the hydrography and associated biological phenomena in the Arabian Sea, off Pakistan. *In* "Mar Geology and Oceanography of the Arabian Sea and Coastal Pakistan" (B. U. Haq and J. Milliman, Eds.), pp. 271–303.

Banse, K. (1987). Seasonality of phytoplankton chlorophyll in the central and northern Arabian Sea. *Deep-Sea Res.* **34**, 713–723.

Banse, K. (1990a). Does iron really limit phytoplankton production in the offshore subarctic Pacific? *Limn. Oceanogr.* **35**, 772–775.

Banse, K. (1990b). Remarks on the oceanographic observations off the east coast of India. *Mahasagar-Bull. Natl. Inst. Oceanogr. India* **23**, 75–84.

Banse, K. (1992). Grazing, temporal changes of phytoplankton concentrations, and the microbial loop in the open sea. *In* "Primary Productivity and Biogeochemical Cycles in the Sea" (P. G. Falkowski and A. D. Woodhead, Eds.), pp. 409–440. Plenum, New York.

Banse, K. (1994). On the coupling of hydrography, phytoplankton, zooplankton and settling organic particles offshore in the Arabian Sea. *Proc. Indian Acad. Sci. (Earth Planet. Sci.)* **103**, 125–161.

Banse, K. (1996). Low seasonality of low concentrations of surface chlorophyll in the Subantarctic Water Ring: Underwater irradiance, iron, and grazing. *Prog. Oceanogr.* **37**, 241–291.

Banse, K., and English, D. C. (1993). Revision of satellite-based phytoplankton pigment data from the Arabian Sea during the northeast monsoon. *Mar. Res.* **2**, 83–103.

Banse, K., and English, D. C. (1994). Seasonality of CZCS phytoplankton pigment in the offshore oceans. *J. Geophys. Res.* **99**, 7323–7345.

Banse, K., and McClain, C. R. (1986). Winter blooms of phytoplankton in the Arabian Sea as observed by the Coastal Zone Colour Scanner. *Mar. Ecol. Prog. Ser.* **34**, 201–211.

Barber, R. T. (1988). Ocean basin ecosystems. *In* "Concepts of Ecosystem Ecology" (L. R. Pomeroy and J. J. Alberts, Eds.), pp. 171–193. Springer-Verlag, Berlin.

Barber, R. T., and Smith, R. L. (1981). Coastal upwelling ecosystems. *In* "Analysis of Marine Ecosystems" (A. R. Longhurst, Ed.), pp. 31–68. Academic Press, San Diego.

Barry, R. G., and Chorley, R. J. (1982). "Atmosphere, Weather and Climate." Methuen, New York.

Bathman, U. V., Scarek, R., Klaas, C., Dubischar, C. D., and Smetacek, V. (1997). Spring development of phytoplankton biomass and composition in major water masses of the Atlantic sector of the Southern Ocean. *Deep-Sea Res. II* **44**, 51–68.

Bauer,. S., Hitchcock, G. L., and Olson, D. B. (1991). Influence of monsoonally-forced Ekman dynamics upon surface layer depth and plankton biomass distribution in the Arabian Sea. *Deep-Sea Res.* **38**, 531–553.

Bayliff, W. H. (1980). Species synopses of biological data on 8 species of scombroids. *Int. Am. Trop. Tuna Commun. Spec. Rep.* **2**, 1–530.

Beklemishev, K. V. (1969). "Ekologiya: Biogeografiya Pelagiali" (Ecology and Biogeography of the Open Ocean). Nauka, Moscow.

Ben-Avraham, Z., and Emery, K. O. (1973). Structural framework of the Sunda Shelf. *Bull. Am. Petrol. Geol.* **57**, 2323–2366.

Berger, W. M., Fischer, K., Lai, C., and Wu, G. (1987). Ocean productivity and organic carbon flux. *Scripps Inst. Oceanogr.,* ref. 87-10, 1–67.

Bernal, P. A., Robles, F. L., and Rojas, O. (1983). Variabilidad fisica y biologica en la region meridional del sistema corrientes Chile-Peru. *FAO Fisheries Rep.* **291**, 683–712.

Berrit, G. R. (1961). Observations de surface le long des lignes de navigation. *Cah. Ocanogr. ORSTOM* **10**, 715–729.

Binet, D. (1983). Phytoplankton et production primaire des régions côtières à upwellings saisonières dans le Golfe de Guinée. *Océanogr. Trop.* **18**, 331–355.

Blackburn, M. (1962). An oceanographic study of the Gulf of Tehuantapec. *Spec. Sci. Rep.-Fish.* **404**, 1–28.

Blackburn, M., Laurs, R. M., Owen, R. W., and Zeitzschel, B. (1970). Seasonal and areal changes in standing stocks of phytoplankton, zooplankton and micronekton in the eastern tropical Pacific. *Mar. Biol.* **7**, 14–31.

Blanchot, J., Le Borgne, R., Le Bouteiller, A., and Rodier, M. (1992). Effect of ENSO events on the distribution and abundance of phytoplankton in the western tropical Pacific Ocean along 165°E. *J. Plankton Res.* **14**, 137–156.

Blanton, J. O., Atkinson, L. P., Pietrafesa, L. J., and Lee, T. N. (1981). The intrusion of Gulf Stream water across the continental shelf due to topographically induced upwelling. *Deep-Sea Res.* **28**, 393–405.

Blasco, D., Estrada, M., and Jones, B. H. (1981). Short time variability of phytoplankton populations in upwelling regions—The example of northwest Africa. *In* "Coastal Upwelling" (F. A. Richards, Ed.), pp. 339–347.

Bohle-Carbonell, M. (1989). On the variability of the Peruvian upwelling system. *In* "The Peruvian Upwelling System: Dynamics and Interactions" (D. Pauly, Ed.), pp. 14–44. ICLARM, Manila.

Boisvert, W. E. (1967). Major currents in the north and south Atlantic Oceans. *NavOceano Trans.* **193**, 1–92.

Bonilla, J., Bugden, W. J., Zafiriou, O., and Jones, R. (1993). Seasonal distribution of nutrients and primary productivity on the eastern continental shelf of Venezuela as influenced by the Orinoco River. *J. Geophys. Res.* **98**, 2245–2257.

Booth, B. C., Lewin, J., and Postel, J. R. (1993). Temporal variation in the structure of autotrophic and heterotrophic communities in the subarctic Pacific. *Prog. Oceanogr.* **32**, 57–99.

Borstad, G. A. (1982). The influence of the meandering Guiana Current and Amazon River discharge on surface salinity near Barbadoes. *J. Mar. Res.* **40**, 421–434.

Bradford, J. M. (1983). Physical and chemical oceanography of the west coast of New Zealand. *N. Z. J. Mar. Freshwater Res.* **17**, 71–81.

Bradford, J. M., and Chapman, B. (1988). *Nyctyphanes australis* and an upwelling plume in western Cook Strait, New Zealand. *N. Z. J.. Mar. Freshwater Res.* **22**, 237–247.

Bray, N. A. (1988). Water mass formation in the Gulf of California. *J. Geophys. Res.* **93**, 9223–9240.

Brink, K. H. (1983). The near-surface dynamics of coastal upwelling. *Prog. Oceanogr.* **12**, 223–257.

Brinton, E. (1962). The distribution of Pacific euphausiids. *Bull. Scripps Inst. Oceanogr.* **8**, 51–270.

Brock, J. C., and McLain, C. R. (1992). Interannual variability in phytoplankton blooms observed in the northwestern Arabian Sea during the southwest monsoon. *J. Geophys. Res.* **97**, 733–750.

Brock, J. C., McLain, C. R., Luther, M. E., and Hay, W. W. (1991). The phytoplankton bloom in the northwestern Arabian Sea during the southwest monsoon of 1979. *J. Geophys. Res.* **96**, 20623–20642.

Brock, J. C., McClain, C. R., and Hay, W. W. (1992). A southwest monsoon hydrographic climatology for the northwest Arabian Sea. *J. Geophys. Res.* **97**, 9455–9465.

Brock, J. C., Sathyendranath, S., and Platt, T. (1993). Modelling the seasonality of subsurface light and primary production in the Arabian Sea. *Mar. Ecol.-Prog. Ser.* **101**, 209–221.

Brockman, C., Fahrbach, E., Huyer, A., and Smith, R. L. (1980). The poleward undercurrent along the Peru coast: 5–15°S. *Deep-Sea Res.* **27**, 847–856.

Brown, C. W. (1995). Global distribution of coccolithophore blooms. *Oceanography* **8**, 59–60.

Bruce, J. G. (1979). Eddies off the Somali coast during the southwest monsoon. *J. Geophys. Res.* **84**, 7742–7748.

Bruce, J. G., Kerling, J. L., and Beatty, W. H. (1985). On the North Brazilian eddy field. *Prog. Oceanogr.* **14**, 57–63.

Bubnov, V. A. (1972). Structure and characteristics of the oxygen minimum in the southeastern Atlantic. *Oceanology* **12**, 193–201.

Bucklin, A., Frost, B. W., and Koscher, T. D. (1995). Molecular systematics of six *Calanus* and three *Metridia* species (Calanoida: Copepoda). *Mar. Biol.* **121**, 655–664.

Burkill, P. H., Mantoura, R. F. C., and Owens, N. J. P. (Eds.) (1993a). Biogeochemical cycling in the northwest Indian Ocean. *Deep-Sea Res. II* **40**, 643–849.

Burkill, P. H., Leakey, R. J. G., Owens, N. J. P., and Mantoura, R. F. C. (1993b). *Synechococcus* and its importance to the microbial food web of the northwestern Arabian Sea. *Deep-Sea Res.II* **40**, 773–782.

Burkov, V. A., and Pavlova, P. Y. V. (1980). Description of the eddy field of the California current. *Oceanologia* **20**, 272–278.

Butler, J. N., Morris, B. F., Cadwallader, J., and Stoner, A. W. (1983). Studies of Sargassum and the Sargassum community. *Bermuda Biol. Station Spec. Publ.* **22**, 1–85.

Campbell, J. W., and Aarup, T. (1992). New production in the North Atlantic derived from seasonal patterns of surface chlorophyll. *Deep-Sea Res.* **39**, 1669–1694.

Capone, D. G., Zehr, J. P., and Paerl, H. W. (1997). Trichodesmium, a globally significant marine cyanobacterium. *Science* **276**, 1221–1229.

Cardon, K. L., Gregg, W. W., Costello, D. K., Haddad, K., and Prospero, J. M. (1991). Determination of Saharan dust radiance and chlorophyll from CZCS imagery. *J. Geophys. Res.* **96,** 5369–5378.

Carmack, E. C. (1990). Large-scale physical oceanography of polar regions. *In* "Polar Oceanography. Part A: Physical Science," pp. 171–221. Academic Press, London.

Carpenter, E. J. (1989). Nitrogen fixation by marine Oscillatoria. *In* "Nitrogen in the Marine Environment" (E. J. Carpenter and D. G. Capone, Eds.), pp. 65–104.

Carr, M.-E., Oakey, N. S., Jones, B., and Lewis, M. R. (1992). Hydrographic patterns and vertical mixing in the equatorial Pacific along 150°W. *J. Geophys. Res.* **97,** 611–626.

Carton, J. A., and Huang, B. (1994). Warm events in the tropical Atlantic. *J. Phys. Oceanogr.* **24,** 888–903.

Caspers, H. (1957). Black Sea and Sea of Azov. *Mem. Geol. Soc. Am.* **67,** 801–889.

Chan, K. M. (1970). Seasonal variation of hydrologic properties in the northern South China Sea. *In* "The Kuroshio" (J. C. Marr, Ed.), pp. 143–162. East-West Centre, Honolulu.

Chang, F. H., Vincent, W. F., and Woods, P. H. (1992). Nitrogen utilisation by size-fractionated phytoplankton assemblages associated with an upwelling event off Westland, New Zealand. *N. Z. J. Mar. Freshwater Res.* **26,** 287–301.

Chapman, P., and Shannon, L. V. (1985). The Benguela ecosystem. II. Chemistry and related processes. *Oceanogr. Mar. Biol. Annu. Rev.* **23,** 183–251.

Chavez, F. P. (1989). Size distribution of phytoplankton in the central and eastern tropical Pacific. *Glob. Biogeochem. Cycl.* **3,** 27–35.

Chavez, F. P., Barber, R. T., and Sanderson, M. P. (1989). The potential primary production of the Peruvian upwelling ecosystem: 1953–1984. *ICLARM Stud. Rev.* **15,** 1–351.

Chavez, F. P., Buck, K. R., Coale, K. H., Martin, J. H., DiTullio, G. R., Welschmeyer, N. A., Jacobson, A. C., and Barber, R. T. (1991). Growth rates, grazing, sinking and iron limitation of equatorial Pacific phytoplankton. *Limn. Oceanogr.* **36,** 1816–1833.

Chelton, D. B. (1982). Large-scale response of the California Current to forcing by the wind stress curl. *Rep. California Coop. Fish. Invest.* **23,** 130–148.

Chelton, D. B., Bernal, P. A., and McGowan, J. A. (1982). Large-scale interannual physical and biological interaction in the California Current. *J. Mar. Res.* **40,** 1095–1125.

Chen, Z., Zheng, Y., and Huang, W. (1992). The effects of a Kuroshio frontal eddy on the biochemical structure of the East China Sea. *In* "Essays on the Investigation of the Kuroshio 4," pp. 115–124. China Ocean Press, Beijing.

Chisholm, S. W. (1992). Phytoplankton size. *In* "Primary Productivity and Biogeochemical Cycles in the Sea" (P. G. Falkowski and A. D. Woodhead, Eds.), pp. 213–237. Plenum, New York.

Chisholm, S. W., and Morel, F. M. M. (Eds.) (1992). What controls phytoplankton production in nutrient-rich areas of the open sea? *Limn. Oceanogr.* **36** (Spec. Symp.).

Chung, S. P., Gardner, W. D., Richardson, M. J., Walsh, I. D., and Landry, M. D. (1996). Beam attenuation and micro-organisms: Spatial and temporal variations in small particles along 140°W during the 1992 JGOFS EqPac transects. *Deep-Sea Res. II* **43,** 1205–1226.

Church, J. A. (1987). East Australian Current adjacent to the Great Barrier Reef. *Aust. J. Mar. Freshwater Res.* **38,** 671–683.

Church, J. A., Cresswell, G. R., and Godfrey, J. S. (1989). The Leeuwin Current. *In* "Poleward Flows along the Eastern Ocean Boundaries" (S. J. Neshyba, C. N. K. Mooers, R. L. Smith, and R. T. Barber, Eds.), pp. 230–253. Springer-Verlag, Berlin/New York.

Citeau, J., Bergès, J. C., Demarcq, H., and Mahé, G. (1988). The watch of ITCZ migrations. *Trop. Ocean-Atmos. Newslett.* **45,** 1–3.

Coachman, L. K. (1986). Circulation: Water masses and fluxes on the southeastern Bering Sea shelf. *Cont. Shelf Res.* **5,** 22–108.

Coachman, L. K., Kinder, T. H., Schumacher, J. D., and Tripp, R. B. (1980). Frontal systems of the southeast Bering Sea shelf. *In* "Second International Symposium on Stratified Flows. Trondheim. June 1980" (T. Carstens and T. McClimans, Eds.), pp. 917–933. Tapir, Trondheim.

Coale, K. H., *et al.* (1996). A massive phytoplankton bloom induced by an ecosystem-scale iron fertilization experiment in the equatorial eastern Pacific Ocean. *Nature (London)* **383,** 495–501.

Codispoti, L. A., Dugdale, R. C., and Minas, H. J. (1982). A comparison of the nutrient regimes off northwest Africa, Peru and Baja California. *Rapp. Proc. Verb. Réun. ICES* **180,** 184–201.

Colborn, J. G. (1975). "The Thermal Structure of the Indian Ocean," pp. 1–173. Univ. of Hawaii Press, Honolulu.

Colebrook, J. M. (1976). Trends in climate of the North Atlantic Ocean over the past century. *Nature (London)* **263,** 576–577.

Colebrook, J. M. (1979). Continuous plankton records: Seasonal cycles of phytoplankton and copepods in the North Atlantic Ocean and the North Sea. *Mar. Biol.* **51,** 23–32.

Colebrook, J. M. (1982). Continuous plankton records: Seasonal variations in the distribution and abundance of plankton in the North Atlantic Ocean and the North Sea. *J. Mar. Res.* **4,** 435–462.

Colebrook, J. M. (1986). Environmental influences on long-term variability in marine plankton. *Hydrobiology* **142**, 309–325.

Colin, C., Gonella, J., and Merle, J. (1987). Equatorial upwelling at 4°W during the FOCAL programme. *Océanol. Acta* **Sp. 6**, 39–49.

Comiso, J. C., McLain, C. R., Sullivan, C. W., Ryan, J. P., and Leonard, C. L. (1993). Coastal zone color scanner pigment concentrations in the Southern Ocean and relationships to geophysical surface features. *J. Geophys. Res.* **98**, 2419–2451.

Conkright, M. E., Levitus, S., and Boyer, T. P. (1994). "NOAA World Ocean Atlas, NESDIS 1 (Nutrients)," pp. 1–150. U.S. Department of Commerce, Washington, DC.

Conover, R. J. (1981). Nutritional strategies for feeding on small suspended particles. *In* "Analysis of Marine Ecosystems" (A. R. Longhurst, Ed.), pp. 363–396. Academic Press, London.

Conover, R. J. (1988). Comparative life histories in the genera *Calanus* and *Neocalanus* in high latitudes of the northern hemisphere. *Hydrobiology* **167/168**, 127–142.

Cooney, R. T. (1986). The seasonal occurrence of *Neocalanus cristatus*, *N. plumchrus* and *Eucalanus bungii* over the shelf of the northern Gulf of Alaska. *Cont. Shelf Res.* **5**, 541–553.

Cornillon, P. (1986). The effect of the New England seamounts on Gulf Steam meandering. *J. Phys. Oceanogr.* **16**, 386–389.

Coste, B., Le Corre, P., Minas, H. J., and Morin, P. (1988). Les éléments nutritifs dans le bassin occidental de la Méditerranean. *Océanol. Acta Spec.*, 87–94.

Cromwell, T. (1953). Circulation in a meridional plane in the central equatorial Pacific. *J. Mar. Res.* **12**, 196–213.

CSIRO (1969). Seasonal variations in the Indian Ocean at 110°E. *Aust. J. Mar. Freshwater Res.* **20**, 1–90.

Cullen, J. J., Lewis, M. R., Davis, C. O., and Barber, R. T. (1992). Photosynthesis characteristics and estimated growth rates indicate grazing is the proximate control of primary production in the equatorial Pacific. *J. Geophys. Res.* **97**, 639–654.

Cury, P., and Roy, C. (1987). Upwelling et pêche des éspèces pélagiques côtières de la Côte d'Ivoire. *Océanol. Acta* **10**, 347–357.

Cushing., D. H. (1982). "Climate and Fisheries," pp. 1–373. Academic Press, San Diego.

Cushing D. H. (1989). A difference in structure between ecosystems in strongly stratified waters and those that are rarely stratified. *J. Plankton Res.* **11**, 1–13.

Cushing, D. H., and Dickson, R. R. (1966). The biological response in the sea to climatic change. *Adv. Mar. Biol.* **14**, 1–122.

Dadou, I., Garcon, V., Anderson, V., Flierl, G. R., and Davis, C. S. (1996). Impact of the North Equatorial Counter Current meandering on a pelagic ecosystem: A modelling approach. *J. Mar. Res.* **54**, 311–342.

Dandonneau, Y., and Charpy, L. (1985). An empirical approach to the island mass effect in the south tropical Pacific Ocean based on sea surface chlorophyll concentrations. *Deep-Sea Res.* **32**, 707–721.

Dandonneau, Y., and Gohin, F. (1984). Meridional and seasonal variations of the sea surface chlorophyll concentration in the southwestern tropical Pacific. *Deep-Sea Res.* **31**, 1377–1394.

Das, K., Gouveia, A. D., and Varma, K. K. (1980). Circulation and water characteristics on isanosteric surfaces in the northern Arabian Sea during February–April. *Ind. J. Mar. Sci.* **9**, 156–165.

Deacon, G. (1984). "The Antarctic Circumpolar Ocean." Cambridge Univ. Press, Cambridge, UK.

de Baar, M. A., *et al.* (1991). The ecology of the Friesian Front. *Int. Council Exp. Sea C. M.* **L25**, 1–12.

de Baar, H. J. W., de Jong, J. T. M., Bakker, D. C. E., Loescher, B. M., Vet, C., Bathmann, U., and Smetacek, V. (1995). Importance of iron for plankton blooms and CO_2 drawdown in the Southern Ocean. *Nature (London)* **373**, 412–415.

de Baar, H. J. W., van Leeuwe, M. A., Scharek, R., Goeyens, L., Bakker, K. M. J., and Fritsche, P. (1997). Nutrient anomalies in *Fragilariopsis* blooms, iron deficiency and the N/P ratio of the Antarctic Ocean. *Deep-Sea Res. II* **44**, 229–260.

de Beaufort, L. F. (1951). "Zoogeography of the Land and Inland Waters." Sidgewick & Jackson, London.

de la Mare, W. K. (1997). Abrupt mid-twentieth century decline in Antarctic sea-ice extent from whaling records. *Nature (London)* **389**, 57–60.

Delecluse, P., Servain, J., Levy, C., Arpe, K., and Bengtsson, L. (1994). On the connection between the 1984 Atlantic warm event and the 1982–83 El Niño. *Tellus* **46A**, 446–464.

Dickey, T., *et al.* (1993). Seasonal variability of bio-optical and physical properties in the Sargasso Sea. *J. Geophys. Res.* **98**, 865–898.

Dickey, T., *et al.* (1994). Bio-optical and physical variability in the subarctic North Atlantic Ocean during the spring of 1989. *J. Geophys. Res.* **99**, 22541–22556.

Dickie, L., and Trites, R. (1983). The Gulf of St. Lawrence. *In* "Ecosystems of the World, 26: Estuaries and Enclosed Seas," pp. 403–425. Elsevier, Amsterdam.

Dickson, R. R. (1983). Global summaries and inter comparisons; Flow statistics from long-term current meter moorings. *In* "Eddies in Marine Science" (A. R. Robinson, Ed.), pp. 278–328.

Dickson, R. R., Kelly, P. M., Colebrook, J. M., Wooster, W. S., and Cushing, D. H. (1988). North winds and production in the eastern North Atlantic. *J. Plankton Res.* **10**, 151–169.

Dietrich, G. (1964). Oceanic polar front survey. *Res. Geophys.* **2**, 291–308.

Dietrich, G., Kalle, K., Krauss, W., and Siedler, G. (1970). "General Oceanography." Wiley, New York.

DiTullio, G. R., and Smith, W. O. (1996). Spatial patterns in phytoplankton biomass and pigment distributions in the Ross Sea. *J. Geophys. Res.* **101**, 18467–18477.

Dodge, J. D., and Marshall, H. G. (1994). Biogeographic analysis of the armored dinoflagellate *Ceratium* in the North Atlantic. *J. Phycol.* **390**, 905–922.

Dodimead, A. J., Favorite, F., and Hirano, T. (1967). Review of the oceanography of the North Pacific. *Bull. Int. North Pacific Fish. Commun.* **13**, 1–195.

Donaghey, P. L., et al. (1991). The role of episodic atmospheric nutrient inputs in the chemical and biological dynamics of oceanic ecosystems. *Oceanography* **4**, 62–70.

Dorgham, M. M., and Moftah, A. (1989). Environmental conditions and phytoplankton distribution in the Arabian Gulf and Gulf of Oman. September 1986. *J. Mar. Biol .Assoc. India* **31**, 36–53.

Dorman, C. E., and Palmer, D. P. (1981). Southern California summer coastal upwelling. *In* "Coastal Upwelling" (F. A. Richards, Ed.), pp. 44–56. American Geophysical Union, Washington, DC.

Dower, K., and Lucas, M. I. (1993). Photosynthesis–irradiance relationships and production associated with a warm-core ring shed from the Agulhas Retroflection south of Africa. *Mar. Ecol.-Prog. Ser.* **95**, 141–154.

Dowidar, N. M. (1983). Primary production in the Red Sea off Jeddah. *In* "Marine Science in the Red Sea; Proceedings of the Conference Celebrating the 50th Anniversary of the Al Ghardaqa Marine Biology Statistics" (A. F. Latif, A. R. Bayoumi, and M. Thompson, Eds.), pp. 160–170.

Dubischar, C. D., and Bathman, U. V. (1997). Grazing impact of copepods and salps on phytoplankton in the Atlantic sector of the Southern Ocean. *Deep-Sea Res. II* **44**, 415–434.

Duce, R. A., and Tindale, N. W. (1991). Atmospheric transport of iron and its deposition in the ocean. *Limn. Oceanogr.* **38**, 1715–1726.

Duce, R. A., *et al.* (1991). The atmospheric input of trace species to the world ocean. *Global Biogeochem. Cycles* **5**, 193–259.

Ducklow, H. W., and Harris, R. P. (1993). Topical studies in oceanography—JGOFS, the North Atlantic Bloom Experiment. *Deep-Sea Res. II* **40**, 1–641.

Duing, W., Ostapoff, F., and Merle, J. (1980). "Physical Oceanography of the Tropical Atlantic during GATE." Univ. of Miami Press, Miami, FL.

Dunbar, M. J. (1979). The relation between the oceans. *In* "Zoogeography and Diversity of Plankton" (S. van der Spoel and A. C. Pierrot-Bults, Eds.), pp. 112–125. Bunge.

Duncan, C. P., Schladow, S. G., and Williams, W. G. (1982). Surface currents near the Greater and Lesser Antilles. *Int. Hydrogr. Rev.* **59**, 67–78.

Dupouy, C., Petit, M., and Dandonneau, Y. (1988). Satellite-detected cycanobacteria bloom in the southwest tropical Pacific: Implications for oceanic nitrogen fixation. *Int. J. Remote Sensing* **9**, 389–396.

Edwards, F. J. (1987). Climate and oceanography. *In* "Key Environments—Red Sea" (A. J. Edwards and S. M. Head, Eds.), pp. 46–69.

Eilertsen, H. C., Tande, K. S., and Hegseth, E. N. (1989a). Potential of herbivorous copepods for regulating the spring phytoplankton bloom in the Barents Sea. *Rapp. Proc.-Verb. Cons .Int. Explor. Mer.* **188**, 154–163.

Eilertsen, H. C., Tande, K. S., and Taasen, J. P. (1989b). Vertical distributions of primary production and grazing in arctic waters (Barents Sea). *Polar Biol.* **9**, 253–260.

Eldin, G. (1983). Eastward flows of the South Equatorial Central Pacific. *J. Phys. Oceanogr.* **13**, 1461–1467.

El Sayed, S., and Taguchi, S. (1981). Primary production and standing crop of phytoplankton along the ice-edge in the Weddell Sea. *Deep-Sea Res.* **28**, 1017–1032.

Emery, W. J., and Meincke, J. (1986). Global water masses: Summary and review. *Océanol. Acta* **9**, 383–391.

Emery, W. J., Lee, W. G., and Magaard, L. (1984). Geographic and seasonal distributions of Brunt–Väisälä frequency and Rossby radii in the North Pacific and North Atlantic. *J. Phys. Oceanogr.* **14**, 294–317.

Enfield, D. B. and Cid, L. (1990). Statistical analysis of ENSO over the last 500 years. *TOGA Notes Fall 1990* **1**, 1–4.

English, T. S. (1961). Some biological oceanographic observations in the central Polar Sea. *Res. Pap. Arct. Inst. North Am.* **13**, 1–80.

Eppley, R. W., Stewart, E., Abbott, M. R., and Heymen, U. (1985). Estimating ocean primary production from satellite chlorophyll. Introduction to regional differences and statistics for the Los Angeles Bight. *J. Plankton Res.* **7**, 57–70.

Errhif, A., Razouls, C., and Mayzaud, P. (1997). Composition and community structure of copepods in the Indian Ocean sector of the Antarctic Ocean during the end of the austral summer. *Polar Biol.* **17**, 418–430.

Estrada, M., and Margalef, R. (1988). Supply of nutrients to the Mediterranean photic zone along a persistent front. *Océanol. Acta* **9**, 133–142.

Estrada, M., Marrasé, C., Latasa, M., Berdelet, E., Delgado, M., and Riera, T. (1993). Variability of the deep chlorophyll maximum in the northwest Mediterranean. *Mar. Ecol.-Prog. Ser.* **92**, 289–300.

Evseenko, P. (1987). Reproduction of Peruvian jack mackerel in the southern Pacific. *J. Ichthy.* **27**, 151–160.

Fan, K. L. (1982). A study of the water masses of the Taiwan Strait. *Acta Oceanol. Taiwan* **13**, 140–153.

Fanning, K. A. (1991). Nutrient provinces in the sea: Concentration ratios, reaction rate ratios and ideal covariation. *J. Geophys. Res.* **97**, 5693–5712.

Farmer, et al. (1993). Effects of low and high river flow on the underwater light field of the eastern Caribbean basin. *J. Geophys. Res.* **98**, 2279–2288.

Fasham, M. J. R. (Ed.) (1984). "Flows of Energy and Materials in Marine Ecosystems: Theory and Practice." Plenum, New York.

Favorite, F. (1974). Flow into the Bering Sea through Aleutian Island passes. *In* "Oceanography of the Bering Sea" (D. W. Hood and E. J. Kelly, Eds.), pp. 3–38. Univ. of Alaska Press, Fairbanks.

Favorite, F., Dodimead, A. J., and Nasu, K. (1976). Review of the oceanography of the North Pacific, 1960–1971. *Bull. Int. North Pacific Fish. Commun.* **33**, 1–187.

Fedoseev, A. (1970). Geostrophic circulation of surface waters on the shelf of northwest Africa. *Rapp. Proc.-Verb. Réun. Cons. Int. Explor. Mer.* **159**, 30–37.

Feldman, G. C., *et al.* (1989). Ocean color: Availability of the Global Data Set. *Eos* **70**, 634–641.

Fernadez, E., and Pingree, R. D. (1996). Coupling between physical and biological fields in the North Atlantic subtropical front southeast of the Azores. *Deep-Sea Res. II* **43**, 1369–1393.

Fiedler, P. C. (1994). Seasonal climatologies and variability of the eastern tropical Pacific surface waters. *Tech. Rep. U.S. Natl. Mar. Fish. Service* **109**, 1–65.

Fiekas, V., Elken, J., Müller, T. J., Aitsam, A., and Zenk, W. (1992). A view of the Canary Basin thermocline circulation in winter. *J. Geophys. Res.* **97**, 12495–12510.

Findlater, J. (1969). A major low-level air current near the Indian Ocean during the northern summer. *Q. J. R. Meteor. Soc.* **95**, 362–380.

Fleminger, A., and Hulseman, K. (1977). Geographic range and taxonomic divergence in the North Atlantic. *Calanus. Mar. Biol.* **40**, 233–248.

Fonseca, T. R., and Farias, M. (1987). Estudio del proceso de surgencia en la costa chilena utilizando percepcion remota. *Invest. Pesq. (Chile)* **34**, 33–46.

Forbes, A. M. G. (1984). The contributions of local processes to seasonal hydrography of the Gulf of Carpentaria. *Océanogr. Trop.* **19**, 193–201.

Foster, T. D. (1984). The marine environment. *In* "Antarctic Ecology, Vol. 2" (R. M. Laws, Ed.), pp. 345–372. Academic Press, San Diego.

Fraga, F. (1981). Upwelling off the Galicia coast, Northwest Spain. *In* "Coastal Upwelling" (F. A. Richards, Ed.), pp. 176–182. American Geophysical Union, Washington, DC.

Froneman, P. W., Pakhomov, E. A., and Perissinotto, R. (1997). Dynamics of the plankton communities of the Lazarev Sea during seasonal ice melt. *Mar. Ecol. Prog. Ser.* **149**, 201–214.

Furnas, M. J., and Mitchell, A. W. (1996). Pelagic primary production in the Coral and southern Solomon Seas. *Mar. Freshwater Res.* **47**, 695–706.

Furuya, K., Hasumoto, H., Nakai, T., and Nemoto, T. (1986). Phytoplankton in the subtropical Convergence during the austral summer: Community structure and growth activity. *Deep-Sea Res.* **33**, 621–630.

Gabric, A. J., Hoffenberg, P., and Broughton, W. (1990). Chlorophyll distribution in the Great Barrier Reef. *Austr. J. Mar. Freshwater Res.* **41**, 313–324.

Gabric, A. J., Garcia, L., van Camp, L., Nykjaer, L., Eifler, W., and Schrimpf, W. (1993). Offshore export of shelf production in the Cap Blanc giant filament as derived from CZCS imagery. *J. Geophys. Res.* **98**, 4697–4712.

Ganssen, G., and Kroon, D. (1991). Evidence for Red Sea surface circulation from oxygen isotopes of modern surface waters and planktonic foraminiferan tests. *Palaeocean* **6**, 73–82.

Garner, D. M. (1967). Hydrology of the south-east Tasman Sea. *Bull. N. Z. Dep. Sci. Ind. Res.* **181**, 1–40.

Garrison, D. L. (1991). An overview of the abundance and role of proto zooplankton in Antarctic waters. *J. Mar. Systems* **2**, 317–331.

Garzoli, S. L., and Katz, E. J. (1983). The forced annual reversal of the Atlantic North Equatorial Countercurrent. *J. Phys. Oceanogr.* **13**, 2082–2090.

Garzoli, S., and Richardson, P. L. (1989). Low-frequency meandering of the Atlantic North Equatorial Countercurrent. *J. Geophys. Res.* **89**, 2079–2090.

Geyer, W. R., *et al.* (1996). Physical oceanography of the Amazon shelf. *Cont. Shelf Res.* **16**, 575–616.

Gibbs, C. F. (1991). Nutrients and plankton distribution at a shelf-break front in the region of the Bass Strait Cascade. *Austr. J. Mar. Freshwater Res.* **42**, 201–217.

Gibbs, R. K. (1980). Wind-controlled coastal upwelling in the western equatorial Atlantic. *Deep-Sea Res.* **27**, 857–866.

Gieskes, W. W. C., Kraay, G. W., Nontji, A., Setipermana, D., and Sutomo (1990). Monsoonal differences in primary production in the eastern Banda Sea. *Neth. J. Sea Res.* **25**, 473–483.

Glorioso, P. (1987). Temperature distribution related to shelf-sea fronts on the Patagonian shelf. *Cont. Shelf Res.* 7, 27–34.

Glover, D. M., and Brewer, P. G. (1988). Estimates of wintertime mixed layer nutrient concentration in the North Atlantic. *Deep-Sea Res.* **35**, 1525–1546.

Glover, D. M., Wrobleski, J. S., and McClain, C. R. (1994). Dynamics of the transition zone in CZCS ocean colour in the North Pacific during oceanographic spring. *J. Geophys. Res.* **99**, 7501–7511.

Godfrey, J. S., Vaudrey, D. J., and Hahn, S. D. (1986). Observations on the shelf-edge current south of Australia, winter 1982. *J. Phys. Oceanogr.* **16**, 668–679.

Godfrey, J. S., Hirst, A. C., and Wilkin, J. (1993). Why does the Indonesian throughflow appear to originate from the North Pacific? *J. Phys. Oceanogr.* **23**, 1087–1098.

Gonzalez-Rodriguez, E. (1994). Yearly variation in primary productivity from Cabo Frio, Brazil. *Hydrobiology* **294**, 145–156.

Gonzalez-Rodriguez, E., Valentin, J. L., André, D. L., and Jacob, S. A. (1992). Upwelling and downwelling at Cabo Frio (Brazil): Comparing biomass and primary production responses. *J. Plankton Res.* **14**, 289–306.

Gordon, A. L., and Bosley, K. T. (1991). Cyclonic gyre in the tropical south Atlantic. *Deep-Sea Res.* **38**(Suppl. 1), S323–S343.

Gordon, A. L., and Greengrove, C. L. (1986). Geostrophic circulation of the Brazil–Falkland confluence. *Deep-Sea Res.* **33**, 573–585.

Gould, W. J. (1985). Physical oceanography of the Azores front. *Prog. Oceanogr.* **14**, 167–190.

Gouriou, Y. (1993). Mean circulation of the upper layers of the western equatorial Pacific Ocean. *J. Geophys. Res.* **98**, 22495–22520.

Gradinger, R., Weiss, T., and Pillen, T. (1992). The significance of picoplankton in the Red Sea and the Gulf of Aden. *Bot. Mar.* **35**, 245–250.

Graham, N. E. (1994). Decadal-scale climate variability in the tropical and North Pacific during the 1970's and 1980's. *Clim. Dyn.* **10**, 135–162.

Gran, H. H. (1931). On the conditions for the production of plankton in the sea. *Rapp. Proc.-Verb. Réun. Cons. Int. Explor. Mer.* 7, 343–358.

Griffiths, D. A., and Middleton, J. H. (1992). Upwelling and internal tides over the inner New South Wales continental shelf. *J. Geophys. Res.* **97**, 14389–14405.

Griffiths, R. C. (1968). Physical, chemical and biological oceanography of the entrance of the Gulf of California, spring of 1960. *U.S. Fish Wildlife Serv. Spec. Sci. Rep. Fish.* **573**, 1–47.

Griffiths, R. W., and Pearce, A. F. (1985). Instability and eddy pairs on the Leeuwin Current south of Australia. *Deep-Sea Res.* **32**, 1511–1534.

Groendahl, F., and Hernroth, L. (1986). Vertical distribution of copepods in the Eurasian part of the Nansen Basin. *Natl. Mus. Can. Syllogeus* **58**, 311–320.

Guan, B. (1984). Major features of the shallow water hydrography in the East China Sea and Huanghai Sea. *In* "Ocean Hydrodynamics of the Japan and East China Seas" (T. Ichiye, Ed.), pp. 1–14.

Guillen, O. (1980). The Peru Current system. 1—Physical aspects. *In* "Scientific Exploration of the South Pacific" (W. S. Wooster, Ed.), pp. 185–216. National Academy of Science, Washington, DC.

Gulland, J. (Ed.) (1971). "The Fish Resources of the Ocean." Food and Agricultural Organization, Rome.

Gunther, E. R. (1936). A report on oceanographic investigations in the Peru Coastal Current. *Discovery Rep.* **13**, 109–276.

Guo, Y. (1992). Ecological studies on phytoplankton in the northern East China Sea. *Int. Symp. Mar. Plankton* **37**, 756.

Halim, Y. (1984). Plankton of the Red Sea and the Arabian Gulf. *Deep-Sea Res.* **31**, 969–982.

Halliwell, G. R., Ro, Y. J., and Cornillon, P. (1991). Westward-propagating SST anomalies and baroclinic eddies in the Sargasso Sea. *J. Phys. Oceanogr.* **21**, 1664–1680.

Hamilton, L. J. (1992). Surface circulation in the Tasman and Coral Seas: Climatological features derived from bathythermograph data. *Aust. J. Mar. Freshwater Res.* **43**, 793–822.

Hansell, D. A., Goering, J. J., Walsh, F. F., McRoy, C. V. P., Coachman, L. K., and Whitledge, T. E. (1989). Summer phytoplankton production and transport along the shelf break in the Bering Sea. *Cont. Shelf Res.* **9**, 1085–1104.

Hardy, A. (1967). "Great Waters." Collins, London.

Hardy, J., Hanneman, A., Behrenfeld, M., and Horner, R. (1996). Environmental biogeography of near-surface phytoplankton in the southeast Pacific Ocean. *Deep-Sea Res.* **43**, 1647–1660.

Harris, G., Nielsson, C., Clementson, L., and Thomas, D. (1987). The water masses of the East Coast of Tasmania. *Aust. J. Mar. Freshwater Res.* **38**, 569–590.

Harris, G. P., Davies, P., Nunez, M., and Myers, G. (1988). Interannual variability in climate and fisheries in Tasmania. *Nature (London)* **333**, 754–757.

Harris, G. P., Griffiths, F. B., Clementson, L. A., Lyne, V., and van der Doe, H. (1991). Seasonal and interannual variability in physical processes, nutrient cycling and the structure of the food chain in Tasmanian shelf waters. *J. Plankton Res.* **13**, 109–131.

Harrison, P. J., Fulton, J. D., Taylor, F. J. R., and Parsons, T. R. (1983). Review of the biological oceanography of the Straits of Georgia. *Can. J. Fish. Mar. Sci.* **40**, 1064–1094.

Harrison, W. G., Li, W. K. W., Smith, J. C., Head, E. J. H., and Longhurst, A. R. (1987). Depth profiles of plankton, particulate matter and microbial activity in the eastern Canadian arctic during summer. *Polar Biol.* **7**, 208–224.

Hart, T. J., and Currie, R. I. (1960). The Benguela Current. *Discovery Rep.* **31**, 123–298.

Harvey, H. W., Cooper, L. H. N., Lebour, M., and Russell, F. S. (1935). Plankton production and its control. *J. Mar. Biol. Assoc. UK* **20**, 407–441.

Hastenrath, S. (1985). "Climate and Circulation of the Tropics." Reidel, Dordrecht.

Hastenrath, S. (1989). The monsoonal regimes of the upper hydrospheric structure of the tropical Indian Ocean. *Atmos.-Ocean.* **27**, 478–507.

Hastenrath, S., and Lamb, P. J. (1977). "Climatic Atlas of the Tropical Atlantic and Eastern Pacific Oceans." Univ. of Wisconsin Press, Madison.

Hastenrath, S., and Lamb, P. J. (1979). "Climatic Atlas of the Indian Ocean. Part 1: Surface Climate and Atmospheric Circulation," pp. i–xv, 97 charts. Univ. of Wisconsin Press, Madison.

Hastenrath, S., and Merle, J. (1987). Annual cycle of subsurface thermal structure in the tropical Atlantic Ocean. *J. Phys. Oceanogr.* **17**, 1518–1538.

Hays, G. C. (1996). Large-scale patterns of diel vertical migration in the North Atlantic. *Deep-Sea Res.* **43**, 1601–1616.

Hayward, T. L. (1987). The nutrient distribution and primary production in the central North Pacific. *Deep-Sea Res.* **34**, 1593–1627.

Head, E. J. H., and Harris, L. R. (1985). Physiological and biochemical changes in *Calanus hyperboreus* from Jones Sound during the transition from summer feeding to overwintering conditions. *Polar Biol.* **4**, 99–106.

Head, S. M. (1987). Introduction to the Red Sea. *In* "Key Environments—Red Sea" (A. J. Edwards and S. M. Head, Eds.), pp. 1–21.

Heath, R. A. (1981). Oceanic fronts around New Zealand. *Deep-Sea Res.* **28**, 547–560.

Heath, R. A. (1985). A review of the physical oceanography of the seas around New Zealand—1982. *N. Z. J. Mar. Freshwater Res.* **19**, 79–124.

Heath, R. A., and Bradford, J. M. (1980). Factors affecting phytoplankton production over the Campbell Plateau. *J. Plankton Res.* **2**, 169–181.

Hebert, D., Moum, J. N., and Caldwell, D. R. (1991). Does ocean turbulence peak at the equator? *J. Phys. Oceanogr.* **21**, 1690–1698.

Hedgepeth, J. W. (1957). Treatise on marine ecology and palaeoecology. *Mem. Geol. Soc. Am.* **67**, 1–1296.

Heinrich, A. K. (1962a). The life histories of plankton animals and seasonal cycles of plankton communities in the oceans. *J. Cons. Int. Explor. Mer.* **27**, 15–24.

Heinrich, A. K. (1962b). On the production of copepods of the Bering Sea. *Int. Rev. Ges. Hydrobiol.* **47**, 465–469.

Hellerman, S. (1967). An updated estimate of the wind stress on the world ocean. *Mon. Weath. Rev.* **95**, 607–614. (See also correction in *Mon. Weath. Rev.* **96**, 63–74)

Hempel, G. (1985). Antarctic food webs. *In* "Antarctic Nutrient Cycles and Food Webs" (W. R. Siegried, P. R. Condy, and R. M. Laws, Eds.), pp. 266–270. Springer-Verlag, Berlin.

Herbland, A., and Voituriez, B. (1977). Production primaire, nitrate et nitrite dans l'Atlantique tropicale. *Cah. ORSTOM Sér. Océanogr.* **15**, 47–55.

Herbland, A., and Voituriez, B. (1982). Vitesses verticales et production potentielle dans l'upwelling de Mauretanie en Mars, 1973. *Rapp. Proc.-Verb. Réun. Cons. Int. Explor. Mer.* **180**, 131–134.

Herbland, A., Le Borgne, R., Le Bouteiller, A., and Voituriez, B. (1983). Structure hydrologique et production primaire dans l'Atlantique tropical orientale. *Océanogr. Trop.* **18**, 249–293.

Herbland, A., Le Bouteiller, A., and Raimbault, P. (1985). Size structure of phytoplankton biomass in the equatorial Atlantic Ocean. *Deep-Sea Res.* **32**, 819–836.

Herbland, A., Le Bouteiller, A., and Raimbault, P. (1987). Does the nutrient enrichment of the equatorial upwelling influence the size structure of phytoplankton of in the Atlantic Ocean? *Océanol. Acta Spec. Publ.* **6**, 115–120.

Herman, A. W. (1983). Vertical distribution patterns of copepods, chlorophyll and production in Baffin Bay. *Limnol. Oceanog.* **28**, 709–719.

Herman, A. W. (1984). Vertical copepod aggregation and interactions with chlorophyll and production on the Peru Shelf. *Cont. Shelf Res.* **3**, 131–146.

Herman, A. W. (1989). Vertical relationships between chlorophyll, production and copepods in the eastern tropical Pacific Ocean. *J. Plankton Res.* **11**, 243–261.

Hernandez-Guerra, M. A., Aristegui, J., and Canton, M. (1993). Phytoplankton pigment patterns in the Canary Islands area as determined with CZCS data. *Int. J. Remote Sensing* **14**, 1431–1437.

Hirota, Y. (1987). Vertical distribution of euphausiids in the western Pacific Ocean. *Bull. Jpn. Sea Regul. Fish. Res. Lab.* **37**, 175–224.

Hirsche, H.-J. (1991). Distribution of dominant calanoid species in the Greenland Sea during late fall. *Polar Biol.* **11**, 351–362.

Hisard, P. (1980). Observations de réponses de type "El Nino" dans l'Atlantique tropical orientale Golfe de Guinée. *Océanol. Acta* **3**, 69–78.

Hobson, L. A., and Lorenzen, C. J. (1972). Relationship of chlorophyll maximum to density structure in the Atlantic Ocean and Gulf of Mexico. *Deep-Sea Res.* **19**, 297–306.

Holligan, P. M. (1981). Biological implications of fronts on the northwest European shelf. *Philos. Trans. R. Soc. London A* **302**, 547–562.

Holligan, P. M., Viollier, M., Dupouy, C., and Aiken, J. (1983). Satellite studies on the distributions of chlorophyll and dinoflagellate blooms in the western English Channel. *Cont. Shelf Res.* **2**, 81–96.

Hopkins, T. L. (1982). The vertical distribution of zooplankton in the Gulf of Mexico. *Deep-Sea Res.* **29**, 1069–1083.

Horne, E., and Petrie, B. (1986). Mean position and variability of the sea surface temperature front east of the Grand Banks. *Atmos. Ocean.* **26**, 321–328.

Horne, E. P. W., Loder, J., Harrison, W., Mohn, R., Lewis, M., Irwin, B., and Platt, T. (1989). Nitrate supply and demand at the Georges Bank tidal front. *Sci. Mar.* **53**, 145–158.

Houghton, R., and Colin, C. (1986). Thermal structure along 4°W in the Gulf of Guinea during 1983–1984. *J. Geophys. Res.* **91**, 11727–11739.

Houghton, R. W. (1981). Equatorial upwelling in the Gulf of Guinea as seen by SST maps. *In* "Recent Progress in Equatorial Oceanography" (J. P. McCreary, Ed.), pp. 249–269. Scientific Committee for Oceanographic Research.

Houghton, R. W. (1983). Seasonal variations of the subsurface thermal structure in the Gulf of Guinea. *J. Geophys. Res.* **13**, 2070–2080.

Houghton, R. W. (1989). Influence of local and remote wind forcing in the Gulf of Guinea. *J. Geophys. Res.* **94**, 4816–4828.

Houghton, R. W. (1991). The relationship of sea surface temperature to thermocline depth at annual and interannual time scale in the tropical Atlantic Ocean. *J. Geophys. Res.* **96**, 15173–15185.

Houghton, R. W., and Mensah, M. A. (1978). Physical aspects and biological consequences of Ghanaian coastal upwelling. *In* "Upwelling Ecosystems" (R. Boje and M. Tomczak, Eds.), pp. 167–180.

Houry, S., Dombowsky, E., de Mey, P., and Minster, J.-F. (1987). Brunt–Väisälä frequency and Rossby radii in the South Atlantic, *J. Phys. Oceanogr.* **17**, 1619–1626.

Hsueh, Y., Wang, J., and Chern, C. S. (1992). The intrusion of the Kuroshio across the continental shelf northeast of Taiwan. *J. Geophys. Res.* **97**, 14323–14330.

Hughes, P., and Fiuza, A. F. G. (1982). Observations of a bottom mixed layer in the coastal upwelling area off Northwest Africa. *Rapp. Proc.-Verb. Réun. Cons. Int. Explor. Mer.* **180**, 75–82.

Humphrey, G. F., and Kerr, J. D. (1969). Seasonal variation in the Indian Ocean along 110°E. (III. Chlorophyll-a and -c). *Aust. J. Mar. Freshwater Res.* **20**, 55–64.

Hunter, J. R., and Sharp, G. D. (1983). Physics and fish populations: Shelf sea fronts and fisheries. *Food Agric. Org. U.N. Fish. Rep.* **291**, 659–682.

Huyer, A. (1976). A comparison of upwelling events in two locations: Oregon and Northwest Africa. *J. Mar. Res.* **34**, 532–545.

Ilahude, A., Komas, G., and Mardans (1990). Hydrology and productivity of the northern Arafura Shelf. *Neth. J. Sea Res.* **25**, 573–583.

Iselin, C. O'D. (1936). A study of the circulation of the western North Atlantic. *Papers Phys. Oceanogr. Meteorol.* **4**, 1–104.

Isemer, H. J., and Hasse, L. (1987). "The Bunker Climate Atlas of the North Atlantic Ocean. 2. Air–Sea Interactions." Springer-Verlag, Berlin.

Isobe, A., Tawara, S., Kaneko, A., and Kawano, M. (1994). Seasonal variability in the Tsushima Warm Current. *Cont. Shelf Res.* **14**, 23–35.

Ivanenkov, V. N., and Zemlyanov, I. V. (1985). Oxygen and carbon production during photosynthesis in the Okhotsk Sea. *Riboprod. Dal'Nevost. Mor.* **110**, 151.

James, M. R. and Wilkinson, V. H. (1988). Biomass, carbon ingestion and ammonium excretion by zooplankton in western Cook Strait, New Zealand. *N. Z. J. Mar. Freshwater Res.* **22**, 249–257.

Jerlov, N. G. (1964). Optical classification of ocean water. *In* "Physical Aspects of Light in the Sea," pp. 45–49. Univ. of Hawaii Press, Honolulu.

Jitts, H. R. (1969). Seasonal variation in the Indian Ocean along 110°E. IV Primary production. *Aust. J. Mar. Freshwater Res.* **20**, 65–75.

Jochem, F. J., and Zeitzschel, B. (1993). Productivity regime and phytoplankton size structure in the tropical and subtropical North Atlantic in spring 1989. *Deep-Sea Res. II* **40**, 495–521.

Jochem, F. J., Mathot, S., and Quéguiner, B. (1995). Size-fractionated primary production in the open Southern Ocean in austral spring. *Polar Biol.* **15**, 381–392.

Johannessen, O. M. (1986). Brief overview of the physical oceanography. *In* "The Nordic Seas" (B. G. Hurdle, Ed.), pp. 103–127. Springer-Verlag. Berlin.

Johns, B., Rao, A. D., and Rao, G. S. (1992). On the occurrence of upwelling along the east coast of India. *East Coast Shelf Sci.* **35**, 75–90.

Johns, W. E., Lee, T. N., Schott, F., Zantopp, R. J., and Evans, R. H. (1990). The North Brazil Current retroflection: Seasonal structure and eddy variability. *J. Geophys. Res.* **95**, 22103–22120.

Johnson, D. R., Fonseca, T., and Sievers, H. (1980). Upwelling in the Humboldt Current near Valparaiso, Chile. *J. Mar. Res.* **38**, 1–16.

Joint, I. R., and Williams, R. (1985). Demands of the herbivore community on phytoplankton production in the Celtic Sea in August. *Mar. Biol.* **87**, 297–306.

Jones, E. P., Nelson, D. M., and Treguer, P. (1990). Chemical oceanography. *In* "Polar Oceanography Part B: Chemistry, Biology, and Geology" (W. O. Smith, Jr., Ed.), pp. 407–476. Academic Press, New York.

Joyce, T. M., and Wiebe, P. H. (Eds.) (1992). Warm core rings: Interdisciplinary studies of the Kuroshio and Gulf Stream Rings. *Deep-Sea Res.* **39**, S1–S417.

Kabanova, J. G. (1968). Primary production in the Indian Ocean. *Oceanology* **8**, 214.

Kamykowski., D., and Zentara, S.-J. (1990). Hypoxia in the world ocean as recorded in the historical data set. *Deep-Sea Res.* **37**, 1861–1874.

Karl, D. M., Letelier, R., Hebel, D., Tupas, L., Dore, J., Christian, J., and Winn, C. (1995). Ecosystem changes in the North Pacific subtropical gyre attributed to the 1991–92 El Niño. *Nature (London)* **373**, 230–234.

Katz, E. J. (1987). Seasonal response of the sea surface to the wind in the equatorial Atlantic. *J. Geophys. Res.* **92**, 1885–1893.

Kawaguchi, S., and Takahashi, Y. (1996). Antarctic krill eat salps. *Polar Biol.* **16**, 479–481.

Kawai, H. (1972). Hydrography of the Kuroshio Extension. *In* "Kuroshio" (H. Stommel and K. Yoshida, Eds.), pp. 235–341. East–West Center, Hawaii.

Kawai, H. (1979). Rings south of the Kuroshio and their possible roles in the transport of the intermediate salinity maximum and in the formation of tuna fishing grounds. *Proc. 4th CSK. Symp.*, 250–263.

Kawai, H., and Saitoh, S. I. (1986). Secondary fronts, warm tongues and warm streamers of the Kuroshio Extension system. *Deep-Sea Res.* **33**, 1487–1507.

Kielhorn, W. V. (1952). The biology of the surface zone zooplankton of a boreo-Arctic ocean area. *J. Fish. Res. Bd. Can.* **9**, 223–264.

Kinder, T. H. (1983). Shallow currents in the Caribbean Sea and Gulf of Mexico as observed with satellite-tracked drifters. *Bull. Mar. Sci.* **33**, 239–246.

Kinder, T. H., Heburn, G. W., and Green, A. W. (1985). Some aspects of the Caribbean circulation. *Mar. Geol.* **68**, 25–52.

Kinkade, C., Marra, J., Langdon, C., Knudson, C., and Ilahudi, A. G. (1997). Monsoonal differences in phytoplankton biomass and production in the Indonesian Seas: Tracing vertical mixing using temperature. *Deep-Sea Res.* **44**, 581–592.

Kitano, K. (1979). Recent developments in the studies of the warm rings off Kuroshio, a review. *Proc. 4th CSK Symp.*, 243–251.

Knox, R. A., and Anderson, D. (1985). Recent advances in study of low latitude ocean circulation. *Prog. Oceanogr.* **14**, 259–317.

Krause, W. (1986). The North Atlantic Current. *J. Geophys. Res.* **91**, 5061–5074.

Krey, J., and Babenard, B. (Eds.) (1976). "Atlas of the IIOE; Phytoplankton Production." Institut für Meereskund an der Universitet Kiel, Kiel.

Krupatkina, D. K., Finenko, Z. Z., and Shalapyonok, A. A. (1991). Primary production and size-fractionated structure of the Black Sea in the winter–spring period. *Mar. Ecol.-Prog. Ser.* **73**, 25–31.

Kullenberg, G. (1983). The Baltic Sea. *In* "Ecosystems of the World, 26: Estuaries and Enclosed Seas" (B. H. Ketchum, Ed.), pp. 309–336. Elsevier, Amsterdam.

LaFond, E. C., and LaFond, K. G. (1968). Studies of oceanic circulation in the Bay of Bengal. (1968). *Bull. Natl. Inst. Sci. India.* **38**, 164–183.

Lagerloef, G. S. E. (1992). The Point Arena Eddy: A recurring summer anticyclone in the California Current. *J. Geophys. Res.* **97**, 12557–12568.

Lagler, D. M., and O'Brien, J. J. (1980). "Atlas of Tropical Pacific Wind-Stress Climatology, 1971–1980." Florida State Univ. Press, Tallahassee.

Le Borgne, R. (1981). Zooplankton production in the eastern tropical Atlantic: Net growth and P:B in terms of carbon, nitrogen and phosphorus. *Limn. Oceanogr.* **32**, 905–918.

Le Boutiller, A., Blanchot, J., and Rodier, M. (1992). Size distribution patterns of phytoplankton in the western Pacific: Towards a generalisation for the tropical Pacific ocean. *Deep-Sea Res.* **39**, 805–823.

Le Fèvre, J. (1986). Aspects of the biology of frontal systems. *Adv. Mar. Biol.* **23**, 163–299.

Le Fèvre, J., and Frontier, S. (1988). Influence of temporal characteristics of physical phenomena on plankton dynamics, as shown by N. W. European marine ecosystems. *In* "Towards a Theory of Biological–Physical Interactions in the World's Ocean" (B. J. Rothschild, Ed.), pp. 245–272. Kluwer, Dordrecht.

Legeckis, R. (1987). Satellite observations of a western boundary current in the Bay of Bengal. *J. Geophys. Res.* **92**, 12974–12978.

Legeckis, R. (1988). Upwelling off the Gulfs of Panama and Papagaya in the tropical Pacific during March, 1988. *J. Geophys. Res.* **93**, 15845–15849.

Legendre, L., *et al.* (1992). Ecology of sea–ice biota. 2—Global significance. *Polar Biol.* **12**, 429–444.

Lepley, L. K., Haar, S. P., and Hendrickson, J. R. (1975). Circulation in the Gulf of California from orbital photographs and ship investigations. *Cienc. Mar.* **2**, 86–93.

Levitus, S. (1982). "Climatological Atlas of the World Ocean", NOAA Professional Papers No. 13, pp. 1–173. U.S. Government Printing Office, Washington, DC.

Levitus, S., and Boyer, T. P. (1994a). "World Ocean Atlas 1994, I. Nutrients," NOAA Atlas NESDIS 1, pp 1–150. U.S. Department of Commerce, Washington, DC.

Levitus, S., and Boyer, T. P. (1994b). "World Ocean Atlas 1994, IV Temperature," NOAA Atlas NESDIS 4, pp 1–117. U.S. Department of Commerce, Washington, DC.

Levitus, S., Burgett, R., and Boyer, T. P. (1994). "World Ocean Atlas 1994, III Salinity," NOAA Atlas NESDIS 3, pp 1–99. U.S. Department of Commerce, Washington, DC.

Li, B., and Fei, Z. (1992). Distribution and characteristics of chlorophyll in the northern East China Sea in 1989. *In* "Essays on the Kuroshio 4," pp. 165–172. China Ocean Press, Beijing.

Li, W. K. W. (1995). Composition of ultraphytoplankton of the central North Atlantic. *Mar. Ecol.-Prog. Ser.* **122**, 1–8.

Li, W. H. W., and Dickie, P. (1984). Rapid enhancement of heterotrophic but not photosynthetic activities in Arctic microbial plankton at mesobiotic temperatures. *Polar Biol.* **3**, 217–226.

Li, W. K. W., and Wood, M. (1988). Vertical distribution of North Atlantic ultraphytoplankton: Analysis by flow cytometry and epifluorescent microscopy. *Deep-Sea. Res.* **35**, 1615–1638.

Lighthill, M. J. (1969). Dynamic response of the Indian Ocean to the onset of the southwest monsoon. *Philos. Trans. R. Soc. London A* **265**, 45–93.

Lim, L. C. (1975). Record of an offshore upwelling in the Southern China Sea. *J. Prim. Ind.* **3**, 53–61.

Lindstrom, E., et al. (1987). The Western Equatorial Pacific Ocean Circulation study. *Nature (London)* **330**, 533–537.

Liu, K. K., Gong, G. C., Shyu, S. Z., Pai, S. C., Wel, C. L., and Chao, S. H. (1992). Response of Kuroshio upwelling to the onset on the northeast monsoon in the sea north of Taiwan. *J. Geophys. Res.* **97**, 12511–12526.

Loder, J. W., and Greenberg, D. A. (1986). Predicted positions of tidal fronts in the Gulf of Maine region. *Cont. Shelf Res.* **6**, 397–414.

Loder, J. W., and Platt, T. (1985). Physical controls on phytoplankton production at tidal fronts. *In* "Proceedings of the 19th European Marine Biology Symposium" (P. E. Gibbs, Ed.), pp. 3–21. Cambridge Univ. Press, Cambridge, UK.

Loeb, V., Siegel, V., Holm-Hansen, O., Hewitt, R., Fraser, W., Trivelpiece, W., and Trivelpiece, S. (1997). Effects of sea-ice extent and krill or salp dominance on the Antarctic food web. *Nature (London)* **387**, 897–900.

Loeng, H. (1991). Features of the physical oceanographic conditions of the Barents Sea. *Polar Biol.* **10**, 5–18.

Lohrenz, S. E., Cullen, J. A., Phinney, D. A., Olson, D. B., and Yentsch, C. S. (1993). Distribution of pigments and primary production in Gulf Stream meanders. *J. Geophys. Res.* **98**, 14545–14560.

Longhurst, A. R. (1964). The coastal oceanography of Western Nigeria. *Bull. lnst. franç. Afr. Noire. A* **26**, 337–402.

Longhurst, A. R. (1966). The pelagic phase of Pleuroncodes planipes in the California Current. *Rep. California Coop. Fish. Invest.* **11**, 142–154.

Longhurst, A. R. (1967). Vertical distribution of zooplankton in relation to the eastern Pacific oxygen minimum. *Deep-Sea Res.* **14**, 1535–1570.

Longhurst, A. R. (1985a). Relationship between diversity and the vertical structure of the upper ocean. *Deep-Sea Res.* **32**, 1535–1570.

Longhurst, A. R. (1985b). The structure and evolution of plankton communities. *Prog. Oceanogr.* **15**, 1–35.

Longhurst, A. R. (1991). Role of the marine biosphere in the global carbon cycle. *Limn. Oceanogr.* **36**, 1507–1526.

Longhurst, A. R. (1993). Seasonal cooling and blooming in tropical oceans. *Deep-Sea Res.* **40**, 2145–2165.

Longhurst, A. R. (1995). Seasonal cycles of pelagic production and consumption. *Prog. Oceanogr.* **36**, 77–167.

Longhurst, A. R., and Harrison, W. G. (1989). The biological pump: Profiles of plankton production and consumption in the open ocean. *Prog. Oceanogr.* **22**, 47–123.

Longhurst, A. R., and Head, E. J. H. (1989). Algal production and variable herbivore demand in Jones Sound, Canadian High Arctic. *Polar Biol.* **9**, 281–286.

Longhurst, A. R., and Pauly, D. (1987). "The Ecology of Tropical Oceans." Academic Press, Orlando, FL.

Longhurst, A. R., and Williams, R. (1979). Materials for plankton modelling: Vertical distribution of Atlantic zooplankton in summer. *J. Plankton Res.* **1**, 1–28.

Longhurst, A. R., and Williams, R. (1992). Carbon flux by seasonal vertical migrant copepods is a small number. *J. Plankton Res.* **14**, 1495–1509.

Longhurst, A. R., and Wooster, W. S. (1990). Abundance of oil sardine and upwelling on the southwest coast of India. *Can. J. Fish. Aquat. Sci.* **47**, 2407–2419.

Longhurst, A. R., Sameoto, D., and Herman, A. (1984). Vertical distribution of arctic zooplankton in summer: Eastern Canadian archipelago. *J. Plankton Res.* **6**, 137–168.

Longhurst, A. R., et al. (1989). Biological oceanography in the Canadian High Arctic. *Rapp. Proc.-Verb. Réun. Cons. Int. Explor. Mer.* **188**, 80–89.

Longhurst, A. R., Sathyendranath, S., Platt, T., and Caverhill, C. M. (1995). An estimate of global primary production in the ocean from satellite radiometer data. *J. Plankton Res.* **17**, 1245–1271.

Lukas, R., and Lindstrom, E. (1991). The mixed layer of the western equatorial Pacific Ocean. *J. Geophys. Res.* **96**, 3343–3357.

Lutjeharms, J. R. E. (1985). Location of frontal systems between Africa and Antarctica: Some preliminary results. *Deep-Sea Res.* **32**, 1499–1509.

Lutjeharms, J. R. E. (1988). On the role of the East Madagascar Current as a source of the Agulhas Current. *S. Afr. J. Sci.* **84**, 236–238.

Lutjeharms, J. R. E., and Connell, A. D. (1989). The Natal Pulse and inshore countercurrents off the South African east coast. *S. Afr. J. Sci.* **85**, 533–534.

Lutjeharms, J. R. E., and Meeuwis, J. M. (1987). The extent and variability of south-east Atlantic upwelling. *S. Afr. J. Mar. Sci.* **5**, 51–62.

Lutjeharms, J. R. E., and van Ballegooyen, R. C. (1984). Topographic control in the Agulhas Current system. *Deep-Sea Res.* **31**, 1321–1338.

Lutjeharms, J. R. E., Bang, N. D., and Duncan, C. P. (1981). Characteristics of currents east and south of Madagascar. *Deep-Sea Res.* **28**, 879–899.

Lutjeharms, J. R. E., Walters, N. M., and Allanson, B. R. (1985). Oceanic frontal systems and biological enhancement. *In* "Antarctic Nutrient Cycles and Food Webs" (W. R. Siegfried, P. R. Condy, and R. M. Laws, Eds.), pp. 11–21. Springer-Verlag, Berlin.

Lutjeharms, J. R. E., Catzel, R., and Valentine, H. R. (1989). Eddies and other boundary phenomena of the Agulhas Current. *Cont. Shelf Res.* **9**, 597–616.

MacIsaac, J. J., Dugdale, R. C., Barber, R. T., Blascoe, D., and Packard, T. T. (1985). Primary production in an upwelling center. *Deep-Sea Res.* **32**, 503–529.

Mackintosh, N. A. (1937). The seasonal circulation of the Antarctic macro plankton. *Discovery Rep.* **16**, 365–412.

Madhupratap, M., and Haridas, P. (1990). Zooplankton, especially calanoid copepods, in the upper 1000 m of the south-east Arabian Sea. *J. Plankton Res.* **12**, 305–321.

Madhupratap, M., Kumar, S. P., and Bhattathiri, P. M. A. (1996). Mechanisms of the biological response to winter cooling in the northeastern Arabian Sea. *Nature (London)* **384**, 549–552.

Maenhaut, W., Raemdonck, H., Selen, A., van Grieken, R., and Winchester, J. W. (1983). Characterization of the atmospheric aerosol over the eastern Equatorial Pacific. *J. Geophys. Res.* **88**, 5353–5364.

Mann, C. R. (1967). The termination of the Gulf Stream and the beginning of the North Atlantic Current. *Deep-Sea Res.* **14**, 337.

Mann, K. H., and Lazier, J. R. N. (1991). "Dynamics of Marine Ecosystems." Blackwell, Boston.

Mantoura, F., Law, C. S., Owens, N. J. P., Burkill, P. H., Woodward, A. M. S., Howland, R. J. M., and Llewellyn, C. A. (1993). Nitrogen biogeochemical cycling in the northwestern Indian Ocean. *Deep-Sea Res. II* **40**, 651–671.

Mao, H. L., Hu, D. X., Zhao, B. R., and Ding, Z. X. (1986). A cyclonic eddy in the northeastern China Sea. *Stud. Mar. Sinica* **27**, 21–31.

Margalef, R. (1994). Through the looking glass; How marine phytoplankton appears through the microscope when graded by size and taxonomically. *Sci. Mar.* **58**, 87–101.

Marra, J. (1982). Variability in surface chlorophyll-a at a shelf-break front (New York Bight). *J. Mar. Res.* **40**, 575–591.

Martin, J. H., and Fitzwater, S. E. (1988). Iron deficiency limits phytoplankton growth in the northeast Pacific subarctic. *Nature (London)* **331**, 341–343.

Martin, J. H., Gordon, R. M., Fitzwater, S., and Broenkow, W. W. (1989). VERTEX: Phytoplankton studies in the Gulf of Alaska. *Deep-Sea Res.* **36**, 649–680.

Martin, J. H., Fitzwater, S. E., Gordon, R. M., Hunter, C. N., and Tanner, S. J. (1993). Iron, primary production and C–N flux studies during the JGOFS North Atlantic Bloom Experiment. *Deep-Sea Res. II* **40**, 1115–1134.

Mascarenas, A. S., Miranda, L. B., and Rock, N. J. (1971). A study of the oceanographic conditions in the region of Cabo Frio. *In* "Fertility of the Sea" (J. D. Costlow, Ed.), pp. 285–297. Gordon & Breach.

Matano, R. P., Schlax, M. G., and Chelton, D. B. (1993). Seasonal variability in the South Atlantic. *J. Geophys. Res.* **98**, 18027–18035.

Mathew, B. (1982). Studies on upwelling and sinking in seas around India. PhD thesis, University of Cochin.

McCarthy, J. J., and Nevins, J. L. (1986). Utilisation of nitrogen and phosphorus by primary producers in warm core ring 82-B following deep convective nixing. *Deep-Sea Res.* **33**, 1773–1788.

McClain, C. R. (1993). Review of major CZCS applications: U.S. case studies. *In* "Ocean Colour: Theory and Applications in a Decade of CZCS Experience" (V. Barale and P. Schlittenhardt, Eds.), pp. 167–188.

McClain, C. R., and Firestone, J. (1993). An investigation of Ekman upwelling in the North Atlantic. *J. Geophys. Res.* **98**, 12327–12339.

McClain, C. R., Esaias, W. E., Felmna, G. C., Elrod, J., and Endres, D. (1990). Physical and biological processes in the North Atlantic during the First GARP Global Experiment. *J. Geophys. Res.* **95C**, 18027–18048.

McClain, C. R., Arrigo, K., Tai, K.-S., and Turk, D. (1996). Observations and simulations of physical and biological processes at OWS P. 1951–1980. *J. Geophys. Res.* **101**, 3697–3713.

McCreary, J. P., Picaut, J., and Moore, D. W. (1984). Effects of remote annual forcing in the eastern tropical Atlantic Ocean. *J. Mar. Res.* **42**, 45–81.

McCreary, J. P., Kundu, P. K., and Molinari, R. L. (1993). A numerical investigation of dynamics, thermodynamics and mixed-layer processes in the Indian Ocean. *Prog. Oceanogr.* **31**, 181–244.

McCreary, J. P., Kohler, K. E., Hood, R. R., and Olson, D. B. (1996). A four component ecosystem model of biological activity in the Arabian Sea. *Prog. Oceanogr.* **37**, 193–240.

McGill, D. A. (1973). Light and nutrients in the Indian Ocean. *In* "The Biology of the Indian Ocean" (B Zeitzschel, Ed.), pp. 53–102. Springer-Verlag, Berlin.

McGillicuddy, D. J., Jr., and Robinson, A. R. (1997). Eddy-induced nutrient supply and new production in the Sargasso Sea. *Deep-Sea Res.* **44**, 1427–1450.

McGowan, J. A. (1971). Oceanic biogeography of the Pacific. *In* "The Micropaleontology of the Oceans" (S. M. Funnell and W. R. Riedl, Eds.), pp. 3–74. Cambridge Univ. Press, Cambridge, UK.

McGowan, J. A., and Williams, P. M. (1973). Oceanic habitat differences in the North Pacific. *J. Exp. Mar. Biol. Ecol.* **12**, 187–212.

McMurray, H. F., Carter, R. A., and Lucas, M. I. (1993). Size-fractionated phytoplankton production in western Agulhas Bank continental shelf waters. *Cont. Shelf Res.* **13**, 307–329.

McRoy, C. P., and Goering, J. J. (1974). The influence of ice on the primary productivity of the Bering Sea. *In* "Oceanography of the Bering Sea" (D. W. Hood and E. J. Kelly, Eds.), pp. 403–421. Univ. of Alaska Press, Fairbanks.

McRoy, P. C., Hood, D. W., Coachman, L. K., Walsh, J. J., and Goering, J. J. (1985). Processes and resources of the Bering Sea shelf (PROBES): The development and accomplishments of the project. *Cont. Shelf Res.* **5**, 5–21.

Melillo, J. M., McGuire, A. D., Kicklighter, D. W., Moore, B., Vorosmarty, C. J., and Schloss, A. L. (1993). Global climate change and terrestrial net primary production. *Nature (London)* **363**, 234–240.

Mensah, M. A. (1974). The reproduction and feeding of the marine copepod Calanoides carinatus in Ghanaian waters. *Ghana J. Sci.* **14**, 167–191.

Menzel, D. W., and Ryther, J. H. (1960). The annual cycle of primary production in the Sargasso Sea off Bermuda. *Deep-Sea Res.* **6**, 351–367.

Merle, J. (1980). Variabilité thermique annuelle et interannuelle de l'oceean Atlantique equatorial est. *Océanol. Acta* **3**, 209–220.

Merle, J. (1983). Seasonal variability of subsurface thermal structure in the tropical Atlantic. *In* "Hydrodynamics of the Equatorial Ocean" (J. C. J. Nihoul, Ed.), Oceanography Series No. 36. Elsevier, Amsterdam.

Merle, J., and Arnault, S. (1985). Seasonal variability of the surface dynamic topography in the tropical Atlantic Ocean. *J. Mar. Res.* **43**, 267–288.

Metcalf, W. G., and Stalcup, M. C. (1967). Origin of the Atlantic equatorial undercurrent. *J. Geophys. Res.* **72**, 4954–4975.

Michaels, A. F., and Knap, A. H. (1996). Overview of the U.S. JGOFS Bermuda Atlantic Time Series study. *Deep-Sea Res. II* **43**, 157–198.

Mihnea, P. E. (1997). Major shifts in the phytoplankton community (1980–1994) in the Romanian Black Sea. *Océanol. Acta* **20**, 119–129.

Miller, A. J., Cayan, D. A., Barnett, T. P., . Graham, N. E., and Oberhufer, J. M. (1994). The 1976–77 climate shift of the Pacific Ocean. *Oceanography* **7**, 21–26.

Miller, A. R. (1983). Mediterranean Sea; Physical aspects. *In* "Ecosystems of the World, 26: Estuaries and Enclosed Seas" (B. H. Ketchum, Ed.), pp. 219–238. Elsevier, Amsterdam.

Miller, C. B., Frost, B., Batchelder, H., Clemens, M., and Conway, R. (1984). Life histories of large grazing copepods in a subarctic ocean gyre. *Prog. Oceanogr.* **13**, 201–243.

Miller, C. B., Frost, B. W., Wheeler, P. A., Landry, M. R., Welschmayer, N., and Powell, T. M. (1991). Ecological dynamics in the subarctic ecosystem: A possibly iron limited ecosystem. *Limn. Oceanogr.* **36**, 1600–1615.

Milliman, J. D., and Meade, R. H. (1983). World-wide delivery of river sediment to the oceans. *J. Geol.* **91**, 1–21.

Millot, C. (1987). Circulation in the Western Mediterranean Sea. *Océanol. Acta* **10**, 143–150.

Millot, C. (1992). Are there differences between the largest Mediterranean Seas? *Bull. Inst. Océanogr. Monaco* **11**, 3–25.

Mills, E. L. (1989). "Biological Oceanography, an Early History, 1970–1960." Cornell Univ. Press, Ithaca, NY.

Minas, H. J., and Minas, M. (1992). Net community production in "HNLC" waters of the tropical and Antarctic oceans: Grazing v. iron hypothesis. *Océanol. Acta* **15**, 145–162.

Minas, H. J., and Nival, P. (1988). Pelagic Mediterranean oceanography. *Océanol. Acta* **9**, 1–250.

Minas, H. J., Codispoti, L. A., and Dugdale, R. C. (1982). Nutrients and primary production in the upwelling region off northwest Africa. *Rapp. Proc.-Verb. Réun. Cons. Int. Explor. Mer.* **180,** 148–183.

Minas, H. J., Coste, B., Le Corre, P., Minas, M., and Raimbault, P. (1991). Biological and geochemical signatures associated with the water circulation in the western Alboran Sea. *J. Geophys. Res.* **90,** 8755–8771.

Mitchell, B. G., Brody, E. A., Yeh, E. N., McLain, C. R., Comiso, J., and Maynard, N. G. (1991a). Meridional zonation of the Barents Sea ecosystem. *Polar Biol.* **10,** 147–161.

Mitchell, B. G., Brody, E. A., Holm-Hansen, O., McLain, C., and Bishop, J. (1991b). Light limitation of phytoplankton biomass and macronutrient utilisation in the Southern Ocean. *Limn. Oceanogr.* **36,** 1662–1677.

Mittelstaedt, E. (1983). Large-scale circulation along the coast of Northwest Africa. *Rapp. Proc.-Verb. Réun. Cons. Int. Explor. Mer.* **180,** 50–57.

Mittelstaedt, E. (1991). The ocean boundary along the northwest African coast: Circulation and oceanographic properties at the sea surface. *Prog. Oceanogr.* **26,** 307–355.

Miya, M., and Nishida, M. (1997). Speciation in the open ocean. *Nature (London)* **389,** 803–804.

Molinari, R. L., and Morrison, J. (1988). The separation of the Yucatan Current from the Campeceh Bank and the penetration of the Loop Current into the Gulf of Mexico. *J. Geophys. Res.* **93,** 10645–10654.

Mooers, C. N. K., and Robinson, A. R. (1984). Turbulent jets and eddies in the California Current and inferred cross-shore transports. *Science* **223,** 51–53.

Mooers, C. N. K., Flagg, C. N., and Boicourt, W. C. (1978). Prograde and retrograde fronts. *In* "Ocean Fronts in Coastal Processes" (M. J. Bowman and W. E. Esaias, Eds.), pp. 43–86. Springer-Verlag, Berlin.

Moore, D. W., Hisard, P., McCreary, J., Merle, J., O'Brien, J., Picaut, J., Verstraete, J. M., and Wunsch, C. (1978). Equatorial adjustment in the eastern Atlantic. *Geophys. Res. Lett.* **5,** 637–640.

Moreira da Silva, P. C. (1971). Upwelling and its biological effects in southern Brazil. *In* "Fertility of the Sea" (J. D. Costlow, Ed.), pp. 469–478. Gordon & Breach, New York.

Morel, A., and Berthon, J.-F. (1989). Surface pigments, algal biomass profiles and potential production of the euphotic layer: Relationships reinvestigated in view of remote sensing applications. *Limn. Oceanogr.* **34,** 1545–1562.

Motoda, S., and Minoda, T. (1974). Plankton of the Bering Sea. *In* "Oceanography of the Bering Sea" (D. W. Hood and E. J. Kelly, Eds.), pp. 207–241. Univ. of Alaska Press, Fairbanks.

Mueller, J. L., and Lang, R. E. (1989). Bio-optical provinces of the Northeast Pacific Ocean: A provisional analysis. *Limn. Oceanogr.* **34,** 1572–1586.

Muench, R. D. (1970). The physical oceanography of the northern Baffin Bay region. *Rep. Arch. Inst. North Am.* **1,** 1–150.

Muench, R. D. (1990). Mesoscale phenomena in the Polar Oceans. *In* "Polar Oceanography. Part A. Physical Science," pp. 223–285. Academic Press, San Diego.

Mulhearn, P. J. (1987). The Tasman front: A study using satellite infrared imagery. *J. Phys. Oceanogr.* **17,** 1148–1155.

Müller-Karger, F. E., McClain, C. R., and Richardson, P. L. (1988). The dispersal of the Amazon's water. *Nature (London)* **333,** 56–59.

Müller-Karger, F. E., McClain, C. R., Fisher, T. R., Esaias, W. E. V., and Varela, R. (1989). Pigment observations in the Caribbean Sea: Observations from space. *Prog. Oceanogr.* **35,** 23–64.

Murray, J. W. (1991). Black Sea oceanography. *Deep-Sea Res.* **38**(Suppl. 2A), 1–1266.

Murray, J. W., and Izdar, E. (1989). The 1988 Black Sea oceanographic expedition: Overview and new discoveries. *Oceanography* **2,** 15–21.

Nagata, Y., Yoshida, J., and Shin, H. R. (1986). Detailed structure of the Kuroshio Front and the origin of the water in warm-core rings. *Deep-Sea Res.* **33,** 1509–1526.

Nathansohn, A. (1909). Beiträge zur Biologie des Planktons. II. Vertikalzirkulation und Planktonmaxima im Mittelmeer. *Int. Rev. Ges. Hydrobiol.* **2,** 580–632.

Nelson, G., and Hutchings, L. (1983). The Benguela upwelling area. *Prog. Oceanogr.* **12,** 333–356.

Nemoto, T., and Harrison, W. G. (1981). High latitude ecosystems. *In* "Analysis of Marine Ecosystems" (A. R. Longhurst, Ed.), pp. 95–126. Academic Press, San Diego.

Neumann, G. (1969). The Equatorial Undercurrent in the Atlantic Ocean. *Proc. Symp. Oceanogr. Fish. Res. Trop. Atlantic UNESCO,* 33–44.

Niiler, P. P. (1977). One dimensional models of the seasonal thermocline. *In* "The Sea" (E. D. Goldberg, Ed.), pp. 97–115. Wiley, New York.

Nitani, H. (1972). The beginning of the Kuroshio. *In* "Kuroshio" (H. Stommel and K. Yoshida, Eds.), pp. 129–164. East–West Center, Hawaii.

Nowlin, W. D., and Klinck, J. M. (1986). The physics of the Antarctic Circumpolar Current. *Rev. Geophys. Space Phys.* **24,** 469–491.

Obata, A., Ishizaka, J., and Endoh, M. (1996). Global verification of critical depth theory for phytoplankton bloom with climatological in situ temperature and satellite ocean colour data. *J. Geophys. Res.* **101,** 20657–20667.

Odum, E. P. (1971). "Fundamentals of Ecology." Saunders, Philadelphia.

Oguz, T., LaViolette, P. E., and Unluata, U. (1992). The upper layer circulation of the Black Sea: Its variability as inferred from hydrographic and satellite observations. *J. Geophys. Res.* **97**, 12569–12584.

Okkonen, S. R. (1992). Shedding of an anticyclonic eddy from the Alaskan Stream. *Geophys. Res. Lett.* **12**, 2397–2400.

Olson, D. B., Hitchcock, G. L., Fine, R. A., and Warren, B. A. (1993). Maintenance of the low-oxygen layer in the central Arabian Sea. *Deep-Sea Res. II* **40**, 673–685.

O'Reilly, J. E., and Busch, D. A. (1984). Phytoplankton primary production on the northwestern Atlantic shelf. *Rapp. Proc.-Verb. Réun. Cons. Int. Explor. Mer.* **183**, 255–268.

Orr, A. P. (1933). Scientific report of the Great Barrier Reef Expedition, 1928–29. *Bull. Br. Mus. (Natl. History)* **2**, 37–86.

Osborne, J., Swift, J., and Flinchem, E. P. (1992). Ocean atlas for the Macintosh, SIO No. 92-29. Scripps Institution of Oceanography, La Jolla, CA.

Osgood, K. E., and Checkley, D. M. (1997). Seasonal variations in a deep aggregation of Calanus pacificus in the Santa Barbara Basin. *Mar. Ecol. Prog. Ser.* **148**, 59–69.

Ottens, J. J. (1991). Planktonic foraminifera as North Atlantic water mass indicators. *Océanol. Acta* **14**, 123–140.

Oudot, C. (Ed.) (1987). "Observations physico-chimiques et biomasse vegetale dans l'océan Atlantique equatorial. Programme PIRAL." Off. Res. Sci. Techn. Outre-Mer., Paris.

Oudot, C., and Morin, P. (1987). The distribution of nutrients in the equatorial Atlantic: Relation to physical processes and phytoplankton biomass. *Océanol. Acta* **SP6**, 121–130.

Owen, R. W. (1981). Fronts and eddies in the ocean. *In* "Analysis of Marine Ecosystems" (A. R. Longhurst, Ed.), pp. 197–233. Academic Press, San Diego.

Owens, N. J. P., Burkill, P. H., Mantoura, R. F. C., Woodward, E. M. S., Bellan, I. E., Aiken, J., Howland, R. J. M., and Llewellen, C. C. (1993). Size-fractionated primary production and nitrogen assimilation in the N. W. Indian Ocean. *Deep-Sea Res. II* **40**, 697–710.

Paffenhöfer, G.-A. (1983). Vertical zooplankton distribution on the northeast Florida shelf. *J. Plankton Res.* **5**, 15–34.

Paluskiewicz, T., Atkinson, L. P., Posmentier, E. S., and McClain, C. R. (1983). Observations of a Loop Current frontal eddy intrusion onto the West Florida continental shelf. *J. Geophys. Res.* **88**, 9639–9651.

Parrish, R. (1989). The South Pacific oceanic horse mackerel (*Trachurus picturatus murphyi*) fishery. *In* "The Peruvian Upwelling System: Dynamics and Interactions" (D. Pauly, P. Muck, J. Mendo, and I. Tsukuyama, Eds.), pp. 321–331. IMARPE/ GTZ/ ICLARM, Lima/Frankfurt/Manila.

Partos, P., and Piccolo, M. C. (1988). Hydrography of the Argentine continental shelf between 38° and 42°S. *Cont. Shelf Res.* **8**, 1043–1056.

Pauly, D., and Tsukuyama, I. (1987). The Peruvian anchoveta and its upwelling ecosystem: Three decades of change. *ICLARM Stud. Rev.* **15**, 1–351.

Pauly, D., Muck, P., Mendo, J., and Tsukuyama, I. (1989). The Peruvian upwelling system: Dynamics and interactions. *ICLARM Conf. Proc.* **18**, 1–438.

Pearce, A. F., and Gründling, M. L. (1982). Is there a seasonal variation in the Agulhas Current? *J. Mar. Res.* **40**, 177–184.

Peeken, I. (1997). Photosynthetic pigment fingerprints as indicators of phytoplankton biomass and development in different water masses of the Southern Ocean during austral spring. *Deep-Sea. Res. II* **44**, 261–282.

Peinert, R., Bachman, U., von Bodungen, B., and Noji, T. (1989). Impact of grazing on spring phytoplankton growth and sedimentation in the Norwegian Channel. *Mitt. Geol. Palaeontol. Inst. Hamburg* **62**, 149–164.

Peterson, R. G., and Stramma, L. (1991). Upper-level circulation in the South Atlantic. *Prog. Oceanogr.* **26**, 1–73.

Peterson, R. G., and Whitworth, T. (1989). The subantarctic and polar fronts in relation to deep water masses through the southwestern Atlantic. *J. Geophys. Res.* **94**, 10817–10838.

Peterson, W. T., Miller, C. B., and Hutchinson, A. (1979). Zonation and maintenance of copepod populations in the Oregon upwelling zone. *Deep-Sea Res.* **26**, 467–494.

Petit, D., and Courties, C. (1976). *Calanoides carinatus* sur le plateau continentale congolaise II. *Cah. ORSTOM. Sér. Océanogr.* **14**, 177–199.

Philander, S. G., and Rasmusson, E. M. (1985). The southern oscillation and El Niño. *Adv. Geophys.* **28A**, 197–215.

Philander, S. G. H. (1978). Variability of the tropical oceans. *Dyn. Atmos. Oceans* **3**, 191–208.

Philander, S. G. H. (1979). Upwelling in the Gulf of Guinea. *J. Mar. Res.* **37**, 23–33.

Philander, S. G. H. (1985). Tropical oceanography. *Adv. Geophys.* **28**, 461–477.

Philander, S. G. H., and Chao, Y. (1991). On the contrast between the seasonal cycles of the equatorial Atlantic and Pacific Oceans. *J. Phys. Oceanogr.* **21**, 1399–1406.

Philander, S. G. H., and Pacanowski, R. C. (1981). The oceanic response to cross equatorial winds with application to coastal upwelling in low latitudes. *Tellus* **33**, 201–210.

Philander, S. G. H., Hurlin, W., and Pacanowski, R. C. (1987). A model of the seasonal cycle in the tropical Pacific Ocean. *J. Phys. Oceanogr.* **17**, 1987–2002.

Philander, S. G. H., Gu, D., and Halpern, D. (1996). Why the ITCZ is mostly north of the equator. *J. Climate* **9**, 2958–2972.

Picaut, J. (1981). Propagation of the seasonal upwelling in the Gulf of Guinea. *In* "Recent Progress in Equatorial Oceanography" (J. P. McCreary, D. W. Moore, and J. M. Witte, Eds.), pp. 271–280. Scientific Committee for Oceanographic Research.

Piccioni, A., Gabriele, M., Salusti, E., and Zambianchi, E. (1988). Wind-induced upwellings off the southern coast of Sicily. *Océanol. Acta* **9**, 309–314.

Pingree, R. D. (1975). The advance and retreat of the thermocline on the continental shelf. *J. Mar. Biol. Assoc. UK* **55**, 965–974.

Pingree, R. D., and Griffiths, D. K. (1978). Tidal fronts on the shelf seas around the British Isles. *J. Geophys. Res.* **83**, 4615–4622.

Pingree, R. D., and Griffiths, D. K. (1980). A numerical model of the M2 tide in the Gulf of St. Lawrence. *Océanol. Acta* **3**, 221–225.

Pingree, R. D., and Mardell, G. T. (1981). Slope turbulence, internal waves and phytoplankton growth at the Celtic Sea shelf edge. *Philos. Trans. R. Soc. London. A* **302**, 663–682.

Pingree, R. D., Pugh, P. R., Holligan, P. M., and Forster, G. R. (1975). Summer phytoplankton blooms and red tides along tidal fronts in the approaches to the English Channel. *Nature (London)* **258**, 672–677.

Pingree, R. D., Mardell, G. T., Holligan, P. M., Griffiths, D. K., and Smithers, J. (1982). Celtic Sea and Armorican current structure and the vertical distributions of temperature and chlorophyll. *Cont. Shelf Res.* **1**, 99–116.

Pitelka, F. A. (1941). Distribution of birds in relation to major biotic communities. *Am. Midl. Nat.* **25**, 113–157.

Platt, T., and Sathyendranath, S. (1988). Oceanic primary production: Estimation by remote sensing at local and regional scales. *Science* **241**, 1613–1620.

Platt, T., Woods, J., Sathyendranath, S., and Barkman, W. (1994). Primary production, respiration and stratification in the ocean. *J. Geophys. Res.* **102**, 12765–12787.

POEM Group (1992). General circulation of the eastern Mediterranean. *Earth Sci. Rev.* **32**, 285–309.

Prata, A. J., and Wells, J. B. (1990). A satellite sensor image of the Leeuwin current, Western Australia. *Int. J. Remote Sensing* **11**, 173–180.

Price, N. M., Ahner, B. A., and Morel, F. M. (1994). The equatorial Pacific Ocean: Grazer-controlled phytoplankton populations in an iron-limited ecosystem. *Limn. Oceanogr.* **39**, 520–534.

Quadfasel, D., and Cresswell, G. R. (1992). A note on the seasonal variability of the South Java Current. *J. Geophys. Res.* **97**, 3685–3688.

Radenac, M.-H., and Rodier, M. (1996). Nitrate and chlorophyll distributions in relation to thermohaline and current structures in the western tropical Pacific during 1985–1989. *Deep-Sea Res. II* **43**, 725–752.

Ragoonaden, V., Babu, R., and Sastry, J. S. (1987). Physico-chemical characteristics and circulation of waters in the Mauritius–Seychelles ridge zone. *Ind. J. Mar. Sci.* **16**, 184–191.

Rahmstorf, S. (1992). Modelling ocean temperatures and mixed-layer depths in the Tasman Sea off the South Island, New Zealand. *N. Z. J. Mar. Freshwater Res.* **26**, 37–51.

Rao, D. P., and Sastry, J. S. (1981). Circulation and distribution of some hydrographical properties during late winter in the Bay of Bengal. *Mahasagar* **14**, 1–15.

Rao, R. R. (1986). Cooling and deepening of the mixed layer in the central Arabian Sea during MONSOON-77: Observations and simulations. *Deep-Sea Res.* **33**, 1413–1424.

Rao, R. R., Molinari, R. L., and Festa, J. F. (1989). Evolution of the climatological near-surface thermal structure of the tropical Indian Ocean. *J. Geophys. Res.* **94**, 10801–10815.

Read, J. F., Pollard, R. T., Morrison, A. I., and Symon, C. (1995). On the southerly extent of the ACC in the southeast Pacific. *Deep-Sea Res. II* **42**, 933–954.

Reid, J. L. (1962). On the circulation, phosphate-phosphorus content and zooplankton volumes in the upper part of the North Pacific Ocean. *Limn. Oceanogr.* **2**, 287–306.

Reid, J. L. (1964). A transequatorial Atlantic oceanographic section in July, 1963 compared with other Atlantic and Pacific sections. *J. Geophysic. Res.* **69**, 5205–5215.

Reid, J. L., Brinton, E., Fleminger, A., Venrick, E. L., and McGowan, J. A. (1978). Ocean circulation and marine life. *In* "Advances in Oceanography" (H. Charnock and G. Deacon, Eds.), pp. 65–130. Plenum, New York.

Revelante, N., and Gilmartin, M. (1994). Relative increase of larger phytoplankton in a subsurface maximum in the Adriatic Sea. *J. Plankton Res.* **17**, 1535–1562.

Reverdin, G., and Fieux, M. (1987). Sections in the Indian Ocean—Variability in the temperature structure. *Deep-Sea Res.* **34**, 601–626.

Rey, F. (1990). P/I relationship in natural phytoplankton populations in the Barents Sea. *Polar Res.* **10**, 105–116.

Richardson, P. L. (1976). Gulf Stream rings. *Oceanus* **19**, 65–68.

Richardson, P. L. (1980). Anticyclonic eddies generated near the Corner seamounts. *J. Mar. Res.* 38, 673–686.

Richardson, P. L. (1983). Gulf Stream rings. *In* "Eddies in Marine Science" (A. R. Robinson, Ed.), pp. 19–45. Springer-Verlag, Berlin.

Richardson, P. L. (1985). Drifting derelicts in the North Atlantic, 1883–1902. *Prog. Oceanogr.* 14, 463–483.

Richter, C. (1994). Regional and seasonal variability in the vertical distribution of mesozooplankton in the Greenland Sea. *Ber. Polarforsch.* 154, 1–87.

Riley, G. A. (1942). The relationship between vertical turbulence and the spring flowering of diatoms. *J. Mar. Res.* 5, 67–87.

Riley, G. A. (1957). Phytoplankton of the north central Sargasso Sea. *Limn. Oceaogr.* 2, 252–267.

Ring Group (1981). Gulf stream cold core rings: Their physics, chemistry and biology. *Science* 212, 1091–1100.

Robinson, A. R., and Malanotte-Rizzoli, P. (1993). Physical oceanography of the eastern Mediterranean. *Deep-Sea Res. II* 40, 1073–1332.

Robinson, C. L. K., Ware, D. M., and Parsons, T. R. (1993). Simulated annual plankton production in the northeastern Pacific Coastal Upwelling Biome. *J. Plankton Res.* 15, 161–183.

Robinson, M. K., Bauer, R. A., and Schroeder, E. H. (1979). "Atlas of the North Atlantic Ocean Monthly Mean Temperatures and Salinities of the Surface Layer." U.S. Naval Oceanographic Office, Washington, DC.

Robles, F., Alarn, E., and Ulloa, A. (1980). Water masses in the northern Chilean zone. *In* "Proceedings of the Workshop on the Phenomenon Known as El Niño," pp. 83–174. UNESCO, Paris.

Rochford, D. J. (1975). Nutrient enrichment of east Australian coastal waters. II. Laurieton upwelling. *Aust. J. Mar. Freshwater Res.* 26, 385–397.

Rochford, D. J. (1986). Seasonal changes in the distribution of Leeuwin Current waters off southern Australia. *Aust. J. Mar. Freshwater Res.* 37, 1–10.

Roden, G. I. (1964). Oceanographic aspects of the Gulf of California. *Mem. Am. Soc. Petr. Geol.* 3, 30–58.

Roden, G. I. (1970). Aspects of the mid-Pacific Transition Zone. *J. Geophys. Res.* 75, 1097–1109.

Roden, G. I. (1975). On North Pacific temperature, salinity, sound velocity and density fronts and their relation wind and energy flux fields. *J. Phys. Oceanogr.* 5, 557–571.

Rodriguez, J., Jimenez, F., Bautista, B., and Rodriguez, V. (1987). Planktonic biomass spectra during a winter production pulse in Mediterranean waters. *J. Plankton Res.* 9, 1183–1194.

Roemmich, D., and McGowan, J. (1994). Climatic warming and the decline of zooplankton in the California Current. *Science* 267, 1324–1326.

Roemmich, D. (1981). Circulation of the Caribbean Sea: A well-resolved inverse problem. *J. Geophys. Res.* 86, 7993–8005.

Rosa, H., and Laevastu, T. (1959). Comparison of biological and ecological characteristics of sardines and related species—A preliminary study. *Proc. World Meeting Biol. Sardines Related Species* 2, 523–534.

Rosen, D. E. (1975). Doctrinal biogeography. *Q. Rev. Biol.* 50, 69–70.

Rougerie, F., and Henin, C. (1977). Coral and Solomon Seas in the austral summer monsoon. *Cah. Orstom Oceanogr.* 15, 261–278.

Rougerie, F., and Rancherie, J. (1994). The Polynesian South Ocean: Features and circulation. *Mar. Pollut. Bull.* 29, 14–25.

Roy, C. (1990). Les upwellings: Le cadre physique des pecheries côtieres ouest-africains. *In* "Pécheries Ouest-Africaines" (P. Curie and C. Roy, Eds.), pp. 38–66. ORSTOM, Paris.

Rudjakov, J. A. (1997). Quantifying seasonal phytoplankton oscillations in the global offshore ocean. *Mar. Ecol.-Progr. Ser.* 146, 225–230.

Russell, F. S., Southward, A. J., Boalch, G. T., and Butler, E. I. (1971). Changes in biological conditions in the English Channel off Plymouth during the last half-century. *Nature (London)* 234, 468–470.

Ryther, J. H., Menzel, D., and Corwin, N. (1967). Influence of the Amazon River outflow on the ecology of the western tropical Atlantic. *J. Mar. Res.* 25, 69–83.

Saetre, R., and Jorge da Silva, A. (1984). The circulation of the Mozambique Channel. *Deep-Sea Res.* 31, 485–508.

Saijo, Y., Kawamura, T., Iizka, S., and Nozawa, K. (1970). Primary production in Kuroshio and its adjacent area. *Proc. 2nd CSK Symp. Tokyo,* 169–175.

Salas de Leon, D. A., and Monreal-Gomez, M. A. (1986). The role of the Loop Current in the Gulf of Mexico fronts. *In* "Marine Interfaces Hydrodynamics" (J. C. J. Nihoul, Ed.), Oceanography Series No. 42, pp. 295–300. Elsevier, Amsterdam.

Saltzman, J., and Wishner, K. (1997a). Zooplankton ecology in the eastern tropical Pacific oxygen minimum zone above a seamount. I. General trends. *Deep-Sea Res. II* 44, 907–930.

Saltzman, J., and Wishner, K. (1997b). Zooplankton ecology in the eastern tropical Pacific oxygen minimum zone above a seamount. II. Vertical distribution of copepods. *Deep-Sea Res. II* 44, 931–954.

Sambrotto, R. N., Goering, J. J., and McRoy, C. P. (1984). Large yearly production of phytoplankton in the western Bering Strait. *Science* 225, 1147–1150.

Sameoto, D., Herman, A., and Longhurst, A. R. (1986). Relations between the thermocline, meso-

and microzooplankton, chlorophyll-a and primary production distributions in Lancaster Sound. *Polar Biol.* **6**, 53–61.

Santamaria-del-Angel, E., Alvarez-Borrego, S., and Müller-Karger, F. E. (1994). Gulf of California biogeographic regions based on CZCS imagery. *J. Geophys. Res.* **99**, 7411–7421.

Sarma, V. V., and Aswanikumar, V. (1991). Subsurface chlorophyll maxima in the northwestern Bay of Bengal. *J. Plankton Res.* **13**, 339–352.

Sathyendranath, S., and Platt, T. (1994). New production and mixed-layer physics. *Proc. Ind. Acad. Sci. (Earth Planet. Sci.)* **103**, 177–188.

Sathyendranath, S., Gouveia, A. E., Shetye, R., Ravindran, P., and Platt, T. (1991). Biological control of surface temperature in the Arabian Sea. *Nature (London)* **349**, 54–56.

Sathyendranath, S., Longhurst, A. R., Caverhill, C. M., and Platt, T. (1995). Regionally and seasonally differentiated primary production in the North Atlantic. *Deep-Sea Res.* **42**, 1773–1802.

Satoh, H., Tanaka, H., and Koike, T. (1992). Light conditions and photosynthetic characteristics of the deep chlorophyll maximum in Solomon Sea. *Jpn. J. Phycol.* **40**, 135–142.

Saur, J. F. T. (1980). Surface salinity on the San Francisco–Honolulu route, June 1966–December 1975. *J. Phys. Oceanogr.* **10**, 1669–1680.

Schnack, S. B., Smetacek, V., von Bodungen, B., and Stegman, P. (1985). Utilisation of phytoplankton by copepods in Antarctic waters during spring. *In* "Marine Biology of Polar Regions" (J. S. Gray and M. E. Christiansen, Eds.), pp. 65–83. Wiley, New York.

Schnack-Schiel, S. B., Hagen, W., and Mizdalski, E. (1991). Seasonal comparison of *Calanoides acutus* and *Calanus propinquus* in the Weddell Sea. *Mar. Ecol. Prog. Ser.* **70**, 17–27.

Schnack-Schiel, S. B., Thomas, D., Dieckmann, G. S., Eicken, H., Gradinger, R., Spindler, M., Weissenberger, J., and Beyer, K. (1995). Life cycle strategy of the Antarctic copepod *Stephos longiceps*. *Prog. Oceanogr.* **36**, 45–75.

Schroeder, E. H. (1965). Average monthly temperatures in the North Atlantic. *Deep-Sea Res.* **12**, 323–343.

Schott, F. (1983). Monsoon response of the Somali Current and associated upwelling. *Prog. Oceanogr.* **12**, 357–381.

Schumacher, J. D., and Reed, R. K. (1983). Interannual variability in abiotic environment of Bering Sea and Gulf of Alaska. *In* "From Year to Year" (W. S. Wooster, Ed.), pp. 111–133. Univ. of Washington Press, Seattle.

Schumann, E. H. (1982). Inshore circulation of the Agulhas Current off Natal. *J. Mar. Res.* **40**, 43–55.

Segerstråle, S. (1957). Baltic Sea. *Mem. Geol. Soc. Am.* **67**, 751–800.

Servain, J., Picaut, J., and Merle, J. (1982). Evidence of remote forcing in the equatorial Atlantic Ocean. *J. Phys. Oceanogr.* **12**, 457–463.

Sevrin-Reyssac, J. (1993). Phytoplancton et production primaire dans les eaux marines ivoiriennes. *In* "Environnement et Ressources Aquatiques de Côte d'Ivoire" (P. Le Loeuff, E. Marchal, and J.-B. Amon Kothias, Eds.), pp. 151–166. ORSTOM, Paris.

Shah, N. H. (1973). Seasonal variation of phytoplankton pigments in the Laccadive Sea off Cochin. *In* "Biology of the Indian Ocean" (B. Zeitzschel, Ed.), pp. 175–185. Chapman & Hall, London.

Shannon, L. V. (1985). The Benguela ecosystem. Part 1. Physical features and processes. *Oceanogr. Mar. Biol. Annu. Rev.* **23**, 105–182.

Shannon, L. V., and Pillar, S. C. (1986). The Benguela ecosystem. Part III. Plankton. *Oceanogr. Mar. Biol. Annu. Rev.* **24**, 65–170.

Shannon, L. V., Agenbag, J. J., and Buys, M. E. L. (1987). Large and mesoscale features of the Angola/Benguela Front. *S. Afr. J. Mar. Sci.* **5**, 11–34.

Sharma, V. V., and Aswanikumar, V. (1991). Subsurface chlorophyll maxima in the northwestern Bay of Bengal. *J. Plankton Res.* **13**, 339–352.

Shaw, P. T. (1992). Shelf circulation off the southeast coast of China. *Rev. Aquat. Sci.* **6**, 1–28.

Sherman, K., and Alexander, L. M. (1989). Biomass yields and geography of large marine ecosystems. *Am. Assoc. Adv. Sci. Selected Symp.* **111**, 1–493.

Sherman, K., Jaworski, N. A., and Smayda, T. J. (1996). "The Northeast Shelf Ecosystem: Assessment, Sustainability and Management." Blackwell Sci., Oxford.

Shibata, A. (1983). Meander of the Kuroshio along the edge of continental shelf in the east China sea. *Umi To Sora* **58**, 113–120.

Shih, C. T. (1979). East–west diversity. *In* "Zoogeography and Diversity in Plankton" (H. van der Spoel and I. Peirrot-Bults, Eds.), pp. 87–102. Bunge, Utrecht.

Shinn, E. A. (1987). Carbonate coastal accretion in an area of longshore transport. Northeast Qatar. *In* "Key Environments—Red Sea" (A. J. Edwards and S. M. Head, Eds.), pp. 179–191.

Shirtcliffe, T. G. L., Moore, M. I., Cole, A. G., Viner, A. B., Baldwin, R., and Chapman, B. (1990). Dynamics of the Cape Farewell upwelling plume. *N. Z. J. Mar. Freshwater Res.* **24**, 555–568.

Shukla, J. (1987). Interannual variability of monsoons. *In* "Monsoons" (J. S. Fein and P. L. Stephens, Eds.), pp. 399–463. Wiley, New York.

Sieburth, J. M., and Davis, P. G. (1984). The role of heterotrophic nanoplankton in nurturing and grazing planktonic bacteria. *Annu. Inst. Océanogr.* **58**, 285–296.

Siedler, G., Zangenberg, N., and Onken, R. (1992). Seasonal changes in the tropical Atlantic circulation: Observation and simulation of the Guinea Dome. *J. Geophys. Res.* **97**, 703–715.

Sievers, H. A., and Silva, N. (1975). Water masses and circulation in the southeastern Pacific Ocean. *Cienc. Technol. Mar.* **1**, 7–67.

Silva Sandoval, N., and Neshyba, S. (1977). Surface currents off the southern coast of Chile. *Cienc. Technol. Mar.* **3**, 37–42.

Simpson, J. H. (1981). The shelf sea fronts: Implications of their existence and behaviour. *Proc. R. Soc. London A* **302**, 531–546.

Sinclair, M. (1988). "Marine Populations." Washington Univ. Press, Seattle.

Smetacek, V., *et al.* (1984). Seasonal stages characterizing the annual cycle of an inshore pelagic system. *Rapp. Proc.-Verb. Réun. Cons. Int. Explor. Mer.* **183**, 126–135.

Smetacek, V., Scharek, R., and Nthig, E.-M. (1990). Seasonal and regional variation in the pelagial and its relationship to the life history of krill. *In* "Antarctic Ecosystems. Ecological Change and Conservation" (K. R. Kerry and G. Hempel, Eds.), pp. 103–114. Springer-Verlag. Berlin.

Smetacek, V., Scharek, R., Gordon, L. I., Eicken, H., Fahrbach, E., Rohardt, G., and Moore, S. (1992). Early spring phytoplankton blooms in ice platelet layers of the southern Weddell Sea, Antarctica. *Deep-Sea Res.* **39**, 153–168.

Smetacek, V., *et al.* (1997). Ecology and biochemistry of the Antarctic Circumpolar Current during austral spring: A summary of SO JGOFS ANT X/6 of R. V. Polarstern. *Deep-Sea. Res. II* **44**, 1–22.

Smetacek, V. H., and Passow, U. (1990). Spring bloom initiation and Sverdrup's critical depth model. *Limn. Oceanogr.* **35**, 228–234.

Smith, P. C., and Petrie, B. (1982). Low-frequency circulation at the edge of the Scotian shelf. *J. Phys. Oceanogr.* **12**, 28–46.

Smith, P. C., and Sandstrom, H. (1986). Shelf edge processes. *Bedford Inst. Oceanogr. Rev.* **1986**, 40–46.

Smith, R. C. (1981). Remote sensing and depth distribution of chlorophyll. *Mar. Ecol.-Prog. Ser.* **5**, 359–361.

Smith, R. L. (1981a). Circulation patterns in upwelling regimes. *In* "Coastal Upwelling" (F. W. Richards, Ed.), pp. 13–35. American Geophysical Union, Washington, DC.

Smith, R. L. (1981b). A comparison of the structure and variability of the flow field in three coastal upwelling systems: Oregon. northwest Africa and Peru. *In* "Coastal Upwelling" (F. W. Richards, Ed.), pp. 107–118. American Geophysical Union, Washington, DC.

Smith, R. L., and Bottero, J. S. (1977). On upwelling in the Arabian Sea. *In* "A Voyage of Discovery" (M. Angel, Ed.), pp. 291–304. Pergamon, London.

Smith, R. L., Huyer, A., Godfrey, J. S., and Church, J. A. (1991). The Leeuwin Current off Australia, 1986–1987. *J. Phys. Oceanogr.* **21**, 323–345.

Smith, S. L. (1984). Biological indications of active upwelling in the northeastern Indian Ocean and a comparison with Peru and northwest Africa. *Deep-Sea Res.* **31**, 951–967.

Smith, S. L., and Schnack-Schiel, S. B. (1990). Polar zooplankton. *In* "Polar Oceanography," pp. 527–597. Academic Press, London.

Smith, W. O. (1987). Phytoplankton dynamics in marginal ice zones. *Oceanogr. Mar. Biol. Annu. Rev.* **25**, 11–38.

Smith, W. O., and Brightman, R. I. (1991). Phytoplankton photosynthetic response during the winter–spring transition in the Fram Strait. *J. Geophys. Res.* **96**, 4549–4554.

Smith, W. O., and DeMaster, D. J. (1996). Phytoplankton biomass and productivity in the Amazon River plume: Correlation with seasonal and river discharge. *Cont. Shelf Res.* **16**, 291–319.

Smith, W. O., and Nelson, D. M. (1985). Phytoplankton biomass near a receding ice-edge in the Ross Sea. *In* "Antarctic Nutrient Cycles and Food Webs" (W. R. Siegfried, Ed.), pp. 70–77. Springer-Verlag, Berlin.

Smith, W. O., and Sakshaug, E. (1990). Polar phytoplankton. *In* "Polar Oceanography Part B: Chemistry, Biology, and Geology" (W. O. Smith, Ed.), pp. 477–525. Academic Press, New York.

Smith, W. O., Codispoti, L. A., Nelson, D. M., Manley, T., Buskey, E. J., Niebauer, H. J., and Cota, G. F. (1991). Importance of *Phaeocystis* blooms in the high-latitude ocean carbon cycle. *Nature (London)* **352**, 514–516.

Soloviev, A., and Lukas, R. (1997). Observation of large diurnal warming events in the near-surface layer of the western equatorial Pacific warm pool. *Deep-Sea Res.* **44**, 1055–1076.

Sorokin, Y. (1983). The Black Sea. *In* "Ecosystems of the World, 26: Estuaries and Enclosed Seas" (B. H. Ketchum, Ed.), pp. 253–292. Elsevier, Amsterdam.

Sournia, A. (1994). Pelagic biogeography and fronts. *Prog. Oceanogr.* **34**, 109–120.

Springer, A., and McRoy, C. P. (1993). The paradox of pelagic food webs in the northern Bering Sea—III. Patterns of primary production. *Cont. Shelf Res.* **13**, 575–599.

Sprintall, J., and Tomczak, M. (1992). Evidence of the barrier layer in the surface layer of the tropics. *J. Geophys. Res.* **97**, 7305–7316.

Stabeno, P. J., and Reed, R. K. (1992). A major circulation anomaly in the western Bering Sea. *Geophys. Res. Lett.* **19**, 1671–1674.

Stabeno, P. J., and Reed, R. K. (1994). Circulation in the Bering Sea basin observed by satellite-tracked drifters: 1986–1993. *J. Phys. Oceanogr.* **24,** 848–864.

Steuer, A. (1933). Zur planmassigen Erforschung der geographischen Verbreitung des Haliplankton, besonders der Copepoden. *Zoogeogr. Int. Rev. Comp. Caus. Anim. Geogr.* **1,** 269–302.

Stramma, L. (1992). The South Indian Ocean Current. *J. Phys. Oceanogr.* **22,** 421–430.

Strass, V. H., and Woods, J. D. (1988). Horizontal and seasonal variation of density and chlorophyll profiles between the Azores and Greenland. *In* "Towards a Theory of Biological–Physical Interactions in the Worlds Oceans" (B. J. Rothschild, Ed.), pp. 113–136. Reidel, Dordrecht.

Strass, V. H., and Woods, J. D. (1991). New production in the summer revealed by the meridional slope of the deep chlorophyll maximum. *Deep-Sea Res.* **38,** 35–56.

Su, J. L., Guan, B. X., and Jiang, J. Z. (1990). The Kuroshio. Part 1: Physical features. *Oceanogr. Mar. Biol. Annu. Rev.* **28,** 11–71.

Sugimoto, T., and Tameishi, H. (1992). Warm-core rings, streamers and their role on the fishing ground formation of Japan. *Deep-Sea Res.* **39,** S183–S201.

Sund, P., Blackburn, M., and Williams, F. (1980). Tunas and their environment in the Pacific Ocean. *Oceanogr. Mar. Biol. Annu. Rev.* **18,** 23–56.

Sür, H. I., Ozoy, E., and Unlüata, U. (1994). Boundary current instabilities, upwelling, shelf mixing and eutrophication processes in the Black Sea. *Prog. Oceanogr.* **33,** 249–302.

Sverdrup, H. U. (1947). Wind driven currents in a baroclinic ocean with application to the equatorial currents in the Pacific Ocean. *Proc. Natl. Acad. Sci. USA* **33,** 219–303.

Sverdrup, H. U. (1953). On the conditions for vernal blooming of the phytoplankton. *J. Cons. Perm. Int. Explor. Mer.* **18,** 287–295.

Swallow, J. C. (1984). Some aspects of the physical oceanography of the Indian Ocean. *Deep-Sea Res.* **31,** 639–650.

Swallow, J. C., and Fieux, M. (1982). Historical evidence for two gyres in the Somali Current. *J. Mar. Res.* **40,** 747–755.

Swart, V. P., and Gonzalves, J. W. (1983). Episodic meanders in the Agulhas Current. *Rep 5th Natl. Oceanogr. Symp. S. Afr.* **79,** 161–172.

Swift, J. H. (1986). The Arctic waters. *In* "The Nordic Seas" (B. G. Hurdle, Ed.), pp. 129–153. Springer-Verlag, Berlin.

Szekielda, K.-H. (1978). Eolian dust into the northeast Atlantic. *Oceanogr. Mar. Biol. Annu. Rev.* **16,** 11–41.

Taft, B. (1972). Characteristics of the flow of the Kuroshio south of Japan. *In* "Kuroshio" (H. Stommel

and K. Yoshida, Eds.), pp. 165–216. East–West Center, Hawaii.

Takahashi, M., Nakai, T., Ishimaru, T., Hasamotu, H., and Fujita, Y. (1985). Distribution of the SCM and its nutrient-light environment in and around the Kuroshio off Japan. *J. Oceanogr. Soc. Jpn.* **41,** 73–80.

Takenouti, A. Y., and Ohtani, K. (1974). Currents and water masses in the Bering Sea: Japanese work. *In* "Oceanography of the Bering Sea" (D. W. Hood and E. J. Kelly, Eds.), pp. 39–58. Univ. of Alaska Press, Fairbanks.

Taniguchi, A., and Kawamura, T. (1972). Primary production in the western tropical and subtropical Pacific Ocean. *Proc. 2nd CSK Symp. Tokyo,* 159–168.

Tett, P. (1981). Modelling phytoplankton production at shelf sea fronts. *Philos. Trans. R. Soc. London A* **302,** 605–615.

Thomas, W. H. (1972). Nutrient inversion in the southeastern tropical Pacific Ocean. *U.S. Fish. Bull.* **70,** 929–932.

Thomas, W. H. (1978). Anomalous nutrient–chlorophyll relations in the offshore eastern tropical Pacific Ocean. *Deep-Sea Res.* **37,** 327–335.

Thompson, R. E. (1981). Oceanography of the British Columbia coast. *Can. Spec. Publ. Fish. Aquat. Sci.* **56,** 1–291.

Thompson, R. E. (1987). Continental shelf scale model of the Leeuwin Current. *J. Mar. Res.* **45,** 813–827.

Thunell, R., Pride, C., Ziveri, P., Müller-Karger, F., Sancetta, C., and Murray, D. (1996). Plankton response to physical forcing in the Gulf of California. *J. Plankton Res.* **18,** 2017–2026.

Tianming, L., and Philander, S. G. H. (1996). On the annual cycle of the eastern tropical Pacific. *J. Climate* **9,** 2986–2998.

Tomczak, M., and Godfrey, J. S. (1994). "Regional Oceanography." Pergamon, Elmsford, NY.

Toole, J. M., and Schmidt, R. W. A. F. (1987). Small-scale structure in the Northwest Atlantic sub-tropical front. *Nature (London)* **327,** 47–48.

Townsend, D. W. (1992). An overview of the oceanography and biological production. *In* "The Gulf of Maine," Regional Synthesis Series No. 1, pp. 5–25. U. S. NOAA.

Townsend, D. W., Christenson, J. P., Berman, T., Walline, P., Scheller, A., and Yentsch, C. S. (1988). Near-bottom chlorophyll maxima in southeastern Mediterranean shelf waters. *Océanol. Acta* **9,** 235–244.

Townsend, D. W., Keller, M. D., Sieracki, M. E., and Ackeson, S. G. (1992). Spring phytoplankton blooms in the absence of vertical water column stratification. *Nature (London)* **360,** 59–62.

Traganza, E. D., Conrad, J. C., and Breaker, L. C. (1981). Satellite observations of a cyclonic upwelling system and giant plume in the California Current. *In* "Coastal Upwelling" (F. A. Richards, Ed.), pp. 228–241. American Geophysical Union, Washington, DC.

Tranter, D., Leach, G. S., and Airey, D. (1983). Edge enrichment in an ocean eddy. *Aust. J. Mar. Freshwater Res.* **34**, 665–680.

Tranter, D. J. (1979). Seasonal studies of a pelagic ecosystem (110°E). *In* "Biology of the Indian Ocean" (B. Zeitzschel, Ed.), pp. 487–520. Chapman & Hall, London.

Tranter, D. J., and Leech, G. S. (1987). Factors influencing the standing crop of phytoplankton on the Australian Northwest Shelf seaward of the 40 m isobath. *Cont. Shelf Res.* **7**, 115–133.

Tranter, D. J., Carpenter, D. J., and Leech, G. S. (1987). The coastal enrichment effect of the East Australia Current eddy field. *Deep-Sea Res.* **33**, 1705–1728.

Tréguer, P., and Jacques, G. (1992). Dynamics of nutrients and phytoplankton and fluxes of carbon nitrogen and silicon in the Antarctic Ocean. *Polar Biol.* **12**, 149–162.

Trees, C. C., and El-Sayed, S. Z. (1986). Remote sensing of chlorophyll concentrations in the northern Gulf of Mexico. *Proc. Int. Soc. Optic. Eng.* **637**, 328–334.

Trotte, J. R. (1985). Phytoplankton floristic composition and size-specific photosynthesis in the eastern Canadian arctic. MSc thesis, Dalhousie University, Nova Scotia.

Tsuda, A., Furuya, K., and Nemoto, T. (1989). Feeding of micro- and macro-zooplankton at the subsurface chlorophyll maximum in the subtropical North Pacific. *J. Exp. Mar. Biol. Ecol.* **132**, 41–52.

Uda, M. (1963). Oceanography of the subarctic North Pacific. *J. Fish. Res. Bd. Can.* **20**, 119–179.

Umatani, S., and Yamagata, T. (1991). Response of the eastern tropical Pacific to meridional migration of the ITCZ: The generation of the Costa Rica Dome. *J. Phys. Oceanogr.* **21**, 346–363.

van Aken, H. M., Quadfasel, D., and Warpakowski, A. (1991). The arctic front in the Greenland sea during February, 1989. *J. Geophys. Res.* **96**, 4739–4750.

van Camp, L., Nyjkaer, L., Mittelstaedt, E., and Schlittenhardt, P. (1991). Upwelling and boundary circulation off northwest Africa as depicted by infra-red and visible satellite observations. *Prog. Oceanogr.* **26**, 357–402.

van der Spoel, S., and Heyman, R. P. (1983). "A Comparative Atlas of Zooplankton." Springer-Verlag, Berlin.

Vedernikov, V. I., and Demidov, A. B. (1991). Primary production and chlorophyll in deep regions of the Black Sea. *Oceanology* **33**, 193–199.

Veldhuis, M. J. W., Kraay, G. W., van Bleijsweijk, J. D. L., and Baars, M. A. (1997). Seasonal and spatial variability in phytoplankton biomass, productivity and growth in the northeastern Indian Ocean; The southwest and northwest monsoon, 1992–1993. *Deep-Sea Res. II* **44**, 431–449.

Venrick, E. L. (1974). Recurrent groups of diatoms species in the North Pacific. *Ecology* **52**, 614–625.

Venrick. E. L. (1988). The vertical distribution of chlorophyll and phytoplankton species in the North Pacific central environment. *J. Plankton Res.* **10**, 987–988.

Venrick, E. L. (1989). The lateral extent and characteristics of the North Pacific Central environment at 35°N. *Deep-Sea Res.* **26**, 1153–1178.

Venrick, E. L. (1991). Mid-ocean ridges and their influence on the large-scale patterns of chlorophyll and production in the North Pacific. *Deep-Sea Res.* **38**, S83–S102.

Venrick, E. L., McGowan, J. A., and Mantyla, A. W. (1973). Deep maxima of photosynthetic chlorophyll in the Pacific Ocean. *Fish. Bull.* **71**, 41–52.

Venrick, E. L., McGowan, J. A., Cayan, D. R., and Hayward, T. L. (1994). Climate and chlorophyll: Long-term trends in the central North Pacific Ocean. *Science* **238**, 70–72.

Veth, C., Peeken, I., and Scharek, R. (1997). Physical anatomy of fronts and surface waters in the ACC near the 6°W meridian during austral spring 1992. *Deep-Sea. Res. II* **44**, 23–50.

Vethamony, P. V., Babu, R., and Kumar, M. R. R. (1987). Thermal structure and flow patterns around Seychelles Islands during austral autumn. *Ind. J. Mar. Sci.* **16**, 179–183.

Villareal, T., Altabet, M. A., and Culver-Rymsza, K. (1993). Nitrogen transport by vertically migrating algal mats in the North Pacific Ocean. *Nature (London)* **363**, 709–712.

Vincent, W. F., Howard-Williams, C., Tildsley, P., and Butler, E. (1991). Distribution and biological properties of oceanic water masses around the South Island. *N. Z. J. Mar. Freshwater Res.* **25**, 21–42.

Vinogradov, M. E., and Voronina, N. M. (1961). Influence of the oxygen deficit on the distribution of plankton in the Arabian Sea. *Okeanologia* **1**, 670–678.

Vinogradov. M. E., Flint, V., and Shushkina, E. A. (1985). Vertical distribution of mesozooplankton in the open area of the Black Sea. *Mar. Biol.* **89**, 95–107.

Voituriez, B. (1981).The equatorial upwelling in the eastern Atlantic. *In* "Recent Progress in Equatorial Oceanography" (J. P. McCreary, Ed.), pp. 229–247. Scientific Committee for Oceanographic Research.

Voituriez, B. (1982). Les variations saisonières des courants equatoriaux 40°W du Golfe de Guinée. *Océanogr. Trop.* **18**, 185–199.

Voituriez, B., and Herbland, A. (1981). Primary production in the tropical Atlantic Ocean mapped from oxygen values of EQUALANT 1 and 2. *Bull. Mar. Sci.* **31**, 853–863.

Voituriez, B., and Herbland, A. (1982). Comparison des systèmes productifs de l'Atlantique tropical est: Domes thermiques, upwellings cotières et upwelling équatorial. *Rapp. Proc.-Verb. Réun. Cons. Int. Explor. Mer.* **180**, 114–130.

Voorhuis, A. D., and Bruce, J. G. (1982). Small-scale surface stirring and frontogenesis in the subtropical convergence of the North Atlantic. *J. Mar. Res.* **40**, 801–821.

Vukovich, F. M. (1988). Loop Current boundary variations. *J. Geophys. Res.* **93**, 15585–15591.

Vukovich, F. M., and Maul, G. A. (1985). Cyclonic eddies in the eastern Gulf of Mexico. *J. Phys. Oceanogr.* **15**, 105–107.

Vukovich, F. M., and Waddell, E. (1991). Interaction of a warm core ring with the western slope in the Gulf of Mexico. *J. Phys. Oceanogr.* **21**, 1062–1074.

Walker, G. T., and Bliss, E. W. (1932). World weather. *Mem. R. Meteorol. Soc.* **4**, 53–84.

Walsh, J. J., Dieterle, D. A., Meyers, M. B., and Müller-Karger, F. E. (1989). Nitrogen exchange at the continental margin: A numerical study of the Gulf of Mexico. *Prog. Oceanogr.* **23**, 245–301.

Wang, X., and Zhu, B. (1991). Discussion of the subsurface chlorophyll maximum forming in the continental shelf and Kuroshio area of the South China Sea. *In* "Essays on the Investigation of the Kuroshio 3," pp. 297–304. China Ocean Press, Beijing.

Wang, Y. (1983). The mudflat coast of China. *Can. J. Fish. Aquat. Sci.* **40**, 160–171.

Ware, D. M., and McFarlane, G. A. (1989). Fisheries production biomes in the northeast Pacific Ocean. *Can. Spec. Publ. Fish. Aquat. Sci.* **108**, 359–379.

Warner, A. J., and Hay, G. C. (1994). Sampling by the Continuous Plankton Recorder Survey. *Prog. Oceanogr.* **34**, 235–256.

Wassman, P., Peinert, R., and Smetacek, V. (1991). Patterns of production and sedimentation in the boreal and polar Northeast Atlantic. *Polar Biol.* **10**, 209–228.

Watts, J. C. D. (1972). The occurrence of the thermocline in the shelf areas south of Hong Kong. *In* "Proceedings of the 2nd Symposium Kuroshio" (K. Sugawara, Ed.), pp. 43–46. Saikon, Tokyo.

Weber, L. H., and El-Sayed, S. (1987). Contributions to the net, nano- and picoplankton standing crop and primary production in the Southern Ocean. *J. Plankton Res.* **9**, 973–994.

Weaks, M. (1984). Upwelling in the Gulf of Oman. *NOAA Oceanogr. Month. Sum.* **4**, 13.

Wefer, G., and Fischer, G. (1991). Annual primary production and export flux in the Southern Ocean from sediment trap data. *Mar. Chem.* **35**, 597–614.

Weikert, H. (1984). The vertical distribution of zooplankton in relation to habitat zones in the area of the Atlantis II deep, Red Sea. *Mar. Ecol.-Prog. Ser.* **8**, 129–143.

Weikert, H. (1987). Plankton and pelagic environment. *In* "Key Environments—Red Sea" (A. J. Edwards and S. M. Head, Eds.), pp. 90–111.

Wells, J. T. (1983). Dynamics of coastal fluid muds in low moderate and high tide range environments. *Can. J. Fish. Aquat. Sci.* **40**, 130–142.

Welschmeyer, N. A., Strom, S., Goericke, R., DiTullio, G., Belvin, M., and Petersen, W. (1993). Primary production in the subarctic Pacific Ocean: Project SUPER. *Prog. Oceanogr.* **32**, 101–135.

Wetsteyn, F. J., Ilahude, A. G., and Baars, M. A. (1990). Nutrient distribution in the upper 300 m of the eastern Banda Sea during and after the upwelling season. *Neth. J. Sea Res.* **25**, 449–464.

Wheeler, P. A., and Kokkinakis, S. A. (1990). Ammonium recycling limits nitrate use in the oceanic subarctic Pacific. *Limn. Oceanogr.* **35**, 1267–1278.

Wiebe, P. H., and McDougall, T. J. (1986). Warm core rings. *Deep-Sea Res.* **33**, 1455–1922.

Williams, R., and Conway, D. V. P. (1988). Vertical distribution and seasonal numerical abundance of the Calanidae in oceanic waters southwest of the British Isles. *Hydrobiology* **167/168**, 259–266.

Williams, R., and Robinson, A. (1973). Primary production at OWS "I" in the North Atlantic. *Bull. Mar. Ecol.* **8**, 115–121.

Wolanski, E. M., Carpenter, D. J., and Pickard, G. L. (1986). The CZCS views the Bismarck Sea. *Annu. Geophys. B Terr. Planet. Phys.* **4**, 55–62.

Wolf, K.-U. (1991). Meridionale variabilitet des physikalischen und plantologischen Jahreszyklus. *Ber. Inst. Meeresk. Univ. Kiel.* **203**, 1–118.

Wolf, K.-U., and Woods, J. D. (1988). Lagrangian simulation of primary production in the physical environment—The deep chlorophyll maximum and thermocline. *In* "Towards a Theory of Biological–Physical Interactions in the Worlds Ocean" (B. J. Rothschild, Ed.), pp. 51–70. Reidel, Dordrecht.

Wood, A. M., Sherry, N. D., and Huyer, A. (1996). Mixing of chlorophyll from the Middle Atlantic Bight into the Gulf Stream at Cape Hatteras in July, 1993. *J. Geophys. Res.* **101**, 20579–20593.

Woods, J. D. (1984). The Warmwatersphere of the northeast Atlantic: A miscellany. *Ber. Inst. Meeresk. Kiel.* **128**, 1–39.

Woods, J. D. (1988). Mesoscale upwelling and primary production. *In* "Towards a Theory of Biological–Physical Interactions in the Worlds Ocean" (B. J. Rothschild, Ed.), pp. 1–29. Reidel. Dordrecht.

Woods, J. D., Wiley, R. L., and Briscoe, M. G. (1977). Vertical circulation at fronts in the upper ocean. *In*

"A Voyage of Discovery" (M. Angel, Ed.), pp. 253–276. Pergamon, Oxford.

Wooster, W. S., Schaefer, M. B., and Robinson, M. K. (1967). Atlas of the Arabian Sea for fishery oceanography, IMR No. 67-12, pp. 1–39, 143 figs. Institute of Marine Resources, La Jolla, CA.

Wooster, W. S., Bakun, A., and McLain, D. R. (1976). The seasonal upwelling cycle along the eastern boundary of the North Atlantic. *J. Mar. Res.* **34,** 131–141.

Worthington, L. V. (1986). On the North Atlantic circulation. *Johns Hopkins Oceanogr. Stud.* **6,** 1–110.

Wroblewski, J. S. (1989). A model of the spring bloom in the Atlantic and impact on ocean optics. *Limn. Oceanogr.* **34,** 1563–1571.

Wroblewski, J. S., Sarmiento, J. L., and Flierl, G. R. (1988). An ocean basin scale model of plankton dynamics in the North Atlantic. *Global Biogeochem. Cycles* **2,** 199–218.

Wyllie, J. G. (1966). Geostrophic flow of the California Current at the surface and at 200 meters. *CalCOFI Atlas* **4,** 1–288.

Wyrtki, K.(1961). Physical oceanography of Southeast Asian waters. *NAGA Rep. (Scripps Inst.)* **2,** 11–195.

Wyrtki, K. (1962). The upwelling in the region between Java and Australia during the south-east monsoon. *Aust. J. Mar. Freshwater Res.* **13,** 217–225.

Wyrtki, K. (1963). The horizontal and vertical field of motion in the Peru Current. *Bull. Scripps Inst.* **8,** 313–346.

Wyrtki, K. (1973a). An equatorial jet in the Indian Ocean. *Science* **181,** 262–264.

Wyrtki, K. (1973b). Physical oceanography of the Indian Ocean. *In* "Biology of the Indian Ocean" (B. Zeitzschel, Ed.), pp. 18–36. Springer-Verlag, Berlin.

Wyrtki, K., and Kilonsky, B. (1984). Mean water and current structure during the Hawaii-to-Tahiti Shuttle Experiment. *J. Phys.Oceanogr.* **14,** 242–252.

Yamamoto, T. (1986). Small-scale variations in phytoplankton standing stock and productivity across the oceanic fronts in the Southern Ocean. *Mem. Natl. Inst. Polar Biol.* **40,** 25–41.

Yamamoto, T., and Nishizawa, S. (1986). Small-scale zooplankton aggregations at the front of a Kuroshio warm-core ring. *Deep-Sea Res.* **33,** 1729–1740.

Yan, X.-H., Ho, C.-R., Zheng, Q., and Klemas, V. (1992). Temperature and size variabilities of the Western Pacific warm pool. *Science* **258,** 1643–1645.

Yentsch, C. S. (1965). Distribution of chlorophyll and phaeophytin in the open ocean. *Limn. Oceanogr.* **11,** 117–147.

Yentsch, C. S. (1974). The influence of geostrophy on primary production. *Tethys* **6,** 111–118.

Yentsch, C. S. (1982). Satellite observation of phytoplankton distribution associated with large-scale oceanic circulation. *Sci. Stud. North Atlantic Fish. Org.* **4,** 53–59.

Yentsch, C. S. (1990). Estimates of "new" production in the mid-North Atlantic. *J. Plankton Res.* **12,** 717–734.

Yentsch, C. S., and Garside, J. C. (1986). Patterns of phytoplankton abundance and biogeography. *UNESCO Tech. Papers Mar. Sci.* **49,** 278–284.

Yentsch, C. S., and Phinney, D. A. (1985). Rotary motions and convection as a means of regulating primary production in warm core rings. *J. Geophys. Res.* **90,** 3237–3248.

Yoder, J. A., McClain, C. R., Feldman, G. C., and Esaias, W. (1993). Annual cycles of phytoplankton chlorophyll concentrations in the global ocean: A satellite view. *Global Biogeochem. Cycles* **7,** 181–193.

Young, R. W., *et al.* (1991). Atmospheric iron inputs and primary productivity: Phytoplankton responses in the North Pacific. *Global Biogeochem. Cycles* **5,** 119–134.

Zevenboom, W., and Wetsteyn, F. J. (1990). Growth limitation and growth rates of pico- phytoplankton in the Banda sea during two different monsoons. *Neth. J. Sea Res.* **25,** 465–472.

Zhabin, I. A., Zuenko, Y. I., and Yurasov, G. I. (1990). Satellite-revealed surface cool patches in the northern part of the Sea of Okhotsk. *Issled. Zemli. Kosm.* **5,** 25–28.

Zheng, Y., Guo, B., Tang, Y., Xiu. S., and Nakamura, Y. (1992). Observation of a Kuroshio frontal eddy in the East China Sea. *In* "Essays on the Investigation of the Kuroshio 4," pp. 23–32. China Ocean Press, Beijing.

Zijlstra, J. J., Baars, M. A., Tijssen, S. B., . Witten, J. I. J., Ilahude, A. G., and Hadiksumah (1991). Monsoonal effects on the hydrography in the eastern Banda Sea and northern Arafura Sea, with special reference to vertical processes. *Neth. J. Sea Res.* **25,** 431–447.

INDEX

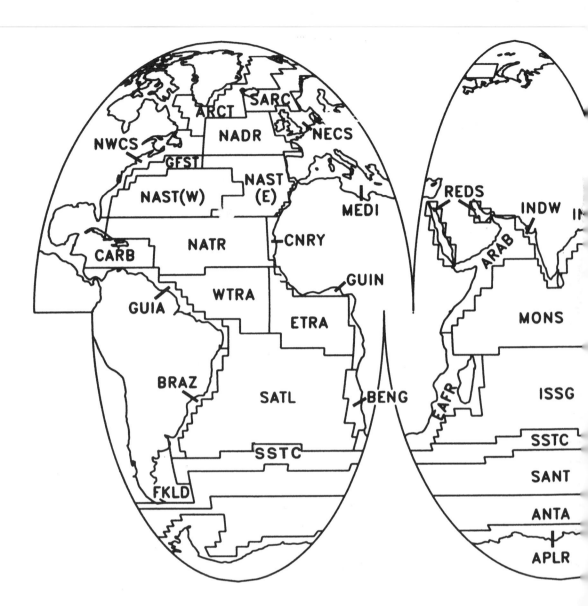

ANTARCTIC POLAR BIOME
ANTA   Antarctic Province
APLR   Austral Polar Province

ANTARCTIC WESTERLY WINDS BIOME
SANT   Subantarctic Water Ring Province
SSTC   South Subtropical Convergence Province

ATLANTIC COASTAL BIOME
BENG   Benguela Current Coastal Province
BRAZ   Brazil Current Coastal Province
CNRY   Eastern (Canary) Coastal Province
FKLD   Southwest Atlantic Shelves Province
GUIA   Guianas Coastal Province
GUIN   Guinea Current Coastal Province
NECS   Northeast Atlantic Shelves Province
NWCS   Northwest Atlantic Shelves Province

ATLANTIC POLAR BIOME
ARCT   Atlantic Arctic Province

BPLR   Boreal Polar Province
SARC   Atlantic Subarctic Province

ATLANTIC TRADE WIND BIOME
CARB   Caribbean Province
ETRA   Eastern Tropical Atlantic Province
NATR   North Atlantic Tropical Gyral Province
SATL   South Atlantic Gyral Province
WTRA   Western Tropical Atlantic Province

ATLANTIC WESTERLY WINDS BIOME
GFST   Gulf Stream Province
MEDI   Mediterranean Sea, Black Sea Province
NADR   North Atlantic Drift Province
NAST   North Atlantic Subtropical Gyral Province

INDIAN OCEAN COASTAL BIOME
ARAB   Northwestern Arabian Upwelling Province
AUSW   Australia-Indonesia Coastal Province